EMERGING PERS

ON

FLORA NWA

EMERGING PERSPECTIVES
ON
FLORA NWAPA:

Critical and Theoretical Essays

EDITED BY MARIE UMEH

Africa World Press, Inc.

P.O. Box 1892

Trenton, NJ 08607

P.O. Box 48

Asmara, ERITREA

Africa World Press, Inc.

P.O. Box 1892
Trenton, NJ 08607

P.O. Box 48
Asmara, ERITREA

Book design: Jonathan Gullery
Cover design: Linda Nickens

Photo credits: Front cover photo of Flora Nwapa courtesy of Nina E. Mba.
Back cover photo of Flora Nwapa by Sabine Jell-Bahlsen.

This book is composed in Veljovik Book and Nueva.

Library of Congress Cataloging-in-Publication Data

Emerging perspectives on Flora Nwapa: critical and theoretical essays/ edited by Marie Umeh.
    p. cm.
    Includes bibliographical references and index.
    ISBN 0-86543-514-6. -- ISBN 0-86543-515-4 (pbk.)
    1. Nwapa, Flora. 1931–1993--Criticism and interpretation. 2. Women and literature--Nigeria--History--20th century. 3. Nigeria--In literature. I. Umeh, Marie.
PR9387.9.N933Z65   1997
823--dc21
                         97-14708
                          CIP

DEDICATION

TO ALL THE ENTERPRISING
WOMEN OF THE WORLD

CONTENTS

PART TWO: (RE)CASTING GENDER RELATIONS: NIGERIAN WOMEN IN THE COLONIAL AND POST-COLONIAL EXPERIENCE

Photograph Section

PART THREE: AESTHETICS AND POETICS IN NWAPA'S CANON

PART FOUR: THE BODY POLITIC: LOCATING FEMALE SEXUALITY

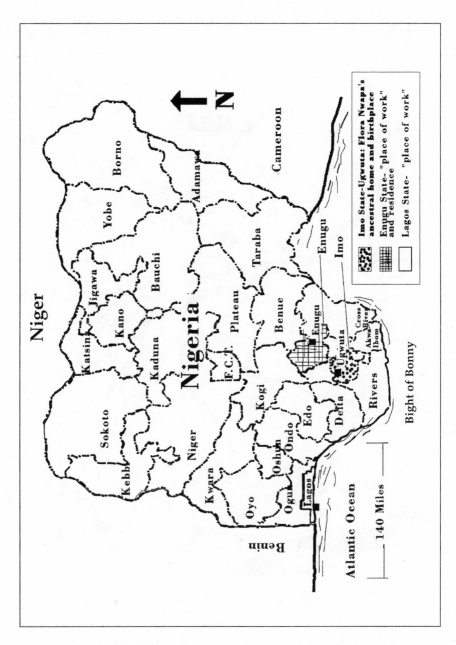

Map of Nigerian states where Flora Nwapa lived and worked

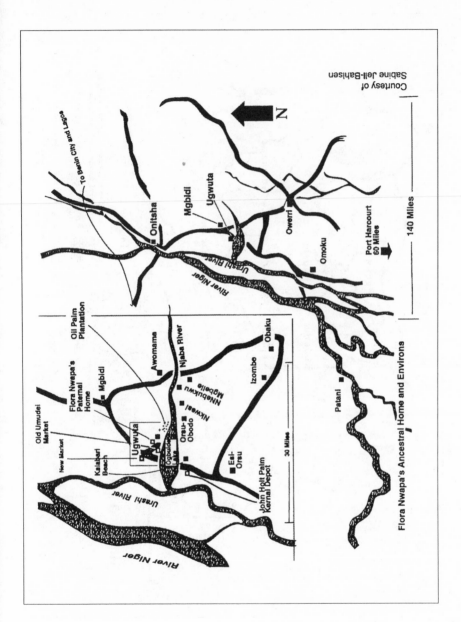

Map showing Flora Nwapa's ancestral home and environs

ACKNOWLEDGEMENTS

Books usually claim that they are the work of one author when they are the collected efforts of many persons. In this vein, I wish to acknowledge and express much gratitude for the camaraderie of colleagues and friends in completing this project. I owe my greatest debt to Ezenwa-Ohaeto, whose enthusiasm and professionalism added to the quality of the anthology. I am especially grateful to him for distributing the flyers and collecting articles and manuscripts in Nigeria on my behalf and writing the introduction to parts of the book. Also of invaluable help was Chimalum Nwankwo, who, when visiting Nigeria, cheerfully distributed letters and collected the manuscripts of some of the contributors, and Shivaji Sengupta and Ernest N. Emenyonu for accepting to write the introductions to parts of the anthology. A debt of gratitude is also owed to Chief Gogo Nwakuche, Flora Nwapa's husband, Uzoma Gogo Nwakuche, Flora Nwapa's son, and Ejine Nzeribe, Flora Nwapa's daughter, for their invaluable assistance and support.

I have also benefited from the collegiality of Theodora Akachi Ezeigbo, who not only collected and sent to me some of the photographs in the book, but who also sent to me the map of Nigeria, some articles and books published in Nigeria. A heart-felt "thank you" also goes to Sabine Jell-Bahlsen, who created the map of Ugwuta for the book. Sabine also provided the photograph of Flora Nwapa on the back cover of the book and the one of the Ugwuta priestess in the book. Additionally, of immense help was Nina E.

Mba who gave me her personal photographs of Flora Nwapa to include in the anthology and who informed me that Flora Nwapa was a published playwright as well as a novelist. To Njideka Nwapa, who advised and assisted me in contacting the members of the Nwapa-Nwakuche family in Nigeria, and Sherry Taylor Gibson, who professionally typed the manuscripts of some of the contributors whose diskettes were destroyed in transit, I am also very grateful.

Many thanks are due as well to the members of the New York Circle for Theory of Literature and Criticism, namely Edward Davenport, Peggy Escher, Victor Hapuarachchi, Fawzia Aszal-Kahn, Mary Cross, Shivaji Sengupta, and Stephen Paul Miller, for the dynamic sessions throughout the years. For her rigorous intellect and theoretical insights, and for being my best friend, I thank Chikwenye Okonjo Ogunyemi.

Over the years my husband, Davidson, and children—Akuaza, Ikechukwu, Uchenna, Chizoba and Ugochukwu—sustained me throughout my research projects; I am especially grateful to them for their love, understanding, and maturity.

PREFACE

Dear reader, you will find this anthology, *Emerging Perspectives on Flora Nwapa*, a definitive text on Flora Nwapa and her works. Although she cannot physically hold and thumb through these pages, we all certainly know that she has paved the way to this well-wrought scholarship. Nwapa has brought a new literary canon from Nigeria and presented it to the world for its perusal; and this anthology examines her works and the characters in her works. Each contribution celebrates the woman who stems from the female archetype, Ogbuide, the water spirit: strong, confident, independent, secure, and divine. This picture of the new African woman remains the legacy of Flora Nwapa, and the first part of this volume explores and addresses the idea of womanhood. As a major twentieth-century writer, Nwapa departs from the previous male-dominated literary predilection of women transmitted by Achebe, Soyinka, et al. In their works, their literary penchant is to provide the female character with all the amenities of life. But the woman's role in their literature is simply to be an adjunct to the man. The woman is socially ill-equipped to occupy an equal place in the society. She possesses little individuality and is powerless to protect herself from the "slings and arrows of outrageous fortune."[1] In Nwapa's works, however, her women play the central part. They are not just cuddly creatures without a future especially if they have no husband. Instead, Nwapa's women are endowed with an energy, a vitality, an intelligence, and an imagination that make them, *sine qua non*, the strong protagonists of her novels.

The contributions to this anthology are, therefore, solid and represent the critical scrutiny of the manifold dimensions of Nwapa's works. Her concerns are gender specific. However, she does not denigrate the male. In *The Lake Goddess*, for example, Mgbada (Joseph) is husband and partner, not predator nor malingerer. He, too, is the giver of life and is accepted and respected as such. But soon the male fades out of the picture and the reader watches the development of Ona, a girl with a special gift. She becomes Ogbuide incarnate, the one who heals and leads; and as she develops, she displays all the evidence and grace of the independent woman. As the water spirit, Ona becomes the aspiration of all women.

Emerging Perspectives on Flora Nwapa is, indeed, a fitting tribute to Flora Nwapa. As you read these pages you will notice that the works are not just contributions to the literary canon and to the literary world but that they extend themselves to other disciplines. Each work examines and re-examines the social construction of our society as well as the role of women in the artistic creation of Flora Nwapa.

Chester Mills

NOTES

1.   William Shakespeare, *Hamlet*. Act III, Scene 1.

FOREWORD

Nina E. Mba

To be asked to write the foreword to this profound and perceptive study commemorating Flora Nwapa is an honor. I applaud the diversity of the contributors, well-balanced as it is between Nigerians ("at home" and "abroad") and non-Nigerians (in Nigeria and elsewhere), thus mirroring Flora Nwapa's dual commitment to international sisterhood and cultural nationalism.

To count myself—an Australian woman married to a Nigerian—among Flora's most intimate friends is a treasured privilege. My first "sighting" took place in July, 1968 in Ugwuta, Biafra, where I, an evacuee from Port Harcourt, had taken refuge. I was amazed one day to see a red, roofless sports car speed down the main road, raising a dust storm behind it, driven by a woman wearing a scarf over her hair and dark glasses. What an incongruously glamorous vision she presented: for that moment the war seemed far away. I later learned how notorious Flora was for fast driving and how much she loved sports cars!

Flora and I did not actually meet until 1971, at the Enugu nursery which our 'war babies,' Uzoma Nwakuche and David Mba, attended. Happily, they remained classmates through primary to the end of secondary school and are good friends today. We soon, however, discovered many other interests in common besides our children, and built up an enduring rapport. We shared our respective discoveries about the international women's movement and the historical political and literary dimensions of feminism—Flora through her exposure to feminists and feminist literature on her visits abroad, and I through my doctoral research on Nigerian women.

Flora Nwapa was one of the very first Nigerian women to stand up and proclaim universal feminist values. On the 9th of March 1973, at a speech to the women's society at the University of Nigeria, Nsukka, she condemned the negative attitude of many women to their sisters—what she described as the *Le-le-muo, obu nwanyi ibem-o* (look it is just a woman like me) syndrome. She concluded: "The most valuable aspect of the new women's movement is the rediscovery that women can think and work together and find common ground for action . . . only when we ourselves can treat one another as full human beings worthy of other women's trust and respect can we be fully liberated."

Flora and I worked together when I was the editor at Nwamife Publishers, who first published *This is Lagos and Other Stories* and *Never Again*. Unfortunately Nwamife experienced various set backs and it was Flora's frustration at this situation, coupled with the termination of her position as Commissioner in the East Central State government with the coup of 1975, that led her and her husband, Chief Gogo Nwakuche, to set up Tana Press. I am pleased that Flora's publishing career is examined in this volume because it was a very important part of her life and Tana Press was on its way to becoming the first feminist press in Nigeria.

Although Flora Nwapa's novels have been studied extensively, her children's literature has been neglected and therefore the chapter on it in this collection is particularly valuable. As a historian I would have liked to see in this book more attention paid to the society (beyond pre-colonial Ugwuta) in and of which Flora Nwapa wrote, for she was very much a product of certain historical conditions in colonial and post-colonial Nigeria. That aside, altogether this volume presents a well-rounded portrayal of Flora Nwapa whom Marie Umeh rightly describes as "a complex figure and a private, dignified individual."[1] Like Virginia Woolf, Flora Nwapa cultivated the principle of anonymity and would have sympathized with Woolf's philosophy, "I must be private, secret, as anonymous and submerged as possible in order to write."[2] Behind her public image, Flora kept her real persona known only to her closest friends and family, but projected through certain characters in her novels.

At a talk on African women in literature in 1973, Flora Nwapa described Efuru as "all grace and grandeur, hardly erring and taking the vicissitudes of life in her stride." That would be an apt epitaph for Flora Nwapa herself.

NOTES

1. Marie Umeh, "Signifying(g) The Griottes: Flora Nwapa's Legacy of (Re)Vision and Voice." *Research in African Literatures* 26.2 (Summer, 1995):121.
2   Frances Spalding in a book review of James King's book, *Virginia Woolf* (C.H. Hamilton, 1994) quotes from the book. See the *London Sunday Times* (28 August 1994):7.

FOREWORD

Chukwuemeka Ike

We invited Flora Nwanzuruahu Nwapa Nwakuche, O.O.N., to join the Board of Trustees of the Nigerian Book Foundation in recognition of her stature as an author, her expertise in book related matters, and her versatility, qualities which the young non-governmental, non-profit organization needed in its mission to provide Nigeria with a vibrant indigenous book industry. She accepted the invitation with enthusiasm, and contributed to the success of the inaugural meeting of the Board.

When the Board assembled in Lagos in October, 1993 for the second meeting, Flora's eldest daughter, Ejine, sent word that Flora was in the hospital at Enugu. I drove to Flora's Enugu residence on Tuesday, October 19 to invite her to Ndikelionwu to convalesce in a quiet, rural environment. Amede, her second daughter who received me, rushed into my open arms and held me tight, shivering:

"Where's mummy?" I promptly asked, my pulse quickening.
"Mummy died on Saturday," she muttered.

Flora's departure (on 16 October 1993) was that sudden and unanticipated. Many thought it need not have happened, but it did happen, and we watched her mortal remains laid to rest on November 20, 1993 on the expansive grounds of her husband's manorial Oguta home.

Flora made history as Nigeria's (and Africa's) first published female novelist in English, and followed it up with an impressive list of published tiles, including novels, short stories, books for children, and poetry. Although the two titles, which launched her as a novelist, were published in the internationally acclaimed *Heinemann African Writers Series*, she established her own publishing company, Flora Nwapa Books Ltd., and backed it up with a printing press, Tana Press Ltd. This was an act of great courage, uncharacteristic of her contemporaries. The publishing outfit enabled her to advance the cause of children's literature in Nigeria.

Flora drew people to herself not only because she was an author, but also because she was a personable human being: simple, warm hearted (with an infectious laughter), always in fashion. She was also down to earth, serious minded, and loyal.

Most writers present at her funeral affirmed that Flora would live for all time. *Emerging Perspectives on Flora Nwapa* is one major effort in this direction. The book takes a broad sweep, paying tribute to Flora, exploring and interpreting major issues in her writing, and examining the background which must have inspired and informed them. It is significant that the selection of contributors transcends national, racial, and ethnic boundaries. I am particularly delighted that the Editor resisted any temptation to make it an all women affair. There is no doubt that Flora wrote primarily about women. She did not, however, write exclusively for women, neither did she live in a world exclusive to women. It would, therefore, be a disservice to her memory for female or male scholars and readers to see continuing interest in her works as an affair for women only.

The choice of the title, *Emerging Perspectives on Flora Nwapa*, suggests that the international literary discourse on Flora is opening up, not closing. I believe it is so because many of the ideas Flora canvassed on womanhood in her writing are universal and timeless.

I warmly commend Marie Umeh and Africa World Press, Inc. for putting out the most comprehensive and informative study ever published so far on Flora Nwapa.

Chukwuemeka Ike
President
Nigerian Book Foundation
P. O. Box 1132
Awka, Anambra State, Nigeria

INTRODUCTION:
Historicizing Flora Nwapa

Marie Umeh

In my native Onitsha, the highest tribute we may pay someone is to say that the person has, possesses, manifests a quality we call *awele*. What is *awele*? It is not wealth but a sense of richness, it is the sense of gracious well being; it is a sense of being in harmony with the world, with one's God and with nature; it is a sense of elitism, of seemingly effortless superiority; it is a sense of contentment, of fulfillment from having done one's best. Flora Nwapa Nwakuche *nwelu awele.* *

—Ajie Ukpabi Asika

Flora is gone and we all have to say "adieu." But she left behind an indelible mark. No one will ever write about Nigerian Literature in English without mentioning her. She will always be the departure point for female writing in Africa. And African publishing will forever owe her a debt. But above all, her contribution to the development of women in Nigeria, nay in Africa, and throughout the world is what she will be best remembered for. *

—Ken Saro-Wiwa

Nwa n'amarǫ nne-ya adị'apụ ya nzute. (A child who does not know the mother does not run out to welcome her.)[1]

                                        —Igbo Proverb

PART I: FLORA NWAPA'S ROOTS AND FOREMOTHERS

It is no mere accident of history that Flora Nwapa became one of the world's most prolific and illustrious women writers in the second half of the twentieth century. Over a span of twenty-seven years, she published six novels, nine children's books, three plays, two collections of short stories, a book of poems and innumerable essays. In addition to her literary accomplishments, Flora Nwapa was the founder and managing director of Tana Press, the first publishing company in Africa to be owned and operated by a woman. Her special mission in this venture was to give women a voice and to publish books for Nigerian children.

    Firstly, Flora Nwapa was born in the right place and at the right time. Africa is credited as the birthplace of the "first" great female monarchs and exceptionally vigorous inhabitants.[2] Ivan Van Sertima in his book *Black Women in Antiquity* asserts that ancient African regents "were not only masters of the state but masters of the spiritual capitals as well" (6). So naturally, Flora Nwapa's foundation was as majestic as Mount Kilimanjaro and as ancient as Tana Lake, the source of the Nile, the longest river in Africa.[3] Because the family is the backbone of African nation-states, the child is nurtured to cultivate acceptable behavior and virtues such as humility, obedience, patience, hard work, empathy, flexibility, reciprocity, and honesty. Elechi Amadi in his book, *Ethics in Nigerian Culture,* cites the following as other important cardinal virtues in the formation of the African personality: "wholesome human relations among people; a respect for elders; community fellow-feelings; a live-and-let-live philosophy; altruism and hospitality."[4] Sylvia Leith-Ross buttresses this position in her anthropological work, *African Women,* when she identifies the Igbo women's personality as ambitious, courageous, self-reliant, hardworking and independent (337). Niara Sudarkasa, in her book *The Strength of Our Mothers,* identifies seven fundamental principles that have traditionally guided African families and communities, namely respect, restraint, responsibility, reciprocity, reverence, reason and reconciliation"(xxii). Like the typical African child, Flora Nwapa was brought up to have good character and never to lose sight of the sense of dignity and truth, no matter what the vicissitudes in life, for fear of losing the essence and value of life. This moral force is not only crucial for the maintenance of a healthy body and soul, but also for the stability and progress of one's group.

In pre-colonial Igboland, the cultivation of yam was the work of men, who also provided land, built houses, and hunted wild game. Most of the other crops were grown and cultivated by women. African women also marketed their surplus agricultural products to buy essential commodities. Trade naturally became a major occupation of women, who did not have to engage in extended trade to get rich.[5] Chieka Ifemesia reports that seventeenth century Igboland maintained a viable export-import trade and that Aro people, the middlemen of the hinterland trade, transacted long-distance trade in gold, timber, ebony, camwood, ivory, implements, iron, tin, kola nuts, foodstuffs, slaves, and other products of the inland farm and forest, which they carried down to the coastal areas. Overseas merchandise and local fish and salt were purchased from the coastal areas in return. Ifemesia goes on to demonstrate how the traders traveling in caravans "organized trade routes which ran from south to north and east to west of the Igbo country and beyond. Routes passing through areas known for their political stability were usually frequented by large numbers of Aro traders" (25). The historian K. Onwuka Dike states in his book, *Trade and Politics in the Niger Delta 1830-1885,* that "in the Delta the middleman was his own master and capitalist and never permitted intercourse between Europeans and producers." Citing Livingstone, Dike goes on to note that the Delta middlemen were strict protectionists who allowed "no trade with white or black except what passed through their hands, at their own price; and each tribe on the river coast, did the same with its next inland neighbour" (103). Leslye Amede Obiora credits innate African intelligence, courage, drive, efficiency, cooperation, networking, and diplomacy when she concurs with Robert W. Harms that it was African initiative that forged trade routes seventeen hundred miles into the interior and developed the machinery for mercantile activities. African traders could not have undertaken this initiative had not the basic institutions and concepts necessary for trade and capitalist activity already existed.[6] Cheikh Anta Diop in *Precolonial Black Africa* also points to the sophistication of African economies when he states that "the sale of goods [was] strictly regulated; there were fixed market days and fixed prices on goods . . . foreign nationals had their own quarters in which they could live in the utmost security with their goods"[7] (1987:133). Before and during European contact with Africa, there existed in Africa homogenous, egalitarian communities in possession of tropical plenitude, a consistent system of continuous traditions, and a disparate distribution of wealth and structures of stratification which sustained the healthy environment that marked everyday life (Obiora, 1993:218). "In most parts of Nigeria," Bolanle Awe maintains, "trade has traditionally been an important activity for both rural and urban women and they were not left out of the new international trade" (ix).

Women gifted in trade who exploited their rich cultural heritage attained recognition and gained status in their communities. Victor Uchendu attests to this phenomenon when he asserts:

> A woman may be highly ranked because of the position of a distinguished son, her success in trading, or her position in the women's title society, irrespective of her marriage order. Indeed, some women are ranked higher than their husbands because of their distinctions in other spheres. Wealthy women marry in their own right, found big compounds, and play the role of social father quite effectively. (86)

Acholonu adds in her book, *Motherism,* that "a rich woman, an educated woman, enlightened woman who is outspoken, hardworking and fearless can hardly expect to be looked down upon by any member of her society."[8] She goes on to point out that "rich and enterprising women are better placed on the social ladder than poor unenterprising men, without prejudice to the individual customs of African communities" (45).

Although the Portuguese were the first Europeans to trade in the riverine town of Oguta, when the British took-over in 1785 the United African Company (UAC) opened a flourishing trading post.[9] UAC-Nigeria provided the ubiquitous market women in Oguta with the opportunity to acquire more wealth. Flora Nwapa, in a paper "Priestesses and Power among the Riverine Igbo" delivered at the Schomburg Center for Research in Black Culture in 1991, confirmed this fact when she read: "In Nigeria's economic and administrative history, trade was considered a woman's job and prominent women among the riverine Igbo sought to enhance their influence with foreign traders."[10] Nwapa's hallmark achievements, which were the beginning of her many milestones, were clearly rooted in her hometown Oguta, a small fishing and trading community which became the center of the palm kernel produce trade in the nineteenth century. Historically, Oguta's industrious market women made a great impression on many research scientists who visited, including Sylvia Leith-Ross, who observed the following:

> The palm-kernel trade also belongs to the women, but I know of only one town where palm-oil is bought and sold on a large scale by women. This is Oguta, where a remarkable figure, Madam Ruth as she is known, dominates the community. A veritable Amazon among traders, she now sweeps through the market-place like a ship in full sail; stands, in rose-pink velvet and purple head-tie, on the bank of Oguta Lake, shouting commands to her women bent over the paddles as they shoot the oil-cask laden canoes across the lake;or weary after the day's work climbs the steep lane to her home, dangling her Yale latch-key. There are others like her but she is the acknowledged leader and nowhere else in Owerri Province did I meet any such outstanding feminine figure in the world of commerce. (343)

Like their neighbors, riverine women were responsible for feeding their husbands and children and providing them with other necessities. However, women in the riverine area became financially independent and economically buoyant through trade.[11] "In Onitsha and Oguta little farming was done by women," contends the historian Nina Mba, "especially those from royal lineages" (29).

Apart from these solid beginnings, a second factor which assured Flora Nwapa's success in the post-colonial world was her distinguished family lineage. Flora Nwapa's mother, Lady Martha Onyema Nwapa was a teacher. She later accumulated wealth as a distributor of georges and coral beads, which earned her the title "The Coral Queen" (Chikwe, 19). Christopher I. Nwapa, Flora Nwapa's father, worked for UAC-Nigeria, and was clearly one of their well-known, and successful palm produce agents. The British colonizers while bringing their trading companies and expatriates for expanding UAC-Nigeria, who traded with local indigenes, also brought in Christian missionaries who established churches and schools. Second generation Christian converts of the Church Missionary Society (CMS), Christopher and Martha Nwapa were instru-

5

mental in building the Christian congregation in the wealthy and strategically sited town of Oguta.[12] The fact that Flora Nwapa's grandparents, Chief Onumonu Uzoaru and his wife, the awesome Madam Ruth, brought the Anglican Church to Oguta—when many people did not join the church or receive a Western education— made Flora Nwapa, who was born into this prominent and wealthy Christian family on January 13, 1931 in Oguta, a privileged child. At that time, the missionaries were responsible for expanding Christianity and Western education through their schools and missionary activities, with the aim of creating a strong middle class who would later become leaders in their society (Amadiume, 46). So her attraction to Western education began at an early age and lasted a lifetime. The first daughter of six children—five girls and an only son (Christopher Fine Nwapa)—Flora who attended the local missionary school and was a student *par excellence,* served as a fine role model for her younger siblings. "Education was seen as the gateway to economic opportunities by the Ibo people," according to the historian Elizabeth Isichei, "who realized that knowledge was power and that it could command a good salary" (1973, 154).

PART II: LINKING THE PAST AND THE PRESENT

A child of the post-colonial world, Flora Nwapa attended Central School, Oguta (1936), Archdeacon Crowther Memorial Girls School, Elelenwa (1944), C. M. S. Girls School, Lagos (1949-50), and Queen's College, Lagos (1951). According to Ogbu U. Kalu, missionary education was designed to cultivate European values in Nigerian children, and in the schools emphasis was placed on English, grammar, dictation, composition, debates, and drama (537). Flora also gained some teaching experience at Priscilla Memorial Grammar School in Oguta (1952), which was founded by her uncle, Chief Richard Nzimiro, the first Mayor of Port Harcourt. Post-colonial Nigerian "girls were taught the basics of modern life," Ifi Amadiume asserts, "including how to dress like ladies, how to straighten their hair and how to use the make-up in vogue" (46). Influenced by paternal and maternal aunts and uncles who were highly educated and studied abroad, Flora successfully passed her School Certificate Examinations and General Certificate of Education (London) and entered the University College, Ibadan (1953), where she earned a B. A. degree in English, History, and Geography, four years later (1957). Eventually she travelled abroad to study at the University of Edinburgh, Scotland, where she obtained a Diploma in Education

(1958). Flora was, therefore, well-groomed, academically and socially, to grace the halls of power and play a prominent role in a new Nigeria which was in the process of gaining political independence from British rule in October, 1960, spearheaded by Dr. Nnamdi Azikiwe, who later became the first President of Nigeria, and the Owelle of Onitsha.[13]

Upon returning home from Scotland, Flora served as an education officer at Calabar in Nigeria (1958), before she taught English, history, and geography at Queen's College in Enugu (1959). It was at this point in her life that she gave birth to her first child—and first daughter—Ejine (1959), by Gogo (Chu) Nzeribe (deceased), a trade unionist and one-time secretary to the Trade Union Congress of the West African sub-region. From 1962 to 1967, Flora served as an assistant registrar at the University of Lagos. While working in Lagos in 1965, Flora garnered a Ford Foundation and U. S. A. State Department grant and subsequently studied Public Relations, Fund Raising, and Alumni Administration at Northwestern University (Illinois). A British Council grant sent her abroad again to study University Administration in the United Kingdom. Following her return to Lagos, Flora became an active member of the Society of Nigerian Authors. Chinua Achebe was the President and Flora Nwapa the Secretary of this literary organization.[14] In an interview with Austa Uwechue, Flora admits that her association with Chinua Achebe in this organization, gave her the courage to show him her script. Achebe quickly read her first novel, liked it, put a title on it, and sent it off to Heinemann for publication (9). It was significant that Chinua Achebe inspired Flora Nwapa to write and encouraged her to publish her work. Achebe is thus rightly dubbed "The Father of Nigerian Writing in English" for the role he played in recognizing talent and putting into print such aspiring writers as Flora Nwapa and Ngũgĩ wa Thiong'o, to name only two who later became "big shots" in the world literary canon. This was the beginning of Flora Nwapa's literary career, which she developed through hard work and her heritage of excellence, because Flora continued to write and subsequently published not only more novels, but also short stories, essays, children's books, poetry, and plays. Even though she married Gogo Nwakuche, a successful industrialist, during the Nigerian Civil War in 1967 and had two children, Uzoma (1969) and Amede (1971), her passion for writing proved to be her "ultimate seduction."[15] Charlotte Chandler, in her interview conversations with some of the most famous people in the world, learned that their "obsession" for their art and

"drive for success" made them the *crème de la crème*. Chandler concluded:

> They all followed strong drives toward goals not
> always known, and found—themselves. All were
> able to work at what they wanted to do and to do it
> successfully. All were willing to risk everything
> when the chances for success seemed small. All
> gave everything to their work and received every-
> thing from it. All had luck, but none was just a lot-
> tery winner. All began by defining their work, and
> then were defined by *it*. (3)

This is very true of Flora Nwapa, who initially just wanted to tell
the story of a dynamic woman, Efuru, who took her fate into her
own hands. But in the process, she has given to the literary world
issues and theories which now transcend her local beginnings. Her
subsequent works have defined Flora Nwapa more than she could
define herself. For example, the turning point in Flora Nwapa's
writing was *One Is Enough*. From this novel onward, her tone
acquired greater militancy. This may have been a result of her contacts with Western feminists who invited her to several conferences
abroad at this stage in her career. Buchi Emecheta in an interview
with Adeola James stated that "the feminist movement in England
brought Flora Nwapa and Ama Ata Aidoo here [London] recently.
At various conferences they always make sure that Black women
are well represented" (43). Florence Stratton, in *Contemporary
African Literature and the Politics of Gender,* theorizes that Nwapa's
novel, *One is Enough,* was inspired by Buchi Emecheta's novel, *The
Joys of Motherhood*. Stratton further points out that there is "an on-
going dialogic relation between Nwapa's and Emecheta's novels,
one in which the authorial roles of precursor and successor are not
fixed but are interchangeable" (118–119). Flora Nwapa's development as a writer and "achieved plural identities," to use Christine
Sylvester's term, can then be attributed to a multiplicity of female
voices on the African continent, in the diaspora and in the world
at large.

PART III: CONTINUING THE TRADITION
OF HER MOTHERS

After the civil war, because of her career as a teacher, her fame as
Nigeria's first female novelist, and her reputation as an excellent

administrator, Ajie Ukpabi Asika, the Governor of East Central State (1970-75), appointed Flora Nwapa Nwakuche the Commissioner for Health and Social Welfare in 1970 and later for Lands, Survey and Urban Development from 1971 to 1975 in East Central State. It was Flora Nwapa's involvement in Nigerian government and politics for five and a half years that led her in the essay, "Women in Politics" to assert that "The days are gone when Nigerian women [are] looked upon as mere objects for decoration. The Nigerian woman has given every indication that, given the opportunity, she can successfully and gracefully take an active part in politics." Nwapa's short story, "The Campaigner," and her play, *The First Lady,* both demonstrate the significant roles Nigerian women continue to play in twentieth-century Nigerian politics. And in her essay, "Priestesses and Power among The Riverine Igbo," she credits the former First Lady of Nigeria, Dr. (Mrs.) Maryam Babangida (1985-1993), for utilizing the traditional dual-sex political system for female empowerment in the 1990s through the Better Life Programme for Rural Women inaugurated in Abuja in September, 1987 (11).[16]

Certainly, the DNA heredity factor which can be traced to her foremother—the Madam Ruth of Oguta—was another factor which explains Nwapa's innovation, competitive spirit, and penchant for excellence. "Excellent writing," Bernth Lindfors states emphatically, "is the only passport for literary canonization" (75). Flora Nwapa, who believed in excellence and hard work, was a trailblazer who has entered the world literary canon as one of the most vigilant historicizers of the Nigerian female experience from the precolonial to the post-colonial eras. "As the materfamilias of Nigerian women's fiction, she has launched a comprehensive wo/man palava," Chikwenye Okonjo Ogunyemi contends, "and bequeathed a rich, literary legacy to us by prominently featuring characters drawn from the collective psyche, especially the Osun-like Uhamiri, her fictionalized version of Mammywata or Ogbuide, the Ugwuta Woman of the Lake, who invariably presides over her texts" (132-33). From *Efuru* (1966) right up to *The First Lady [A Play]*, Nwapa, through her womanist poetics, limns her commitment to centering women from their marginal identities as objects rather than as subjects in Nigerian written literatures. Juma, one of the protagonists in the play, "Two Women in Conversation," mirrors the strong, long-suffering, self-sacrificing, and even celibate African mother who more often than not independantly cares for her children. By interrogating Chinua Achebe and his brothers, Cyprian

Ekwensi, Chukwuemeka Ike, and Wole Soyinka, to name only four Nigerian male writers who painted one-dimensional images of African women as appendages to the male in Nigerian fiction, Nwapa dismantled the woman-as-object trope and constructed the woman-as-subject metaphor, thus inaugurating the African woman's literary tradition of illustrious torchbearers who are carrying on her "legacy of (re)vision and voice." In an interview with Ezenwa-Ohaeto, Flora Nwapa states:

> From my childhood I lived among very strong women: my two grandmothers and their co-wives. They were strong women. In my native Oguta it was the women who first started trading with the foreign companies. Trading was the job of the woman in the riverine areas. So the women gained economic power through trade before the men, and all this influenced my writing and that is why I project women as great achievers. I did not see women as second-class citizens. (14)

This is exactly what the author and critic Chinweizu wants to hear as he asserts that not only are African women dynamic leaders in commerce, but that they are also far from powerless in postcolonial African society (11). "That women operate by methods which often differ from those available to men," he goes on to say, "does not in any way mean that women are bereft of power . . . the hand that rocks the cradle is the hand that rules the world" (15).

Another area in which Flora Nwapa made her mark and displayed her valor was in the field of publishing. With her husband, Chief Gogo Nwakuche, she founded Tana Press Limited and Flora Nwapa Books, in 1977, thus becoming Africa's first woman to own and operate a publishing company.[17] In addition to running the business and reading all the manuscripts that came in for publishing, she had to her literary credit a total of six adult novels and many unpublished manuscripts. Apart from publishing *Efuru* (1966) and *Idu* (1970), she wrote *Never Again* (1975), *One Is Enough* (1981), *Women Are Different* (1986), and *The Lake Goddess* (forthcoming); two collections of short stories, *This Is Lagos and Other Stories* (1971) and *Wives at War and Other Stories* (1975); a book of poetry, *Cassava Song and Rice Song* (1986); and three plays, *The First Lady* (1993), *Two Women in Conversation* (1993), and *The Sychophants* (1993). Like Cyprian Ekwensi, who was the first Nigerian author to write books for children, Flora Nwapa wrote nine books for her own growing children: *Emeka—Driver's Guard* (1972), *Mammywater* (1979), *My Tana Colouring Book* (1979), *My Animal Colouring Book* (1979), *The Miracle Kittens* (1980), *Journey to Space and Other Stories,* which includes the story, "The Magic Cocoyam" (1980), *The Adventures of Deke* (1980), *My Animal Number Book* (1981), and *My Tana Alphabet Book* (1981). Since its inception, Nwapa's publishing company has produced approximately twenty-five titles, which she marketed at conferences, in the tradition of her foremothers, as she traveled around the world. In 1992, she successfully launched her publishing debut with Africa World Press in the United States, which reprinted and distributed her novels and two collections of short stories.[18] Many of Flora Nwapa's manuscripts were left unpublished largely because of the high cost of publishing in Nigeria.[19] Nevertheless, it is hoped that her children, who are presently managing Tana Press Limited in Enugu, will continue her legacy by seeing that all her creative works are put in print. Her son, Uzoma Nwakuche, informed me that just before Flora Nwapa was to begin a teaching appointment as a Visiting Professor in North Carolina (USA), she marked off a tract of land for a Flora Nwapa Foundation to be built in Oguta as a center of learning.[20] The Flora Nwapa Foundation would enable scholars and students of African literature and African studies to read and study her journals, letters, published and unpublished manuscripts, short stories, plays, photographs, and essays, as well as Oguta lore and culture.[21]

PART IV: FLORA IN HER GLORY

Without question, Flora Nwapa was the most celebrated woman writer of post-colonial Nigeria. Internationally, her literary and publishing achievements have placed her picture on the cover of *Africa Woman* magazine (London) in 1977, along with a revealing cover story where Nwapa reminisces about her childhood days in Oguta with her kith and kin, as well as about her university, writing, and nation-building activities in post-independent Nigeria. Other European periodicals, namely the magazines *West Africa* and *Africa Now* and London newspaper *The Guardian,* included pictures and interviews with her and critical reviews of her books and publishing activities. Her first novel, *Efuru,* was translated into French in 1986 and into Dutch in 1991. In 1991 after Flora Nwapa was in the United States to present a Keynote address, "Nwanyibu: Womanbeing and African Literature," at the 17th Annual African

Literature Association Conference at Loyola University in New Orleans, Louisiana, Flora was inducted into PEN (Poets, Essayists, Novelists) International. The following year after Flora Nwapa gave her Keynote address, "Women and Creative Writing in Africa," during the Plenary Session at Obioma Nnaemeka's "First International Conference on Women in Africa and the African Diaspora: Bridges Across Activism and the Academy," at the University of Nigeria, Nsukka, she traveled to Nairobi, Kenya to join a committee of distinguished writers chosen to select the winner of the Commonwealth Writers Prize for 1992. Flora Nwapa's *oeuvres* are part of the school curriculum in many countries in Africa, Europe, and North America. Many post-graduate theses, doctoral dissertations, professional journals, anthologies, and scholarly books around the world engage in her woman-centered discourse. Flora Nwapa, herself a product of two worlds—traditional and western—has successfully woven together an authentic theory of Igbo/Nigerian/African woman herstories. For this reason the journals, *Research in African Literatures (RAL)* in the United States, and the forthcoming *Journal of Women's Studies in Africa (JOWSA)* in Port Harcourt, Nigeria, have both dedicated entire volumes in honor of one of Africa's "Firsts" and foremost African women writers.

In 1978, Flora Nwapa acquired the chieftancy title *Ogbuefi*, which literally means "killer of a cow." In Oguta, unlike most other Igbo societies, a woman of means, integrity, and good standing in the eyes of the elders, known as the "Ndi Igbu" Society, can take the chieftancy title, *Ogbuefi*. It is important to note that the same rules apply to the men who take the title of *Ogbuagu*, which literally means "killer of a lion." According to Lolo Kema Chikwe, Lady Mary Nzimiro, "a veritable Amazon and merchant Princess of Igboland and Nigeria" was another Oguta woman who had the prestigious title of *Ogbuefi* (23). The Nigerian Association of University Women had also honored her with the Certificate of Merit Award in 1980. Additionally, Flora Nwapa's twin interests—creative writing and publishing—won her from the Nigerian Federal Government's former Head of State, President Shehu Shagari (1979-1983), the coveted Officer of the Order of the Niger (OON) Award in 1983. This honor is similar to the recognition President Clinton gives to distinguished Americans. Flora Nwapa's outstanding career also earned her the University of Ife's Merit Award for Authorship and Publishing in 1985. In 1987, the Solidra Circle in Lagos and the Octagon Club in Owerri honored her with two distinguished awards. And the former Governor Amadi Ikweche of Imo State in

1992 gave her a medal of honor for her service to the community. Flora Nwapa died from pneumonia at the University of Nigeria Teaching Hospital, Enugu in 1993.

Although Flora Nwapa, the receptacle and reservoir of a sea of love that never dries, did not live to receive the Honorary Doctoral Degree that I was planning to nominate her for, there are many ways to crown a life and career as exemplary as Flora's: hence, this commemorative volume. The essays in this anthology have been commissioned and organized to reflect Flora Nwapa's aesthetic practice and gender-specific discourse. Scholars from around the world (re)present a wide variety of interdisciplinary approaches to Nwapa's *oeuvres;* they cover all of her literary and poetic landscapes. The negative criticism of her works and the unappreciative approach to assessing them by earlier "critics" have given way to more truly critical and theoretical evaluations of her writing. The *linchpin* of the articles gathered in this anthology is Nwapa's central ideology: female autonomy, self-fulfillment, and economic independence.

Emerging Perspectives on Flora Nwapa: Critical and Theoretical Essays begins with a short biographical sketch which situates Flora Nwapa in the world literary canon. This is followed by a section titled *"Tribute(s) to Flora Nwapa,"* which eulogizes the author/poet/playwright upon her untimely death in 1993. The essays in *PART ONE—Igbo Women: Culture and Literary Enterprise,* explore the impact of Igbo cultural traditions on women in Nwapa's novels. The myth of Uhamiri, indeed the entire history and culture of Oguta, are used by Nwapa to assess women's condition in Igboland. Theodora Akachi Ezeigbo's essay, "Myth, History, Culture, and Igbo Womanhood in Flora Nwapa's Novels," shows how Flora Nwapa deconstructed the Uhamiri myth and other cultural paradigms and reconstructed history from women's perspectives, thus making a case for female empowerment. Sabine Jell-Bahlsen in "Flora Nwapa and Uhammiri/Ogbuide, The Lake Goddess: An Evolving Relationship" analyses how Nwapa's relationship to Oguta's Lake Goddess, Uhammiri/Ogbuide, has evolved over the years. According to Jell-Bahlsen, Nwapa, in her earlier works, challenged the core values of Oguta society by deconstructing local beliefs in the Lake Goddess's power to grant women their greatest desire, children. However, in her novel, *The Lake Goddess,* Nwapa's criticism has shifted towards questioning the merits of foreign values and gifts introduced to Oguta via Christianity, and reevaluating local beliefs about the Lake Goddess and her requirements and gifts to women. Teresa U. Njoku's article, "The Mythic World in Flora Nwapa's Early Novels," examines the myths in *Efuru* and *Idu* and shows how they give shape to her art as well as deepen literary meaning. Florence Stratton, in "'The Empire, far flung': Flora Nwapa's Critique of Colonialism," demonstrates the unique way in which Flora Nwapa challenged the colonial view of Igbo women by reclaiming African women's unique cultural past so glaringly denied them in colonial historiography. Gay Wilentz, in "Not Feminist but Afr*a*centrist: Flora Nwapa and the Politics of African Cultural Production," sees Flora Nwapa as an Afr*a*centrist, rather than as a feminist. Through a discussion of a "critical incident" in three of Nwapa's novels—*Idu, One Is Enough,* and the posthumous *The Lake Goddess*—Wilentz exposes the Afr*a*-center of Nwapa's creative works. Ada Uzoamaka Azodo's essay, "*Efuru* and *Idu:* Rejecting Women's Subjugation," argues that Flora Nwapa recaptured the communal values of Oguta people in her early novels. According to Azodo, a feminist reading of the texts shows that Flora Nwapa is protesting against the secondary position of women in Nigerian colonial society. Ezenwa-Ohaeto's essay, "Breaking Through: The Publishing Enterprise of Flora Nwapa," discusses the various aspects of Nwapa's activity as a publisher. Ezenwa-Ohaeto stresses the dynamism of Nwapa's publishing enterprise which enabled her to overcome the tribulations associated with such enterprises in societies with harsh economic conditions. He concludes that Tana Press and Flora Nwapa Books illustrate and exemplify an effective utilization of limited resources and relevant ethical standards to pioneer literary and educational developments.

PART TWO—(Re)casting Gender Relations: Nigerian Women in the Colonial and Post-Colonial Experience, focuses on the ways in which Flora Nwapa claimed more room for women by overturning patriarchal stereotypes of women. Susan Arndt, in "Applauding A 'Dangerous Luxury': Flora Nwapa's Womanist Re-Interpretation of The *Ifo* about The 'Handsome Stranger,'" argues that Nwapa supported the ideas of a free choice of marriage partners for women and condemned the Igbo cultural assumption that the payment of a bride price guaranteed the success of a marriage. Mary E. Modupe Kolawole, in "Space for the Subaltern: Flora Nwapa's Representation and Re-presentation of Heroism," contends that the novelist's gender awareness correlated to a growth in consciousness and a progressive attempt to come to terms with social change. According to Kolawole, Nwapa proved that African women are mouthpieces for their gender and their race. Ada Uzoamaka Azodo, in "Nigerian Women in Search of Identity: Converging Feminism and Pragmatism," asserts that Flora Nwapa's later writings reached a new height as she gave pragmatic pieces of advice to Nigerian women in search of identity in complex modern times. Emelia C. Oko, in "Woman, The Self-Celebrating Heroine: The Novels of Flora Nwapa," shows how Flora Nwapa captured in her depictions of daughters the resonance of the Oguta people's riverine culture and their celebration of difference as woman. According to Oko, in Oguta society, woman is not marginal; she is central. Adeline Apena, in "Bearing the Burden of Change: Colonial and Postcolonial Experiences in Flora Nwapa's *Women Are Different,*" contends that the challenges of modern women in Nigeria's colonial and postcolonial society demonstrate the contradictions of double patriarchal experiences of traditional and modern Nigeria. According to Apena, Nwapa's sense of pragmatism and determination to succeed is demonstrated in the women's strategies of adjustment reflecting the changing cultural environment. She concludes that while these strategies enabled modern women to gain economic power and independence, they did not necessarily ensure emotional stability. Tess Onwueme, in "Shifting Paradigms of Profit and Loss: Men in Flora Nwapa's Fiction," argues that in *One Is Enough,* Nwapa stages a contest between the prospective winners and losers, between men and women, and it is abundantly signified that men are the huge losers while women are the huge winners in the race for power. In this regard, the critic insists that by consciously disclosing Izu's experiences of dislocation and marginality imposed on him by Amaka, Nwapa skillfully shows the "other" tormented face behind the mirror—man. Chimalum Nwankwo, in *"The Lake Goddess:* The Roots of Nwapa's Word," demonstrates how Flora Nwapa's last novel, with its return to the mythological foundations of Oguta, offers us a window into what her politics would have been if she had grown as a writer in the directions which the first two novels *Efuru* and *Idu,* appeared to have been headed. According to Nwankwo, Nwapa's final novel acknowledges the resilience and strength of Igbo culture in the face of the onslaught of Western cultural imperialism.

The essays in PART THREE—Aesthetics and Poetics in Nwapa's Canon, analyze the stylistic and metaphoric qualities in Nwapa's *oeuvres.* Ernest N. Emenyonu, in "Portrait of Flora Nwapa As A Dramatist," examines Nwapa's dramatic works; "The Sychophants," "Two Women in Conversation," and "The First Lady," which were unknown to the literary world until after her death in 1993. He argues that Nwapa in her plays seems to have rediscovered herself; for the plays show her at her best in terms of her mastery of subject matter, perfect control of her medium and an excellent blending of the artistic attributes of social relevancy, communicative competence and imaginative vigor. Emenyonu concludes that had Flora Nwapa lived longer, she would have achieved the global excellence and recognition of a playwright which she never quite did as a novelist. Jane Bryce, in "Evading Canonical Constraints: The 'Popular' As An Alternative Mode—The Case of Flora Nwapa," addresses the peculiarities of Flora Nwapa's relationship to the canon and the critical establishment, in light of the 'paradigmatic status' of her first novel, *Efuru,* and the general neglect of her later, self-published works, *One Is Enough* and *Women Are Different.* Obododimma Oha, in "'To the root/of dear Cassava': Rhetorical Style in Flora Nwapa's Poetry," examines the classical tropes in Nwapa's only collection of poems, *Cassava Song* and *Rice Song,* and identifies the relationship of her linguistic resources to sexual politics.

In "Never A Gain? A Critical Reading of Flora Nwapa's *Never Again,*" Obododimma Oha exposes the negative effects of pro-war propaganda in Biafra on personal security and interpersonal relationships. Oha shows how Nwapa's anti-war rhetoric was weakened by the technique of appealing to emotions. Naana Banyiwa Horne's essay, "Flora Nwapa's *This Is Lagos:* Valorizing the Female Through Narrative Agency," engages in a structural and thematic analysis of the stories in Nwapa's collection, *This Is Lagos and Other Stories.* Using Mieke Bal's theory of narrative, Horne explores the process through which Nwapa constructed the elements comprising the narratives which constituted her texts. Theodora Akachi Ezeigbo, in her essay "Vision and Revision: Flora Nwapa and the Fiction of War," appraises the artistic tools with which Nwapa presented the theme of survival in her war fiction. Ezeigbo also focuses on Nwapa's use of the autobiographical and realistic modes and the images and metaphors that illuminate Nwapa's ideas. In "Orality and Metaphoric Dichotomy of Subject in the Poetry of Flora Nwapa's *Cassava Song and Rice Song,*" Ezenwa-Ohaeto argues persuasively that the unrecognition of Nwapa's poetry as an interesting work of literature should be redressed. He identifies the orality in the technique and content of the poetry, as well as the artistic dichotomy of subject that Nwapa uses to project a relevant vision of her society. Ezenwa-Ohaeto insists rightly that Nwapa's poetry revises prevalent gender notions and humanistically reflects on the nature of life and society.

PART FOUR—The Body Politic: Locating Female Sexuality, discusses the centrality of female sexuality in Nwapa's novels, *Efuru, One Is Enough,* and *The Lake Goddess.* The erotic relationships between the Woman of the Lake and her priestesses are critically explored by Ifi Amadiume in "Religion, Sexuality, and Women's Empowerment in Nwapa's *The Lake Goddess.*" Amadiume explores the problematic sexuality that is often linked to Mammywater spirit possession. Linda Strong-Leek, in "The Quest for Spiritual/Sexual Fulfillment in Flora Nwapa's *Efuru* and *The Lake Goddess,*" analyzes the relevance of female circumcision—psychological and physiological—in Nwapa's fictional Igbo communities, while discussing the physical, psychological, and spiritual ramifications of the life-altering event. She also focuses on how the main characters in the novels find redemption and peace only after they retreat from the traditional roles of wife and mother and become worshippers of Uhamiri, the Lake Goddess. Shivaji Sengupta in his essay, "Desire, The Private and The Public in Flora Nwapa's *Efuru* and *One Is Enough,*" discusses the transition of Flora Nwapa's protagonists from the private to the public domain, and the roles desire and sexuality play in this transition. Sengupta posits that Nwapa locates herself in feminist criticism by taking a stand, different from that of the Western "first world feminists," by enjoining the public to the private rather than reversing the roles of the private and the public.

PART FIVE—(Re)Configurations of Child Symbolism, is a study of Flora Nwapa's body of children's literature. Julie Agbasiere, in her essay, "Conquerors of the Universe: Flora Nwapa's Kiddies on the Move," shows how Flora Nwapa gives her child-characters ample space to assume responsible positions and solve problems. Agbasiere also assesses the child-character as an important factor in harnessing the forces of nature and in giving meaning to the universe. In her essay, "Flora Nwapa's Legacy to Children's Literature," Ifeoma Okoye pays tribute to Flora Nwapa for her innovative and pedagogical insights in her children's book, *Mammywater.* Through a careful study of *Mammywater,* Okoye demonstrates the ways in which the author delineates characters and advances plot in a variety of question and answer techniques. Ernest N. Emenyonu, in his essay, "Flora Nwapa's Writings for Children: Visions of Innocence and Regeneration," shows through close textual analyses of *Emeka—Driver's Guard, Mammywater* and "The Magic Cocoyam," Nwapa's concerns as a mother writing for growing children. He states that Nwapa's commitment is towards the restoration of good moral values and its provision through the deliberate characterization of role models for contemporary youth. Marvie Brooks, in her essay, "'Either Pound in the Mortar or Pound on the Ground': Didacticism in Flora Nwapa's Children's Books," analyzes Flora Nwapa's use of ritual, folktales, proverbs, custom, characterization, and plot in five of her children's books. The ways in which each illustrator uses artistic insight to complement Flora Nwapa's themes, characters, setting, and plot are also examined.

PART SIX—Conversation(s) With Flora Nwapa, contains three interviews which Flora Nwapa granted in 1988, 1989, and 1992, respectively. Sabine Jell-Bahlsen's "An Interview with Flora Nwapa" reveals the novelist's ambivalencies both towards Oguta society's goddess worship and her own Christian background. Through a series of skillful and sustained questions by the interviewer, Nwapa reveals her vacillation towards traditional African practices because of her Christian upbringing and Western education. Theodora Akachi Ezeigbo's interview, "A Chat With Flora Nwapa," throws light on many issues relating to Flora Nwapa's creativity, approach to writing, and, above all, her views on what constitutes good and bad criticism in literary studies. Unimpressed by some of the past critical opinions of her works, she bares her mind on the destructive criticism some critics indulge in. In Flora Nwapa's interview with Marie Umeh, "The Poetics of Economic Independence for Female Empowerment: An Interview with Flora Nwapa," the author traces her literary and publishing career to the Oguta griottes (who inspired her with stories during her youth) and to her mentor, Chinua Achebe, who recommended her first book-length manuscript, *Efuru,* to Heinemann Educational Books, London, for publishing in 1966.

Many literary bards have gone without honor in their own time and place, but Flora Nwapa had the good fortune of being recognized in her own lifetime for her innovative thinking and extraordinary courage, her creative writing and publishing prowess. Public and private photos of Flora Nwapa have been included in this anthology to document and "provide powerful statements" about important persons and events who played critical roles in her life. As Flora Edouwaye Kaplan posits, "photographs prove useful tools in linking memory and oral tradition to recorded events, customs and persons."[22] In addition to the visual evidence of Flora Nwapa's literary and publishing entrepreneurship in post-independent Nigeria, my comprehensive chronology of the author's life and creative works and Brenda F. Berrian's massive bibliography of the literary criticism that Flora Nwapa's texts have inspired, point to Flora Nwapa, the "Mother of African Women's Writing," as a vessel of creation and a nurturer of young creative writers. Just as Chinua Achebe "begat" many sons, Flora Nwapa "begat" many daughters, who today carry on her legacy in various genres—the novel, the essay, the short story, children's books, poetry, and drama—in Africa and in other parts of the world. I agree with Niara Sudarkasa that "the strength of Black American women is rooted in African societies" (179). Flora Nwapa, the doyenne of African women writers, was my own literary mother who nurtured me to growth. Her novels captured my imagination and thus gave me a voice. Since our first meeting at Syracuse University in 1973, our professional friendship blossomed. Her novels inspired me and I used them as texts in my classes; her books for children played a significant role in the development of my own children, as Flora Nwapa became a household name. Flora and I continued to meet at her lectures held at conferences and book fairs in Nigeria and the United States. Her female characters were often the subject of many of my critical and theoretical essays. Many of us remember her and think of her as the proverbial Igbo child remembers a most loving mother. How can one ever forget the role she has played in the lives of so many of her literary daughters—Buchi Emecheta, Ifeoma Okoye, Zaynab Alkali, Chikwenye Okonjo Ogunyemi, Yemi Mojola—to name only a few of us who are committed to passing on her "poetics of economic independence for female empowerment?" In her transition to ancestorhood, we continue to be renewed, keeping in mind the words from Birago Diop's poem "Spirits":

> Listen to things
> More often than beings,
> Hear the voice of fire,
> Hear the voice of water.
> Listen in the wind,
> To the bush that is sighing:
> This is the breathing of ancestors,
> Who have not gone away
> Who are not under earth
> Who are not really dead. (152-4)

Acknowledgements

I am very grateful to Davidson C. Umeh, Ernest N. Emenyonu, Chimalum Nwankwo and Mrs. Njideka Ibuaka, née Nwapa, for their voices from Igboland.

Notes

*Both Ajie Ukpabi Asika and Ken Saro-Wiwa's quotes are taken from the obituary, *The Woman Called Flora: Professor Flora Nwakuche (Née Nwapa) 1931-1993*, written by her family. These excerpts reflect the high regard the Nigerian community held for Flora.

1. This proverb is taken from Chieka Ifemesia's book, *Traditional Humane Living Among The Igbo: An Historical Perspective*, p. 117. The proverb points to the importance of reciprocity between mother and child. It cautions the mother to nurture her child to growth. The child must also acknowledge the mother's tender loving care. The mother gives love and reaps constant respect, honor, and appreciation from her child.

2. In a conversation with Flora Nwapa's son, Uzoma Nwakuche, he informed me that Tana Press was named after a river in East Africa. *Webster's New Geographical Dictionary* states that the Tana River is situated in Kenya, East Africa. The dictionary points out that there is also a Tana/Tsana Lake in Northern Ethiopia, p. 1181. Whether Tana Press is named after Tana River or Tana/Tsana Lake is left for Nwapa's biographers to tell us.

3. The first woman—the earliest known human ancestor—is "Lucy" the African mother of all mankind. See Van Sertima's *Black Women in Antiquity*, p.5.

4. Elechi Amadi cites from J. A. Sofola's book, *African Culture and the African Personality*. See Amadi's book, *Ethics in Nigerian Culture*, p. 52.

5. A class of women were allowed to go on extended-trade but they were either spinsters or barren women. Ernest N. Emenyonu cited this fact in a paper he presented, "Nigerian Women Entrepreneurship: Myth or Reality?" at the Workshop of the Nigerian Association of Women's Entrepreneurships, Port Harcourt Civic Centre, River State, November 21, 1995.

6. L. Amede Obiora in her essay, "Reconsidering African Customary Law," credits Robert W. Harms' book, *River of Wealth, River of Sorrow*. New Haven: Yale UP, 1981.

7. Cheikh Anta Diop, in *Precolonial Black Africa*, deconstructs western hegemony in African trade and economy when he contends that African societies were economically developed before European intrusion.

8. Catherine Acholonu presents an Afrocentric theology of female power and economic influence in African societies in her theoretical work, *Motherism*.

9. The United African Company (UAC) is one of the early trading companies the British brought to Nigeria during the colonial era and it has survived until today. See A.C. Izuagha's Essay, "Oguta and the Novels of

Flora Nwapa: A Bibliographic Approach."

10. Nwapa also makes references in this essay to the powerful Queen of the Onitsha market, who directly controls all trading activities.

11. *Nigerian Women Mobilized,* Nina Mba's interviews with Oguta women, offers authentic oral testimonies on post-colonial women's experiences.

12. The historian Elizabeth Isichei, like Nina Mba and Chieka Ifemesia, et al., also points to Oguta women's unique and autonomous cultural heritage and their role in spreading Christianity in Iboland. See her book, *The Ibo People and the Europeans.* London: Faber, 1973.

13. Nnamdi Azikiwe (1904-1996), the Owelle of Onitsha, and the first President of the Federal Republic of Nigeria (1963-1965), led the campaign for the political emancipation of his people.

14. In an interview with *Quality Weekly* magazine team, Flora Nwapa shares this information with Segun Adeleke, Bose Adeogun and Ben Nwanne. See *Quality Weekly,* August 23, 1990, Vol. 6, No. 8, p. 34.

15. This is the title of Charlotte Chandler's book, *The Ultimate Seduction,* which reveals the basic concepts and insights of successful people who found in their work supreme satisfaction. According to Chandler, "the ultimate seduction is not about sex, but about people who found work the ultimate satisfaction (3)."

16. Nwapa's essay, "Priestesses and Power Among The Riverine Igbo," is published in Flora Kaplan's book, *Queens, Queen Mothers, Priestesses and Power: Case Studies in African Gender.*

17. Katherine Frank states that Flora Nwapa is Africa's first woman publisher. See *Africa Now* (May, 1983): 61-62.

18. In 1992, Mr. Kassahun Checole, President of Africa World Press, published Flora Nwapa's novels, *Never Again, One Is Enough, Women Are Different,* and her two collections of short stories, *This Is Lagos and Other Stories* and *Wives at War and Other Stories.* In a conversation with Mr. Checole in 1996, he informed me that both he and Uzoma Nwakuche are negotiating to publish the novel, *The Lake Goddess,* jointly with Tana Press Limited.

19. Ezenwa-Ohaeto's essay in this volume, "Breaking-Through: The Publishing Enterprise of Flora Nwapa," gives the history of Tana Press Limited and Flora Nwapa Books and Company.

20. Uzoma Nwakuche, personal interview with Marie Umeh, 13 July 1994.

21. Uzoma Nwakuche, in memory of his mother Flora Nwapa, is soliciting assistance and support for the Flora Nwapa Foundation. For more information see his advertisement in *Research in African Literatures* 26.2(1995):251.

22. Flora S. Kaplan in her essay, "Some Uses of Photographs in Recovering Cultural History at the Royal Court of Benin, Nigeria," gives a scintillating analysis of the ways in which photographs supplement and support ethnographic study. See *Visual Anthropology,* 1990, 317-341.

Works Cited

Acholonu, Catherine O. *Motherism: The Afrocentric Alternative to Feminism.* Owerri, Nigeria: Afa, 1995.

Adeleke, Segun, Bose Adeogun, and Ben Nwanne. "Quality Interview with Flora Nwapa: Some ANA Members Are Crazy." *Quality Weekly* 6.8(August 23, 1990): 28-35.

Afigbo, A. E., ed. *Groundwork of Igbo History.* Lagos: Vista, 1992.

——. "The Igbo Under British Rule." Ed. M. J. C. Echeruo, and E. N. Obiechina, *Igbo Traditional Life, Culture and Literature.* Owerri, Nigeria: Conch, 1971.

——. ed. *The Image of the Igbo.* Lagos: Vista, 1991.

——. "Women in Nigerian History." Ed. M.O. Ijere, *Women in Nigerian Economy.* Enugu: Asena, 1991.

Amadi, Elechi. *Ethics in Nigerian Culture.* Ibadan: Heinemann Educational Books, 1982.

Amadiume, Ifi. "Beyond Cultural Performance: Women, Culture and the State in Contemporary Nigerian Politics." Ed. David Parkin, Lionel Caplan and Humphrey, *The Politics of Cultural Performance.* Providence: Berghahn, 1996, 41-59.

Asante, Molefi Kete & Abarry S. Abu. *African Intellectual Heritage: A Book of Sources.* Philadelphia: Temple University Press, 1996.

Awe, Bolanle, ed. *Nigerian Women in Historical Perspective.* Lagos/Ibadan: Sankore/Bookcraft, 1992.

Azikiwe, Nnamdi. *Choose Independence in 1960 or More States: A National Challenge.* Onitsha: Varsity, 1959.

Chandler, Charlotte. *The Ultimate Seduction.* New York: Doubleday, 1984.

Chikwe, Kema. *Women and New Orientation: A Profile of Igbo Women in History.* Owerri: Prime Time, 1994.

Chinweizu. *Anatomy of Female Power. A Masculinist Dissection of Matriarchy.* Lagos: Pero, 1990.

Clarke, John Henrik. "New Introduction." *The Cultural Unity of Black Africa.* Chicago: Third World Press, 1978. i-xv.

Dike, K. Onwuka. *Trade and Politics in the Niger Delta: 1830-1885.* Oxford: Oxford University Press, 1956.

Diop, Birago. "Spirits." Ed. Ellen Conroy Kennedy, *The Negritude Poets.* New York: Thunder's Mouth Press, 1975, 152-154.

Diop, Cheikh Anta. *The Cultural Unity of Black Africa.* Chicago: Third World Press, 1978.

——. *Precolonial Black Africa.* Brooklyn, New York: Lawrence Hill, 1987.

Egbufor, Ada. "Flora Nwapa: Africa's First Lady of Letters." *African Profiles International.* March-April, 1993, 34.

Emenyonu, Ernest N. "Nigerian Women Entrepreneurship: Myth or Reality?" Paper presented at the Workshop of the Nigerian Association of Women's Entrepreneurships. Port Harcourt Civic Centre, River State, November 21, 1995.

Ezenwa-Ohaeto. "Flora Nwapa (†1993): Interview with Ezenwa-Ohaeto (April 1993)." *ALA Bulletin* 19.4(Fall, 1993): 12-18.

Frank, Katherine. "Flora Nwapa, Africa's First Woman Publisher." *Africa Now* 25 (May, 1983):61-62.

Harms, Robert W. *River of Wealth, River of Sorrow: The Central Zaire Basin in the Era of the Slave and Ivory Trade, 1500-1891.* New Haven: Yale, 1981.

Ifemesia, Chieka. *Traditional Humane Living Among The Igbo: An Historical Perspective.* Enugu: Fourth Dimension, 1979.

Isichei, Elizabeth. *A History of Christianity in Africa: From Antiquity to the Present.* Trenton, New Jersey: Africa World Press, 1995.

——. *The Ibo People and the Europeans.* London: Faber, 1973.

Izuagha, A.C. "Oguta and the Novels of Flora Nwapa: A Biographical Approach." A paper presented at the 13th Annual International Conference on African Literature and the English Language, University of Calabar, Nigeria, May, 1994.

James, Adeola. *In Their Own Voices: African Women Writers Talk.* London: Heinemann, 1990.

Kalu, Ogbu U. "Education and Change in Igboland, 1857-1966." Ed. A. E. Afigbo, *Groundwork of Igbo History.* Lagos: Vista, 1992, 522-541.

Kaplan, Flora Edouwaye S., ed. *Queens, Queen Mothers, Priestesses, and Power: Case Studies in African Gender.* Vol. 810. New York: The New York Academy of Sciences, 1997.

——. "Some Uses of Photographs in Recovering Cultural History at the Royal Court of Benin, Nigeria." *Visual Anthropology* 3(1990): 317-341.

Leith-Ross, Sylvia. *African Women: A Study of the Ibo of Nigeria.* New York: Frederick A. Praeger, 1965.

Lindfors, Bernth. *Long Drums and Canons: Teaching and Researching African Literatures.* Trenton, New Jersey: Africa World Press, 1995.

Mba, Nina Emma. *Nigerian Women Mobilized: Women's Political Activity in Southern Nigeria, 1900-1965.* Berkeley, University of California, 1982.

Ngũgĩ wa Thiong'o. *Moving the Centre: The Struggle for Cultural Freedoms.* Portsmouth, New Hampshire: Heinemann, 1993.

Nwakuche, Uzoma Gogo. Personal Interview with Marie Umeh. 13 July 1994.

Nwapa, Flora. *The First Lady [A Play].* Enugu: Tana, 1993.

——. "Priestesses and Power Among The Riverine Igbo." Ed. Flora Edouwaye S. Kaplan, *Queens, Queen Mothers, Priestesses and Power: Case Studies in African Gender.* Vol. 810. New York: New York Academy of Sciences, 1997.

——. "Two Women in Conversation." *Conversations [Plays].* Enugu: Tana, 1993, 70-144.

——. "Women in Politics." *Présence Africaine.* 141.1(1987): 115-121.

Nzeribe, Ejine, Uzoma Nwakuche and Amede Nwakuche. *The Woman Called Flora: Professor Flora Nwakuche (Née Nwapa) 1931-1993* [Obituary Program]. November 20, 1993, Oguta, Nigeria.

Obiora, L. Amede. "Reconsidering African Customary Law in Africa." *Legal Studies Forum* 17.3 (1993):217-252.

Ogunyemi, Chikwenye Okonjo. *Africa Wo/man Palava: The Nigerian Novel By Women.* Chicago: University of Chicago, 1996.

Stevenson, Arthur J., ed. *Webster's New Geographical Dictionary.* Springfield, MA: G. & C. Merriam, 1980, 1181.

Stratton, Florence. *Contemporary African Literature and the Politics of Gender.*

London: Routledge, 1994.

Sudarkasa, Niara. *The Strength of Our Mothers: African & African American Women And Families: Essays and Speeches.* Trenton, New Jersey: Africa World Press, 1996.

Sylvester, Christine. "African and Western Feminisms: World-Traveling the Tendencies and Possibilities." *SIGNS* 20.4 (Summer, 1995): 941-969

Umeh, Marie. "Signifyin(g) The Griottes: Flora Nwapa's Legacy of (Re)Vision and Voice." *Research in African Literatures.* 26.2(Summer, 1995): 114-123.

Uwazurike, Chudi. "Azikiwe—'*The Renascent African*' Takes His Farewell." *African Profiles International.* June-July, 1996, 52-57.

——. "Flora Nwapa—Tributes & Testimonials." The Nigerian House Conference Hall, New York, New York, November 21, 1993.

Uwechue, Austa. "Flora Nwakuche, née Nwapa, A Former Cabinet Minister and One of Africa's Leading Women Writers Talks to Austa Uwechue." *Africa Woman.* 10(July/August, 1977):8-10.

Van Sertima, Ivan, ed. "The African Eve: Introduction and Summary." *Black Women in Antiquity.* New Brunswick: Transaction, 1985.

Tribute(s) to Flora Nwapa

The acknowledgement of the human essence, the illumination of the individual's achievements and the indication of future influences are part of the features of tributes. Thus, the various writers of these tributes centralize both the person of Flora Nwapa and the context of her life. By relating their works to the cultural context, continental context and especially human context, the writers of these tributes clearly stress that a life of achievement possesses wider ripples beyond the borders of the locality of birth place. And for Flora Nwapa, this is a deserved tribute from respected writers and academics.

The tribute poems here clearly portray Flora Nwapa's influence and impact on the literary development in Africa as worthy of celebration. Chimalum Nwankwo in "Ivory Befits Her Ankles" uses the image of ivory which adorns the ankles of titled women in some parts of Igbo society as a metaphor for grounding Nwapa's successful literary life. The figure of Nwapa as a worthy ancestor is vividly delineated in the second stanza of Nwankwo's poem as Nwapa is praised for living, learning, mothering, writing, weeping, laughing, and more importantly loving "with our burden of life." Thus the poet perceives Nwapa as having lived her life in the service of her people. This same sense of loss in which is subsumed a feeling of gratitude for Nwapa's life is also present in the poem by Fehintola Mosadomi which memorably presents her brief encounter with Nwapa. Mosadomi is convinced that rather than time functioning as the great healer that it will be the late writer's literary legacy that will assuage the pain of her death.

Similarly, Naana Banyiwa Horne and Ezenwa-Ohaeto in their

poems capture the sense of loss associated with Nwapa's death. Horne in the poem, "Demirifa Due," perceives a unique development in Nwapa's death which she symbolizes as the way the "sun and the rain" vie for attention like "the way Nwapa's literary children often vie for (her) love." This symbol clearly reflects on exhibition of appreciation in the fact that Nwapa made indelible literary contributions for the poem adds that Nwapa is "the sunshine that stayed/our season of frost." The same symbol of greatness is used by Ezenwa-Ohaeto in his poem, "The Trees at Home are Falling Down" to illustrate the fact of Nwapa's death. Ezenwa-Ohaeto perceives Nwapa rightly as a "lamp bearer in the darkness" of social tergiversations—a role that situates her achievements as a writer. However, this poet widens the lament in the poem to question that regularity of such losses of human lives in Africa for he insists that "the gods must change this pad/with which we carry our world." The association of Nwapa's death in Ezenwa-Ohaeto's poem with the idea of a falling giant tree is not an overstatement for Buchi Emecheta and Molara Ogundipe-Leslie echo that same impression.

Buchi Emecheta in her tribute essay, "*Nwanyi oma, biko nodu nma,*"[1] recollects an intellectually emphatic relationship with Nwapa. Emecheta acknowledges Nwapa's literary influence and her subsequent relationship with her as successful literary colleagues. Emecheta's testimony goes to substantiate what many of those who knew Flora Nwapa had accepted as her warmth, her humaneness, and her quiet dignity. In her own tribute, "Dirge to Flora Nwapa," Molara Ogundipe-Leslie also testifies to that warmth and elegant personality. Using a Yoruba dirge chant, Ogundipe-Leslie's poetic lament captures the many facets of Flora Nwapa as woman, person, pioneer and now ancestor. Adopting the poetic device of invocation, she ends her dirge with the haunting words: "You who have gone where they go, departed sister, who has become rain water flowing; rain water is healing, energizing, fructifying. Fructify our lives that we may continue to produce as you did throughout your life. . . ."

This section ends with a tribute essay by Ernest N. Emenyonu who highlights the significant accomplishments and literary interventions of Flora Nwapa. He identifies Nwapa's relevant thematic occupations that include casting insight on contemporary Nigerian politics, exploring domestic feuds and parochial malice, centralizing women affairs, condemning undeserved humiliation and dehumanization of womanhood, as well as stressing issues of self-development. More importantly, Emenyonu argues that Flora

Nwapa, in spite of her achievements remained a warm, dignified, unassuming, gentle and articulate woman. His closing statement that Nwapa "showed the light to other women" and that through her literary works she enabled African women to acquire a "distinct image and identity" reaffirms the indelible nature of Nwapa's life. Such deserving tributes stress the importance of not only Flora Nwapa but also this collection of essays.

Ezenwa-Ohaeto

1. The English translation is "Beautiful woman, please farewell."

Ivory Befits Her Ankles

Chimalum Nwankwo

Because she was of water
When the last wave heaved
Her dainty ankles answered

She lived with our burden
She learned with our burden
She mothered with our burden
She wrote with our burden
She wept with our burden
She laughed with our burden
She loved with our burden

And before the last wave rolled
She could still the water
Streaming from the lake goddess
Before the last forest of the seven forests
Before the seventh river of the seven rivers
Before the last step of her dainty ankles
She could still the water
Streaming from the lake goddess

The songs of songbirds at the festival
The dirges of fire at the funerals
The ripples of mirth at low ebb
The whirlpool of wrath at hightide
She could still the water
Streaming to the lake goddess

She could still the water
Streaming to the lake goddess

Ivory befits her ankles
Ivory befits her ankles
Ivory befits her ankles

Nwayi oma, biko nodu nma[1]

Dear Flora,

I can't believe you're gone. Your going makes death so near. What is it like up there or down here, in the air or in the water—wherever your soul is now? How I wish you could give us, all your friends, a hint to let us know where you are and what you are up to. Are you still laughing your mischievous laugh and saying, "Oh, it's all politics"?

It seems like only yesterday when you telephoned me in 1979 to say you wished to speak to me! What a shock that was. Since childhood, I have heard of your family's name—Nwapa. Nwapa, made more famous by your politician uncle, Golden Water (Star Beer) Nwapa. They say he loved his Nigerian Star Beer. Then when he passed away, we started hearing and reading about you as one of those early Nigerian young women born with silver spoons in their mouths.

As a young girl, I virtually devoured all the books I could lay my hands on written by women, whilst nurturing the hope of writing one day. When I came to England, borrowed a copy of *Efuru* from my local library, and read and re-read it, my mind was made up. Since you had written this, I could start writing as well. You became my model. I was not born with a silver spoon in my mouth, but I was not lacking in ambition. You left the literary field and went into Nigerian politics. I lost interest in your progress. I never liked politicians and still don't respect them that much.

During this time, I was going through hell (*In the Ditch*) and came across *Idu*. Then I started writing in earnest. When my first book was burned by my husband, I did not give up. Flora Nwapa could do it, so could I. My husband said I was living in cloud cuckoo land. I had not then seen you in the flesh, but you were influencing me. Then the civil war came, and I bought Achebe's *Girls at War*. I could not push my way to make him sign my copy. I must have looked a sight with two babies in a buggy and three toddlers holding onto my skirt, and the day was wet, and the big man was surrounded by many loud and inspiring Igbo scholars.

I thought about you then, wondering where you were. The anger at the humiliation, coupled with the frustration of the civil war, made me bank on the diary that later became *In the Ditch*. And it was published. All that time I thought you were permanently lost to the literary circle and had joined the Nigerian political vanguard.

I met Dr. Brenda Berrian in May '79 and she told me she was going to Nigeria. Imagine my surprise when she wrote to say she went to your office in Enugu and saw my picture on the wall with the list of all my publications proudly displayed! All I was thinking was "Flora Nwapa knew about me!" And our mutual friend and wonderful woman from Pittsburgh, America was determined to make us meet. She succeeded.

You came to London and telephoned. I promised nervously that I would come and pay you a visit, because I thought my house with my boisterous family would not be good enough to welcome somebody like you. Whilst I was mopping my kitchen floor and wondering what to wear for the visit, you phoned again, you could no longer wait, you were coming right away. I started running round in circles like a rat on a bed of hot coal. I gave millions of instructions to my children to behave, to disappear, to wash their faces. I was still holding the mop when you came in a black cab. Few people visited us in black cabs in those days. My Alice refused to budge and kept staring at you, and you gave her that huge chocolate box you bought for all of them, and she said she liked you and sat on your lap. You looked at my face, I looked back at you and I was rewarded with that famous Flora's special laughter—with your tongue sticking out a little, head thrown back, one hand adjusting your head tie, and this beautiful uninhibited sound coming from your mouth like someone singing. That moment you disarmed me with your naturalness. I think I became one of your friends. I knew you became my big sister in the literary field. You followed me everywhere, and I followed you too.

When you saw me in Calabar you said, "Pity Nigeria did not value you." This was because I left you in my suite at the Savoy in Frankfurt, when I was a guest speaker there. I knew then, and I think you suspected, that most of us Nigerian writers would have to spend our writing lives in exile. Dear Sister and gentle woman, I am missing you a lot. Jetting round the world and staying in very, very expensive hotels used to be double joy, because I knew that at the end of the conference, we would stay a day or two behind, spending our hard-earned money the way we liked, reading and discussing all the new books and conference papers and of course making jokes about some pompous writers.

One innocent Igboman wrote an appreciation of you recently in one of our weeklies. He said you're not a Feminist, not even a feminist with a small "f." How those of us who knew you well had laughed. How was he to know that most of your last visits to London were funded by the Feminist movement, and that your last talk at the Feminist Bookshop in Upper Street, London, had a strong cultural Feminist message? How was he to know that what we African women writers resented in the Feminist movement was the fact that the name was from the West, and that we still cherish our families, that we value most of our relevant cultures? This is why some of us claim to call it by other names: Cultural Feminism or Feminism with a small "f," some say Womanism and some Universal Womanism, but at the end of the road, we are all working towards the same end—the dignity of the Woman.

I think you are having the last laugh over all this. That bell-like laughter, with one foot in the air and your elegantly arranged headtie thrown backwards, saying, "Let our men believe what they like, Buchi, what does it really matter? It's all politics."

We all miss you. *"Nwayi oma, biko nodu nma."*

Dr. Buchi Emecheta

1. The title in Igbo literally means in English, "Beautiful woman, please farewell."

Your Golden Words

Dear Flora,
Did you receive my last letter?

Our brief meeting
under the sunny skies
of Nsukka
the site of WAAD,
the vision of OBI
A happy moment.

Under the shady roof
where lay your body
of knowledge and experience
smiling at me, saying
"take me home."
A moment of joy.

Thereafter, your soft thin voice,
your cool smiling face
beaming their lights of happiness
through my ears and eyes
penetrating the loudness of New York
through the serenity of Hamden-Sydney.

The echo of your words
from Nigeria
to Virginia
whispering to me
"You've not seen
the last of me.
May be gone,
but here to stay."

Little did I know
that my first glance
of you
was my last.
Little did I know

that your words of wisdom
would be kept brief.
Your sudden departure,
my miriads of goosebumps.
Sister Flora
you were loved
and still are.
A loss to me and to all
but not
forever.

Memories of you
shall remain
at the corner
of my heart
from Nsukka
with love
to Hampden-Sydney
with tears.

Time will not heal
the pain of your death,
your legacy
will.

Sleep on,
beautiful angel
sleep on.
Your golden words shall glow
over the dim shadow of life.

'Til we meet again,
You are missed.

Sincerely,

Fehintola Mosadomi

Dirge to Flora Nwapa[*]

O d'agbara
O d'agbara
Flora Nwapa d'agbara ojo
K'o ya'le mo

(translation)
She has become flowing water
She has become flowing water
Flora Nwapa became flowing rain water
She will not stop by houses anymore

No, her essence is no longer housed in flesh. She will no longer have mundane human needs. She need not look for shelter, food and clothing in human houses. She is beyond stopping to say hello at our houses where we receive her with or without gladness. She has eternal well-being now with the ancestors.

But we still need her spirit to visit our houses, particularly our creative houses. We need the grace of that most elegant of women; elegant in speech and appearance. Oh, Flora Nwapa, big sister, why did you leave so early? We cannot call you an old *ogbanje (abiku)*. Sixty years are quite respectable in our continent of shorter lives. But it is still too early, as it always is, for those who are treasured. Blame it, not on history. Blame it on collapse. The collapse of every- thing basic at home. The collapse of our medical institutions, our communications system, our schools, our ethics, our... how many more shall we name in our country, wrecked by those who promised us a better Nigeria.

Who would have thought you would leave so soon, my elegant sister, laughing now among the ancestors. You, a well-born woman who bore herself again into well-deserved fame, national and inter- national. Niece and intellectual daughter of the great A.C. Nwapa, early Nigerian nationalist and honored gentleman; sister to C. Fine, our fine college mate at the great, old Ibadan University; you, lead- ing light from the Archbishop Crowther Memorial Girls' College, fondly known by us and throughout Nigeria as Elelenwa; early woman graduate of the University at Ibadan; university administrator and minister of State; writer, wife, mother and trusty friend. May the ancestors embrace you.

Woman of many firsts! First Nigerian woman novelist to be published in English and in the U.K.; first African woman to set up her own publishing house and publish different kinds of books; first great seer who perceived that African women need to be justly represented in fictional art; woman who knew so early that African women were not just male-oriented sexual beings but multifaceted and *human* beings - strong and weak, sweet tempered and tempestuous, stupid and brilliant; redoubtable and self-assured, resourceful, tireless, loving or quarrelsome; sexual beings too, but only just people. May your creativity empower us.

Who would have thought it would be so soon, Flora, when I introduced you to a respectful audience at Rutgers University only last year, the Spring of 1993, when you read a new, rich paper on ways of looking at African women? Who would have thought you would join the ancestors so soon when we were laughing together about phases in the life of a woman at the reception Africa World Press held for you in Seattle (ASA, 1992) when that publishing house reproduced some of your major fiction? Who would have thought you were leaving?

Beloved older colleague who left behind much writing, enjoy yourself among the ancestors (if they allow women to sit among them!). Laugh and dance yourself translucent as you loved to dance here, agile as a maiden. Tell the ancestors to send some light to our brothers here. Tell them to put some balm on their souls, so that they give women writers the attention they do not hesitate to give to mediocre male writers.

Tell the ancestors to tell the brothers here that all women need not write *Things Fall Apart* before their work is given critical attention. Remind the brothers that your *Efuru* was published with enthusiasm by our great, older brother and author of *Things Fall Apart* who had the depth of perception to see (as genius would) the independent value of *Efuru*; a brother who knows in his great, respected and graceful soul that women's writing need not be judged by works written by men.

You who have gone where they go, departed sister, who has become rain water flowing, rain water is healing, energizing, fructifying. Fructify our lives that we may continue to produce as you did throughout your life. Flow into our lives and houses like rainwater, sweeten and heal our thirst for creativity and knowledge, plead with the ancestors to unite for positive action in Africa, the searching minds of women and men, who would be humble; you, most gracious of women, Flora Nwapa Nwakuche.

Molara Ogundipe-Leslie
1/13/94
Fort Wayne, Indiana

*1st version published in the *ALA Bulletin* 20, 1 (1994) 14-15.

Demirifa Due

(for Mamas Selina Kunene and Flora Nwapa)

Demirifa due
Demirifa due
Demirifa due
Due
Due
Demirifa Due!

It rained today Mama
My tears were not enough
to fill up the rivers
to speed you on
your journey to *Asamando*
to join our *Nananom*
So I called upon the heavens
to open up and weep rain

The bounty of the heavens
Mama matched your bounty
to us your children
The heavens dropped
tears in torrents
washing the skies blue
in celebration of your passing on

The sun and the rain
vied for your attention today
Mama the way your children
often vie for your love

The sun blazed an eternity
shining bright and blue
causing everything to shimmer
in intimation of your mystery
The sun caused everything to stand
still for your eternity

The rain fell profuse and steady
despite the blazing sun
filling up the river
the river to *Nsaman pow mu*
to make your transit smooth
Today was a wonder to behold
Mama the wonder of you
You the sunshine that stayed
our season of frost

Demirifa due
Demirifa due
Demirifa due
Due
Due
Demirifa due!

Naana Banyiwa Horne

The Trees at Home
Are Falling Down

(For Flora Nwapa)

What Hurricane came to the clan?

The news from home
Sends the heart jumping.

In the forest at home
the trees are falling down,
In the farm at home
the plants are turning yellow.

Which wind blew into the house?
Now from the armpits of pain
We scratch out joys
We snatch out hopes,

The gods must change this pad
with which we carry our world
For the unlucky one is related to the prodigal.

Even in a foreign land
the wail of an ambulance
tells how it is at home
When the trees fall down.

But why is pain so sharp?
When the news comes from home
recalling for us faces we farewelled
With smiles that promised reunion.

The hurricane came to the clan
The wind blew into our house
If we fall must we reach ground?

The news from home
sends the heart jumping
It falls bruised and bleeding
And tears start flowing.

It hurts most
When a lamp bearer is recalled
And one cannot find a path in the dark
Or find the path in the dark with no lamp.

It hurts most
When the trees fall down
The desert overtakes the land
The sun blazes on bare heads
And we are far from shelter.

It hurts most
When death has not been kind.

Ezenwa-Ohaeto

Flora Nwapa: A Pioneer African Female Voice Is Silenced!

Nigerians, Africans, and indeed people the world over who have been in one way or the other associated the African Literature heard the rude shock on Saturday, October 16, 1993—the announcement of the passing away of Mrs. Flora Nwakuche, known to the literary world as Flora Nwapa. Flora Nwapa has a unique place in the history of African Literature of the 20th century.

She was Nigeria's first female novelist and one of the earliest self-confident female voices on the continent of Africa at a time when African writing was dominated by and seen as the exclusive preserve of men and men alone.

In 1966, Flora Nwapa published her first novel, entitled *Efuru*. This singular act, which drew relatively little attention at home, soon blossomed into a feat of international significance. By the publication of *Efuru*, Nigeria had produced in Flora Nwapa its first female novelist. The significance of this novel, however, was more than the ordinary historical fact of a pioneer work; for in this novel Flora Nwapa created female characters that were at once unique and unconventional. The typical female characters created by male writers before this point were docile, subservient, and mindless women who were only seen but never heard. They existed only for the services they provided inside the home for their children and husbands. They did not count in any significant matter affecting community or national development.

Flora Nwapa's heroine, Efuru, was an uncharacteristic artistic phenomenon. She had her own mind. She had her own personality. She distinguished herself by her industrious nature. She had administrative acumen and economic prudence. Whatever the merits and demerits of Flora Nwapa's first novel, its major significance lay in its uncompromising desecration of the myth of female irrelevance in national development. Flora Nwapa's subsequent works followed this pattern of giving the African female characters voice and identity; imagination and intellect, wit and wisdom in human

41

affairs—no matter how offensive and loathsome to the typical (para-noid) Nigerian readers. After Flora Nwapa's debut, Nigerian and, other, African female authors found their feet and found the vocal resonance which they needed in their voices to repudiate what had become traditionally and fashionably a chauvinistic approach by male writers writing about the female human condition.

The setting of Flora Nwapa's major novels is her own home town, Oguta, which she has placed indelibly on the world literary map. Flora Nwapa focused on the culture and life-ways of the Igbo, specifically, the fishing and farming residents of Oguta, who found occupation and pleasure in the Oguta Lake and to whom the fantasies of "The Woman of the Lake" are a reality. This is the realism which became the special feature of Flora Nwapa's art. Her style was a distinct one. She introduced the reader to a problematic life situation such as barrenness, infidelity, or envy. This problematic situation soon took on an aura of tragedy. In the process, Flora Nwapa explored in depth the beliefs, aspirations, failures, and successes of the people whose life-ways created the particular human condition depicted in the novel. Flora Nwapa in her idiosyncratic manner always brought female closeness and intuition into the theme in a way that would best be understood by Nigerian women in particular and all women everywhere in general.

In the nearly three decades that Flora Nwapa almost dominated the African female literary scene, she brought variety and versatility into her writings. She wrote on the Nigerian Civil War. She wrote on the frivolity of contemporary Nigerian Politics. She wrote on domestic feuds and parochial malice. She wrote on women's affairs. She wrote on undeserved humiliation and the dehumination of African womanhood. Above all, she wrote about Oguta Lake and the Oguta people. She wrote about the Igbo and the Igbo people. She wrote about Nigerian and the Nigerian people.

No one can write an authentic history of African Literature or indeed Nigerian literary history today without giving Flora Nwapa a prominent place because without her, African women writers would have continued to remain in the shadows of male writers. Flora Nwapa was a resounding success on various fronts. To many she was an accomplished educator. To others she was an astute administrator being one of the first women to be appointed a Commissioner in the Eastern Nigerian Government. To literary scholars, she was one of the first few female writers on the African continent who had the vision to combine imaginative creativity with business enterprise when she established a viable publishing company—the Tana Press. To children and all mothers, Flora Nwapa was a kind-hearted, humane, Christian mother and friend. A woman who had social concern for the less privileged and the despised in the contemporary society.

The death of Flora Nwapa is an irreparable loss on the Nigerian literary scene because she had steadily grown from an experimental novelist to a self assured author who had begun to take full mastery of the art of story telling. There may never be again a Flora Nwapa in the Nigerian literary environment; a woman with an articulate voice, a woman with a sense of dignity; a woman with firmness yet unsurpassed in her gentleness of demeanor; a woman who like the female protagonists in her fiction believed in meritocracy; a woman who believed that success naturally stemmed from hard work. A woman who took nothing for granted. A woman of vision. An innovator. A catalyst. An inspirer. A woman who was never ashamed to proclaim the dignity of labor. Flora Nwapa until her death was a model of womanhood both for the young and the aged. She gave all she had to humanity. She gave all she had to Nigeria. She gave all she had to womanhood. But she took nothing from all these! She has left for posterity over a dozen books, each of them with a clear message. On the Nigerian Civil War, she told us *Never Again*. On the permissiveness of our biased polygamous environment, she cautioned us, *One Is Enough*. On the bizarre ecstacies of city life, she beamed onto our spectacle *This Is Lagos*. On the incontrovertible dynamism of the Nigerian womanhood in the face of challenges, she projected for us, *Wives at War*. On the battle of the sexes, and who plays what roles, where and when, she admonished us that *Women Are Different*.

Nigeria, therefore, has lost one of her best female voices. Africa has lost one of her most confident female personalities, and the world has lost a crucial link between male and female imaginative art of our times. Oguta has lost her greatest liaison and public relations advertiser with the outside world. But we will always remember Flora Nwapa, the author of *Efuru*; Flora Nwapa, the author of *Idu*; and Flora Nwapa, a woman who showed the light to other women and after whose literary creativity, African womanhood and/or feminism acquired a distinct image and identity.

We mourn her for our sake and for the sake of posterity.

Adieu, Flora.

May God give you eternal repose.

—**Ernest N. Emenyonu**

PART ONE
IGBO WOMEN:
CULTURE AND LITERARY
ENTERPRISE

Ugwuta, the setting of Flora Nwapa's novels, *Efuru, Idu, Never Again,* and *The Lake Goddess,* is one of the few towns in Igboland with a woman at the helm of village affairs. In fact, it is only in Ugwuta that Nigerian women break and share the prestigious kola nut and the men will eat.[1] Today in Ugwuta, the oldest person in the village who is highly regarded and called upon to solve difficult problems is a titled woman, addressed as the "Ogene Nwanyi."[2] This prominent place accorded to Ugwuta women stems from women's role in the palm produce trade of the United African Company (UAC) which was established by the British who came to Ugwuta in 1889.[3] Ugwuta women were the first group of people to buy and sell the British products such as biscuits, pots, clothes and tobacco along with their palm kernels. Ugwuta men, who were primarily farmers and fishermen, were not eager to trade with the foreigners in the beginning. Hence, the centrality of intelligent, rich, wealthy, industrious, and confident women in Nwapa's *oeuvre.* Furthermore, the myth of the beautiful, wealthy and proud

Woman of the Lake also influenced Flora Nwapa in her creation of the formidable Uhamiri a.k.a. Ogbuide. Uhamiri/Ogbuide is a female deity whom Ugwuta indigenes deeply respect, if not boldly worship in the face of Christianity. Ugwuta people believe that *Ogbuide,* the lake goddess, is the source of life, wealth and fortune— the one who protects them from all evil and misfortune.

Before the impact of Christianity/Westernization in Igboland, Ugwuta women made their mark in the society as verbal wordsmiths as well as wealthy traders. The local artists were notable for telling moonlight stories, educating children, and entertaining the community through the medium of proverbs, riddles, folktales, songs, chants, and dance, as well as singing praise songs at traditional marriage, birth, and burial ceremonies. In interviews with countless journalists and scholars, Flora Nwapa insists that her literary imagination and creativity were sparked by the women in her grandparents' village, where she often spent her holidays and for the first time heard stories about Mammywater. Additionally, in another interview, Flora Nwapa says that as a child she was an avid listener "to a lot of stories, moonlight stories told by the women in Ugwuta."[4]

The grandchild and daughter of Western-educated parents, Flora Nwapa was a descendant of one of Ugwuta's first Christian families. So, naturally, she went to the local missionary school, where she learned the art of reading and writing. Flora Nwapa later furthered her Western education at the University College, Ibadan in 1953, and was the classmate of Ukpabi Asika.[5]

The seeds of Nwapa's literary enterprise was, therefore, planted in Ugwuta by the community of ubiquitous traders, born fabulists, traditional grandparents and Western-educated family members, who were wealthy palm kernel traders, as well as astute educators. The Ugwuta community was, therefore, one of those special communities in that status and recognition were not biologically based. Ugwuta society, it appears, subscribed to Victor Uchendu's view that "a child who washes his/her hands, eats with elders."[6] Indeed, the unit of analysis is the individual. Nwapa straddled both traditional African and colonial and post-colonial worlds successfully. And she was able to reinscribe Ugwuta oral traditions and Ugwuta's unique cultural heritage into a woman-centered discourse.

The essays in Part One, "Igbo Women: Culture and Literary Enterprise," demonstrate how Flora Nwapa as a story teller transformed her art into the written word—how the existence of formidable, wealthy and socially relevant women in Nigerian history have provided role models for young women to look up to and the community to acknowledge.

Theodora Akachi Ezeigbo, in "Myth, History, Culture and Igbo Womanhood in Flora Nwapa's Novels," explores women's conditions in Igboland in Nwapa's novels, *Efuru* and *Idu*. According to Ezeigbo, by deconstructing the Uhamiri myth and other cultural paradigms and by reconstructing history from the women's perspective, Nwapa makes a case for women's empowerment and establishes the right of women to autonomy, self-fulfillment and economic independence. The centrality of Ugwuta women to the cultural and socioeconomic life of their community is also demonstrated in the discourse of Sabine Jell-Bahlsen, in her essay, "Flora Nwapa and Uhammiri/Ogbuide, the Lake Goddess: An Evolving Relationship." Here, she contends that in the novel, *The Lake Goddess*, Nwapa recognizes the Goddess Uhammiri as a traditional recourse for Ugwuta women to maintain their identity and powers against the conflicting demands of Ugwuta society and beliefs as well as the increasing pressures of foreign values and Christian demands. Similarly, Teresa U. Njoku in "The Mythic World in Flora Nwapa's Early Novels" demonstrates the ways in which Flora Nwapa draws from Ugwuta's mythic reservoir to reveal the supremacy and resilient nature of woman. This gives shape to her art and generates a new tradition in which women become the center of focus.

It is the distortion of Igbo women's true heritage of dynamic female ancestors that is explored by Florence Stratton, in "'The Empire, far flung': Flora Nwapa's Critique of Colonialism." Stratton shows how the author deconstructs the colonial trope of "primitive" and "underdeveloped" African women through her portrayal of Efuru and Ona as intelligent, rational beings, who are able to analyze their oppressive and stifling predicaments and choose courses which liberate and empower them.

Gay Wilentz takes this examination of conflicting (re)configurations of African women further in her essay: "Not Feminist but Afracentrist: Flora Nwapa and the Politics of African Cultural Production." She explores the implications of Nwapa's works within the context of African women's cultural production. According to Wilentz, Nwapa's resistance to uncomplicated views of women's experience and her commitment to tell the tale of women's complex roles in traditional and contemporary Nigerian society informs her works. In addition, the controversial views of feminism is highlighted by these critics. Thus, Ada Azodo in "*Efuru* and *Idu*:

Rejecting Women's Subjugation," is convinced that Flora Nwapa's contribution to Nigerian feminism in both books is very significant for engendering criticism of African writing by an African woman writer. According to Azodo, for the first time in Nigerian literary history, Nwapa posits a set of gynocentric stories, in contrast to writings by Nigerian male novelists of the day.

A summative view of Flora Nwapa's accomplishments and her ability to make her vision practical emerges in Ezenwa-Ohaeto's essay, "Breaking Through: The Publishing Enterprise of Flora Nwapa," which discusses the various aspects of Nwapa's activity as a publisher. Ezenwa-Ohaeto stresses the dynamism of Nwapa's publishing enterprise which enabled her to overcome the tribulations associated with such business adventures in societies with harsh economic conditions. He concludes that Tana Press Limited and Flora Nwapa Books and Company illustrate and exemplify an effective utilization of limited resources and relevant ethical standards to pioneer literary and educational developments.

With the publication of *Efuru* in 1966, Flora Nwapa was the first African woman artist to write a full-length novel about a woman and to name the book after the heroine. With the stroke of her pen, Nwapa initiated the development of African women's literature with independent-minded, savvy, and successful female characters who were mirrors of her Ugwuta female kith and kin, whom she witnessed when she was growing up. With the launching of Tana Press Limited and Flora Nwapa Books and Company in 1977, Nwapa also established herself as Nigeria's first woman to actively publish and distribute her books to promote a national and international awareness and understanding of the African personality. As an intellectual female artist, energetic entrepreneur, and Nigeria's literary ambassador to the world, Flora Nwapa will go down in world history as an illustrious woman who achieved extraordinary feats.

Marie Umeh

Notes

1. In a greater part of Igboland, women do not break the kola nut. Kola nut is a status symbol and male preserve; it symbolizes peace, love, happiness, trust and friendship. However, according to Flora Nwapa, prominent, influential women in the Ugwuta area of Igboland have this right. Perhaps it is because of the power and influence of the mystical Lake Goddess, Ogbuide.

2. In a conversation with Uzoma Gogo Nwakuche, Flora Nwapa's son, in July, 1994 in New York, I was informed that the oldest person in Ugwuta at the helm of village affairs is a woman, the "Ogene Nwanyi," who holds the seat of power.

3. In an interview with *Quality Weekly* magazine's staff, Segun Adeleke, Bose Adeogun and Ben Nwanne, in Lagos, Flora Nwapa imparts this information. See *Quality Weekly* 6.8(August 23, 1990): 34.

4 In an interview with Marie Umeh in Scarsdale, New York, Flora Nwapa spoke of her early roots which lead to her position as one of Africa's foremost women writers. See "The Poetics of Economic Independence for Female Empowerment: An Interview with Flora Nwapa," *Research in African Literatures* 26.2 (1995):22-29.

5 Ukpabi Asika was the Administrator of East Central State in 1970 when he appointed Flora Nwapa as Commissioner of Health and Social Welfare and Commissioner of Lands, Survey and Urban Development.

6. Victor C. Uchendu makes this statement in his book, *The Igbo of Southeast Nigeria*. Portsmouth, NH: Heinemann, 1991.

Myth, History, Culture, and Igbo Womanhood in Flora Nwapa's Novels

Theodora Akachi Ezeigbo

I. INTRODUCTION

One of the most interesting developments in the criticism of African Literature is the serious and sustained evaluation and re-evaluation of African women's writing by scholars, particularly feminist critics. In recent years, Flora Nwapa's works have been given primary focus in local and international seminars and conferences[1] as well as critical works written by both local and foreign scholars.[2] Nwapa has been variously referred to as "the mother of the African novel by women," "the doyenne of African female writers," and "ancestor and literary foremother."[3]

I identify with the current research efforts based on Nwapa's *oeuvres* and the insights that pervade the best of these critical evaluations. However, I believe that all of us who are engaged in the profession of literary criticism and find ourselves making value judgements on texts must first of all be knowledgeable in the author's cultural background as well as the intricate issues raised in the text we are evaluating. As Obioma Nnaemeka has succinctly put it, "Our job as literary critics should be to open texts up to possibilities by addressing the complex issues in them, however contradictory they may be; often those contradictions lead us to where the meanings are" (98).

Coming from the same cultural tradition that informs Nwapa's works, it is not difficult for me to observe and comprehend the paradoxical and sometimes contradictory depiction given to Igbo traditions and social experience in her works. However, one should not lose sight of the relationship between art and life, between imagination and reality as one reads Nwapa's novels, so as not to fall into the trap of perceiving her fictional universe as a replica of the Ugwuta[4] community; for, as we shall see in due course, Nwapa invents as well as recreates Ugwuta community in her novels. I have read books and articles in which critics made erroneous statements about the concept of motherhood[5] or the levirate system of marriage in African cultures. In my view, rather than uncritical condemnation of these institutions or traditions, a more acceptable approach would be for the critic to examine their relevance or irrelevance in the text as well as their positive or negative impact on the characters directly concerned. Take, for instance, the levirate arrangement in marriage, which is always condemned in feminist criticism of African literature. We know that this arrangement was introduced in many African societies as a welfare scheme for widows, especially young widows who could not stand on their own economically. But it should be noted that this form of marriage was never forced on women, as widows who objected to it were allowed their freedom.[6] Most readers would agree that Ramatoulaye in Mariama Bâ's *So Long a Letter* (1980) had no use for this form of marriage after the death of Modou Fall, especially as the new husband would be no one but the obnoxious and insensitive Tamsir. In the same way, if the wealthy and competent Idu had lived after the death of Adiewere in Nwapa's *Idu* (1970), she would have had every reason to refuse such an arrangement with her lazy and foolish brother-in-law Ashiodu, and nobody would insist. Consequently, I disagree with Marie Umeh's argument that Idu's "death shortly after her spouse's demise is therefore a protest against the traditional custom of levirate" ("Finale: Signifyin(g)..." 118). But let us assume that things had been the other way round and it was Ashiodu who died and Adiewere had lived. Would Ogbenyeanu (Ashiodu's wife) not have rejoiced if she had had to become Adiewere's wife by levirate arrangement? The point being made here is that some of the traditions or customs that were instituted in traditional African societies were actually put in place to cater to those who found themselves either isolated, traumatized, or destabilized by the vicissitudes of life. Whether these traditions remain relevant today in the face of prevailing circumstances or social change is another matter altogether. But the critic must show a deep understanding of the forces that operate in society which give rise to certain peculiar developments and expectations in such a society.

In this study we will be examining the forces in Igbo cultural tradition—with particular reference to the Ugwuta community—which have shaped the life and destiny of women in Flora Nwapa's early novels. How has she recreated the myth, history, and culture of Ugwuta in the novels? We would like to theorize on the societal forces as well as mechanisms that have shaped the lives of Igbo women in the home and outside the home. A close reading of Nwapa's early novels reveals that, in theory, women are excluded from wielding power and influence in patriarchy, but in practice they have their own significant spheres of influence and they do pull strings for things to happen. Socio-economic activities and some political actions in Nwapa's fictional universe are set in motion by women and they dictate the pace at which these activities are organized and executed. Considering the diversity of women's socio-economic and political activities, Nwapa's readers may well wonder what is left for men to do. How could men fit into or operate in such an environment that appears to be completely 'woman-manned'? Nwapa's women not only have a voice, but their voice is loud and clear.

The question, then, is: has she arrogated more powers to these women than they would normally be allowed in the Igbo world of Ugwuta which she delineates? These issues and hypothetical questions will be examined in the rest of the paper. Our analysis will be based, specifically, on Nwapa's first two novels, *Efuru* (1966) and *Idu* (1970) with references to *One Is Enough* (1981) and the other novels, as occasion demands. Naturally, we will also draw freely from our own knowledge of Ugwuta traditions and of Igbo culture and world-view to provide a balanced analysis.

Flora Nwapa's *Efuru* and *Idu* are set in her home town of Ugwuta, where she grew up and began her education. The two novels depict life of traditional Ugwuta society and embody the history, culture and spirit of the community. They are set in the early

20th century when British colonial administration and Christianity had already been established in Ugwuta and other parts of Igboland, though their impact is little felt in the fictional world of Nwapa's novels. The society portrayed in the novels was largely traditional, with the worship of the local deities, Uhamiri, Okita and Utuosu still intact and traditional mores strictly observed. Sometimes, in the conversations of the women, we are given a hint of the activities of the church and a few other reminders of the colonial experience.[7]

II. NWAPA'S MYTHIC IMAGINATION

Nwapa's theory of female existence and situation in her early fiction is based on the practical experience of her life in Ugwuta as she grew up. The mystical influence of the "beautiful blue Ugwuta lake" which the community depended upon for food, transportation, and for life sustenance was decisive in Nwapa's mythopoiea when she began to create her women-centered fiction. Her mythic imagination derives its force from the spiritual being that controls this body of water—Uhamiri (variously called Ogbuide and Woman of the Lake), the powerful female deity worshipped by Ugwuta people. Judging from the complex portraiture given to Uhamiri in Nwapa's works, there is no doubt that, for the novelist, when a writer makes use of a myth from her culture, her role is to breathe life into it through the resources of her imagination. The myth of Uhamiri constitutes an important symbol in practically all Nwapa's major writing, especially the two earliest novels and *Never Again* (1975), which is based on the Nigerian Civil War. Uhamiri is the central and controlling image and represents the feminine principle in *Efuru* and *Idu*. In the war novel, *Never Again,* the Woman of the Lake becomes the influential, powerful figure and deity to which the people run to for refuge. The Ugwuta people believe that the routing of the invading Federal Forces, who came into the town through the Lake, was accomplished by Uhamiri.

Uhamiri is the epitome of female power, independence, beauty, and accomplishment. Nwapa gives the water goddess a larger-than-life portrait and invests her with some attributes and roles that constitute a deviation from the norm. She reinvents the Uhamiri myth and claims the female deity for womanhood. Uhamiri is Ugwuta woman *par excellence*—she is married, beautiful, hardworking, rich, assertive, and independent. There is every indication that Nwapa modifies the myth to suit her purpose of

54

empowering women and uplifting them. Contrary to established tradition, in which the worshippers of Uhamiri are both male and female, Nwapa modifies this practice in her novels, allowing only women to function as devotees and worshippers of the deity.

In *Efuru* and *The Lake Goddess* (forthcoming)—the last of her novels, which will be published posthumously—Nwapa creates the impression that Uhamiri prefers celibacy for her worshippers. Efuru's total devotion to the worship of Uhamiri does not begin until she ends her marriage and leaves her husband for good. Similarly, Ona, the female protagonist in *The Lake Goddess* can find peace and devote her life of service to the goddess only after she has given up her husband and children for her higher calling. Why does Nwapa preclude, in the novels, the knowledge that men can also be priests to the goddess? Can this be part of her design to claim the goddess for womanhood? It is interesting to note that Idemmiri,[8] the other well-known female water deity, and other female deities such as Agwazi, Ogwugwu, Isigwu, Omaku, Oda, and Nguma, from other parts of Igboland, are worshipped by priests and not by priestesses.

Another modification Nwapa makes in the myth of Uhamiri is to project

> the idea that Uhamiri does not give children to
> women: Can she give me children?...She cannot
> give me children, because she has not got children
> herself...she was happy, she was wealthy. She was
> beautiful. She gave women beauty and wealth but
> she had no child. She had never experienced the joy
> of motherhood, why then did the women worship
> her? (165, 221)

But according to Sabine Jell-Bahlsen, information she gathered and oral histories collected in her research in and around Ugwuta testify to the fact that men *and* women pray to Uhamiri for children and such prayer is usually answered (31-2).

Why does Nwapa associate Uhamiri with barrenness and charge her with denying children to women, particularly her worshippers? Indeed why is an issue made of the goddess's so-called childless state? For it is not common for Igbo female deities to be directly associated with procreation or child-rearing—not even those who have husbands or those whose images connote sensuality.[9] Such an association is limited to the gift of children to their people. In the interviews I held with a number of priests who worshipped female deities in the area where I come from Uga—in Aguata Local Government Area of Anambra State, I found out that most of the goddesses have no husbands, except one—Idemmiri, whose husband is Nkwo,[10] but there is no talk about children. Isigwu, Omaku, Oda, Agwazi, Nguma, and Ogwugwu[11] (the other female deities) have neither husbands nor children. If Igbo female deities are not usually expected to experience maternity,[12] why does Nwapa make Efuru, at the end of the novel, say that Uhamiri "had never experienced the joy of motherhood?" Why should she?[13] Sabine Jell-Bahlsen comments that in a talk she had with Nwapa, the novelist admitted that "she did not know why she ended *Efuru* with these words"(36).

Nwapa appears to have no apologies to give for some of the contradictions that are thrown up by not only her deconstruction of the myth of Ogbuide but also the new attributes she attaches to the figure of the goddess. Her reasons for the ambivalent end to her *magnum opus* will remain forever a matter for speculation by critics. However, the point that stands out clearly, to my mind, as very germane to Nwapa's deconstruction and reconstruction of the Uhamiri myth is her overriding desire to recreate and remodel the goddess to achieve her purpose of providing myth-ic, psychological as well as practical explanations for female exis-tence and condition in her Ugwuta cultural tradition. The novelist probably conceived the idea that if Uhamiri was barren, then that could provide a good explanation for barrenness among her hero-ines, especially those who are Uhamiri's priestesses or protegees. After all, it is logical to insist that the goddess could not give what she did not have. If this was Nwapa's line of thought, then her vision was somewhat flawed, being based on the wrong premise; and in addition, the novelist has succeeded in casting a shadow on the credibility of a goddess who either fails to reward her wor-shippers with the greatest desire nurtured by women in the Ugwuta community—the experience of being mothers, the pos-session of children—or who insists on celibacy, as in the case of Ona in *The Lake Goddess*. One might wonder whether Uhamiri has been just or fair to her devotees.[14] But if the heroines have problems with procreation, they are, however, rewarded and com-pensated in other important ways by their mentor. They are usu-ally very beautiful, wealthy women who stand out when compared with others in terms of self-awareness and strength of character. Efuru and Idu are good illustrations and so is Amaka in *One Is Enough*.

III. Nwapa As An Historical Novelist

Apart from her use of a mythic framework to deepen the portrayal of her major female characters, Nwapa also adopts a historical approach in exploring female experience in her society. This is easily seen in the chronological scheme of her five novels which explore female experience from the early colonial period (in a society that was still largely traditional) to the post-civil war period and contemporary Nigeria. The study of the historical vision of many African writers and their works has been a major preoccupation of critics (particularly male) of African literature.[15] The novels of five writers—Chinua Achebe, Ngũgĩ wa Thiong'o, Ayi Kwei Armah, Ousmane Sembene and Yambo Ouologuen—have been examined as historical interpretations of the writers' societies, from prehistory and traditional society to colonial, post-colonial, and contemporary periods. No such study has, to our knowledge, focused on the interpretation of history by African women writers. Does this mean that no woman writer has grappled with the history of her society in fictional terms?

It is, however, not difficult to see why no such study has been done. History has been a record of male experience, written by men and from a male perspective. This is the case in most societies, not Africa alone. History is not usually written about women, as they are assumed not to be national figures, the stuff about whom history is made. Even though such remarkable personalities like Queen Amina of Zaria, Moremi of Ile-Ife, Emotan of Benin and others existed, little is said or known about them. The dearth of research into the lives of African women by historians has also affected the recreation of women from the historical perspectives by African women writers. When female writers recreate history in fiction, the account they give is mostly that of women's history and the product is not regarded as historical novel "proper" by the dominant (male) culture. David Daiches, for instance, defines the historical novel "proper" as "an attempt to use a historical situation to illustrate some aspect of man's fate which has importance and meaning quite apart from that historical situation"(35).

However, a close examination of Nwapa's fiction reveals that her imagination is historical in the sense that her novels represent the exploration of women's experience in society from the past to contemporary times—a corollary to what Achebe and Ngũgĩ have done from the male perspective. As I stated in another study, Nwapa has explored the many faces of the truth of Igbo traditional society as well as contemporary society and, by so doing, she has "demonstrated her remarkable understanding of history, historical process and individual and corporate psychology"(159).

The African male novelists wrote for various reasons, including the need to put the record of African history right; the need to preserve the African cultural tradition and make it known to Africans and their friends as well as detractors; and the need to interpret the historical process from the past to the present as it impacts on African peoples. On her part, Nwapa believes that women's history must be recreated in fiction by African female writers. In a paper she read at Iowa (USA) in 1984, Nwapa had this to say:

> The Nigerian male writers fail to elevate women to their rightful plane. They overlook the safeguards with which custom surround her: the weight of feminine opinion, the independence of her economic position, the power she wields by the mere fact that she holds the pestle and the cooking pots. They fail to see all these things because they are men and are influenced by the colonial administration's Victorian type prejudices against women. (14)

She further stated that her "heroines are the ordinary women in the African countryside, often enmeshed in the customary beliefs of her society: beliefs which oftentimes drive (the) heroines to a point of rigidity"(11).

Nwapa's primary aim, therefore, is the restoration of the dignity of the African woman and the appreciation of her true nature, her role, and her contribution to her society—which were played down or neglected in the works of the male writers. This objective is to be accomplished through the recreation of the woman in traditional society as well as contemporary times. Her focus is on the interaction between people, especially women, in a communal set-up and the portrayal of human beings as they grapple with personal dilemmas and problems of communal living. Each person's individuality is subsumed in the lives of others and an individual's problem becomes an issue that interests the whole community. Nwapa does not grapple with colonialism or even nationalism; neither does she pointedly explore post-independence corruption which formed the thematic thrust of many an African male writer's creative effort, especially those who wrote in the Sixties. Rather, she concentrates on women's lives and women's culture, on domestic history as well as economic history as they touch on women's lives. Consequently, many important scenes, especially in her early works, are dramatized in the home, in the stream, and in the marketplace, which are the centers of women's social and economic activities.

Nwapa's strongest and most memorable female characters, drawn from *Efuru* and *Idu,* are types of Ugwuta women she knew when she was growing up, women of her grandmother's generation. Efuru, Ajanupu, Idu, Ojiugo, Nwasobi, and Uzoechi are such women who evoke nostalgic feelings about strong independent women of the past, in the mind of not only Nwapa but every reader who shares Nwapa's historical and cultural background, and even the reader whose only source of knowledge or information is what is read from books on Igbo history and tradition. However, as Nwapa's works grew in number, her vision of women in Nigeria's historical process became more ambivalent, complex, and uncompromising, especially as colonization gave way to independence and as the period of post-independence yielded to the civil war and postwar era.

The Igbo woman moved from the countryside to the urban area and experienced its dislocations and contradictions as well as its "liberating" influences from inhibitions and certain social responsibilities. Consequently, the modern women in *One Is Enough* (1981) and *Women Are Different* (1986) become more intolerant of socio-cultural constraints such as polygyny, unfaithfulness in marriage, arranged marriages and abandonment. Amaka in *One Is Enough* leaves Obiora, her insensitive and faithless husband, and goes at it alone in the city of Lagos. Dora in *Women Are Different* survives abandonment by Chris to become rich; and when he returns she accepts him back but puts him in his place—a secondary position where he is a mere figurehead. *Women Are Different* is beyond the scope of this paper; but it is relevant to Nwapa's interpretation of history. Indeed, a wide expanse of Nigeria's history is explored in the novel, set as it is in the period between the Forties and the Seventies, the colonial period to the postwar years. But the characters are essentially not operating under the influence of Igbo cultural tradition and are, therefore, not conditioned by its norms. Rather it is Western influence and the Christian religion, together with the conflicts both generate, that influence as well as determine the fate of women characters like Rose, Dora, Agnes, and the rest of them.

IV. IGBO WOMEN AND CULTURE IN NWAPA'S EARLY NOVELS

Nwapa's deconstruction of the myth of Uhamiri and her reconstruction of women's history are aspects of her overall design to
plumb the deep mysteries and the complex systems of beliefs and
values of Igbo culture and the pervasive influence they have on
women. The definition of culture that serves my purpose here is
the one provided by Kroeber and Kluckholn:

> Culture consists of patterns, explicit and implicit, of
> and for behavior acquired and transmitted by sym-
> bols constituting the distinctive achievement of
> human groups, including their embodiments in arti-
> facts; the essential core of culture consists of tradi-
> tional (i.e., historically derived and selected) ideas
> and especially their attached values.... (522)

In the early novels, Nwapa situates her women solidly in Igbo culture, which shapes their destiny and is equally shaped by their
activities and responses to the environment. It is, however, Ugwuta
traditions that, specifically, come into focus in the novels. *Efuru*
and *Idu* explore Ugwuta customs that characterize group behavior
and habits that characterize individual behavior. The way of life of
the community and their value and belief system are determined
by the social environment. Women's culture features most prominently in the interaction of women in the home and outside the
home, in specific locations such as the village streams, market-
place, funeral ceremony, age-grade meeting grounds, and play-
grounds. The patterns of life that gain Nwapa's primary attention
are those that relate to women, such as marriages, childbirth, motherhood, storytelling, and commercial activities. Reading these novels is an experience in the study of womanhood and women's
culture in Igboland in general and in Ugwuta in particular. Nwapa's
searchlight is focused on the true experience of women and she
explores issues such as the quality of women's daily lives, the conditions in which they lived and still live, in which they worked and
continue to work, the ages at which they married and procreated,
their roles in the family and in society. What we have in these novels is a woman-centred analysis of Nwapa's society and a powerful impression of the difference between women's social reality
vis-a-vis that of men.

The importance of social institutions such as the family, marriage, age-grade formations, birth and burial rites, especially in the

lives of women, is highlighted in *Efuru* and *Idu*. The family members are knit together by the bond of kinship; and the extended family system holds sway. In the novels, there is a sharing of joy and sorrow by family members and the phrase "be your brother's keeper," which applies strictly to Ugwuta people, is strongly evident in the novels.

The figure of the mother looms very large in Igbo family life.[16] It is the mother or the woman of the house on whom the foundation of the home squarely rests.[17] Her abilities or inadequacies would affect the home positively or negatively. Idu is the prop on which everyone depends and she is the one constantly making decisions, carrying them out and attending to a large proportion of the demands of the business enterprise jointly owned by her and Adiewere, her husband. Idu and Adiewere are responsible for the impoverished family of Ashiodu, Adiewere's dependent brother; but it is Idu who determines the nature of the help to be given and prompts that it be given at appropriate times. True, Igbo patriarchal set-up in Nwapa's novels suppresses women, but women are also shown to exercise economic power, possess some measure of independence, and attain other achievements, like becoming priestesses and taking traditional titles.

An important role women play in the family is the upbringing and nurturing of children. This role is often seen as limiting women and confining and domesticating them. It is also seen as a distraction from other higher roles or callings; hence, Ona has to give up her family and the responsibility of caring for them in order to devote her life to the worship of Uhamiri in *The Lake Goddess*. But Nwapa also explores the fulfillment and satisfaction Ugwuta women derive from taking care of their children. In *Efuru* Nwapa constantly refers to the proper upbringing of children, especially girls who are expected to become wives and mothers. Amede, Efuru's mother-in-law and her friend, Omirima, express disappointment with Ogea, Efuru's maidservant. Amede complains:

'It is that silly girl, Ogea. She washed my wrappers and all of them will have to be washed again because there is still black soap on all of them. How is it that a grown-up girl like that is not able to wash clothes properly? How can she live in a man's house?' (181)

Omirima replies:

'That's what I keep on saying, children of these days

are no good. How men of today marry them is what
I cannot understand.' (181)

On another occasion, Omirima deplores the thoughtlessness of the
younger generation who fish in the Lake on Orie day, a day devoted to the worship of Uhamiri:

> 'That's why our Uhamiri is angry with us. The chil-
> dren of these days have polluted the Lake. I saw
> three girls—all school girls, on Orie day, going to
> fish. I scolded them. You are responsible for the
> poverty of this town, I told them. I took a cane and
> chased them. But did they listen to me? Of course,
> they did not. As I was returning from the market in
> the evening, I saw them returning from their fish-
> ing. That's what they learn in school—to disobey
> their elders.' (195)

In the novels, the incursion of Western traditions and the Christian
religion are constantly alluded to as polluting the traditional culture, interfering with strict and proper upbringing of children, and
introducing negative traits to the town. In *Idu* prostitution is identified as an alien practice which has trapped some of the young
women. Uhamiri is said to frown at it and punish those that indulge
in it. Afagwu and Obiaku are reported to have lost all the property they acquired in the "profession" and return home broken;
Obiaku, in addition, loses her sanity.

The concern shown by mothers as well as other elders in the
community underscores how seriously they take the socialization
of young people to proper behavior. Idu constantly gives expression to the negative traits she observes in her younger sister,
Anamadi, and tries to correct them and discipline the girl. There
is a proper code of behavior expected to guide children's conduct
which will usher them into responsible adulthood. These concerns
are always expressed by women and they enforce the code. Women
are, therefore, the custodians of tradition which they guard—sometimes, to a ridiculous extent, like when Ajanupu insists that
Nkonyeni should greet her and Efuru more warmly rather than
casually, as she had done (179), and the time Omirima censures
her daughter-in-law for trying to protect her children from yaws
by sending Omirima's daughter away (174). Folktales are often
used in the novels not only to entertain children but primarily to
teach them good manners, wisdom, and Ugwuta customs. A good

example is the story narrated to the children by Eneke in *Efuru* about "the woman whose daughter disobeyed her and as a result was married to a spirit"(106).

The marriage institution as it impacts on women's lives receives no less in-depth treatment in Nwapa's fiction than the family. The two are closely related. In Igbo culture, the most important reason for marriage is procreation. Even in marriages where love is the main attraction that brings couples together, the desire to have children is always the ultimate goal. This is the reason why most marriages, including those built on affection, crumble or are seriously threatened when they are not blessed with children. What is responsible for this ever-present, or enduring desire of every Igbo woman—married or unmarried—to have a child?[18]

After a thorough examination of Nwapa's fiction, I have come to the conclusion that marriage and procreation[19] are given so much emphasis because both are indispensable in creating new family units and in increasing the population of the family or lineage. Children are greatly valued in Ugwuta society and each marriage is expected to produce many siblings, both male and female (with a preference for male). Nwapa is reflecting in her work the situation as it exists in her society. Nwasobi expresses this expectation succinctly in *Idu* when she says, "What we are all praying for is children. What else do we want if we have children?"(150)

The concern with procreation is not limited to the married couple. Indeed, it is their relatives, friends and neighbors who first express these concerns when the woman has not become pregnant. A year after their marriage, Efuru and Adizua (in her first marriage)—and she and Eneberi (in the second marriage)—are still enjoying the bliss of marital life when gossip spreads about Efuru's barrenness, among her female neighbors, as anxious exchanges are made over her childless state:

> 'Seeing them together is not the important thing,' another said. 'The important thing is that nothing has happened since the happy marriage. We are not going to eat happy marriage. Marriage must be fruitful. Of what use is it if it is not fruitful. Of what use is it if your husband licks your body, worships you and buys everything in the market for you and you are not productive?' (137)

The desire for children in Igbo culture is motivated by both materialistic and spiritual reasons. As Akam puts it, "Who among the

Igbos would like to exist without his kith and kin?"(50). Akam further states that (for Igbo people):

> [N]o line is drawn between the spiritual and the physical. Even life in the hereafter is conceived in materialistic and physical terms...The reward after death is understood as an overflow and a prolongation of the one already started on earth. Thus they [Igbo people] cannot understand a childless man having a retinue of children in the next world. A lazy man who does not cultivate plenty of yams and rear a multitude of fowls, goats and sheep, suffers hunger in the next world. (50)

In the same way, childless people not only endanger continuity in their families, but also deny themselves and their ancestors potential worshippers. Apart from this, children are also valued for the service they render to their parents, especially in their old age.[20]

If motherhood is so vital to the mental health of the African woman,[21] why then does Nwapa punish her heroines with the malignant trauma of childlessness? The pain of infertility is inflicted on her most assertive and independent women in four novels— the eponymous heroines Efuru and Idu, Amaka in *One Is Enough,* and Rose in *Women Are Different.* When they eventually conceive, this occurs with much difficulty and may not always bring total satisfaction, except, perhaps, in the case of Amaka. In *The Lake Goddess,* Ona has three children and has to give them up to worship Uhamiri.

Is it then Uhamiri who is responsible for the unhappy childless state of these women who either worship her or share her attributes—long hair, great beauty, business acumen, great wealth, and independent spirit? Omirima states in Efuru that Uhamiri's worshippers are mostly childless:

> 'How many women in this town who worship Uhamiri have children? .... All right let's count them: Ogini Azogu,' she counted off one finger, ' she had a son before she became a worshipper of Uhamiri. Since then she has not got another child. Two, Nwanyafor Ojimba, she has no child at all. Three, Uzoechi Negenege, no child. They are all over the place. Why do we bother ourselves counting them.' (162)

Efuru's mother had only her, and Efuru herself loses her only child, Ogonim. Uhamiri herself is barren and her childless state justifies the others. Her childlessness and her stupendous wealth have a structural and thematic relevance to the lives and experiences of the heroines.[22]

What is the logic behind Nwapa's complex and sometimes ambivalent attitude to motherhood and marriage in her novels? Consider for instance that though motherhood is given so much prominence, yet one also gets the impression that children alone do not give happiness or self-fulfillment to women, as the experience of Ogbenyeanu illustrates in *Efuru*. Uhamiri is also said to be happy, even though she has no child. Is Nwapa consoling barren women in her culture, whose plight she has highlighted in so many of her works?[23] She appears to be saying there are other factors necessary for happiness. Companionship and love in marriage are as important as motherhood (if not more), as illustrated in *Idu* where the pregnant heroine wills herself to die to be able to rejoin her dead husband, Adiewere, thus leaving her son behind.[24]

In *One Is Enough*, Nwapa becomes even more forthright in her encouragement of childless women to look for other ways of living a self-fulfilled and profitable life. Amaka's self-questionings— "Was a woman nothing because she was unmarried or barren? Was there no other fulfillment for her?"(25)—later change to a strong resolution to divorce her husband and strive to be independent and successful.[25]

The marriage institution is held with great respect in Ugwuta culture and every normal man or woman is expected to get married. Those who remain single are considered abnormal, and parents and other relations show concern when a young man or woman delays marriage. Nwapa sees marriages as a desirable state but she does not believe a woman should remain in a marriage in which she is oppressed or humiliated. In her novels, women are encouraged to strive to be self-reliant and independent. In an unpublished paper entitled, "Sisterhood and Survival: The Nigerian Experience," Nwapa states:

> Sisterhood will survive if we women pay less attention to men and marriage. Please, don't get me wrong. I am married and I have children. But I do not live for my husband alone. I live for him, for my children and for my profession. (6)

If a marriage breaks down, it should not be the end of a woman's life. In Igboland, there is hardly a married woman who does not engage in one income-generating activity or the other. It is often said that Igbo women work too hard. Most young girls, in the past and even today, start making money early by trade or some other activity so that by the time they marry, they have already established a business. In *Efuru* and *Idu,* women engage in trading, farming, and fishing and these activities do not stop them from taking care of their families. Efuru is described as a woman whose hands make money(125) and Idu is also very successful in her palm oil trade. Marriage by itself is not conceived as an unwholesome tradition for women in Nwapa's novels. Rather it is individual partners who cause pain to the other partners and thereby ruin marriages.[26] A marriage fails or succeeds depending on the circumstances that affect it. Apart from childlessness, there are other factors responsible for failure in marriage, like incompatibility, unfaithfulness, neglect, and lack of trust. Efuru's marriage fails while Idu's succeeds because of the type of men they marry. Efuru's husbands are unfaithful, ungrateful and irresponsible. Idu's is loving and kind.

But Efuru survives her failed marriages and gains in strength and stature. She finds fulfillment in her worship of Uhamiri. In addition, her business acumen and strength of character enable her to leave both husbands and continue with her life. Her two husbands were her choice and she blames no one. She adjusts to a life of service to her community and undisturbed worship of the Woman of the Lake. In her ordeal, Efuru has the support of her age-group—a formidable and empowering social organization to which every Ugwuta man or woman belongs and which comes to a member's aid in times of misfortunes or celebrations.

In Igbo culture, divorce[27] is available to both men and women when a marriage breaks down irretrievably. A man or woman can initiate it and it requires a repayment of the bride price in most cases. According to J. U. T. Nzeako (1972):

> Mgbe di na nwunye bi n'ulo wee muta umu, ma o
> bu ha amutaghi nwa wee see okwu, gbasaa, otu
> nwere ike igba ibe ya alukwaghi m. Nke a bu igosi
> na ha achoghi ilu di na nwunye ozo. (30)

> [When a married couple quarrel and separate,
> whether they have children or not, any of the two
> can start divorce proceedings against the other.

> This is an indication that they do not want to
> remain husband and wife any longer. [My transla-
> tion)].

It is Efuru who divorces both husbands. Many people counsel patience in the case of Adizua, including the assertive Ajanupu. But Efuru was solely responsible for her marriage to Adizua—she chose him herself—and feels and considers it her responsibility to terminate it when it becomes burdensome and destructive to her. She remains self-reliant and independent in her action. By divorcing Adizua, Efuru rejects her mother-in-law's needless self-sacrifice and self-abnegation and her acceptance of abandonment and neglect, in firm and clear language:

> Perhaps self-imposed suffering appeals to her
> (Adizua's mother). It does not appeal to me. I know
> I am capable of suffering for greater things. But to
> suffer for a truant husband, an irresponsible hus-
> band like Adizua is to debase suffering. My own suf-
> fering will be noble. (61-2)

And when her second marriage to Eneberi fails and she leaves him for accusing her falsely of committing adultery, Efuru again rejects her doctor friend, Difu's advice to return to her husband:

> If you are my friend, why then do you want me to
> go back to the man who accused me of adultery. You
> don't know the seriousness of the offense. (220)

It should be pointed out that Nwapa also gives some space to explore marriages that succeed. She identifies positive attributes like affection, trust, faithfulness, mutual respect, compatibility and equity as contributing to a peaceful and happy married life. And, of course, having children in the marriage does help too. These are the factors that determine the success or failure of the marriages in the three novels, *Efuru, Idu,* and *One Is Enough.*

Nwapa has also explored fully the aspect of Ugwuta tradition embodied in the linguistic cultural traits noticed in women's speech, which bind the womenfolk together. The way people speak in *Efuru* and *Idu* is so authentic. Women bond together in the novels and congregate in the homes of one another, in the marketplace, the farm, and the stream. One could describe these practices as the psychological aspect of culture insofar as they point to certain processes often identified with women, like the habit of talking, gossiping, and the sentimental but affectionate ways of addressing one another, comic and harmless exclamations that connote familiarity and friendship and, occasionally, comments and statements meant to hurt or correct people's failings or express envy.

Nwapa's women, in the early novels, are ordinary rural dwellers, clothed with dignity and strength. It is her creative delineation of these characters that largely defines her significant linguistic and stylistic contribution to African fiction. Her effective capturing of the speech habits of these women constitutes the "appropriateness of technique" which is commended in Emenyonu's essay, "Who Does Flora Nwapa Write For?":

> ... the way she reaches down to all her characters
> and communicates authentically at all their levels
> not excluding their idiosyncrasies and mannerisms
> which she manifests in their speech patterns to help
> reveal more and more the nature of the characters.
> (31)

Nwapa's perfect control of local color, tonal inflections in dialogue, and rhythms of rural lifestyle in all their complexities are the factors that make *Efuru* and *Idu* remarkable novels. The following illustration, a scene at a village stream, in *Idu,* demonstrates the importance of female bonding, the activities that preoccupy women, and the speech rhythms and mannerisms of Igbo rural women:

> 'Our Uzoechi, did you come to the stream?'
> 'Yes, Our Nwasobi, I came to the stream. Are
> you well?'
> 'I am well. I have come to wash my children's
> dirty clothes. You know I have nobody to help to
> do housework.'
> 'You had a maid, what happened to her?'
> 'She has gone. Her people came for her one
> day, and she went with them; I did not even
> know. She took all the dresses I made for her, and
> left my youngest daughter crying.'
> 'This is bad. Who are her people?'
> 'She is from Ezu,' Nwasobi replied. 'They are
> like that. You cannot rely on them. Look that's
> Idu. Idu, our Idu, are you well?'

'I am very well,' replied Idu. 'I have come to
fetch some water. My husband is not very well
today.'

'What's wrong with him?' Nwasobi asked.

'He had a headache yesterday, so he took some
purgative medicine which I bought from the market. He has had a tummy upset since this morning.'

'Let not our Adiewere be ill. God forbid that our
Adiewere should be ill,' Uzoechi said. (11)

This sure-footedness in the control of language and imagery in
her early novels and her deconstruction of the Uhamiri myth as
well as her use of oral tradition are heightened by her interpretation of women's history and culture.

V. Conclusion

Nwapa's vision of the traditional Igbo woman and the modern
woman underscores women's relentless capacity to survive despite
all odds and their determination to achieve economic independence and a measure of fulfillment as human beings in their community. The strength of her heroines is a manifestation of the
attributes of the mythical figure of the Woman of the Lake, the goddess who provides Nwapa's fiction with structural and thematic
unity and relevance. Obioma Nnaemeka aptly summarizes
Nwapa's achievement in these words:

Nwapa's work is a biography, a collective biography
of beautiful, strong Ugwuta women and their majestic lake. Nwapa's work captures the complexity,
ambiguities and contradictions of her environment
as they are embodied in the force that lies at the bottom of the lake, Uhamiri, the goddess of the crossroads. (104)

The myth of Uhamiri is deconstructed by Nwapa and reinvented
and used to criticize the Ugwuta cultural tradition that values the
woman primarily as an instrument of procreation rather than as
an individual with aspirations to attain self-fulfillment and independence. The image of the goddess is reconstructed to symbolize
the fullest realization of the potential of Igbo womanhood, just as
it also represents the glory and beauty of womanhood. For Nwapa,
the myth, as she interprets it, provides the exemplary model for

women's aspirations, be they economic, political, or social. In her works, the most important attributes of her female protagonists are strength, self-determination, and physical beauty. And to make this point glaringly clear, she deconstructs "motherhood" by giving it a secondary position to women's empowerment and economic independence.

NOTES

1. Since Nwapa's death in October 1993, interest in her works has intensified. Recently, for example, there were three international conferences where her works were given much attention: the May 1994 Annual International Conference on African Literature and the English Language at the University of Calabar, Nigeria; the July 1994 International Conference on Women in Africa and African Literature held at the African Research Institute at La Trobe University in Australia; and the 1995 March ALA Conference at Columbus, Ohio, USA. Articles on her works also appeared in the Special Issue on her of *Research in African Literatures*, Vol. 26. 2, 1995, devoted to her *oeuvre* edited by Chikwenye Okonjo Ogunyemi and Marie Umeh.

2. Such works include Florence Stratton, *Contemporary African Literature and the Politics of Gender* (London and New York: Routledge, 1994) and Helen Chukwuma (ed.), *Feminism in African Literature* (Enugu, Lagos, Abuja: New Generation Books, 1994).

3. See Chikwenye Okonjo Ogunyemi, "Introduction: The Invalid, Dea(r)th, and the Author: The Case of Flora Nwapa, *aka* Professor (Mrs.) Flora Nwanzuruahu Nwakuche," *Research in African Literatures* 26. 2 (1995):5, Marie Umeh, "The Poetics of Economic Independence for Female Empowerment: An Interview with Flora Nwapa," *RAL* 26.2 (1995) 22, and Marie Umeh, "Finale: Signifyin(g) The Griottes: Flora Nwapa's Legacy of (Re)Vision and Voice," *RAL* 26. 2 (1995):26.

4. "Ugwuta," which we retain in this essay, is the name used by the community; the derivative "Oguta" was a colonialist distortion of the real name.

5. In her article "Rewriting History, Motherhood, and Rebellion: Naming an African Women's Literary Tradition," *Research in African Literatures* (1990), Susan Z. Andrade writes: "Efuru's desire to be traditional threatens to subvert the text's manifest assertion of female independence"(105). If "being traditional" refers to "motherhood" as Stratton states in her book (92), then Andrade is mistaken to suggest that "motherhood" subverts female independence, for Efuru is a mother though she lost her only child and there was nothing to show she suffered any disadvantage in her self-assertion nor independence by experiencing motherhood.

6. I interviewed people from different ethnic groups in Nigeria and they told me that though the levirate system of marriage (wife inheritance) operated in their cultures, widows were not forced to accept it. I also talked to people from different parts of Igboland (East and West) and

they confirmed that the arrangement was made only with the widow's consent.

7. These reminders of the threatening new world represented by British administration are adequate for Nwapa's purpose in the two novels. In his article "The Igbo Word in Flora Nwapa's Craft," *RAL* 26.2 (1995):49, Chimalum Nwankwo criticizes Nwapa for not highlighting the conflict between Ugwuta community and the white man at the point of contact. But there is no need for this in the novels, especially as Nwapa's focus is on women and what concerned them directly. It should be noted that the white man had little or no dealings with Igbo women. The occasions when women resisted British administration in Igboland were when they were directly affected, like the 1929 Aba Women's War, which was a struggle against the taxation of women.

8. The spelling of the name varies in different parts of Igboland where this deity is worshipped. Other spellings are "Idemili," "Idemmili," and "Idemiri."

9. Some Igbo goddesses are associated with a beautiful and sensuous figure. *Uhamiri* is said to be very beautiful and feminine—long hair, gorgeous in her dressing, possessing trinkets and ornaments. The goddess Isigwu, worshipped in parts of Anambra State, is imbued with a very maternal image that exudes fecundity; yet she is not associated with giving birth.

10. *Nkwo* is a powerful deity but his wife, *Idemmiri*, who is said to have come from Agbaja (Nnewi) is even more powerful than her husband.

11. Isigwu is also worshipped in other areas of Aguata Local Government, for instance, in Aguluezechukwu. She is the goddess of protection and justice. She punishes offenders and attacks the enemy of her people with swarms of bees. *Ogwugwu* is both female and male in Uga town, depending on the village. In Oka and Umuoru villages, she has both identities; in Umueze, she is female; in Awarasi, *Ogwugwu* is male. *Omaku* brings peace and wealth; *Oda* and *Nguma* give children and protect pregnant women.

12. Ironically, the female deities in Igboland are usually addressed as "mother" by their people and their priests (worshippers).

13. To make *Uhamiri* a mother would have meant overstretching the imagination too far; that is, deconstructing the myth of the Lake Goddess to a ridiculous extent.

14. For instance, did Ona have to leave her husband and three children to take up "the full-time role of diviner and spiritual healer of the community"? as Marie Umeh argues in her article, "Finale: Signifyin(g)..."(115).

15. Two of such studies are: Chidi Ikonne, et al, (eds.), *African Literature and African Historical Experiences* (Ibadan: Heinemann Educational Books, 1991) and Eldred Jones, (ed.), *Myth and History, African Literature Today,* No. 11.

16. So much has been written and said about the "mother" image in African culture both by African scholars and writers. But "mother" as used in this study is simply the woman of the house—the female partner in a marriage.

17. One of the paradoxes in Igbo culture is that, though the man is the head of the family, the woman wields subtle but real power, perhaps by virtue of what Nwapa has referred to as her control of "the pestle and the cooking pot."

18. I have yet to come across an Igbo woman who expresses antipathy towards children or who rejects motherhood. I have seen instances where women who remain single for years suddenly decide to get married, even in middle-age, and in spite of their advanced age, they try to have children. In Nigeria, today, many women decide to have children with or without husbands, especially single women who reach the age of 35 or 40. Some women, even among the educated (degree holders) are accepting polygyny partly for economic reasons and partly to be able to have children of their own; these are women who find it distasteful to have children as single parents.

19. I was struck by the common occurrence of these themes in the works of African women writers. Why is the African woman writer so fascinated by and preoccupied with the themes of motherhood, infertility or barrenness? Is it because these factors are so crucial in determining a 'successful' or 'unsuccessful' marriage? Is it because childlessness is seen as responsible for marriage breakdown, divorce, separation, wife-battering and abandonment? But the irony about the obsession with these themes is that in our culture, most women have children. Only a negligible number of women are barren. Consider, for example, the women writers themselves. I know some of them well, have met some others and read the biodata of a few more—and they are (were) all mothers. Then I look around in my village and discover that very few women are barren; in my extended family, for example, starting with my mother's generation and spreading to my generation and the one next to mine, I cannot see a single barren woman among us. In my town, there are only a few, I am told. I look around at the University of Lagos community where I live and work and I find that all my friends and colleagues have children. The point I am trying to make is that very few women are childless and even fewer can really be described as barren. Most women end up giving birth. This has been confirmed by my medical colleagues who say, according to medical science, very few women are indeed barren. Why then has so much attention been lavished on this female condition in our women-authored texts and even in a reasonable number of male-authored texts? The question is: Will the theme endure? Or has it lost its fascination? Will it take a back seat, even as new areas of women's experience are beginning to receive attention?

20. In a situation where the state does not provide adequate welfare service to old people, it becomes of paramount importance for people to receive this service from their own children or relations. The loneliness suffered by many old people in the West is usually not experienced by the elderly in Igbo culture and in other African cultures because they have children or grandchildren to keep them company and care for them.

21. On a television interview in Lagos on August 26, 1995, on the eve of the Beijing Conference on Women, Professor (Mrs.) Felicia Ekejiuba, the

UNICEF Representative for Africa, said, "For the African woman, motherhood is supreme. Women are asking for economic empowerment so as to be able to take care of their children and they are also calling for equitable distribution of resources and opportunities and the removal of all constraints that obstruct women's development and progress." If motherhood is a basic desire of African women, then the concern of African governments and relations of barren women should be what to do to help these women to have their own children or provide them with alternatives. The most practical option is adoption which alone can solve the problem of childless women as well as discourage men from resorting to polygyny in childless marriages. Our men should be socialized to accept adoption (which, at present is unacceptable to most Nigerian men).

22. There is a generally-held belief in some parts of Nigeria, especially in the East, that water deities (often referred to as Mammywater) offer three gifts to those they favor, from which they are allowed to choose any two. These are children, wealth, and long life. A choice of wealth and long life means a life without children for the one who has so chosen. If we apply this to Nwapa's novels, could Efuru have chosen wealth and long life and was Idu's choice children and wealth?

23. The theme of barrenness spills over to Nwapa's short story collection, *This Is Lagos and Other Stories* (1971); a barren woman in the story "The Child Thief" is arrested for stealing another woman's baby.

24. In an earlier study, "Traditional Women's Institutions in Igbo Society..."(1990), I had questioned the validity of this death in the context of Idu's character and belief and the importance attached to motherhood in the novel. But further reading and research into Ugwuta tradition has brought me a credible explanation. Ugwuta people have a belief called *Di Uwa*. In a paper she presented at the 13th Annual University of Calabar International Conference on African Literature and the English Language in 1994, Mrs. A. C. Izuagba states: "The Oguta belief in *Di Uwa* is reflected in the marriages between Idu and Adiewere and that between Ojiugo and Amarajeme. According to Oguta tradition, before one comes to this planet, there is a place he or she leaves his/her paddle (*amara uwa*) and as he/she does this, he/she takes a decision on what his/her mission on earth will be. It is also at this point couples decide to get married and if on coming here this wish is fulfilled, the consequence is that if either of the couple dies, the other must surely die" (5). It is this belief that Nwapa dramatizes in the lives of the two couples. Idu follows Adiewere; Ojiugo remains alive after Amarajeme but she never recovers from his death, as Idu comments: "... but the death of Amarajeme will forever live with her. She does not forget it, she cannot get over it. Sometimes, Adiewere, one can say that God molded two people in the same way for the purpose of living together, being husband and wife" (206). However, the belief in *Di Uwa* is somewhat deconstructed in the contradiction Nwapa creates in the crumbling of the marriage between Ojiugo and Amarajeme. Ojiugo's strong and irrepressible desire to have a child compels her to run to another man who is able to

get her pregnant, sacrificing her marriage and her true love. Nwapa appears to be here endorsing the primacy of motherhood and negating the strong point made for love and companionship in marriage in the case of Idu and Adiewere. This is what makes Nwapa's attitude to motherhood and marriage rather complex and ambivalent in the novel.

25. The pursuit of economic independence, self-fulfillment and autonomous existence often leads Nwapa's heroines in her later novels to go into excess, abandoning moral and social constraints. Chimalum Nwankwo deplores this development in Nwapa's art in his article when he comments on *One is Enough:* "There is something rankling about the prolificacy of women like Amaka and the unscrupulous Madam Ojei and the Cash madam Club Women"(50).

26. At a seminar in my African Literature graduate class, one of the students commented that "Efuru was beaten by tradition." But I do not agree with this view. Efuru manipulates tradition to suit her—for example, she chooses her husbands and line of business herself and even elopes with her first husband and helps him later to pay her bride price. When she discovers she cannot have children for her second husband, she takes steps to get him a second wife so that she can face her business and her spiritual role of worshipping Uhamiri. If Efuru can be said to be "beaten," it is not tradition but the worthless men she had married that are to blame; but then, Efuru transcends the "beating" and achieves self-fulfillment and contentment in her natal home where she can now devote her service to Uhamiri and her community.

27. The Igbo expression for divorce is *igba alukwaghi* m which, literally translated, means "Publicly rejecting a marriage."

Works Cited

Akam, Uche J. B. "The Mustard Seed." *Uga: An Ecclesial Cell in A Local Church.* Owerri: Assumpta P, 1987.

Andrade, Susan Z. "Rewriting History, Motherhood, and Rebellion: Naming an African Women's Literary Tradition." *Research in African Literatures* 21.1(1990):91-110.

Daiches, David. "Scott's Achievements as a Novelist." Ed. D. D. Devlin. *Walter Scott: Modern Judgements.* London: Macmillan, 1968.

Emenyonu, Ernest. N. "Who Does Flora Nwapa Write For?" *African Literature Today* 7 (1975):28-33.

Ezeigbo, Theodora Akachi. "Traditional Women's Institutions in Igbo Society: Implications for the Igbo Female Writer." *African Languages and Cultures* [London] 3.2(1990):149-165.

Izuagba, A. C."Oguta and the Novels of Flora Nwapa: A Biographical Approach." A paper presented at the 13th Annual International Conference on African Literature and the English Language at the University of Calabar, Nigeria, May 1994.

Jell-Bahlsen, Sabine. "The Concept of Mammywater in Flora Nwapa's Novels." *Research in African Literatures* 26.2(1995):30-41.

Kroeber, A. L. & Kluckhohn, Clyde. *Culture: A Critical Review of Concepts and*

Definitions. Cambridge: Mass. U of Massachusetts P, 1952.

Nnaemeka, Obioma. "Feminism, Rebellious Women, and Cultural Boundaries: Rereading Flora Nwapa and Her Compatriots." *Research in African Literatures* 26.2(1995):80-113.

Nwankwo, Chimalum. "The Igbo Word in Flora Nwapa's Craft." *Research in African Literatures* 26.2(1995): 42-52.

Nwapa, Flora. *Efuru*. London: Heinemann, 1966.

——. *Idu*. London: Heinemann, 1970.

——. *The Lake Goddess*. (Unpublished manuscript)

——. *Never Again*. Enugu: Nwamife, 1975.

——. "Nigeria: The Woman As a Writer." A paper delivered at Iowa, USA, 1984.

——. *One Is Enough*. Enugu: Tana, 1981.

——. "Sisterhood and Survival: The Nigerian Experience." A paper presented at the Second International Feminist Book Fair in Oslo, Norway, 1986.

——. *This is Lagos and Other Stories*. Enugu: Nwamife, 1971.

——. *Women Are Different*. Enugu: Tana, 1986.

Nzeako, J. U. Tagbo. *Omenala Ndi Igbo*. Ikeja: Longman, Nigeria, 1972.

Ogunyemi, Chikwenye Okonjo. "Introduction: The Invalid, Dea(r)th, and the Author: The Case of Flora Nwapa, *aka* Professor (Mrs.) Flora Nwanzuruahu Nwakuche." *Research in African Literatures* 26.2 (1995):1-16.

Stratton, Florence. *Contemporary African Literature and the Politics of Gender*. London and New York: Routledge, 1994.

Umeh, Marie. "Finale: Signifying(g) The Griottes: Flora Nwapa's Legacy of (Re)Vision and Voice." *Research in African Literatures* 26.2(1995):114-123.

——. "The Poetics of Economic Independence for Female Empowerment: An Interview with Flora Nwapa." *Research in African Literatures* 26.2(1995): 22-29.

Ogbuefi, Ezemmiri di Egwu, priestess of the Lake Goddess, Ogbuide/Uhammiri and her husband, the river God Urashi, with her group of fellow worshippers during a ritual performance on Ugwuta lake in 1988.

Flora Nwapa and Uhammiri/ Ogbuide, the Lake Goddess: An Evolving Relationship

Sabine Jell-Bahlsen

Introduction

Flora Nwapa, Nigeria's "First Lady of Letters,"[1] was born and raised in Ugwuta/Oguta,[2] a town located on a beautiful lake in Imo State of Southeastern Nigeria. While Nwapa's writings reveal much about local customs and beliefs, Nwapa was a novelist, not an ethnographer. As an insider of Ugwuta society, the artist/novelist expressed doubts about and ambivalence toward her own culture, its beliefs and values. Nor was Nwapa's understanding of and attitude towards that society, culture, and religion static. Particularly her views on African spirituality, local religious beliefs, and their significance for women have changed and evolved throughout Nwapa's life and in her *oeuvre.*

Ugwuta's pre-eminent local deity is the lake goddess, Uhammiri/Ogbuide.[3] The goddess and her importance to women figure prominently in much of Nwapa's work. In her earlier works, *Efuru* and *Idu,* Nwapa questions the goddess's primary gift to women, i.e. "children," and rather celebrates women's economic empowerment which is locally equally credited to the goddess. Later in life (e.g. in her posthumously published book, *The Lake Goddess*), Nwapa became increasingly critical of foreign intrusions, especially the detrimental impact of Christianity on African spirituality, African identity, and African women. Nwapa personally struggled with the mixed blessings of Christianity and westernization in Igbo society, positioned as she was culturally as a native of Ugwuta and socially within Ugwuta's and Nigeria's upper class and Christian elite.

THE GODDESS OF UGWUTA LAKE AND HER GIFTS

Eze Mmiri di Egwu / Uhammiri / Ogbuide, Eze Nwanyi[4]—the water goddess of Ugwuta Lake in Imo State of Southeastern Nigeria—is awesome. The lake goddess is the mythical mother of the Oru people,[5] as Mgbada, Ona's father explains to his out-of-town son-in-law, Mr. Sylvester, in Flora Nwapa's novel, *The Lake Goddess:*

> "We are the lake people."

> "The lake people?"

> "Yes, lake people. We live on the lake. We call the lake Uhamiri or Ogbuide. We sometimes call her mother. She is the mother of us all."(208)

In real life, Mrs. Onyemuru Uzonwanne, a local ferrywoman from Ugwuta who ferries people and their goods across the lake, had this to say about the goddess:

> "Ogbuide, Eze Nwanyi, the Water monarch, Iyi comes from the moon. Iyi has followers. Iyi killed a person's son and used a ram to pay back. Any person who wants to go well on his way must offer a ram to Her [Ogbuide]. If somebody drowns, you must give Her a ram. Ogbuide has twisted hair [dreadlocks]; she has a big head. Iyi is good. The Woman King is fine [beautiful]. You cannot see her from the outside. She helps her group very well. She helps and feeds the poor. When a poor person is hungry and comes to the waterside, She [Ogbuide] will find food to eat for that person."[6]

The lake goddess, Ogbuide, is *the* major reference point in the lives of the local Oru Igbo of Orsu-Obodo, Ugwuta/Oguta, and a host of smaller towns around Ugwuta Lake and the rivers Njaba and Urashi. The people's lives and cultural activities revolve around the flooding and recessing of these waters, as evidenced in the local farming cycle and timing of cultural activities.[7] The notion of the flexible, fluid, "female" side of the universe is manifested in the water goddess and balances the static side expressed in ancestral male traditions. Ogbuide/Uhammiri and Urashi, the divine pair of water deities, were worshipped locally until and even after the advent of Europeans. The goddess's place in the local pantheon of Igbo deities is indicated in a prayer by the late Obiadinbugha, Uzor Ohaka of Orsu-Obodo, the town's chief hereditary male priest of Uhammiri, in 1979. This is how Obiadinbugha addressed the goddess, Uhammiri, when he offered white chalk to her at her shrine:

> "Ogbu ama[8] ngwa were nzu gi. Nke a bu nwa Ebiri Obua ya biara achu aja na. Nkwa okwere gi, na ifeoma i me ni ya, ya ka o biara. Ngwa were kwa nzu gi. Urashi Ome Igbo bu-di. Ani Orsu, Ani nama Oferi, Nganaga, Ogbu Niyii Omerenwaka, ani du mu uzo echi. Arishi ufojioku oma, Ikenga oma. Nne Beke ngwa were kwa ni nzu o. Obu nwa Ebiri Obua na achu aja naa Iyi kwara ochu, Nne Mmiri Ukwu, Iyi ojii, Odi-ide nna ogo, Nwanyi nwe mmiri, Ndi bu oke osisi, Ndi bu Ngwu, ngwa were kwani nzu. Obu nwa Ebiri biara bia achu aja na, obu ka ife oma kwuru ya. Ifeoma wa, ja ekwuru ga, o nahu afu ya anta-afu onye!"[9]

English Translation:

> "You [Ogbuide], sparkling/glittering one, take your chalk! This is the son of Ebiri Obua. He has come to offer this sacrifice because of the promise he made to you [Ogbuide], regarding the good things you did for him. You take the chalk now! Urashi Ome Igbo[10] you take the chalk! Ani Orsu, [11] Ani Nama Oferi, Úganaga, Ogbu Niyii, Omerenwaka,[12] Ani di mu uzo echi[13] (World/Earth, show me the road back! [i.e. re-incarnation], Nne Beke, [14] please take the chalk [nzu]! It is the son of Ebiri who has

come to offer this sacrifice. River goddess who revenges the dead mother of our great water, goddess of the black water, Odi-ide Nna ogo, [15] woman/goddess who owns the water, those who are big trees, those who are ngwu, you also take chalk. It is the son of Ebiri who has come here. He is already witnessing [good things]. Don't you see the person [the current author] with him? I intreat you!"[16]

Like the people who worship her, the goddess Uhammiri has multiple names and identities. While Uhammiri is her ancient name, Ogbuide is her most popular name today.[17] She is ephemeral, multifaceted, and mysterious, sparkling in her numerous appearances, as does the greyish-blue water of the lake under the cloudy skies of the rainy season. Like the rainbow, Ogbuide has many colors, and she includes and expresses more than one idea. She is more than one, all-embracing and multifaceted, beneficial and awe-inspiring all at once. In the belief of the local Oru people, the lake goddess *gives children*[18] *and wealth*. This is evident in another prayer:

Obiadinbugha:	"Uhammiri Ogbu ama ke oji!"
Group:	"Ise."
O.:	"A si m gi zo ya ndu, si gi zosi ya ike."
Group:	"Ise."
O.:	"Ife oma biani ya. Ife oma bu ego. Ife oma bu nwa. Ife oma bu ndu."
Group:	"Ise."
O.:	"Ngwa gbufo pu ni okpo uzo, ngi bu Uhammiri Ogbu ama biko anya sa anya gba."
Group:	"Ise."
O.:	"Biko na o bu ya bu ife o bia gwa gi na gi ereanahu, na nkena amuru - amu ma ke dikwani na ihu na ya ka."
Group:	"Ise."
O.:	"Biko nu anya sa anya gba. Nke di na ihu ka."
Group:	"Ise."
O.:	"Ya ka obiara bia ekweni gi nkwa si gi jisini ike mebe ni ya ngo. Gbufopu ni okpo uzo omumu ngo, uzo ego, uzo

ndu, ife oma bata ni ya. Onye nfu adina, onye onwu adina na ibe ya. Ekwe na ni onye ga na ya no, bu ndi ocha. Biko ngi, bu Uhammiri ngi kpotara."

Group.:   "Ise."

English translation:

Obiadinbugha:	"Uhammiri Ogbuama, the Sparkling One, here is kolanut!"
Group:	"May it be so!"
O.:	"I am asking you to protect and guide his life, guide and protect him strongly."
Group:	"May it be so!"
O.:	"Let him have good things. Good things is wealth. *Good things is having offsprings* [children. Emphasis supplied]. Good things is life."
Group:	"May it be so!"
O.:	"Now, keep open his way, you, Uhammiri Ogbuama, the Sparkling One, we intreat you. Be highly attentive!"
Group:	"May it be so!"
O.:	"Please, he has come to inform you about how *you* [lake goddess] *have performed wonderfully in delivering this very child* [emphasis supplied]. But then, let the future be even greater!"
Group:	"May it be so!"
O.:	"Please be highly attentive! Let the future be even brighter!"
Group:	"May it be so!"
O.:	"He has come to fulfill his promise so that you will continue doing him favors. Please keep open his road [continue to give him access] to money, his road to *life,* his road also to *offspring* [emphasis supplied]. Let good things come into his house and

> stay with him! Let there be no wasted person, no death in his place! Defend also the European woman in his company, please. For it is you, Uhammiri, who brought her [the present author]."

Group: "May it be so!"[19]

FLORA NWAPA'S AMBIVALENCE TOWARDS UHAMMIRI AND HER GIFTS

In her earliest publications, Flora Nwapa depicted the lake goddess and the significance she held for local people, particularly women, in a slightly different character than that endorsed by the villagers themselves. The heroines of Nwapa's early novels, *Efuru* and *Idu*, both are wealthy but childless, or have only one child, a daughter who dies in childhood. With the exception of *The Lake Goddess* and *One is Enough*, Nwapa's novels focus on highly successful women who accumulate wealth *other than* offspring. This in itself is a problem among the local peoples, whose deepest desire and greatest pride are children. There is a lot of pressure on women to get married and bear as many children as possible as early as possible, before even thinking of making money, entering business, or starting a career. As a result, the childless heroines of Nwapa's novels face difficult lives despite their wealth and successes in business, as they are plagued by unhappy marriages and circumstances beyond their control.

To some readers, Nwapa seems to imply that childlessness is associated with a woman's economic or professional success and/or her devotion to the lake goddess; that worshipping the deity (and individual success) may contradict women's ordinary lives and the joys of motherhood. In *Efuru*, Nwapa seems to suggest that the goddess herself is childless and, as a result, may be unable to grant women children. But Nwapa's message is highly codified, as her criticism is aimed at the values of her own society.

Efuru, in the novel of the same name, apparently believes that Uhammiri does not have children of her own. Efuru may even doubt the goddess's ability to give children to humans. But it is also conceivable that Efuru—and by extension Nwapa herself—was doubting the *benefits* of the divine gift of fertility to women (*Efuru*: 208, 281) and/or the unfounded, albeit widespread charge that successful, wealthy women are neccesarily childless.

> Efuru slept soundly that night. She dreamt of the woman of the lake, her beauty, her long hair and her riches. She had lived for ages at the bottom of the lake. She was as old as the lake itself. She was happy, she was wealthy. She was beautiful. She gave women beauty and wealth but she had no child. She had never experienced the joys of motherhood. Why then did women worship her? (281)

Flora Nwapa was a novelist, not an ethnographer. She had her own concerns and agenda, particularly on women's issues. In pursuit of her cause, she challenged the basic tenor of local beliefs. By questioning the child-giving aspect of the water goddess/ woman, she also questioned the *benefits of children to woman* herself.

Yet, at least in her earlier work, Nwapa herself is somewhat ambivalent on this issue. First of all, Efuru's vision of the childless, wealthy, divine woman is only a dream, and Efuru never explicitly states that the goddess prevents her from conceiving. However, when she finally devotes herself as a worshipper of Uhammiri, she must abstain from sexual relations every fourth day (207), a practice that objectively reduces her chances of conceiving.

Efuru's enemies, who also condone the goddess herself together with her worshippers, charge that only barren women worship Uhammiri. When Efuru becomes a worshipper of Uhammiri, some women gossip behind her back:

> " Do I hear that she now has Uhamiri in her bedroom"?[20] Omirima sneered.

> "That's what I hear. She and her husband plunged into it. I was not consulted."

> "She has spoilt everything. This is bad. How many women in this town who worship Uhamiri have children? Answer me, Amede, how many? ... Your daughter-in-law must be a foolish woman to go into that." (203)

Reading these lines, we must be careful however not to adopt their statement at face value. Nwapa herself introduces doubts: The very characters who voice these charges are quite *negative,* in contrast to the positive heroine Efuru who is now worshipping the goddess. Omirima, the very woman who charges that Uhammiri cannot give children and who consequently condemns Efuru for worshipping

Uhammiri, is the same woman who later *wrongfully* accuses Efuru of adultery and thereby destroys her second marriage. After the death of her child from the previous marriage, this marriage was Efuru's last and only acceptable chance to conceive again. Omirima is known as a malicious character who is not highly respected in her community. Most importantly, Omirima wrongfully accuses Efuru of something *she did not do.* By extension, *Omirima equally wrongfully accuses the lake goddess,* Uhammiri, of not having or giving children, of causing childlessness, or even of killing children.[21]

In one respect, *all* (Ugwuta) people are the goddess's children, as Nwapa tells us in *Efuru:* "Uhamiri, the most beautiful woman, your children have arrived safely, we are grateful to you" (256). But then, Nwapa immediately introduces doubts, as this greeting offered by travellers is exclaimed *in the wrong place,* on the River Urashi. The river god Urashi is the lake goddess's husband. According to Nwapa the couple is constantly quarreling (156/7). Here again, Nwapa's story differs from the ideals expressed in local beliefs: According to local myths, the divine pair embodies the ideal of conjugal harmony and procreation.

THE ANIMAL SYMBOLISM AT PLAY IN NWAPA'S SOCIAL CRITIQUE

In her children's book, *Mammywater,* Nwapa separates Uhammiri from certain animals, namely snakes and crocodiles. But, these animals are more than merely entertaining elements of the story: they carry very specific symbolic meanings in local African mythology. The python in particular is associated with the water goddess as her consort and/or manifestation. According to local myths, the goddess's favorite child is the crocodile. This sacred animal must therefore not be killed, according to Ugwuta custom. In Ugwuta myths, as in much African (e.g. Ashanti) mythology, this semi-acquatic reptile is associated with female fertility. In *Mammywater,* as well as in an interview I conducted with Nwapa in 1988 (in this volume), the author separates Uhammiri from her favorite child, the crocodile, a creature symbolizing *female fertility.* She also disassociates Uhammiri from the python and other aquatic snakes commonly associated with the water goddess. Nwapa writes in *Mammywater:*

> Urashi is fanning himself with a huge 'ifuru' fish, whose red fins almost frighten Deke. There are smaller fishes surrounding him on all sides. Further

84

away from where they sit are huge crocodiles, playing and swimming about joyfully. Even further
away are huge long snakes, one coiling around the
other. They appear as happy and contented as the
crocodiles.

Deke wonders. This is not a good spirit. In the home
of Mammywater, one saw good things all around. It
is   quite   different   here.   He   wonders   why
Mammywater has brought him to this evil place, full
of harmful reptiles. (37)

Snakes, particularly the sacred python, are major symbols of creation, procreation, the indestructible soul, and eternity in African
mythology,[22] in contrast to their negative connotations in Judaeo-
Christian beliefs.[23] In *Efuru*[24] and in the 1988 interview(in this volume), Nwapa removes these creatures and the notions they
symbolize from the lake goddess/ woman, and transfers their association and their symbolism—the desire for children—to the lake
goddess's husband/man, the river god Urashi. In doing so, Flora
Nwapa questions the child-giving aspect and benefits of the water
goddess, Uhammiri, and moreover, transfers the association with,
and desire for, children and fertility from female to male—an idea
highly befitting a patrilinear society.

The water goddess could bring anything, including life, children, money, Europeans, and death. According to Nwapa, the water
goddess could also take children away, as she took Ojuru's daughter in *The Lake Goddess* (48,49), and Deke's sister, Soko, in
Mammywater.[25] She can also take people's lives, health, and wealth
away.[26] The goddess Uhammiri is indeed powerful and awesome.
She cannot be measured on a human scale. The local people
believe with Nwapa(*Mammywater*) that the water goddess *could*
take children away. But while the villagers recognize *two opposite
sides* of the lake goddess and pray to the deity for children, Nwapa
and her heroine, Efuru, question not only the goddess's *ability* to
deliver, but also the *benefits to women of fertility, an obsession of men.*

NWAPA'S EVOLVING PERCEPTION OF THE LAKE GODDESS

From the very beginning, Nwapa voices her critique of local beliefs
expressed in her doubts about the lake goddess—causing infertility, or even taking away children—with caution. In *Efuru* these allegations against the goddess and against the heroine—successful

childless woman—are voiced first of all in the words of a deviant character, and secondly in a dream. In *Mammywater,* Nwapa goes a step further, telling us that the [African] lake goddess "takes away children." Today, this charge is propagated by the churches and by some teachers who deliberately spread scary stories about the "pagan" goddess, in order to discredit who-/whatever competes with Christian believes. Sadly, these rumors led to a "Mermaid scare stampede" in an Enugu school, in 1985, where nine school children died.[27] Nwapa later distanced herself from all of those allegations against the water goddess. In her last, posthumously published book, *The Lake Goddess,* the goddess no longer takes away children, nor are her worshippers necessarily barren women.

Everyone still in touch with Ugwuta's own African deities knows that the vicious allegations of infertility and destruction of children against the lake goddess—and by extension against Efuru, the embodiment of the ambitious /successful / professional woman—are wrong for several reasons:

(1) The divine pair of Uhammiri and Urashi is commonly known to *give* children, as attested by Mrs. Nwammetu Okoroafor, an elderly Ugwuta woman:

> *"Uhammiri na akpota nwa, na enye madu ife nine."*
> "Uhammiri gives children and everything else."[28]

(2) There are also prominent *hereditary male priests* of the goddess in Ugwuta whom Nwapa ignores.[29]

(3) Many of the goddess's women priestesses and followers are married and have children,[30] as does Ona, who was called by the goddess to become her priestess in *The Lake Goddess.*

(4) Most importantly, many of the priests and priestesses of the water deities are also herbalists and renowned healers of both mental and gynaecological disorders. These herbalists, and priests and priestesses of Uhammiri specialize in *helping women to conceive, and/or keep their pregnancies.*[31] Uhammiri's priests/priestesses who are also native doctors ascribe their medical knowledge and healing powers to a gift from the goddess. Thus, it would be highly simplified if not preposterous to assume that worshipping the goddess Uhammiri requires a woman's childlessness, or that "in Igbo cosmology, the deity Uhamiri can grant women wealth but never children," as suggested by Gay Wilentz in *Binding Cultures: African Women Writers and the Diaspora* (16).[32]

The opposite is evident in the lives and oral literature of *ordinary* [33] Ugwuta people. There, both men and women are anxious to live up to society's norms and expectations and to have children. Those who are not immediately blessed with offspring implore above all the lake goddess Uhammiri and her husband, the river god Urashi,[34] for help in begetting and keeping offspring. This is evident from the (above) prayers offered at the goddess's shrine, as well as from the abundance of Ugwuta's songs praising Ogbuide/Uhammiri and her gifts. But there are also women who resist marriage or clitoridectomy, and, by extension, child-bearing, as female circumcision was customarily considered a prerequisite of childbearing in Ugwuta. The association of child-bearing and circumcision is evidenced in a conversation between Ona 's mother, Akpe, and her prospective mother-in-law, Nwafo, in *The Lake Goddess*.

> "Are you sure that you have been circumcised?"
>
> " Yes ma."
>
> "Because if you are not, you will not be able to have a child."
>
> "Mother did not tell me so." Akpe said rather timidly.
>
> "Don't bother about your mother and what she told or did not tell you."
>
> *Silence.*
>
> "Did you hear me?"
>
> "Yes ma. But she did not tell me that if I was not circumcised that I will not have a child."
>
> "Forget about that."
>
> "Yes ma." (23)

Then there are also women who cannot conceive, those who have lost all of their children, and those who have "only" one daughter or who bear twins,[35] women who behave unusually, who cook late at night, have visions, are unusually smart, beautiful, or wealthy, and are for some reason or another struggling with traditional social norms. All of these extraordinary individuals may take recourse in the goddess to find consolation, an emotional outlet, a safe-haven, assistance or empowerment in her worship. As Leslye Obiora and others have pointed out, pre-colonial African culture was not as static as some outside observers have painted it.[36] As elsewhere in the world, there are non-conservative, innovative individuals who challenge and contradict the norms. In Ugwuta, these industrious women are often women who refuse to sit back. Ugwuta women in particular are known to be enterprising and energetic. In *The Lake Goddess,* Mgbada's mother and her prospective mother-in-law recall their travels and trades up and down the waterways.

> "Patani, yes. They have good *kaikai* there. So our women still go there?" "They do. Our women are industrious. You remember when we were young we went to such far away markets. The young ones still do." (9)

Ugwuta women never rest; they always move, accumulate and create. These industrious women are socially and economically mobile. They move similar to their evasive, ever flowing patron, the water goddess who cannot be tied down.

The complex deity, Uhammiri/ Ogbuide/Mammywater, was pre-eminent in the lives of local Ugwuta people until the takeover by Christian missionary forces, the colonial conquest, and the postcolonial struggle and rushes for riches of oil, politics, money, and crime. Catholic missionaries with their negative attitudes towards the female body and sexuality were quick to condemn local beliefs and practices, and especially *female* deities.[37] These prudish and invasive foreigners did not hesitate to intrude into people's most private spheres of life. In Ugwuta, men's and women's bathing areas on the lake are separate. Nevertheless, female ritual nudity, customarily believed to enhance fertility, is today scorned and supressed by local adherents of the ideals of the foreign missionaries, as evident in a conversation I had with an Ugwuta couple:

> Azoga: "Women who have not given birth for the first time are not supposed to go to the lake to bathe with towels or underwear. They are supposed to go stark naked and then bathe and come out ..."
>
> Sabine: "Until they conceived?"
>
> Azoga : "Yes."
>
> Sabine : "So, now people are changing it?"
>
> Azoga : "Now people changed it. So it is one of the problems that is happening now because of all these

intrusions ....There are so many things that have changed because of civilization."

Sabine: "So it's a big conflict?"

Azoga: "Yes. The church says that women should bathe with their bathing suit on."[38]

Foreign missionaries and their local disciples condemn not only the goddess's ritual requirements, but they also undermine the loyalties of her local followers by introducing doubts in the goddess's ability to grant to people the one gift they desired most, children. While these doubts were most forcefully spread by missionaries during the colonial regime, the lake goddess Uhammiri is still prominent among local people in and around Ugwuta—some of whom resist the continued assault on their local beliefs and values, despite both tempting and threatening foreign influences.

The lake goddess is central in the work of her most prominent daughter, Flora Nwapa. Uhammiri is ubiquitous in Nwapa's books, in her visions of womanhood, in woman's complexity, in her many dimensions: her struggles, powers, and mysteries. Like the goddess herself and like her people, Flora Nwapa was also a highly complex person in her background and upbringing, her life and identity, her messages and achievements. Flora Nwapa was bi-lingual in Igbo and English. She was at once a daughter of the goddess, firmly rooted in local life and traditions, *and* a member of the educated Christian elite, trained abroad and highly conversant with European social norms, beliefs, and culture. She was a poet and a novelist; she wrote for adults and for children, for Nigerians and foreigners. Nwapa was also an artist and, at the same time, a business woman, a teacher, and a friend. She was a daughter, a sister, a lover, a wife, a mother, as well as a highly successful professional. A highly cultivated Igbo woman, she mastered seemingly unsurmountable antagonisms in life as she negotiated her multiple powers and identities.[39] Flora Nwapa was just as complex as the goddess herself. Her views of the goddess Uhammiri were not static, and instead have changed as she herself travelled in space and time.

NWAPA'S CHALLENGE AND SOCIAL CRITIQUE AND CHAMPIONSHIP OF WOMEN

In her literary *oeuvres,* Flora Nwapa at once explores, challenges, deconstructs, and restructures her hometown Ugwuta's lake goddess, Uhammiri/ Ogbuide, and her worship, to voice her own artistic concerns. Her critique is both deeply personal, highly social, and universal. Nwapa wrote fiction; she did not construct an ethnographic image of her culture, as an anthropologist would. Nwapa's novels are set in her own society and culture, where she lived for most of her life.[40] As Chimalum Nwankwo has pointed out, whether Nwapa was "steeped in tradition," as indeed she was, is not an issue. Rather we are focusing on how she used her cultural premise to formulate her own "coherent or consistent artistic vision."[41]

Nwapa's vision has changed over the years from an internal criticism of her own society and its beliefs, to a critical perspective on wider issues, including the deconstruction of her own society's values by foreign belief systems. In one of the playful author's children's books, the goddess's name is *"Mammywater."*[42] In this book, Nwapa merges the motifs of childlessness, beauty, and wealth expressed in the deity's Mammywater *aspect* with Ogbuide/Uhammiri—who is traditionally associated with giving children and wealth, but whom Nwapa in her earlier novels *Efuru* and *Idu* and in *Mammywater* deprives of her most precious gift, children. Later, Nwapa's concerns shift from personal conflicts resulting from the dichotomy between raising children and pursuing one's own career in a woman's life, to deeper concerns about the conflicts affecting contemporary Nigerian society in general, and Nigerian women in particular.[43] As a result, Nwapa's image of the lake goddess, Uhammiri/Ogbuide/Mammywater also shifts.

In *Efuru* and *Idu*, Nwapa focuses on the extraordinary woman and on her abilities, interests, and desires that may, at times, conflict with the confines and norms of tradition, the laws of the land "omenala." In Ugwuta, as in other areas of Igboland, there is a strong emphasis on, desire for, and appreciation of having many children. Ugwuta's farmlands are rich. To have many children is highly valued among the peasants who need many hands, "umua-ka," to help with the family farm and enterprise. To the local farmer, having many children is a necessity, even though this may also entail many mouths to feed, an increasingly difficult task in today's changing world. Resenting the "joys of motherhood"[44] is a totally unacceptable, almost unspeakable idea in Oru Igbo culture. Nwapa and others have maintained that in this society, *men* in particular need male children to continue their patrilineages. However, as Nwapa admitted later in life, children are equally important for women, because women who reside with their husbands are customarily buried in their fathers' land by their brothers and their brothers' children to whom these women may not be close. Therefore, a woman's children are ultimately the ones to make sure their mother receives an appropriate burial and "her body is not thrown out into the bush," a fate according to Nwapa[45] feared especially by elderly women who are childless, or whose children have died before their mothers. Children *traditionally increase a family's wealth,* as expressed in the Igbo name, "Kego" [Nwa ka ego— (a) child is better than money]. As a result, producing many [male] children enhances a woman's status.

On the other hand, the riverine Igbo, and Ugwuta women in particular, have long had access to means of wealth other than farming. Trading on the water routes along Ugwuta Lake, the River Urashi, and even up and down the great River Niger, Oshimmiri, is the Oru woman's domain, and a major source of her wealth and economic independence.[46] As in modern life, small children may be a burden to the career and ventures of an enterprising Oru woman, as much as children are their source of joy, happiness, and security in old age. Moreover, children may crown married life. But the death of a child may also torture the mother. A woman's fear of leaving a beloved child behind may tie her to an unhappy marriage, forcing her to endure psychological stress, in a society where the children of a divorcee would normally remain in their father's home. In a place where infant mortality is still high, women not only bear many children to make up for potential losses, but are also particularly prone to the pain of loosing a child, either to illness and death, or to her husband's lineage if she separates from her spouse. Flora Nwapa voices her ambivalency towards the mixed blessing of child-bearing that is invariably coupled with the anxiety of loosing the child in *Mammywater,* where Uhammiri tells her husband, the river god, Urashi:

> " ... I told this boy's mother that her husband had let me down and I must take her daughter, Urashi Great One. She wept and wept. Women, how they love children. But I had to do it to save my people from a certain death." (42)

Nwapa explores the deep seated and perhaps subconscious resentment or fear of younger women of the Janus-faced nature of childbearing, and of the goddess, by deconstructing the image of Uhammiri, the goddess believed to enhance female fertility, and focusing on the awesome aspects of the goddess known as Mammywater.[47] A novelist, not an ethnographer, Nwapa *takes the artistic liberty* to divorce the precolonial unity of wealth and children[48] by questioning some of the goddess's gifts. In both novels, *Efuru* and *Idu,* and in the children's book, *Mammywater,* the lake goddess gives wealth, but *not* children. Nwapa ends *Efuru* with the ominous question, "She had never experienced the joy of motherhood. Why then do the women worship her?"[49] (281) By extension, we may also ask together with Flora Nwapa "Do young women really want children?" Or, "Wouldn't some women prefer to pursue their career before raising children?" Or, "Why do women pursue a career?" As Nwapa pointed out, pursuing both, children and a business, or a career, poses a serious obstacle to women worldwide.[50]

In her later work, Nwapa explores the entwined theme of woman's struggles with social and personal conflicts, under the goddess's auspices. In *Never Again,* the lake goddess saves the people of Ugwuta from complete destruction by the invading Nigerian soldiers, a belief locally upheld to this day.[51] The heroines of Nwapa's post-war novels, *One is Enough* and *Women are Different* are true daughters of Ugwuta's lake goddess: they set out to accumulate wealth.

Expanding the Issue Beyond Local Custom

The wider conflicts of women who cannot live up to society's expectations and must chose between either wealth and a career, *or* children, are only dawning in *Efuru* and *Idu,* as both novels are set in an historical time in which—at least theoretically and according to that society's own ideals—wealth and children were still one, and in which a woman who could not fullfill this requirement was not only unhappy, but also potentially powerful and dangerous. In the days of *Efuru* and *Idu,* a woman who found herself in this conflict between society's expectations of child-bearing and her own desires for a career, or her abilities to conceive and keep a child alive, may have been unusual. But such a woman could still find a niche and worship the goddess without being declared "mad."

The smoldering conflict of interests of the localized individual against imposed power structures soars beyond gender in *The Lake Goddess.* Here, woman's fertility or childlessness is no longer the issue—and the pythons are back in the goddess's domain (74, 183). But other than that, nothing is normal any more. The issue of women's potential personal conflicts resulting from their childlessness in customary society has shifted to the issue of existential and identity crisis affecting everybody and threatening people's very existence and state of mind. The heroine is no longer extraordinary, but crazy, and society around her has gone mad. This madness clearly is the result of foreign intrusions, undermining local custom, cosmology, and religious beliefs that provide the core of the customary social, economic, and emotional foundations of people's lives.[52]

The story takes off not with the heroine, Ona herself, but with an introduction of her father, Mgbada, whose conflicts result from the intrusions of Christianity into his life. Ona 's father struggles to maintain his heritage, identity, and even his name, all of which he must defend against the vicious attacks of the new belief system. Christian missionaries and their fanatic local followers pursue the destruction of Igbo cosmology, religious beliefs, economic foundations, and social organization. The "church people's" pejorative attitude towards Igbo names, culture, and identity is clearly revealed in *The Lake Goddess* in a conversation between Ona's father's mother in-law, Mama Theresa, a fanatic Christian, and his mother, Nwafor/Theresa.

> "Your son, Joseph?"
>
> "Yes, Joseph or Mgbada, what does it matter? I call him Mgbada. I am so used to the name his father gave him when he was born."
>
> "You mustn't continue to call him that pagan name, Theresa. You must call him Joseph. Joseph is his Christian name. Drop that pagan name."
>
> "Yes, my namesake, you are right. I should drop it. I must make an effort."(12)

Nwafor/Theresa knew that "there was no point contradicting Mama Theresa. She knew what she believed in. She (Nwafor) was a Christian, but she appreciated some non-christian traditions"(11). The imported religion creates serious existential and spiritual conflicts for the people of Ugwuta and, moreover, antagonisms within the local community. The resulting pressures—particularly on women—are enormous. Some feel compulsively moved to attend mass every morning despite their already heavy burdens of domestic and business obligations. Others "fall by the wayside. The demands of the new religion were too much" (15). While some manage more or less successfully to combine both worlds in their lives, others are tormented to the brink of loosing their minds. These conflicts, created by the extreme demands of Christianity pitted against local life and spirituality, run like a trail of poison through the entire book, reflecting the painful tensions spread by fanatics who inject their venom into almost every home and family, sparing nobody, and no social level of contemporary Ugwuta.[53]

Ona 's grandmothers differ in the degree to which they follow the Christian camp. But they conspire to fake Akpe's circumcision.[54] As a result of this deception, "the people were happy and satisfied that their tradition was not trampled upon by the new religion" (26). This fake circumcision may also be an allusion to the flexibilty of custom, for as Leslye Obiora has pointed out:

> The old rules continue to bear their pristine names and well-worn features, but inconvenient and harsh aspects are being continually toned down and adapted to new circumstances. (224)

Ona herself is called by the lake goddess in visions and dreams, early in life. But her inclination is brutally suppressed by fanatic missionary teachers. Ona cannot find a local husband and must marry a man from another town. In contrast to Efuru, Ona bears a child nearly every year. Ona has three children in five and a half years, a record in fertility (171). But this also is not "normal," because according to custom, Ugwuta women space the births of their children in a three year rhythm,[55] and would *at most* bear two children in six years. Despite her frequent pregnancies, Ona does not enjoy sex and ponders "Why did men make such a fuss about it?"(170). [56] Although she initially thrives in her marriage for some time, she cannot live happily in her husband's home and returns to her father's house. This separation from her marital home is justified as a request from the goddess Ogbuide (178/179). While all she wants is "to live a normal life," society finally declares her mad[57] (165, 187, 194) and marginalizes her.

The worship of and belief in a goddess that was once basic to the local people's very existence is relegated to the margins of society into the realm of psychic disorders as Christianity flourishes. Ona, however, follows her visions of the goddess and is finally accepted as a healer. Although Nwapa does not tell us what type of ailments Ona treats people for, we must assume that Ona cures gyneacological problems, especially infertility, mental illness, and also a host of minor ailments, as her fellow priestess-healers do.[58] To fully devote herself to the goddess, Ona must finally abandon her husband.[59] This, again, is due to Nwapa's artistic liberty, not a customary requirement for worshipping Uhammiri/Ogbuide. Many of the goddess's real life priestesses are married and have children. Only some are widowed, separated, childless, or have lost all of their children. Nwapa's assumed requirement of ritual chastity must be contextualized with Christian teaching, and could perhaps be regarded as another allusion to a society gone mad, where the community's very existence is threatened by people with strange ideas. Nwapa writes:

> Strange people came to the town with strange ideas. They talked of a God who was born by a woman and who died for the sins of the world. They criticized the religion of the people calling them pagans and heathens. On certain days of the week they congregated; they prayed and sung all day. (*The Lake Goddess* 1, 2).

Ona's mother's late father nearly lost the economic foundation of his very livelihood, his land, as a result of the antagonisms sown by the intruders. He managed to secure his existence only by distancing himself from the church:

> ....When his brother reaped the proceeds of the land belonging to their father without as much as informing him, my late husband called everybody, and the whole clan settled the misunderstanding. It was after that that he stopped taking the church seriously. (44)

Ona 's father, Mgbada, a traditional herbalist and a diviner, tries to strike a balance by combining the two religions. But ultimately, he must drop his teaching career in order to pursue his vocation and uphold his heritage. Right from the beginning, he wonders:

> What kind of religion preaches that one should abandon the worship of our ancestors? Why should these people, who are foreigners for that matter, be concerned about where one goes when one dies? We know that there is life after death and so when we die, we join our ancestors and continue to live. (2)

Mgbada represents customary order against the disruptions and uncertanties of foreign intrusions, beliefs, and social change. He is the only person, apart from the goddess, who never ceases supporting Ona—and by extension woman—and who does not regard her as "mad," but rather as gifted. Even before Ona was born, Mgbada discovered with the aid of another diviner that Ona herself was a gift of Ogbuide. He named her Ona, "gift of the goddess," as in the myth invoked in a song during the deity's worship:

Solo:	Kporo ni mu jebe na mmiri mani anyi mma
Chorus:	...
Solo:	Ona nwa Eze akporona ni mu jebe na mmiri mani anyi mma. (repeat).
Solo:	Nwa mmiri oma, Eze nwa Owu akporo na ni mu jebe, Eze nwa ikuku akporona anyi jebe.
Chorus:	...
Voice:	Egwu adiadi ya ...
Eze Mmiri:	(O na ama odu ya—blowing her ram's horn).
Solo:	Nwa eze osidi, nwa onye oke okporona ni mu jebe ...
Solo:	Ikuku bu onye mma owu. Ona nwa mmiri, nwa di ukpa bia...
Solo:	Ocha esuo anye nkeni, na afo inaye, ma ya ere aria. Eze nwa ikuku bu onye mma owu, Ona nwa ikuku amana mma ukwu...
Chorus:	...
Solo:	Ona nwa Ojobo mmiri eze, ikuku bu anyi egwu, nwa Ojobo onye mmiri oma, a sina ikuku bu anyi egwu ukwu.
Chorus:	...
Solo:	Nwa Owu eku na ubu na agba egede, na egwu na bu egwu ndi ukwu, onye obunna biko kwu be okwu egwu na egwu na bu egwu ndi ukwu.
Chorus:	...
Solo:	Ikuku bia egbuna mu, nwa ojobo mmiri mara mma, Ona nwa Ojobo na egwu di ukwu.
Chorus:	...
Solo:	Egwu nwe Ojobo mmiri na egwu na bu eze ndi ukwu na egwu na bu egwu ndi oma.

Chorus: ...

Solo: Eze nwe oha na achu mmadu, oja achuta mu na ole aka, ome okachie nwa ogaranya ukwu oma eze nwe oha na achm m achu.

Chorus: ...

Solo: Mmiri mara nwa afo oma ikuku bu onye mma owu, i ma anyi na owu na mara na be yi beyi, na ikuku bu anyi egwu ukwu.

 Eze Mmiri: (O na ama odu ya—blowing her ram's horn).

Solo: Egwu na bu onye mma owu egwu na mara na be yi nabe yo, ikuku bu onye mma owu na egwu na mara na be yi nabe yo.

Solo: Asa kwa na nwa oba na ikuku bu onye egwu ukwu asisa asakwa nwa eze na ikuku bu onye egwu ukwu.

Chorus: ...

Chorus: Nye oye oye oye.

Solo: Sa mara, Imari jo i mara mara i mari jo,

Chorus: Ima mma ewo ewo.

Solo: Sa mara, imani jo, i mara mara.

Solo: Ikuku oji ukwu na ere ogunna soro anyi. ...

Chorus: Ikuku amana mma odogwu

Solo: I mara Ona i soro anyi
 I maha Ona, i so ni je i mara mara.

Solo: Oromma ka anyi jebu egwu ...

Chorus: Ojenima, Ojenima, Ojenima, ojenye, egwu, egwu, egwu oma.

Solo: Egwo ike agwm na mu. Mmadu o no kwa ya

Eze mmiri: Eze fo otu.

Chorus: O cha eri ji.

Eze mmiri: Egbu eni.

Chorus: Anu na asu.

Eze mmiri: Egbu eshi.

Chorus: Anu na asu.

Group: Eze mmiri di egwu! Eze Mmiri di egwu!

English translation:

Solo:	Ona the queen has taken me out, the water has done good to us. (Repeat)
Solo:	The good water, queen of 'Owu' has taken me out, the king of cool breeze/air has taken me out.
Chorus:	...
Voice:	There is no music!
Eze mmiri:	(Blowing her horn).
Solo:	The breeze is a good 'Owu', Ona the water who is important, come!
Solo:	The breeze is a fine 'Owu'. Ona the breeze is good.
Solo:	Ona, the daughter of Ojobo, queen of the water, the breeze is a good music, the daughter of Ojobo is a good water and they say that the breeze is music for the big people.
Solo:	The daughter of 'Owu' that is on the shoulder that dances 'Egede' music, this music is for the big people. Let everybody join hand because the music is for the big people.
Solo:	Please come, air, but don't kill me! Ona, the daughter of the Ojobo water that is good, music for the big people.
Solo:	The queen who owns the group, you will rule us well, 'Ome okachie' the rich, big and good, the queen who owns us.
Solo:	The water is good and it has a good heart. The breeze is a good 'Owu' water and this 'Owu' is different, different because the breeze is a big dance.
Eze mmiri:	(Blowing her horn)
Solo:	Don't act like an 'Oba' because the breeze is a person of big music. Don't act like the queen because the breeze is a person of a big dance.
Chorus:	Nye, oye, oye, oye. (End of music).
Chorus:	Sa mara, imeni jo, imara mara (Repeat)

Solo:	The breeze is good, hero.
Chorus:	If you know Ona, you will follow us.
	If you don't know Ona, you will stop following us.
Solo:	Proud person, we are going to dance.
Chorus:	If you know Ona, you will follow us, but if you don't know Ona, you will stop following us.
Voice:	I am tired, please, are the people here? Queen of the group?
Voice:	If there is only one tooth left
Chorus:	It will eat yam.
Eze mmiri:	Kill and divide!
Group:	Meat and fish.
Eze mmiri:	Kill and cook!
Group:	Fish and meat. The Queen of the Water is awesome! *Eze Mmiri di Egwu!*[60]

CONCLUSION

Ona /woman herself (and her fertility) is a gift of the lake goddess; she is not "a problem."[61] The lake goddess has endowed Ona /woman with special gifts besides fertility: Ona has the power to "see" and to heal. In the changing world of contemporary Ugwuta, the goddess—and by extension (African) woman—is constantly pushed back and encroached upon by alien ideas, problems, and forces. Nevertheless, in *The Lake Goddess,* she appears as a source of healing, a power that could cure contemporary society's madness, ultimately pointing to the universal values and spirituality of a local African culture.

Novelist Flora Nwapa has over time at first challenged, deconstructed, and reinvented the gooddess Uhammiri. In her earlier novels, Nwapa has turned a deity who once bestowed life, children, *and* prosperity on her people into an awesome spirit who grants women monetary wealth and emotional relief *instead of* children. Later, in *Never Again,* the goddess saves and preserves life, as she rescues her people from disaster during the Nigerian civil war. While more wealth is accumulated after the war, the goddess receeds to the background in Nwapa's post-war novels, *One is Enough* and *Women are Different.* Towards the end of Nwapa's life and writing career, society crashes as its values collapse in conflict

and madness. The goddess whom Nwapa invoked finally re-sur-
faces in her original glory in *The Lake Goddess* to brighten the path
for healing. Uhammiri's mysteries once again shine, despite the
onslaught of foreign powers and their religions. Nwapa now recognizes the destructive forces of those elements attempting to push
Uhammiri's children into the abyss of derangement. Yet, when the
lake goddess finally appears with her full image restored, the messenger who invoked her has already left the land and joined her
ancestors to "live on."

NOTES

1. Flora Nwapa was first recognized as "Africa's First Lady of Letters" by
Ada Egbufor in *African Profiles International.* March/April 1993: 34.

2. Ugwuta is the Igbo name of an Igbo town named "Oguta" by the British
colonial administration. Ugwuta/Oguta is located in Imo State of
Southeastern Nigeria. The British name and spelling, Oguta, is still in
use by the contemporary Nigerian administration and on road maps.

3. The goddess of Ugwuta Lake is known locally by several different names,
e.g.Uhammiri and Ogbuide. In a similar vein, the people who worship
this goddess use several different names for themselves on the ethnic,
community, and individual levels. See also Sabine Jell-Bahlsen, "Names
and Naming: Instances from the Oru Igbo," *Dialectical Anthropology* 13
(1989): 199-207.

4. The goddess's praise name and form of address is "Eze Nwanyi" (Queen
of Women) or "Eze Mmiri" (Water Monarch). These titles are also given
to the goddess's priestesses. In addition to the lake goddess's many
names, there are different possible spellings of Uhammiri. Nwapa wrote
"Uhamiri," while I am following the spelling of local translators who spell
the goddess's name "Uhammiri." The latter spelling's "mm" fully retains
the Igbo word for water, "mmiri," while Nwapa's spelling is contracted
and may represent a compromise with English. When quoting from
Nwapa's books, I am using the author's spelling and English place names,
e.g., Oguta instead of Ugwuta.

5. Ugwuta/Oguta and her sister towns, Orsu-Obodo, Esi-Orsu, Nnebukwu,
Nkwesi, Mgbele, and Izombe form a group of riverine towns known as
the Oru. The Oru, Omoko and Onisha/Onitsha Igbo are the riverine Igbo.
This group is a division of the Igbo people, an ethnic group of approximately 20 million people in Southeastern Nigeria. On the one hand, the
Oru/Oguta people speak their own dialect of the Igbo language and distinguish themselves through their own cultural peculiarities. On the
other hand, the Oru also share certain cultural and linguistic features
with the Igbo people as a whole. Today, Ugwuta /Oguta is the most
prominent Oru town, while Orsu-Obodo is locally known and recognized
as Oguta's "senior brother" town. I have spent many months of fieldwork
there, since 1978.

6. This interview with Mrs. Onyemuru was conducted in her boat on

Ugwuta lake, in March 1989. A portion of the interview can be seen on camera in my 1991 documentary film, *Mammy Water: In Search of the Water Spirits in Nigeria,* distributed by University of California Extension Center for Media and Independent Learning, Berkeley, Calif.

7. This is evidenced in my ethnographic documentary film, *Owu: Chidi Joins the Okoroshi Secret Society,* distributed by Ogbuide Films, New York, and by the National Black Programming Consortium in Columbus, Ohio. The film features the annual Agugu festival. This festival is the Oru towns' New Year festival, timed in accordance with the moon cycles and the corresponding water levels of Ugwuta lake, the rivers Niger, Urashi, and their tributaries, after the annual flood.

8. Ogbu-ama, the Sparkling/Glittering One, is yet another praisename and a form of addressing the lake goddess. According to linguist Dr. Victor Manfredi, there are several possible translations for the goddess's name, Ogbuide, depending on the intonation of the name. The addition of Pan Kwa with tone markers would be necessary for the following explanation. However, the tone markers could not be printed here.

1.1. "Given the prevalence of agentive deverbal nouns in o-gbu- x [x-kill-er] in Igbo, the word in question is not just directly intelligible, but is actually stereotypical, in the function of a praise name (afa-otutu).

1.2. Support for [this version] is ...the expression ogbu -ama ("sparkling/glittering one") as applied to the goddess herself. Some obscurity (to me) is attached to the morpheme "ama," which the translator seems to have taken as a noun derived from a predicate "ma" or "mma" ("sparkle/glitter"); but assuming that interpretation is correct, the function of the morpheme *gbu* ("cut/kill") is transparently adverbial, so that a more literal rendering of this item would be "S/he of the devastating brighntness." Equivalent collocations occur in everyday Igbo, e.g. *"Oma-gbu-ru onwe ya"* ("S/he is exceedingly beautiful")... in which the item *"gbu"* does not [litterally] denote cutting or killing, unless perhaps metaphorically.

1.3. Indispensable in an etymology for *"O-gbu-ide"* is the meaning of "ide". Williamson's 1972 Onicha dictionary (p.170) has two plausible candidates: *ide* (I) "floodwaters" (HL: high-low) and ide "pillar" (LH: low high)—the latter as in the name Idemmili. The pronunciation of the word "Ogbuide" will decide the matter. In standard Igbo, a complement noun in an agentive nominalization has its initial tone replaced by L(low). Therefore, in o-gbu-x, if x = ide (I) "flood" (HL) the predicted surface tone is *O-gbu-i de* (L-H-LL). That is semantically attractive, because it makes the appelation as a whole hypocharistically mean "S/he who kills with floods" or "S/he who floods excessively." But if x = "pillar," *O-gbu-ide* (LHH) is the expected outcome. To confirm the latter, one must verify that an Ugwuta-speaker pronounces "killer of rats (i.e., oke)" as *O- gbu-oke* (LHH) and not *O-gbu-oke* (LHL). The alternative with "pillar" is semantically difficult, which is why I am hoping the tone is *Ogbu-ide,* (LHL).

2. The Onicha dictionary (p.363) also has ogbu mmili ("deep water").

(The sense of *"gbu* " here is related to the intensifying, adverbial use mentioned in 1.2 above.) Thus, if the second word is "ide" (I) "flood" (HL) and the pronunciation is Ogbuide (LHL) or Ogbuide (LHL), the meaning is "deep floodwater"—very easy to believe." Victor Manfredi, letter to Sabine Jell-Bahlsen, 28 March 1996.

9. Igbo is a tone language with two distinctive tones, high [´] and low [`]. The third is a grammatical (predictable tone). There are two distinct tone-marking conventions currently in use in the language. Green and Igweã (1963) mark all low tones and the downstepped tone, the latter with a macron. Emenanjo (1978) follows the same practice. The convention adopted here is that of Nwachukwu (1983) in which the three tones are marked using only the high and low tone marks. The first syllable of an utterance is marked, and subsequent syllables are left unmarked if they are on the same pitch level as the immediately preceeding tone. Thus, when a high tone is marked immediately after another high tone the second tone is downstepped. Mr. Uzo P. Ihionu has provided the tone markings of the Igbo words in the Igbo quotations throughout this paper. Unfortunately, they could not be printed in the current edition.

10. One of the river god Urashi's praise names, referring to a mythical incident where the deity drowned attacking Igbos from other areas who could not swim.

11. The communal earth goddess, Ani, of the town, Orsu- Obodo.

12. More local deities.

13. The earth goddess of an individual compound.

14. The farm god of the nearest village.

15. A male deity.

16. This prayer was recorded during my first field research in Ugwuta, at the Ihu Uhammiri square of Orsu-Obodo, on December 14, 1978. Translation by Chief Francis Ebiri. The prayer accompanied the sacrifice of a chicken to Uhammiri. This ritual was performed on behalf of the late Chief Ebiri to thank the goddess for the birth of his first son, and also, to ask the goddess for an extension to raise enough money to fulfill a previously given pledge to offer a white ram to the goddess. This pledge was finally fulfilled in June 1979, when Mr. Ebiri offered a white ram to Uhammiri. This event was documented on film by Georg Jell and myself in the film, *Divine Earth - Divine Water, Part I: Sacrifice to the Lake Goddess, Uhammiri.*

17. Ugwuta's main road—as well as several "hotels" are named "Ogbuide" after the town's patron goddess. According to local consultants, the name "Ogbuide" originates from Benin. The name may have been given to the goddess by the strand of Ugwuta's population that migrated to the present location from Benin some 500-800 years ago. This migration is also evidenced in an interview I recorded with Ifunanya (Eze Ugo) Azogo, an Eze Nwanyi of Ugwuta (one of the lake goddess's priestesses) and her husband, Azogo, on October 13, 1988. Ifunanya is the daughter of Ugwuta's hereditary male priest of Uhammiri, Chief Nwokocha Eze Ugo. At the time of the interview, the priest had himself abandoned the goddess in favor of the church, and his daughter, Ifunanya, carried his priest-

hood on. However, Ifunanya shortly afterwards made headlines in a local newspaper reading: "The priestess of Oguta lake is a Christian!"

18. See also: Sabine Jell-Bahlsen, "The Concept of Mammywater in Flora Nwapa's Novels," *Research in African Literatures,* 26.2 (Summer 1995): 30-41.

19. Translation by the late Chief Francis Ebiri.

20. Worshippers of Uhammiri place a clay pot with water from the lake in their room as a shrine. The priestesses also keep the pots of those who cannot take their pots home because they live in a faraway city, or abroad. The installation of a clay-pot shrine in the house of a newly initiated woman can be seen in my documentary film, *Mammy Water: In Search of the Water Spirits in Nigeria.*

21. The idea that Uhammiri/ Mammy Water kills/takes away children is propagated by various churches, in order to discredit the indigineous deities competing with Christianity. See below.

22. Geoffry Parrinder, *African Mythology* (London: Hamlin 1967). Snakes are also prominent in the popular image of Mammy Water known all over West Africa. See also: Sabine Jell-Bahlsen, *Mammy Water: In Search of the Water Spirits in Nigeria.* (documentary film) and Sabine Jell-Bahlsen, "Eze Mmiri di Egwu!—The Water Monarch is Awesome, Re-considering the Mammy Water Myths," in *Queens, Queen Mothers, Priestesses and Power,* Edited by Flora Kaplan.

23. In The Bible and in Christian teaching, as opposed to African mythology, the snake is associated with evil temptation. It represents sexuality as negative, resulting in Adam and Eve's expulsion from paradise.

24. As we shall see later, Nwapa reverses this step in *The Lake Goddess.* There, the pythons are back in Uhammiri 's domain, as the author's concerns have shifted.

25. The theme is also recurrent in Nwapa 's novels, *Efuru* and *Idu.* It may be an expression of local mothers' fears for their small children's life in the face of high infant mortality.

26. According to Mrs. Chinwe Achebe, the supreme water goddess, Nne Mmiri, is believed to destroy the lives of those who do not honor their pre-natal vow to worship her, given on their journey across water from the ancestral world to earth, when the goddess helped them change their destiny. See: Chinwe Achebe, *The World of the Ogbanje* (Enugu: Fourth Dimension Publishers, 1986). See also the documentary film, *Eze Nwata: The Small King.*

27. *The Guardian* (Lagos), Vol.3, No. 863: (Friday, Nov.1, 1985): 1,2.

28. Interview with the late Madame Nwammetu Okoroafor, at Nduka's house in Abatu village, Ugwuta, April 14, 1989.

29. Sabine Jell-Bahlsen, "Female Power: Water Priestesses of the Oru Igbo," in *Sisterhood, Feminisms, and Power,*" edited by Obioma Nnaemeka (Lawrenceville, N.J.: Africa World Press, in press). See also: Sabine Jell-Bahlsen, "Social Integration in the Absence of the State." *Doctoral dissertation* (Ann Arbor, Mich.: University Microfilms, 1980).

30. Sabine Jell-Bahlsen, "Female Power: Water Priestesses of the Oru Igbo " and Jell-Bahlsen, "The Concept of Mammy Water in Flora Nwapa's Novels."

31. As seen in my documentary film, *Mammy Water: In Search of the Water Spirits in Nigeria*. See also: Sabine Jell-Bahlsen, "Mammy Water als Therapie: Weltbild, Ritual und Heilung bei den Oru Igbo in Sdost Nigeria." (Mammy Water as a form of therapy: World view, ritual, and healing among the Oru Igbo of Southeastern Nigeria), *Jahrbuch/yearbook for Ethnopsychotherapy* 1993, Edited by W. Andritzky (Berlin/Germany: Aglaster 1996).

32. One may wonder what " Igbo cosmology " Wilentz is citing, and also, whether this really was Nwapa's message. Wilentz's suggestion that "ironically, although Uhamiri brings fertility to both land and water, she is unable to bring fertility to women"(16) reads like an abomination, or like a church sermon insulting the beliefs of local villagers and what is dear to them, in and around Ugwuta. Wilentz quotes Chinua Achebe's "Uncle Ben's Choice" in her effort to underscore the presumed Igbo cosmological dichotomy between women's wealth and children. I am uncertain, whether Achebe was referring to the goddess Uhammiri in this story. Rather, he may have been inspired by Idemili, the goddess to whom he refers in *Things Fall Apart* and in *Anthills of the Savannah*. In addition, I am assuming that Achebe is referring to the modernized Mammywater *aspect* of either deity in that story, rather than to her ancient cosmological side, evidenced in local mythology and oral literature.

    There are indeed (at least) two different sides/*aspects* of the goddess that may be associated with her two names: 1) Uhammiri/ children and 2) Mammywater/money. As we have seen in the prayers offered by local villagers *(infra)*. The villagers believe that Uhammiri has two sides: she can give and she can take life. She can *give/take* both, wealth *and* children. The two were formerly regarded as one. In the *urban* context, Mammy Water may not be associated with giving children, but rather with money. But that is *not* what Nwapa describes. When Flora Nwapa writes "Mammywater," she actually describes Uhammiri/Ogbuide of Ugwuta Lake in her own *rural* setting. Nwapa was also very specific and explit in merging the two, Uhammiri and Mammywater, when I interviewed her in 1988 (see: infra). Nwapa explicitly stated that Mammywater and Uhammiri/ Ogbuide *are one*, that these are two names for one deity. (Mrs.) Chinwe Achebe *(The World of the Ogbanje)*writes that "Mammywater" is just an English term for the generic Igbo *Nne Mmiri*, the water goddess, who has different local names.

33. Ugwuta, just as other places, has various layers of social reality. There are the ordinary people; there are adherents of African religious beliefs; there are Christians; and there is a wealthy elite, some of whom are highly educated. The latter are the people that a wealthy foreigner is likely to meet and talk to, on a brief superficial visit. It takes a prolonged stay and many visits to get close to the ordinary people and their beliefs.

34. See also the documentary film, *Eze Nwata: The Small King* by Georg Jell & Sabine Jell-Bahlsen.

35. In Ugwuta, as among other Igbo groups, twins were customarily abandoned and their mothers scorned. See also: Victor Uchendu, *The Igbo Speaking Peoples of Southeastern Nigeria* (New York: Holt, Rinehart, and

Winston, 1965). Despite this, Amaka, the powerful female heroine of *One is Enough* decides to keep her twins. Twins also appear in Ona 's dreams as an additional sign of her association with the unusual. Among the Oru, mothers of twins may find a safe haven from social ostracism for themselves and their twin children, and also mental peace, by joining a group led by a priest or priestess of the water deities. See also: Jell-Bahlsen, "Female Power: Water Priestesses of the Oru Igbo."

36. Obiora has pointed out that "customary law was not variant... People adopt traditional norms not only as possessions but also as points of departure for new patterns of thought and conduct, and as constituent elements therein. Hence Gwyn Prins eloquently remarks "tradition is a process—it only lives as it is continually reproduced. It is effervescently vital in its apparent stillness." Leslye Amede Obiora, "Reconsidering African Customary Law," *Legal Studies Forum* XVII.3 (1993): 224. Obiora quotes from Gwyn Prins. "Oral History" *New Perspectives on Historical Writing*, edited by Peter Burke, (1991): 114,115. See also: Stanley Diamond, "The Rule of Law Versus the Order of Custom," *In Search of the Primitive* (New Brunswick, NJ: Transaction Books, 1987): 255-280.

37. A shocking example of the viciousness of these attacks is the recent beheading of the shrine sculptures for the river goddess, Ava, at Ukana in Anambra State, documented in my video, *Ava: A River Goddess Near Nigeria, Nsukka 1978-1992*. See also: Sabine Jell- Bahlsen, "Can the Arts be transformed to Death? Some Instances from Southeastern Nigeria" a paper presented at the African Studies Association Meeting in Toronto 1994 forthcoming in *Anthropology and Theology: New Perspectives,* edited by Frank A. Salamone (in preparation).

38. Interview with Ifunanya (Eze Ugo) Azogo, the Eze Nwanyi (priestess of Uhammiri) of Ugwuta, and her husband Azogo, in Ugwuta, October 13, 1988.

39. Flora Nwapa truly represented the complexities of African womanhood as envisioned by Zulu Sofola in her paper, "Feminism and the Psyche of African Womanhood," presented at the first international conference on *Women in Africa and the African Diaspora,* convened by Obioma Nnaemeka at the University of Nigeria in Nsukka, July 1992.

40. In contrast to other African authors (e.g. Buchi Emecheta who largely lives abroad), Flora Nwapa predominantly lived not only in Nigeria, but largely in Enugu where she was in constant touch with her home, Ugwuta.

41. Chimalum Nwankwo, " The Igbo Word in Flora Nwapa 's Craft," *Research in African Literatures,* 26.2 (Spring 1995): 43.

42. Flora Nwapa, *Mammywater.* (Enugu: Tana Press, 1979). According to (Mrs.) Chinwe Achebe, "Mammywater," with different possible spellings in pidgin English, is "Nne Mmiri" in Igbo, a generic term for the supreme (mother) water goddess of the cross roads (between life and death). See: *The World of the Ogbanje* (Enugu, Nigeria: Fourth Dimension, 1986, p.14-25).

43. See also: Gloria Emeagwali, (ed.). *Women Pay the Price.* (Trenton, N.J.: Africa World Press 1995), and Leslye Amede Obiora, "Reconsidering African Customary Law: Part IV. On Situating Women in the Colonial

Experience," *Legal Studies Forum* XVII.3 (1993): 227-252, and Jell-Bahlsen, Sabine. "Female Power: Water Priestesses of the Oru Igbo."

44. To paraphrase the title of Buchi Emecheta's novel, *The Joys of Motherhood* (New York: Braziller, 1979).

45. According to Nwapa, in times of increasingly scarce town land, this is also a reason for many elderly women to join the church so they could "get a place in the grave yard." (Private conversation, New York, October 1992.) The churches of various denominations are apparently cleverly exploiting these deep- seated fears to solicit membership and thereby increase their own wealth, in Africa as much as in Europe. In several European countries, including Italy, Spain, Germany, and Austria, both catholic and protestant churches collect 10% of peoples' income on top of the income tax from their members with the aid of state authorities. Despite the economic hardship raised by this extra expense on low-income families, many people especially in rural areas are afraid of officially dropping their church membership for fear of being denied a place in the graveyard owned by the church. In addition to gaining followers from exploiting peoples' fears, the leader of a parish of 3,000 souls is nicely rewarded in Nigeria, with a home with running water, electricity, a car, and other luxuries beyond the reach of ordinary rural people.

46. Amadiume has pointed to the differentiation in productivity in different parts of Igboland and to the early importance of local trade between the producers of different crafts and crops. Ifi Amadiume, *Male Daughter, Female Husbands: Gender and Sex in an African Society* (London: Zed Books, 1987). Catherine Acholonu also points to the historical importance and wealth of the powerful women traders of the riverine areas: Catherine Acholonu, *The Igbo Roots of Olaudah Equiano* (Owerri, Nigeria: AFA Publications, 1989).

47. Mammy Water also has her benevolent sides. In Ugwuta, Mammywater priests and priestesses insist that the Mammy Water they worship is identical withUhammiri and Urashi, and also with the (male) River Njaba, the River Niger, Oshimmiri (who is both, a man and a woman), and a host of other rivers. All of these water deities are collectively referred to as "Mammy Water." In contrast to observations in Liberia by the Canadian psychatrist Wintrob, Ugwuta villagers believe that Mammy Water also *heals* madness, rather than merely causing mental disorders. Nwapa alludes to both, the goddess's awesome side that may cause "strange behavior" and to the healing powers ascribed to Uhammiri in *The Lake Goddess*. In addition, and *in contrast to Nwapa's earlier novels*, the goddess's followers are no longer childless in her last book.

48. This artistic liberty is also a bourgeois liberty, not available to the peasants who still need many hands/ children to till the soil, even today. Nwapa also questions the ideal of conjugal harmony as basic productive unit, as she deconstructs the harmonious relation between the divine pair, Uhammiri and Urashi, and lets Mammywater/ Uhammiri constantly quarrel with her husband, Urashi in *Mammywater*. Likewise, in *Efuru*, the divine couple is not on good terms (255).

49. In a similar way, Ona questions the pleasures and benefits of sexuality

to *women*, when she asks herself why *men* cherish sex so much. This question also relates to Catholic thinking.

50. In an interview I conducted with Flora Nwapa in Enugu, in 1988 (infra).

51. This view is also confirmed on camera by an elderly Ugwuta woman, Madame Nwammetu, in my documentary film, *Mammy Water: In Search of the Water Spirits in Nigeria.* Local beliefs in the lake goddess as savior are so strong that they are regarded as a threat to the superiority of Christian clergymen who, as a result, have brandished the idea. See also: S.Jell-Bahlsen, "Interview with Flora Nwapa" (infra).

52. See also: Chikwenye Okonjo Ogunyemi, "Introduction: The Invalid, Dea(r)th, and the Author; The Case of Flora Nwapa, aka Professor (Mrs.) Flora Nwanzuruahu Nwakuche, " *Research in African Literatures* 26. 2 (Summer 1995).

53. Flora Nwapa herself, although highly recognized and wealthy, was not immune to these conflicts and, as a result, refused to honor the repeated urgent requests by a renowned local priestess of Uhammiri to sacrifice a cow to the goddess to thank her for her professional success. When I visited the priestess in July 1992, she asked me to deliver this message to Flora Nwapa, again. This I did when Nwapa came to New York in October that year. Flora Nwapa replied that she could not do this "because of her social standing, her age, and the church." (Personal communication w. Flora Nwapa in New York, Oct. 1992, and with the Eze Mmiri, Mrs. Martha Mberekpe, in Orsu-Obodo, Oguta, II, July, 1992).

54. In Ugwuta, the practice of female circumcision in the form of clitoridectomy was formerly believed to be a necessary pre-requisite to child-bearing. Clitoridectomy was performed on young women as a rite of passage, or at the time of marriage, or latest, before the birth of the first child. Women refer to this practice as "taking the bath," Clitoridectomy in grown-up girls has today been largely abolished. But according to Dr. Ehirim, "a mild form of clitoridectomy" was routinely performed at birth on new-born baby girls born in Ugwuta's general hospital, in 1979.

55. An Oru Igbo woman normally avoids sexual intercourse for two years after a child's birth to ensure that a new pregnancy would not disturb the health of the nursing baby and its mother. Ugwuta women enforce sexual abstinence (and with it, their right to birth- spacing) according to the laws of the earth and the water goddesses. Christians might not adhere to these spiritual rules. During the time of sexual abstinence, some women would encourage their husbands to look elsewhere for sexual partners and/or to marry another wife if possible. After 2 years, the first woman would resume sexual relations with her husband again, and could then become pregnant again. This practice resulted in the spacing of a woman's giving birth only every three years at the most. Some women might also skip another year and conceive even later.The three years of birth-spacing are also reflected in the Oru town's social structure because they form the basis of the town's male and female age grades. Every three years, a new age grade is formed by "the people who were nursed together." See: Sabine Jell-Bahlsen, "Social Integration in

the Absence of the State: A Case Study of the Igbo Speaking Peoples in Southeastern Nigeria," Doctoral dissertation, The New School for Social Research, New York (Ann Arbor, Mich.: University Microfilms, 1980).

56. This idea relates to catholic thinking and also contradicts Ugwuta beliefs, whereby women insist that sexual fulfillment is a pre-requisite to conception.

57. Several forms of "madness" are known and recognized locally in Ugwuta, as is evidenced in a conversation between Mgbeke and Ekecha, two fish sellers in *The Lake Goddess* (111). One particular type of "madness" is today associated with spirit calling, initiation, and ultimately priesthood of the lake goddess, Ogbuide. Many of the goddess's priests and priestesses were at one time or another in their lives believed "mad". One priestess, the Eze Mmiri, revealed to me in an interview: "I did not know what to do. I put wood in the soup and salt in the fire. In the night, the spirits would come and take me to the bush. They showed me leafs that will cure and leafs that will kill. My husband spent money on my treatment." The Ezemmiri, Mrs. Martha Mberekpe, Orso-Obodo, 1989. See also: Sabine Jell-Bahlsen, "Mammy Water als Therapie," and her documentary films, *Mammy Water: In Search of the Water Spirits in Nigeria*, and *Eze Nwata: The Small King*.

58. This is evidenced in several interviews with the priestesses and priests of the water deities. See also the documentary film by Sabine Jell-Bahlsen, *Mammy Water: In Search of the Water Spirits in Nigeria*, and "Mammy Water als Therapie."

59. The requirement of absolute celibacy must be read as an artistic liberty and as Christian inspired fiction. In real life, the priests and priestesses of the lake goddess are not at all forced to leave their spouses, even though some few may chose to do so. Real life worshippers of Uhammiri and Urashi are required to abstain from sexual relations with their spouses/lovers on certain days, but *not permanently*. They are African priestesses, not European nuns. Nor are these women social outcasts as Nwapa implies in *The Lake Goddess*. Several contemporary priestesses of the lake goddess suffered from emotional disorders at one point in their lives, and some may position themselves outside ordinary social norms. But others are highly recognized, depite—or because—of their special social position. Nwapa puts these words in Mgbada's mouth: "Women are better than men in keeping the numerous taboos of our Mother. They are endowed with great patience, and remember, priestesses of Ogbuide *must abstain from sexual relationship with men."* [emphasis added] (*The Lake Goddess:* 216). However, the requirement of total abstinence directly contradicts local practices and ideas of partial abstinence, in a society where the notion of celibacy makes no sense. Nwapa's words therefore seem to reflect largely the prevalent European-catholic doctrines of chastity, and perhaps a phantasy about the lake goddess's worship. See also: S.Jell-Bahlsen, *Mammy Water: In Search of the Water Spirits in Nigeria.*

60. Recorded by Sabine Jell-Bahlsen on an *Orie* day [day of worship of Uhammiri/ Ogbuide) at the house of the Eze Mmiri, Mrs. Martha Mberekpe of Orus-Obodo, Oguta II, on Oct.15, 1988. Translation by the

late chief Francis Ebiri.

61. The notion of "woman as a problem" can be directly related to Christian ideology and influences expressed, for instance, in the biblical notion of "Eve" being responsible for man's expulsion from paradise. In a similar vein, we must note that the western world seems obsessed with the idea of a threat to its resources emanating from the fertility of African and other brown and yellow women. The discourse on "overpopulation" is nearly always focused on Africa and other "third world countries" while Europe and America, who consume most of the world's resources, rarely think of themselves as "overpopulated."

WORKS CITED

Achebe, Chinua. "Uncle Ben's Choice." *Girls at War and Other Stories.* London: Heinemann, 1972.

Achebe, Christie, Chinwe (Mrs.). *The World of the Ogbanje.* Enugu, Nigeria: Fourth Dimension, 1986.

Acholonou, Catherine. *The Igbo Roots of Olaudah Equiano.* Owerri, Nigeria: AFA Publications, 1989.

Amadiume, Ifi. *Male Daughters, Female Husbands: Gender and Sex in an African Society.* London: Zed, 1987.

Diamond, Stanley. "The Rule of Law Versus the Order of Custom" *In Search of the Primitive.* New Brunswick, N.J.: Transaction Books: 1987: 255-280.

Emeagwali, Gloria (ed.). *Women Pay the Price.* Trenton, N.J.: Africa World Press, 1995.

Emecheta, Buchi. *The Joys of Motherhood.* New York: Braziller, 1979.

Emeanjo, Nolue. *Elements of Modern Igbo Grammar.* Ibadan, Nigeria: Oxford University Press, 1978.

Green, M. and G. Igwe. *A Descriptive Grammar of Igbo.* Berlin: Akademie Verlag, 1963.

Jell-Bahlsen, Sabine. *Ava: A River Goddess Near Nsukka, Nigeria 1978-1992.* Video. New York: Ogbuide Films, 1992.

___. "The Concept of Mammy Water in Flora Nwapa's Novels." Research in African Literatures 26.2 (Summer 1995): 30-41.

___."Can the Arts be Transformed to Death? Some Instances from Southeastern Nigeria." Paper presented at the African Studies Association Meeting in Toronto 1994, forthcoming in *Anthropology and Theology: New Perspectives,* edited by Frank A. Salamone (in preparation).

___. "Eze Mmiri di Egwu, The Water Monarch is Awesome: Reconsidering the Mammy Water Myths." *Queens, Queen Mothers, Priestesses and Power: Case Studies in African Gender.* Ed. Flora Kaplan; New York: The New York Academy of Sciences, 1997: 103–134.

___. "Female Power: Water Priestesses of the Oru-Igbo." *Sisterhood, Feminisms, and Power.* Edited by Obioma Nnaemeka. Trenton, N.J.: Africa World Press (forthcoming).

___. "Mammy Water als Therapie: Beobachtungen bei den Oru-Igbo in Südost Nigeria" (Mammy Water as a form of therapy: a case study among the

Oru-Igbo of Southeastern Nigeria) *Jahrbuch 1993 für Ethnopsychotherapie* (1993 Yearbook of Ethnopsychotherapy). Edited by Walter Andritzky. Berlin, Germany: Aglaster 1996.

___. *Mammy Water: In Search of the Water Spirits in Nigeria.* Film/Video. Berkeley: University of California Extension Center for Media and Independent Learning, 1991.

___. "Social Integration in the Absence of the State: A Case Study of the Oru-Igbo." Doctoral dissertation. The New School for Social Research, New York. Ann Arbor, Mich.: University Microfilms, 1981.

Jell-Bahlsen, Sabine and Georg Jell. *Divine Earth—Divine Water (a film trilogy). Part I: Sacrifice to the Goddess of the Lake, Uhammiri.* New York: Ogbuide Films, 1981.

___. *Eze Nwata—The Small King.* New York: Ogbuide Films, 1983.

Nwachukwu, P. *Readings on Igbo Verbs.* Onitsha/ Nigeria: Africana/FEP Press, 1983.

Nwapa, Flora. *Efuru.* London: Heinemann, 1966.

___. *Idu.* London: Heinemann, 1970.

___. *The Lake Goddess.* (forthcoming posthumously).

___. *Mammywater.* Enugu: Tana Press, 1979.

___. *Never Again.* Enugu: Tana Press, 1975; Trenton, N.J.: Africa World Press, 1992.

___. *One is Enough.* Enugu: Tana Press, 1981; Trenton, N.J.: Africa World Press, 1992.

___. *Women Are Different.* Enugu: Tana Press 1986 Trenton, N.J.: Africa World Press, 1992.

Nwankwo, Chimalum. "The Igbo Word in Flora Nwapa's Craft." *Research in African Literatures* 26.2 (Summer 1995): 42-52.

Obiora, Leslye Amede, "Reconsidering African Customary Law" *Legal Studies Forum.* XVII.3 (1993): 217-252.

Ogunyemi, Chikwenye Okonjo. "Introduction: The Invalid, Dea(r)th, and the Author: The Case of Flora Nwapa, aka Professor (Mrs.) Flora Nwazuruahu Nwakuche." *Research in African Literatures* 26.2 (Summer 1995): 1-16.

Parrinder, Geoffrey. *African Mythology.* London: Hamlin, 1967.

Sofola, Zulu. "Feminism and the Psyche of African Womanhood". A paper presented at the first international conference on *Women in Africa and the African Diaspora* convened by Obioma Nnaemka at the University of Nigeria in Nsukka, July 1992.

Uchendu, Victor. *The Igbo of Southeastern Nigeria.* New York: Holt, Rinehart, and Winston, 1965.

Wilentz, Gay. "Flora Nwapa, *Efuru." Binding Cultures: Black Women Writers in Africa and the Diaspora.* Bloomington: Indiana University Press, 1992.

Williamson, Kay. *Igbo-English Dictionary.* Benin City, Nigeria: Ethiope, 1972.

Wintrob, Ronald, M.D. "Mammy Water: Folk Beliefs and Psychotic Elaborations in Liberia." *Canadian Psychiatric Association Journal* 15.2 (April 1970): 143-157.

The Mythic World in Flora Nwapa's Early Novels

Teresa U. Njoku

Great writers have always recognized the value of myth in literature. Myths enable them to control and give shape to their art as
well as deepen literary meaning. Quite early, female writers were
confronted by a lot of prejudice. It was believed that novels written by men were more representative of life, had deeper insight,
and conveyed greater knowledge. Therefore, the female writers
had to explore those areas which male writers ignored or considered irrelevant. Female writers concentrated on female experiences and embodied their ideas in myths. However, Molara
Ogundipe- Leslie criticizes the mythification of women by her fellow female writers.[1] Mineke Schipper supports Ogundipe-Leslie
when she condemns the image of women portrayed in myths.[2] She
believes African traditional myths degrade the status of women.
However, Flora Nwapa uses myth to achieve aesthetic purposes;
they become her means of empowering the woman.

Nwapa is a mother-figure in novel writing in West Africa. Born
into a society where the myth of female inferiority and other negative myths that destroy the female psyche are accepted and
unquestioned, she adopts a type of writing whose "surface designs

conceal or obscure deeper, less accessible (and less socially accept-able) levels of meaning."[3] She adopts this stance in order to "express and to camouflage"[4] her art. Her reason is evident: "readers who find themselves outside of and unfamiliar with the symbolic sys-tems that constitute female experience in women's writings will necessarily dismiss those systems as undecipherable, meaningless or trivial."[5]

In her essay on Nwapa's writings, Maryse Conde states that the author "conveys a very poor impression of her society. Her men are weak and dissolute and irrational. Her women are a formidable gallery of malicious gossipers" (136). She thus believes that Nwapa's artistic effort falls of excellence. What Maryse Conde highlights as the defects of Nwapa's writing are among her strong points; she manipulates her dialogue to reveal not only her characters but also to depict a world in which traditional values have significance. It is to rescue Nwapa's early novels from such a dismissal as Maryse Conde's that Ernest Emenyonu wrote to show for whom and about what Flora Nwapa writes (28-33). Bernth Lindfors also draws atten-tion to the mythic perspective in the criticism of African Literature when he states:

> The interpretative critic who studies traditional ele-ments in contemporary African Literature is more likely to be interested in investigating their artistic functions or their aesthetic and metaphysical impli-cations than in merely validating their existence. He seeks to go beyond the obvious into less accessible regions, sometimes even venturing to use this tool to probe the mysterious inner workings of the human mind. (12-13)

This study posits that Nwapa's earliest novels, *Efuru* (1966) and *Idu* (1969), reveal the deeper reality embodied in myths. Since she comes into a tradition of literary convention already established to express male experience, Nwapa builds a structure of ideas ade-quate for interpreting female experience. She generates a new tra-dition in which women are the center of focus, and draws largely from the mythic reservoir to explore every aspect of female expe-rience. *Efuru* and *Idu* embody references to mythic symbols and themes and, in this way, make Nwapa's earliest novels relevant to her time. In fact, the use of myth is not only diversified in the two novels but also the interpretation that could be given to them. Nwapa's treatment of the mythic helps to articulate her liberal feminism. In *Efuru* and *Idu,* Nwapa seems to agree with David Bidney that myth is "not a conscious creation or invention of an individual but...a product of man's spontaneous expression of emotion and feeling of unity with nature as a whole" (14). Nature abhors inferiority, consequently Nwapa takes recourse to myth to help her re-affirm the unity of women with nature. Nature is limitless while societal conventions make the woman a limited being. Thus, the myths in *Efuru* and *Idu* help Nwapa to conceptualize authentic experiences of women which, before the appearance of her works, had been given a superficial and stereotypical treatment by male novelists.

Nwapa also draws on Jung's theory of "the collective unconscious" which consists of "primordial images for archetypes." This states that "civilized man preserves, though unconsciously, those pre-historical areas of knowledge which he expresses indirectly in myth"(248). This inherited knowledge is the residue of repeated kinds of experiences in the lives of our ancestors, and are buried deep in man's psyche, beneath the suppressed memories belonging to the individual. The mythic elements in Nwapa's novels often seem impossible in real life. This agrees with Cassrier's view that myth is "only real in the sense of its symbolic value to a given society and can only be interpreted allegorically."[6] Thus, Cassrier asserts that "every period declares its truth and is warmly attached to it. Our truth of the moment is often only a myth that does not know it is one, and ... we make myth everyday without knowing it."[7] Since myth can thus be viewed as having different shades of truth to reflect various experiences Nwapa resorts to it to give, according to Eric Fromm, "a message from ourselves to ourselves, a secret language which enables us to treat inner as outer event"(249). She depicts her characters as archetypes and gives archetypal patterns to many of the incidents in the novels. These basic cultural patterns assume a mythic quality because of their permanence in the culture.

II

The mythic world of *Efuru* and *Idu* performs two functions. Firstly, the novelist uses the myths to destroy earlier myths of the society in which thought and imagination "are employed uncritically and are deliberately used to promote social delusion"[8] of women. Such deliberate falsification is employed in the interest of sexism and propaganda and to maintain the status quo in cultural norms and attitudes. One of the myths which Nwapa explodes in her two early novels is that of female inferiority. She does this by projecting female characters who emerge as heroines in a world where heroism is usually associated with men. Another myth in this category is that which regards barrenness as a mark of failure on the part of the woman rather than as an incident of nature. In *Efuru*, other myths include that which state that a baby born with two upper teeth desecrates the land and will cause a poor harvest. Gilbert, Efuru's husband, expresses the false myth that it is a waste of money to send girls to school. For him, preference should be given to boys because girls get married quickly and thus waste the money used in training them. This is why he believes that a woman's education ends in the kitchen. Myths that endanger the personhood of the woman also occur in *Idu*. It is a myth that "when a woman starts with money, children run away"(3). Nwapa shatters this myth because Idu later gets a male child and is expecting another one when she dies. In *Women Are Different and One Is Enough*, Nwapa shows that educating the female is worth the effort.

Nwapa also erects new myths. *Efuru* and *Idu* deal with the adventures of two women, Efuru and Idu respectively. These are embodied in the rites of passage they experience from one stage of life to another. The ceremonies are life-cycle ones and are both social and religious. Each novel is structured on a different layer of mythic patterns. In *Efuru*, Efuru elopes with

> Adizua—a nonentity who is unable to afford the bride-price. But Efuru's hard work enables him to pay it. Although Efuru undergoes a "bath" (circumcision), she experiences an initial barrenness. A dibia who attends to her advises her on what to do and, a year after, she gives birth to a baby girl. Before the death of this child, Efuru's husband elopes to Ndoni with a woman who is inferior to Efuru in all respects. He also fails to show up even when his daughter dies. So, Efuru is forced to abandon him. Gilbert later marries her and she experiences childlessness. An accusation of adultery against Efuru makes this marriage end in divorce.

Similarly, in *Idu*, Idu suffers initial childlessness but later gives birth to a son. The child is born on the day of an eclipse, whose mystery announces his birth. Before the birth, Idu's husband marries another wife. Idu is pregnant with another child when her husband dies. Unable to accept the social norm that imposes the tradition of leviration on the bereaved woman, Idu dies and follows her husband to the spirit world. Thus, the two novels project women who are married, challenged and tried by the different experiences they undergo as women. But they emerge as independent persons in life, or in the case of Idu, in death. The unity between love and death is portrayed in Idu, as well as the power of mutual love to defy social norms and taboos.

The rites in the two novels are the central activities from which myths are derived. The ritual represented in "formalized activity" is related to occasions which express beliefs in mystical beings. The rites are accompanied with action and speech. In this regard, the myth Nwapa uses "suffuses the natural with preternatural efficacy."[9] Myth and ritual "replicate each other; the myth exists on the conceptual level and the ritual on the level of action."[10] Thus, the fundamental mythic pattern is that of rites of passage. Each rite fulfills Arnold Van Gennep's mythic standard: "the standard path of the mythological adventure of the hero is the magnification of the formula represented in the rites of passage: separation-initiation-return"(11).

The first rite of passage the grown up Efuru undergoes in *Efuru* is her "bath." This has to do with the transitional stage to procreation and involves the seclusion of Efuru, and such taboos as visitors sprinkling black substance on their feet before going to her room. The taboo is to prevent pollution of the room and to ensure that her wound is not infected. The taboo also indicates the division between her visitors, who constitute the "profane," and she herself, who is regarded as "sacred." On the cultural level, the circumcision indicates social puberty. Though a physical mutilation of the female, it is regarded as a rite of purification by the society. The final act in the ritual is the outing ceremony in the market—a mythic center—which belongs to the living, the dead, and the supernatural. For instance, we are told that not only ordinary women but also the Woman of the Lake attends the Nkwo Market.

In addition, it is only after the ritual of the payment of the bride-price that Efuru and Adizua feel *really* married. The participation in this rite of union by members of the kindred gives it a collective assent. On the other hand, the rite of libation and the communal sharing of wine and food ensure the future security of Efuru's marriage since the union involves "the mutual transference of personality."[11] Efuru, like her mother, experiences difficulties in procreation. Her first visit to the dibia is fruitless because the *dibia* knows supernaturally that Efuru is menstruating. Though the menstrual taboo defines the woman only biologically, the potency of Efuru's period will render his medicine ineffective if he allows her to get too close. Efuru is thus an agent of invisible power. There is power in her menstruation and the *dibia* fears it despite his mythical powers. Here, Nwapa recreates the male fear of female power which is the origin of the restrictions against the woman. When she and her husband return later, the dibia confirms her fertility and prophesies that she would beget a daughter the following year. He outlines the ritual Efuru has to carry out: she should sacrifice to the ancestors every Afor Day. Communication between the living and the dead which the ritual inaugurates is through the mediation of the *dibia*. The spiritual communion is the root of the myth of the ancestors. The prophecy about Efuru's birth of a child is fulfilled during the *Owu* Festival. Seven market days after her delivery, Efuru goes to the lake and puts her feet in it. On Nkwo day she goes to the market. The *Owu* festival, the cleansing symbolized by dipping her feet in the lake, and the outing in the market have ritualistic tones. The *Owu* festival itself which involves dances and dramatic activities is embedded in myth and celebrates communal existence and protection. Besides, the pregnancy and childbirth are ritually important.

The myth of the *dibia* and his god is here depicted. The *dibia's* utterances show an inborn psychic perception of natural phenomena. The second *dibia* Efuru meets with her father mystically explains the theme of crime and punishment. He is psychically aware of the power of the gods. In addition, the regularity in the number of the incidents and actions of the characters points to their mythic dimensions. As Levi-Strauss has stated, "Duplication, triplication, and quadruplication of elements are constant tendencies in myths. The function of the repetition is to make the structure of myth apparent"(81). Again, it is a *dibia* who deduces the reason for Efuru's dreams of the Woman of the Lake. Efuru's dreams of getting a lot of money from her wares is explained by the *dibia* as indications of her selection as a worshipper of the Woman of the Lake. She would be more wealthy if she keeps the laws of the goddess. The *dibia* also analyses Efuru's position as a worshipper of the Woman of the Lake:

> Now listen to me. *Uhamiri* is a great woman. She is
> a goddess and above all she is very kind to women.
> If you are to worship her, you must keep her taboos.

> Orie day is her great day. You are not to fish on this
> day. You are not to sleep with your husband. You
> have to boil, roast, or fry plantains on Orie days.
> Uhamiri likes plantains very much.... When you go
> to bed, you must be in white on Orie nights. You can
> sacrifice a white fowl to Uhamiri on this day....
> Above all, you will keep yourself holy.... The most
> important thing. You are to buy an earthenware pot.
> Fill it with water from the Lake, and put it at one
> corner of your room. Cover it with a white piece of
> cloth. That's all you have to do. (153-154)

The Woman of the Lake is the center of the ritual and it is repeat-ed and formalized in order to please her. The earthenware pot is the symbol that indicates Efuru's status as the worshipper of the Woman of the Lake. She is identified with the goddess, and the rit-ual enumerated above is the myth of the Woman of the Lake. Because the goddess deifies the water, its spiritual essence is high-lighted. Efuru fulfills the ritual and gets a spiritual essence. The ritual marks her initiation into the cult. The theme of initiation is one of the oldest archetypal motifs.

When Efuru becomes ill later, the dibia also tells her mother-in-law that she must appease the goddess with a prescribed ritual because Efuru had neglected her. When Efuru does not recover after this ritual, her husband accuses her of adultery. She is urged to confess her sins, and be assured of life. At her recovery Efuru calls her age group and tells them about the accusation. Selected members follow her to the shrine of Otuosu where she swears by the goddess to be killed if she is guilty. She is absolved after seven Nkwo market days, which means that the goddess has spared her life because she is innocent. Efuru's oath is "imprecatory"; subse-quently, she returns to her father's house. Her action also shows the double moral standards that exist in patriarchal society. The female is the victim of male moral judgement.

Efuru ends where she begins, in her father's house. The rites of passage she has gone through protect and provide her with health and prosperity. The rituals in the story give validity to the myth of the water goddess and the ancestors. Efuru derives peace in her final situation:

> Efuru slept soundly that night. She dreamt of the
> woman of the lake, her beauty, her long hair and
> her riches. She had lived for ages at the bottom of

the lake. She was as old as the lake itself. She was happy, she was wealthy. She was beautiful. She gave women beauty and wealth but she had no child. She had never experienced the joy of motherhood. Why then did the women worship her? (221)

Efuru gains spiritual elevation because of her association with the supernatural. The deification of the earth results in myths. The myth of the "mermaid" or mamiwater or "Woman of the Lake" derives from the idea that the stream is a source of divinity. Though Efuru is child-less, she is empowered by her position as the worshipper of the Woman of the Lake. She accepts herself as she is—a woman of beau-ty, wealth, and generosity who exercises spiritual authority. Her beauty, like that of the Woman of the Lake, is mythical.

III

Similar rites of passage occur in *Idu. Idu* is the name of a mythical Kingdom which refers to the ancient Benin Kingdom. It features in Igbo folk tales, legends, and myths. In Nwapa's story, the name "Idu" has mythical dimensions: she is a woman of promise who embodies all the values and treasures of Africa.

In *Idu,* whereas Idu marries Adiewere, Ojiugo marries Amarajeme. The Ugwuta belief in *"Di Uwa"*[12] (marriage in the past and present lives) is reflected in the two marriages. It is based on the myth of reincarnation which states that dead people can come back to life. According to the myth of reincarnation, when one is born one leaves one's paddle (*amarauwa*) before journeying to earth. It is while one is doing so that one determines one's goals in life. It is at this point, where the paddles are deposited, that cou-ples decide to get married. If this wish is fulfilled in physical life, it means that the couples will be intimate and, if one of them dies, the partner follows.[13] This is so because the marriage on earth is merely a continuation of the betrothal started in the mystic world. Unaware of the mystic relationship, other characters refer to Idu and Adiewere as "created by God to be husband and wife and to live together"(206). When Idu's husband dies mysteriously, Idu does not first tell anyone but goes to Nwasobi's house. Both eat a kolanut before Idu breaks the news and asks Nwasobi to accom-pany her. The kolanut symbolizes unity and since it is believed to represent life, the eating of the nut is a communion and is accom-panied with a rite. Idu fails to weep for her husband according to custom, or to wear mourning cloths which separate the mourner from the other people and the deceased. She is determined to go with her husband to the land of the dead where they would continue their lives: "Weep for Adiewere? That is not what we agreed on. He has cheated me...leave me alone. I am going with him"(210).

Idu rejects the ritual scraping of the hair in the tradition of her people. Idu starves for twenty-eight days and then demands food. Before she eats, there is the ritual cleansing of her hands. She throws a morsel of the food outside for the dead relations before she eats. She does this in the tradition of her ancestors before her. She then goes to sleep and dies. It must be noticed that Idu's death is caused by the force which saw to her creation and subsequent marriage to Adiewere. The life force is completed by her death. Her friend Nwasobi prepares her corpse for the "journey to the other world" which also is believed to depend on a series of rites of passage, one which involves the crossing of a river. Thus, Idu fulfills the cultural norm of ebibi. The myth of the continuation of life of dead persons provides the emotional tolerance that enables people to say her death is a journey.

The second example of a marriage contracted in another world also resounds with mystic echoes. When Ojiugo leaves Amarajeme, he laments that Ojiugo's loss symbolizes for him the loss of his hands and feet. He refuses to participate in all cultural activities because he regards her as both his wife and sister. He asks for food to be prepared for him and the absent Ojiugo three times every day. At night, he takes all the food and throws it into the lake, washes the plates, and comes back to his house. In his sorrow at his loss of Ojiugo who has given birth to a child for an age mate, Amarajeme sews black clothes for mourning Ojiugo, though she is still alive. In the end, he commits suicide. The society regards his suicide as an abomination; he is thrown into the bush and is not mourned by his age group. His fate is considered unfavorable because he is impotent and does not fulfil the norm. That Amarajeme passes the condition from life to death without rites to cushion the harmful effects of such a transition in a society which protects life and believes that nature cannot be won without ritual ceremonies is serious indeed. Besides, Ojiugo's abandonment exposes his impotency and inadequacy. When Amarajeme earlier in the novel returns to his village, he re-integrates with his age group; this assures him of a heightened mystical stature. His type of death however earns him a funeral that is unceremonious.

Idu refuses to accept that Adiewere is dead; Amarajeme rejects the idea that Ojiugo has in frustration of his impotence, deserted him. When Nwapa makes Idu reject the death of her husband, she is echoing Durkheim and Malinowski and Cassrier that "myth denies the fact of death but affirms the unbroken continuity of life"(13-24). As a result, the rites to separate the living from those they have lost are not enacted. The continuity and cyclic nature of life is here affirmed. Marriage is an event of "profound emotional significance."[14] Though it satisfies man's need for companionship and procreation, Idu refuses to be inherited by Ishiodu because such a relationship is a denigration of an individual in her rights and choices. Therefore, when Idu and Amarajeme cling to their loved ones even when they have lost them, they are merely rehearsing the experience of loyalty in the racial collective unconscious. This is also proved by Ojiugo's dramatic transformation on the death of Amarajeme. "She does not forget or get over it" (206). It is Idu's belief that God molded both Amarajeme and Ojiugo for the purpose of living together.

When Idu gives birth to her son, "white thread was tied round her two wrists and ankles. The baby also had white thread round his wrists and ankles"(85). Again, when Idu's husband dies, Ishiodu puts a thread round Idu's neck, to claim her as his wife because it is the custom of the people. Idu, however, repudiates such a custom which offers a woman no choices as an individual. The white thread symbolizes ownership and while Idu accepts it during her delivery of a child, she rejects its significance in the inheritance of a woman as a piece of property.

The rehearsal of primordial experiences is also recreated in the theme of quest in the two novels. Idu's husband, Adiewere, has a ne'er-do-well for a brother whom he tries to raise. Idu also tries to raise her sister. Both Idu and her husband are engaged in the cyclic handing down of tradition to their juniors. This is itself a continuous activity that spans all cultures. When Ishiodu is detained in Abonema, Idu and her husband journey to the cell to rescue him. Idu and her husband here engage in an action in which the strong and rich rescue the weak in times of trouble. When Idu's sister paddles to *Okporodum* to fetch firewood with her age mates and are later declared lost, they are searched for and brought home by a group of men.

The myth of children who get lost when they go to fetch firewood resounds in traditional narratives. Idu's sister's journey takes her to *Okporodum*—all roads. The rescue enlarges the mythic echoes which derive from the story of the drowning of Amarajeme's first wife when she travels to fetch wood with her own age mates. Search parties are also sent to find Idu's only son when he is temporarily lost. Nwasobi uses the *Ugene* to search for him throughout the village. This is the instrument used in traditional society to give urgent, important messages and information. Nwasobi symbolizes the "crier" of generations and a mother agonizing for a lost child. Nwapa shows the reality of female experience in motherhood, which often is full of anguish rather than the patriarchal idealization of it.

Besides, Adizua's mother in *Efuru* travels with her young son in search of her husband. Though her quest is fruitless, the husband later returns home a broken man but disappears soon afterwards. Adizua disappears like his father and never returns. Efuru is abandoned like her mother-in-law, but unlike her, she does not wait endlessly for an irresponsible and truant husband. Efuru understands that suffering should be "for greater things" and chooses the type that "will be noble."

Similarly, characters are analyzed and assessed through their parents and ancestors. The novels also emphasize blood relationships to show the mythological relationships in the lineages. In *Idu,* Idu's husband follows in the tradition of his father who was not at heart a polygamist. In *Efuru,* Efuru is considered as beautiful as her mother. Efuru's mother is so beautiful that people regard the Woman of the Lake as her mother. Efuru suffers initial barrenness and eventually gives birth to one daughter like her mother. Besides, the woman who causes the outbreak in *Efuru* is said to have followed in the footsteps of her mother. In addition, when Onyemuru's daughter dies, Onyemuru carries away all her daughter's property and leaves nothing for her granddaughter. In this way, she breaks the chain through which a mother bequeaths a heritage to her daughter or granddaughter. That break in the chain and therefore in the mother/daughter bond, makes other villagers see her as a witch. Similarly, women who are owed debts in the novel do not bother themselves unnecessarily because they realize that their children will collect the debts if they do not do so before they die. All the children in the society are also taught to swim by their mothers. The children are taught to swim in a similar manner; the child is thrown into the water where it struggles and then learns to swim. The mother helps the child to enact this ritual of swimming. The ritual here provides the youngster with knowledge of the universe by handing down the tradition to them. In all the incidents elaborated above, there is a celebration of "the mythic condition of mothering."[15]

Efuru in *Efuru* and Idu in *Idu* are more than mere characters. Archetypal figures, they are mothers of all and symbolize care, endurance, and positive mystical roles. As mothers, they imitate and repeat "the primordial act by which life appears in the womb of the Earth."[16] Idu and her husband symbolize the parents of the other villagers. Ishiodu's wife regards Idu and her husband as her parents. Therefore, she is not afraid about her miscarriage since the couple will take care of her children. Idu plays the role of a mother to her husband and Ishiodu's wife and is regarded as a sister by quarrelsome Onyemuru. Idu provides the cloth and coffin for the burial of her business friend, Nwosu, and takes over the education of Nwosu's son. Idu and her husband live in close relationship with others. The myths in the novel emphasize this relationship. Both Efuru and Idu in *Efuru* and *Idu* respectively live fully. Their lives are mythic because "myths involve living, which includes the element of struggle."[17]

Nwapa's mythic world takes readers to the past, the present, the future, and the supernatural world. There is interaction between the living and the dead. The female characters in the two novels fetch water, make fire for cooking, bathe, wash clothes in the lake, go to market, prepare food, trade and care for the home. These are activities which have met the different needs of man throughout human history and are also means of socialization. They carry forward the customs, manners, norms and conventions of Ugwuta people. In this way, the activities express the inner experiences of female living. By their repetitive nature and, therefore, their necessity, Nwapa raises them to the status of myth. Perhaps, originally, these repetitive female activities were accompanied by rituals because they constitute life-sustaining activities. They place women at the center of domestic life, rural economy, and nurturance; they are also associated with primordial emotions and sentiments which reveal the supremacy of womanhood.

Although the society Nwapa projects in *Efuru* and *Idu* is patriarchal and sexist, she gives the culture mythic proportions in order to show the centrality of female experience. To do this, she portrays female characters who are fulfilled as human beings in spite of difficulties in having children. Efuru's and Idu's greatness assume mythic proportions that are not cut down to fit the conventional societal view of women as limited and ignorant beings. Idu undermines the foundations of traditional mores which state that a woman must marry her husband's brother at his death. She chooses to die in order to be free from such a marriage. She exercises the freedom to live and die as a moral being, for whom one marital experience is enough. Like the heroine of mythology, she comes to a sudden end. On Efuru's part, her identity as the priestess of Uhamiri is a passport to the spiritual realm of female experience. She returns to the source of the "mythic world of female powers."[18]

The mythic materials that Nwapa deploys in the novels become, therefore, her first step in "the dialectic of bondage and liberation which the human spirit experiences with its own self made image worlds."[19] Through myths Nwapa structures her own ideals for her culture in order to remodel it for modernity, she heightens her artistic temperament and "legitimizes" female life in written literature.

NOTES

1. Ogundipe-Leslie rejects the mythification of women by female writers such as Ama Ata Aidoo and Flora Nwapa. See her article, "The Female Writer and Her Commitment" in *Women in African Literature Today* 15 (1987):5-13.

2. See Mineke Schipper's "Mother Africa on a Pedestal: the Male Heritage in African Literature and Criticism" in *Women in African Literature Today* 15 (1987):35-54.

3. This expression comes from Sandra Gilbert and Susan Gubar's *The Madwoman in the Attic: The Woman Writer and the Nineteenth Century Literary Imagination* (New Haven: Yale University Press, 1979), 73.

4. Ibid., 81. Before Flora Nwapa started writing, novels written by men had great impact on her people. Nwapa, therefore, had to find indirect means of questioning some of the cultural ideas of her Igbo society without seeming to be radical.

5. See Annette Kolodny's "Dancing through the Minefield: Some Observations on the Theory, Practice and Politics of a Feminist Literary Criticism" in *Feminisms: An Anthology of Literary Theory and Criticism* (Rutgers: Rutgers, 1991), 97-116.

6. See Afam Ebeogu's "Leadership, the Myth-making Imagination and the Nigerian Situation." Mimeograph.

7. See Eric Bardel's "The Mythic" *Dogenes* 7(Summer 1954):37; quoted in Ebeogu, "Leadership, the Myth-making Imagination and the Nigerian Situation." Mimeograph.

8. See David Bidney's "Myth, Symbolism and Truth" in *Myth: A Symposium* (Bloomington: Indiana University Press, 1958): 22. The myth of female inferiority is employed to enthrone sexism in a patriarchal society.

9. See Richard Chase's "Notes on the Study of Myth" in *Myth and Literature* (Lincoln: University of Nebraska, 1966): 121.

10. See Claude Levi-Strauss's "The Structural Study of Myth" in *Myth: A Symposium*, 81-106.

11. This is how Arnold Van Vennep describes the ritual of the union of a couple in marriage in *The Rites of Passage*, 133.
12. This concept is derived from Angela Izuagba's paper, "Oguta and the Novels of Flora Nwapa: A Biographical Approach," presented at the 13th Annual University of Calabar International Conference on African Literature and the English Language (ICALEL), May 3-7, 1994.
13. The *Di Uwa* concept is rooted in the belief in *Chi* and reincarnation. Victor Uchendu captures the idea thus: "At reincarnation, individuals are guided by their *Chi* to make a choice from two parcels that Chineke the Creator presents. Of these two parcels, one is believed to contain the desired social position that the individual predicted in his *ebibi* (pre-incarnation social position that the individual predicts during his lifetime on earth). In their *ebibi*, most Igbo predict long life, intelligence, wealth, 'having mouth,' that is, the power of oratory and wisdom. See Uchendu's *The Igbo of South East Nigeria* (New York: Holt, Rhinehart and Winston, 1965), 16.
14. See Lord Raglan's "Myth and Ritual" in *Myth: A Symposium*, 122-135. Wifehood and motherhood are means of achieving status in traditional society. Idu's action is revolutionary. For a woman to refuse to be inherited when the deceased husband's brother is living to acquire her is unusual.
15. See Maggie Humm's *Feminist Criticism: Women as Contemporary Critics*, (Brighton, Sussex: The Harvester, 1986), 98. In the society depicted in Nwapa's early novels, children acquire almost all the skills which help to sustain life from their mothers. The mother-child bond is thus culturally stronger and deeper.
16. See Mircea Eliade's *Myths, Dreams, and Mysteries* (New York: Harper and Row, 1967), 166; quoted in Shirley Weitz, *Sex Roles: Biological, Psychological, and Social Foundations* (New York: Oxford, 1977), 174-175.
17. See Booth's *African Religions: A Symposium* (New York: Nok Publishers, 1977), 118.
18. See Carolyn Heilbrun's *Towards a Recognition of Androgyny* (New York: W.W. Norton and Co., 1982), 70.
19. See David Bidney's "Myth, Symbolism and Truth" in *Myth: A Symposium*, 8.

WORKS CITED

Bardel, Eric. "The Mythic." *Dogenes* 7 (Summer 1954):37.
Bidney, David. "Myth, Symbolism and Truth" in *Myth: A Symposium*, ed. Thomas Sebeok. Bloomington: Indiana University Press, 1958:3-24.
Booth, Newell S. *African Religions: A Symposium*. New York: Nok Publishers, 1977:118.
Chase, Richard. "Notes on the Study of Myth" in *Myth and Literature*. Ed. John B. Vickery. Lincoln: Univ. of Nebraska Press, 1966:121.
Condé, Maryse. "Three Female Writers in Modern Africa: Flora Nwapa, Ama Ata Aidoo and Grace Ogot." *Présence Africaine* 82 (1972):136.
Ebeogu, Afam. "Leadership, the Myth-Making Imagination and the Nigerian

Situation." In Mircea Eliade, ed., *Myths, Dreams, and Mysteries.* New York: Harper and Row, 1967:166.

Emenyonu, Ernest N. "Who Does Flora Nwapa Write For?" *African Literature Today* 7(1975):28-33.

Gilbert, Sandra M. and Susan Gubar. *The Madwoman in the Attic: The Woman Writer and the Nineteenth Century Literary Imagination.* New Haven: Yale University Press, 1979.

Heilbrun, Carolyn G. *Towards a Recognition of Androgyny.* New York: Norton, 1982.

Humm, Maggie. *Feminist Criticism: Women as Contemporary Critics.* Brighton, Sussex: Harvester Press, 1986.

Izuagba, Angela. "Oguta and the Novels of Flora Nwapa: A Biographical Approach." Paper presented at ICALEL Conference, May 3-7, 1994.

Kolodny, Annette. "Dancing through the Minefield: Some Observations on the Theory, Practice and Politics of a Feminist Literary Criticism." In *Feminism: An Anthology of Literary Theory and Criticism.* Ed. Robyn R. Warhol and Diane Price Herndl. Rutgers: Rutgers University Press, 1991:97-116.

Levi-Strauss, Claude. "The Structural Study of Myth." In *Myth: A Symposium:* 81-106.

Lindfors, Bernth. *Folklore in Nigerian Literature.* New York: African Publishing, 1973:12-13.

Nwapa, Flora. *Efuru.* London: Heinemann, 1966. (All other references within the text will be cited from this edition.)

——.*Idu.* London: Heinemann, 1971.(All other references within the text will be cited from this edition.)

Ogundipe-Leslie, Molara. "The Female Writer and Her Commitment." *Women in African Literature Today* 15 (1987):5-13.

Raglan, Lord. "Myth and Ritual." In *Myth: A Symposium:* 122-135.

Schipper, Mineke. "Mother Africa on a Pedestal: The Male Heritage in African Literature and Criticism." *Women in African Literature Today* 15 (1987):35-54.

Scott, Wilbur S. *Five Approaches to Literary Criticism.* London: Collier Macmillan Publishers, 1962.

Uchendu, V.C. *The Igbo of South East Nigeria.* New York: Holt, Rhinehart and Winston, 1965.

Van Gennep, Arnold. *The Rites of Passage.* London: Routledge and Kegan Paul, 1960.

Weitz, Shirley. *Sex Roles: Biological, Psychological, and Social Foundations.* New York: Oxford University Press, 1977.

"The Empire, Far Flung":
Flora Nwapa's Critique
of Colonialism

Florence Stratton

If on little else, Sylvia Leith-Ross and Flora Nwapa agree on one thing: the considerable power and autonomy of Igbo women. The recipient of a research grant in 1934 to study "the conditions of life of the women of Iboland" (5), Sylvia Leith-Ross provides yet another example of the collusion between scholarship and imperial design. Written in the wake of the 1929 Women's War, or "Aba Riots" as the British (misleadingly) called it, an uprising of tens of thousands of mainly Igbo women against the conditions of colonialism, Leith-Ross's book, *African Women: A Study of the Ibo of Nigeria,* presents the colonial (in this case female) subject as an object of surveillance: "an accurate knowledge concerning women's rights and responsibilities in primitive communities, some idea of their thoughts and the trend of their opinions, are important, especially so to the European legislator, administrator, and educationalist"(21).

African Women is, in all but one respect, a typical colonial text. Complex issues are reduced to simple dichotomous opposites: colonizer/colonized, civilized/uncivilized, modern/primitive. Such hierarchical dualisms are, of course, signifiers for the legitimate right of Europeans to administer, manage, and reform the affairs of "other" people.

Leith-Ross does not perceive her text as contaminated by colonialism. Rather, she considers her own valuations as somehow beyond relations of power. Thus it is only "personal bias," or "the human factor" as she also calls it, that, in her view, constitutes a potential threat to her making detached, value-free observations, and that she feels obliged to acknowledge: "I was conscious all the time that there was a danger of mis-statement, of mis-interpretation due to the fact that my 'type' did not fit in with the Ibo 'type'"(40-41).

As if she were also conforming to the demands of current scholarship, Leith-Ross acknowledges her "positionality" in the colonial hierarchy. Father, brother, husband and Leith-Ross herself have all been in the colonial service. In addition, she has been assisted in carrying out her research by Frederick, Lord Lugard, one of the key players in late nineteenth as well as twentieth century British colonial history, who also wrote the preface to her study. An examination of my own connections with Western imperialism would, perhaps, reveal a difference in degree, but certainly not in kind, from Leith-Ross's. Like her, I, too, have had research grants, in my case to study African literature—an opportunity afforded much more frequently to "First World" than to "Third World" scholars.

There has, of course, been a change in the way such ties with the imperial order are evaluated—a change which casts Leith-Ross's acknowledgments in an ironic perspective. For she considers her relationship to the institution of colonialism to be not a liability, but rather an asset to her study. So the fact that her father "was in command of a sloop at the taking of Lagos"(41) is cited to enhance her credibility.

The change also signals a paradigm shift in humanities and social sciences research: a rejection of claims to objectivity and universality and an admission of bias in the epistemological apparatus. We can never be vigilant enough! For as well as discontinuity, there continues to be continuity between past (that is colonial) research practices, such as Leith-Ross's, and present (neocolonial?) ones, like my own. While Flora Nwapa's critique of colonialism is the main subject of this essay, pointing up some of those continuities will also be part of its project. No doubt many traces of colonial contamination remain in my essay. In Robert Young's words:

"The nightmare of the ideologies and categories of racism continue to repeat upon the living"(28).

As Leith-Ross sees it, the majority of Igbo are incapable of reason. There are, however, a few exceptions, one being a woman named Mary who, though "there were strange gaps in her reasoning," was able, when suitably prompted, to "make a picture of the Empire, far flung, with King George sitting on his throne in the middle of it"(113-114).

Flora Nwapa, who was born in 1931, sets three of her novels— her first two, *Efuru* and *Idu,* and her last, *The Lake Goddess* (to be published posthumously)—in the same period during which Leith-Ross was doing her research, the time of Nwapa's own childhood. While *African Women* has a tendency toward self-deconstruction, reading it from the vantage of Flora Nwapa's novels uncovers further great gaps in Leith-Ross's reasoning. Nwapa's work, however, is much more than simply an act of deconstruction, for it also creates a space for postcolonial aspirations.

Challenging the colonial view of history as a celebration of the glories and triumphs of Empire, Nwapa takes on the task of recovering an Igbo past, a past which is explicitly denied in Leith-Ross's historiography. In countering the dominant European narrative, Nwapa repeatedly turns to the myth of Uhamiri or Ogbuide, a water Goddess. One of Uhamiri's textual functions is to signify the glorious history of Ugwuta, a lake-shore town which provides the setting for a number of Nwapa's novels. The town's founding in the mythic past is associated with the Goddess, as are its victories over enemies, both in precolonial times, as in the conflict with the "warlike people on the Urashi river" (*Lake Goddess,* 49) and much later during the Nigerian Civil War: "Uhamiri heard the pleadings of her people. She did not turn a deaf ear....And she had acted according to the belief of the people. No invader coming by water had ever succeeded in Ugwuta"(*Never Again,* 84).

Nwapa does not, however, in any way idealize precolonial Igbo society. Most notably, her account of the historical events that preceded colonialism contains repeated reference to Igbo involvement in the trans-Atlantic slave trade. Thus in *Efuru,* when the heroine's father dies, the manner in which he has acquired his wealth is revealed in the narrative—an aspect of the past the people of Ugwuta would prefer to forget:

> It was the death of a great man. No poor man could
> afford to fire seven rounds of a cannon in a

day....The cannons were owned by very distinguished families who themselves took part actively
in slave dealing....The booming of thecannons was
announcing the departure of a great son, the last of
the generation that had direct contact with thewhite
people who exchanged their cannons, hot drinks
and cheap ornaments for black slaves. (200-203)1

This is a taboo subject for most Igbo writers. The British also, it
would seem, want to forget that they were ever involved in the
nasty business of slavery. Leith-Ross, at least, suppresses any reference to it. Thus Nwapa's fiction serves as a reminder of the many
perspectives which are so frequently erased from the historical
record.

Leith-Ross also suppresses all knowledge of colonialism as a
system of economic exploitation. Furthermore, the economic analysis she offers—that the price of goods is controlled by imperson-
al market forces—is still being propagated by Western nations,
providing the rationale for current neocolonial relations. Leith-
Ross's Igbo interlocutors know better. Like Walter Rodney, they are
well aware that "[t]he differences between the prices of African
exports of raw materials and their importation of manufactured
goods constitute[s] a form of unequal exchange," and that there is
"no objective economic law which determine[s] that primary produce should be worth so little" (Rodney, 1982:160). "Why does
Government not say that [palm]oil must be sold at the same price
as kerosene?" they demand to know of Leith-Ross (167). Perhaps
her reason is clouded by her place in the colonial order?

> It is a delicate matter, this entry of the nearly untu-
> tored savage into the world market without prepa-
> ration, without economic education. He is...not yet
> enough developed to appreciate the intricacies, the
> financial and technical problems of world trade.
> They bewilder, frighten, or anger him and at pre-
> sent he has no one to teach him, no one to explain
> these mysterious happenings that occur thousands
> of miles away. (62-63)

The demystification of colonial economic relations is one of Nwapa's
projects. Thus, in *Efuru,* there is repeated reference to the unequal
nature of trade between Britain and Nigeria. The specific issue is the
banning of the production of local gin by the colonial government.

As the inhabitants of Ugwuta know very well, the primary purpose of the banning is to create a guaranteed market for a British export and to produce conditions under which British companies can exploit Nigerians. Such restrictions on economic activity also, of course, limit the capacity of Nigeria to make maximum use of its economic potential. The people of Ugwuta are, however, defiant:

> "Why the Government does not allow us to drink our home-made gin, I do not know....Does it know that it cannot stop us from cooking gin; that our people will continue to go to jail instead of giving it up completely. If they must stop us from cooking gin, then the white man's gin and his schnapps should be sold cheap. We sell our gin two shillings or sometimes two shillings and six pence a bottle, and they sell their gin and schnapps for many shillings." (86)

With the same aim of clarifying issues, Nwapa turns her attention in her prose poem "Rice Song" to the Nigerian political economy of the post-independence era. This time the obfuscators are the Nigerian military government and its supporters in the academy. The singer of the song presents herself as an ordinary housewife, one who does not "understand Economists" (71). What her analysis reveals is the dependent, neocolonial nature of the Nigerian political economy.

The focus of Nwapa's attack is on the nation's dependence on imported rice. It is a dependence, she indicates, which is a hangover from the colonial era, for south of the Niger rice was neither grown nor eaten until the British introduced it. It is also a dependence which obstructs national development, depriving vital areas of the economy of hard-earned foreign exchange and depressing local incentives to produce food by undercutting prices local producers must receive to stay in business. Why, then, has it been allowed to entrench itself? This is the crucial questions the poem addresses. Nwapa's answer is the greed of those who have a stake in the rice business: the nation's business sector, as well as members of the academy who, she hints, may be involved in the rice distribution business, and the military government. The poem concludes with a plea to the government to "Ban the importation of Rice"(73).

According to Leith-Ross, all Igbo are greedy; it is an hereditary characteristic. Thus she speaks of "the innate graspingness" of the women she studied (156). Not for a moment does she think of colonialism as "the vilest scramble for loot that ever disfigured the history of human conscience" (Conrad 1926:25). Rather it is the Igbo she has "no great liking for," they are "so entirely dominated by material values"(355).

Is there not something objectionable/hypocritical about middle-class Western women censoring "other(ed)" women for their material values? In any case, commentators on Nwapa's novels routinely take exception to the materialism of her later heroines, *characters who evidently have their author's whole- hearted support and approval.* Perhaps such criticism is merely a version of the common complaint that "they" want what "we" have got. After all, most Western academics take credit cards, automobiles, and home-ownership for granted.

The Igbo sociologist Ifi Amadiume calls Leith-Ross "a Victorian puritan" for her disapproval of "the money-making zeal of Igbo women"(14). As Amadiume demonstrates, the transportation of "Victorian ideology...into Igboland by British missionaries and educationalists" undermined Igbo women's autonomy(136). Nwapa aims in her fiction to rescue Igbo women from colonial domination and to reverse the trend toward increasing inequality between men and women which colonialism inaugurated. Thus she seeks to revalue the materially successful, independent woman and remove the stigma placed on her by the colonizing culture.

"She was so good that whatever she put her hand to money flowed in" (*Efuru,* 149). Any number of binary oppositions are undone by this assertion, including the one which designates the economically able Igbo woman evil (i.e., grasping and greedy). Nwapa's insistence on the kindness and generosity of her heroines is equally subversive and also counters Leith-Ross's characterization of Igbo women as "so money-minded" as to find the notion of "voluntary work....inconceivable"(262). Thus when her elderly neighbor falls ill, Efuru takes her to the hospital, arranges for an operation, and pays all the medical bills. She also helps to nurse her former mother-in-law back to health when she is stricken. Furthermore, in *Idu,* Nwapa compares Igbo women's with Christian charity, to the detriment of the latter. When a madman steals a cloth from the church in order to keep himself warm, a churchman snatches it back, leaving the madman shivering with cold: "[Idu] threw her wrapper to the madman, and ran home naked"(120).

Through the myth of Uhamiri, Nwapa also gives religious sanction to economic prosperity for women. Uhamiri's main function in Nwapa's novels is to affirm the legitimacy of female independence and freedom. The Goddess of an ancient matriarchal religion, she bestows wealth, as well as beauty, on the women who worship her.

On the heroines of her later novels, the ones critics tend to find too materialistic, Nwapa bestows her own authorial blessing. Thus when, in *One Is Enough*, Amaka achieves material success, Nwapa rewards her by fulfilling her fondest dream: to be a mother. Similarly, in *Women Are Different*, Chinwe is upbraided and labelled a prostitute by women of her mother's generation, who, as a result of their missionary education, disapprove of a young single women "possessing a Mercedes Benz...credit cards and so on, without a husband." But she is commended by her author:

> Chinwe had done the right thing. Her generation was doing better than her mother's own....They are saying that women have options. Their lives cannot be ruined because of a bad marriage. They have a choice, a choice to set up a business of their own, a choice to marry and have children, a choice to marry or divorce their husbands. (118-119)

Once again there is a convergence of views, in this case on the subject of prostitution, between the (Victorian? colonial? Western middle-class?) attitude of Leith-Ross and that of at least some latter-day literary critics. According to Leith-Ross, "the morality of the younger generation [of Igbo women] had become increasingly lax"(219). Nor was it "ever real necessity that drove the girls to immoral ways. It was more a glad throwing off of all restraint and a blind grab after money, new clothes and tasteless finery"(220).

Prostitution is a vexed issue in African literary and critical discourse. Along with some critics, men writers also, quite routinely, define it in terms of moral laxity, encoding women as agents of moral corruption. By contrast, moral laxity is replaced by necessity in Nwapa's definition. For her, prostitution is a strategy women have adopted for confronting male domination. Though Uhamiri "hate[s] prostitution," she evidently tolerates it as long as profits from it are reinvested in a more conventional form of trade within a two-to-three year period (*Idu* 120, 195). Nwapa's later novels define prostitution more specifically, as a way for women to regain the freedom and power they lost under colonialism. Thus in *One Is Enough*, Amaka, the product of a mission-school education, wonders, when she is made miserable by her marriage, whether there was not "something traditional which she did not know because she went to school and was taught the tradition of the white missionaries." Amaka's first step in her negotiation of freedom is to discard the values of "[t]he good missionaries [who] had emphasized chastity, marriage and the home"(11). She then puts into practice "all that her mother taught her"(74), using "her bottom power" to achieve wealth and independence (121).

The effect of a colonial education on Igbo women is a much debated issue in Nwapa's fiction. On the one hand, certain of the skills it teaches seem to be necessary for women if they are to keep pace with men economically in their rapidly changing society. On the other hand, the values it transmits have a debilitating effect on Igbo women. Thus a colonial education can be an experience of cultural, and consequently economic, dispossession, one which undermines the self-definition of Igbo women as industrious and prosperous, replacing it with gender norms which define women as dependent. In contrast to Amaka, who, as we have seen, is finally able to break out of the definition of womanhood she had internalized in the course of her missionary education, the heroines of *Women Are Different* remain trapped in a "missionary way of thinking"(116). Their stories thus tell of defeat and humiliation.

Elleke Boehmer (1995:190) speaks of the "colonial state of bereavement" as being frequently textualized as orphanhood or bastardy. Nwapa's metaphor for Igbo women's loss of their cultural heritage is motherlessness. Thus so many of her heroines (Efuru and Idu, as well as Rose and Agnes in *Women Are Different*) have lost their mothers, while in other cases there has been a rupture, brought about by colonialism, in the mother-daughter relationship. The main symbol in Nwapa's novels of anti-colonial feminist resistance is Uhamiri/Ogbuide. Consistently referred to in *The Lake Goddess* as "Our Mother," Uhamiri represents in that novel, as well as in *Efuru,* the repossession by Igbo woman of a feminist culture. Thus the heroines of these novels, both of whom take Uhamiri as their model of womanhood, have the capacity to serve as role models for succeeding generations of women.

"Rice Song," along with its companion piece "Cassava Song," also tells of cultural loss, of the erosion of Igbo culture during colonial occupation. While "Rice Song" might be called a song of blame, "Cassava Song" takes the form of a praise song in honour of "Great Mother Cassava"(2). As a left-over of colonization, rice signifies cultural dispossession. Cassava, on the other hand, emerges as another of Nwapa's symbols of Igbo—and more specifically Igbo women's—resistance to colonial domination. Thus Nwapa recalls the 1929 Women's War against colonial occupation and the carrying of branches of the cassava plant by the women as a symbol of their defiance:

> The women, they stood
> Defiantly they stood
> Armed only
> With Mother Cassava.
> The Colonial Army
> They were afraid
> Of the Women
> According to their tradition.
> Their own women
> Were delicate beings
> They must be
> Very well protected.
> So they were afraid
> Afraid to fire
> Their deadly weapons
> At the formidable women. (41)

The British did, of course, eventually fire on the women, killing more than fifty and injuring many others, which brought an end to the uprising. Amadiume writes of the outcome of the women's struggle that, even though reforms were made, the women's demands that they be given a place in the new judicial and legislative order were ignored, for "they seemed absurd to the British who were ignorant of, or so ethnocentric they could not accept the fact of indigenous women's organizations and their place in the traditional political structure":

> In 1933, the British therefore legitimized a new sexual politics based on a very rigid gender ideology which Igbo men were to manipulate effectively in their monopolization of power in the public spheres during the post-independence period. (140)

Nwapa's evocation of the Women's War is therefore an attempt to reclaim an earlier, more "formidable," as well as self-constituted identity for Igbo women. As an indigenous food crop, one which rice is threatening to replace as a dietary staple, cassava also signifies cultural reconnection. Playing on the literal and figurative meanings of the word, Nwapa urges a return to "roots": "We must go back/To the roots/To your roots/Oh Mother Cassava"(18).

Nwapa's song in praise of cassava thus becomes a song of cultural repossession, the quest for roots being one of the strategies of cultural decolonization she advocates and one which, she makes quite clear, is to be understood as tied to Nigerian economic de-linkage with Western supremacy.

Nwapa's position is not, however, a nativist one. In other words, she is not a proponent of cultural purity or authenticity. This is, in fact, very evident in the two songs I have just been discussing. For she emphatically does not advocate a "ban [on] the eating of Rice" but only on its importation (69), while also urging that incentives be given to farmers so that Nigeria will become self-sufficient in rice production.

Hybridity is, of course, integral to all of Nwapa's work. Like other African writers, she combines Western and African literary modes, bringing together in her fiction the Western realist mode and Igbo forms of narration, especially the folktale and the mythic mode, with the myth of Uhamiri giving shape to a number of her novels. But what is, perhaps, the distinguishing mark of Nwapa's fiction is its submersion of the conventional realist novel's well-made plot in conversation or verbal exchange, which provides her novels with what Elleke Boehmer (1991) refers to as their "self-generating orality" (16).

In other ways, too, Nwapa's novels tell us they are to be read syncretically. For example, in the Ugwuta of *Efuru* and *Idu*, Igbo and Christian beliefs co-exist, and the dibia practices his divination and homeopathic medicine alongside the Western-trained medical doctor. Hybridity, is, however, at the heart of Nwapa's last novel where it acquires thematic status.

Like *Efuru* and *Idu*, *The Lake Goddess* tells of its heroine's quest for her identity. Like the other novels, it too is set in the colonial era, though a little later than, at least, *Efuru*, as is made evident by the explicit intertextuality between the two novels. Speaking of Uhamiri/Ogbuide worship, one of the market women in *The Lake Goddess* eulogizes Efuru: "She gave dignity to the whole cult. She was the leader and every member of the cult from far and near respected her. She was always calm, always spoke in a slow and measured tone. She never raised her voice. She did not claim to do what she had no power to do"(132-133).

Given its later setting, then, it is not at all surprising that colonialism plays a much larger role in the dialogism of *The Lake Goddess* than it does in *Efuru*, and that the quest of its heroine, Ona, is much more for cultural identity than is Efuru's, though like that of her precursor/model, Ona's also encompasses Igbo female identity. Nwapa uses the family to dramatize the social tension created by colonialism. Ona's grandmothers and mother are all practicing Christians, the latter having a tendency towards fanaticism. Her father, on the other hand, both teaches in the Ugwuta primary school and practices divination and Igbo medicine, while her husband is a businessman who deals in Western medicines.

As the novel's protagonist, Ona's main task is to define an Igbo identity during a time of rapid cultural transformation and upheaval. From the outset, she seems to be destined to be a priestess of Uhamiri/Ogbuide, for she has dreams and visions of the water Goddess from early childhood. Believing she is possessed by an evil spirit, her parents send her to a mission school. Ona, however, resists all Western learning and soon drops out of school in order to study herbal medicine with her father and to take up fishing as her vocation. Her parents then arrange for her to marry a prosperous businessman who is a member of the church. Though she is happy enough in her marriage, she eventually abandons her husband and three children to take up her calling as Ogbuide's priestess, a role in which she finds satisfaction and peace, as well as status and power. She does, however, have one regret: that she has not learned to read and write. Hence she advocates "education for our daughters" as well as for sons. For "Ogbuide wants all women to have voices"(241- 242).

Leith-Ross places Igbo women in a double bind. Those who remain uninfluenced by Western culture she labels "primitive." In contrast to her "backward" or "underdeveloped" sisters, however, the "modern" or "sophisticated" woman is lazy and morally lax. Nwapa undoes this colonial trope through her portrayal of such of her heroines as Efuru and Ona as sophisticated, modern women, that is as women who are able to deal with the complex issues of their day. Similarly, in *One Is Enough*, Amaka must recover Igbo feminist values in order to find happiness and prosperity in contemporary postcolonial Nigerian society.

The theme of the primitive and the modern is one which emerges consistently in Western discourse on Africa. In Leith-Ross's case, she uses the categories not only in her analysis of Igbo women, but also to characterize all Igbo people. In her estimation, the Igbo are "the most primitive" of Nigerian peoples. They therefore will have the least trouble adapting to "modern life" for, "[s]tarting at scratch, they have nothing to unload: no institutional religion, no traditions, no caste or other hide-bound social system"(19).

Behind Leith-Ross's and other uses of the primitive/modern dichotomy lies social Darwinism and the notion of the differential development of cultures. Thus Leith-Ross proclaims the Igbo not merely backward, but so backward that they are nearly at the bottom of the evolutionary scale. Scandalously, the notion of a hierarchy of cultures is also one which is still largely taken for granted. In current scholarship, the primitive/modern opposition frequently recurs in the guise of the traditional/modern dichotomy. It makes no difference whether the initial or "lower" term is "primitive" or "traditional." For "modern" means recent, current, contemporary, as opposed to outmoded, outdated—backward. Thus Europe or the West acquires a monopoly on the present, while the West's "others" remain always behind the times, permanently trapped in an earlier evolutionary stage of social development. As Olu Oguibe remarks, "[t]he West may require an originary backwoods," Africa, to be precise, "against which to gauge its progress"(8). Certainly the primitive-traditional/modern dichotomy is important in sustaining notions of Western superiority and hence in legitimizing continued Western interference in African affairs through such instruments of neocolonial control as the IMF (International Monetary Fund) and World Bank.

Such common locutions as "bringing Africa into the Twentieth Century," as well as related dichotomies—developing/developed, Third World/First World—are also based on social Darwinism. The history of the way such terms have been applied to marginalize and derogate "other(ed)" people must be remembered. It continues to be a violent history.

Nwapa's work as a whole constitutes an assault on this particular brand of binary thinking. Most significant, perhaps, is the consistent and repeated inscription of Uhamiri as both an ancient deity and a figure of considerable influence and power in colonial, that is *modern*, Igbo society. In another move, Nwapa often focuses her narratives on the relative merits of Igbo and western beliefs and cultural practices. Sometimes she valorizes Igbo over British values, as is the case, say, in the evaluation of the gender ideologies of the two societies in *Efuru* and *Women Are Different*. More frequently, however, Nwapa displaces the dichotomy, rather than merely reversing it, displacement or effacement of the opposition being her main strategy of postcolonial definition. To return once more to *The Lake Goddess*, its narrative encompasses many comparisons of Igbo and Western cultural practices, including a number of discussions in which parallels are drawn between Christian and Uhamiri worship. What is significant is that the narrative does not seek any oppositional stance, but rather affords equal status to the practices of the two cultures. Thus Ona's father, Mgbada, the wise man of the story, sees little conflict between the two cultures and insists that his society "can blend [its] traditions and cultures with the Christian culture"(218). And as we have already seen, Ona herself is an advocate of cultural syncreticism.

As indicated at the beginning of this essay, Leith-Ross's "study of the Ibo of Nigeria" is a typical colonial text in all but one respect. And that is that the subject of its study are women. As Leith-Ross herself says,

> We have perhaps until recently received our infor-
> mation too exclusively through men, whether act-
> ing as investigators or as informants. Even with the
> highest degree of trained discernment, it is difficult
> for a male investigator to get an accurate impres-
> sion of what goes on in a woman's mind when it is
> revealed to him by another man; and with all pos-
> sible goodwill, it is difficult for that man not to be
> biased by tradition, vanity, or self-importance. (21)

We might note that while Leith-Ross sees gender as a barrier to objectivity in research, she does not view race in the same light. Is this not a view shared by many of her counterparts in the late Twentieth Century? In any case, it is its focus on women which also primarily distinguishes Nwapa's work from that of her male predecessors in African novel writing. Nwapa "extends the bound-aries of the African novel to include the women's side of the com-pound," as Boehmer puts it (1991:14).

However, as we have also seen, Leith-Ross catches Igbo women in the colonial gaze, making them the object of her scrutiny, with a view to finding out how they might best be controlled or man-aged by the colonial authorities. Nwapa, representing her society from her own perspective, reconstitutes an Igbo cultural identity which has been damaged by the colonial experience. Sometimes she even totally excludes the outsider from the scene of her nar-ratives by using, for example, untranslated "foreign" words such as "*ganashi*," or exclusionary expressions such as "our town." What is most notable, however, is that her heroines are self-defining. They are strong, courageous, independent Igbo women.

Following Homi Bhabha, we can also say that Igbo women's resistance to colonial definition can also be discerned in Leith-

Ross's *African Women.* For just as Nwapa's discourse is hybridized in the colonial context, so too is Leith-Ross's in the context of Igbo society. While *African Women* is riddled with contradictions, its crucial moment—that is the moment that most "mark[s] the disturbance of [the text's] authoritative representation" (Bhabha 174)—comes with the collapse of the primitive/modern opposition within its own discourse. Thus Leith-Ross declares the Igbo "modern" in their gender relations:

> This idea of marriage as a business partnership, this appreciation by the husband of his wife's powers, this expectancy that the wife will contribute to the family exchequer, and her own sentiment that her self-respect demands that she should do so, have no counterpart that I can think of until the most recent times and in the most advanced among European countries. (348)2

Even though Leith-Ross's "Empire, far flung" is still very much with us—in terms, for example, of the economic exploitation of "Third World" by "First World" nations, as well as in colonialist patterns of thinking within the Western academy and more generally in Western society—it had cracks in it from the beginning. Flora Nwapa has worked to widen those cracks—a very substantial contribution. Pressure must continue to be exerted.

NOTES

1. See also *Efuru* 176, *Idu* 139-140, and *The Lake Goddess* 239.
2. For other instances of the undoing of the primitive/modern binary oppositions in Leith-Ross's discourse, see 230, 231, and 277.

WORKS CITED

Amadiume, Ifi. *Male Daughters, Female Husbands: Gender and Sex in An African Society.* London: Zed, 1987.

Bhabha, Homi K. "Signs Taken for Wonders: Questions of Ambivalence and Authority under A Tree Outside Delhi, May 1817." *'Race,' Writing and Difference.* Ed. Henry Louis Gates, Jr. Chicago: University of Chicago Press, 1986.

Boehmer, Elleke. *Colonial and Postcolonial Literature: Migrant Metaphors.* Oxford: Oxford University Press, 1995.

——. "Stories of Women and Mothers: Gender and Nationalism in The Early Fiction of Flora Nwapa." *Motherlands: Black Women's Writing from Africa, The Caribbean and South Asia.* Ed. Susheila Nasta. London: Women's Press, 1991.

Conrad, Joseph. *Last Essays.* London: J. M. Dent, 1926.

Leith-Ross, Sylvia. *African Women: A Study of The Ibo of Nigeria.* 1939. New York: Frederick A. Praeger, 1965.

Nwapa, Flora. *Cassava Song and Rice Song.* Enugu: Tana, 1986.

——. *Efuru.* 1966. London: Heinemann, 1978.

——. *Idu.* London: Heinemann, 1970.

——. *The Lake Goddess.* (Page references to unpublished manuscript version.)

——. *Never Again.* 1975. Enugu: Tana, 1986.

——. *One Is Enough.* Enugu: Tana, 1981.

——. *Women Are Different.* Enugu: Tana, 1986.

Oguibe, Olu. "In the 'Heart of Darkness.'" *Third Text* 23 (1993): 3-8.

Rodney, Walter. *How Europe Underdeveloped Africa.* Washington D.C.: Howard Univ.Press, 1982.

Young, Robert J. C. *Colonial Desire: Hybridity in Theory, Culture and Race.* London: Routledge, 1995.

Not Feminist but Afracentrist: Flora Nwapa and the Politics of African Cultural Production

Gay Wilentz

In reconfiguring Flora Nwapa's place as one of the foremothers of contemporary African literature, there has been a great deal of critical discussion concerning her position as a spokesperson for African women and women in general. In this not-always-friendly dialogue, especially within and among African feminists, non-African feminists, academic feminists, womanists, and others, the question of whether Nwapa was a "feminist," however conscribed, tends to surface. This essay is not intended to answer that question (which in some ways is a non-question but rather a labeling, as Lorde has noted[1]); instead, I aim to explore its implications in Nwapa's works within the context of African women's cultural production.

In an article in *Research in African Literatures*' special issue on Flora Nwapa, Obioma Nnaemeka critiques feminist critics of

Nwapa, particularly the non-African kind, and comes up with a definition of a (feminist) critic who is an "'inoutsider' who pays equal attention to cultural contexts and critical theory"(81). She states further: "Our job as literary critics should be to open texts up to possibilities by addressing the complex issues in them, however contradictory they may be; often those contradictions lead to where the meanings are"(98). In examining the contradictions concerning women's experience in Flora Nwapa's novels, I accept my role as "inoutsider" (so dubbed by Nnaemeka in her article), and explore how the confusion over Nwapa's feminism is reflected in the way in which she portrays women's complex role in traditional and contemporary Nigerian society, while situating it within the context of women's cultural production. Since, as critics, we are all in the act of calling names, one might rather see Nwapa as an Afracentrist, rather than a feminist: Her resistance to an uncomplicated view of women's experience and her commitment to tell the tale (as in the patterns and traditions of the orature of the women storytellers before her) informs her works. Through a discussion of a "critical incident" in each of three of Nwapa's novels— *Idu, One Is Enough,* and the posthumous *The Lake Goddess*—I expose the Afra-center of Nwapa's work, and hopefully present another way of reading the importance of Nwapa's inheritance to African literature and women writers/readers.

Before examining Nwapa as Afracentrist, I think it useful to define my feminizing appropriation of "Afrocentrism." Afracentrism as a women-centered affective theory is grounded in the socio-historical realities of women's experience, hinging on women's major role in orally transmitting the values and traditions of their culture(s) to future generations.[2] By examining the dailiness and specificity of women's lives, it resists some of the totalizing notions encoded in Afrocentricity.[3] As a critical perspective, Afracentrism clarifies women's role in the creation of a diaspora culture and what commonalities exist in female modes of production throughout Africa and the diaspora. This critical perspective displaces the essentialist idea of Africa to focus in on the historicity of how generational continuity becomes cultural continuity. It acknowledges connections in what I call "oraliterary" works that resist interpretation in narrow terms, based in the orature of undocumented women storytellers. This approach especially allows for contradictions in the kinds of traditions that women pass on, as well as complications in the mother-daughter bond. How women become citizens in their own right as well as how they are indoctrinated into limited positions in the society is closely tied to the women who trained them into the culture's values and traditions. It is within this context that I examine Flora Nwapa's novels.

It has been well recorded that Nwapa does not perceive herself as a "feminist" in any conventional framework; however, as she related to me in a conversation (now partially documented in the *Dictionary of Literary Biography* entry on Nwapa), her aims as a writer relate to the most basic of feminist concerns: "Flora Nwapa writes stories about women because these stories are familiar to her....I keep saying that I am not conscious of doing more than telling stories about women, but if you insist that I'm trying to prove something, *it is that women are first and foremost human beings*" (personal conversation, 3 June 1984). Not only is this the bottom line of feminism, but it is also a significant statement in relation to an observation of Obioma Nnaemeka's in her article cited earlier. Nnaemeka notes that the problem with Nwapa's "inconsistency" regarding her "feminist identity" has more to do with the problems of what Nnaemeka dubs "Western feminism" (practiced by insiders and outsiders alike) than Flora Nwapa's own position. She states most forcefully: "What has been inconsistent is the way questions of 'feminist identity' have been framed and posed"(84).[4] I raise this extratextual discussion to highlight the framing of Nwapa as a "resisting" feminist. Although I do not agree generally with the limited way in which feminism has been constructed by many "feminist" critics of Nwapa's works, it may serve us well to understand Nwapa's writings if, rather, we critique her work under the rubric of Afracentrism. In the three works discussed in this essay, Nwapa's working of male-female, traditional-modern relations in a changing Nigerian milieu gives us no easy answers. If, in fact, she presents "imperfect pictures" of society in which "there is no reason why proverbially bad things happen to the good" (Nwankwo 1995:49; 48), perhaps that is her intention as an Afracentric writer. Nwapa, as in the manner of a dilemma tale, allows space for the readers of her work to engage with the work, and as she herself states, "I show it and people will make their own judgments" (Wilentz interview; 14 July 1984). As I have argued elsewhere, Afracentric writers displace the binary oppositions encoded in these debates and problematize rather than reify women's position in both traditional and contemporary cultures in their works.[5]

One of the most contested spaces concerning Nwapa's work is the unresolved ending of her first novel, *Efuru* (1966). Although the conflict surrounding motherhood and the role of priestesses to Uhamiri fit the model I see Nwapa developing in her writings, I choose in this essay to examine some of her other work, less extensively discussed in critical circles, beginning with *Idu* (1970). *Idu* and *Efuru* are often placed together as companion works; they are her first two novels and are grounded in an almost precolonial existence in Nwapa's home community of Oguta, Eastern Nigeria. However, even with the contradictions implied in the ending of *Efuru*, it is an optimistic novel, and Efuru, despite her misgivings at becoming a worshipper of Uhamiri, sleeps "soundly" at the end of the novel (208). *Idu*, on the other hand, is somber in tone, reinforced by the number of personal disasters that are foregrounded by an environment of natural calamities and disturbances. For example, there are unusual incidents of violence and robberies, as well as an unexplained fire. This fire appears to be a metonymy of the disturbances in the community as well as the tragedies that befall its members. The community viewpoint is that there "was something very deep behind it all"(98). As the novel unfolds, we witness a community out of balance, where the needs of the individual versus the demands of the community are in constant and everpresent stress. Lloyd Brown notes that there is a "tightly knit, even suffocating communality in Idu's world"(156–57).

The tone and structure of the novel appears, at first, to be in direct opposition to the portrayal of the protagonist Idu and her husband Adiewere's relationship. They are, in most ways, ideal members of the community, even through they flaunt other aspects of the conventions. They are, as the novel opens, a consummate couple: a sum greater than its parts. Nwasobi, Idu's friend, remembers: "You know, Adiewere was not all that magnaminous before he married Idu. Now the two of them are about the kindest couple I know in town"(3). However, as in the earlier novel, a happy marriage cannot take the place of children and, like Efuru, Idu cannot conceive.

Like *Efuru*, this novel calls into question the issue of women's role in the community in terms of the emphasis on childbearing and the "tragedy" of childlessness. Sociologist Ada Mere succinctly states that when, after marriage, "procreation does not happen, that woman has failed an essential life goal" (25 April 1984). In the case of Idu and Adiewere, Idu is much more devastated by the fact of her childlessness than he is. Adiewere is quite satisfied with their life together and feels that children will come in time. However, in the eyes of the community, those insistent voices of the novel, Idu

is holding him back from marrying another wife and therefore rob-
bing them of their right to future generations. The words of the community come through the malicious gossip, Onyemuru, demanding
Idu's confidante Nwasobi's involvement: "How can a man live without children? Wasn't it a woman who bore him in her womb? No,
you must tell Idu to find another wife for her husband"(34). In
Moorings and Metaphors, Karla Holloway comments that Idu's fate
testifies to the "overwhelming significance of biological motherhood
in all aspects of African traditional life"(179). This is evident in both
novels. However, Idu questions the tenets of her society in another way. Idu's relationship with Adiewere also challenges the notion
that consanguinal relationships are more important than conjugal—
in this case, the bond between mother and child. It also shifts the
"blame" from barrenness being solely a woman's problem to the
men who are infertile.

The novel revolves around a pairing of two "happy" couples: Idu
and Adiewere; Ojiugo and Amarajeme. However, both couples,
despite their basic success in both work and community/personal
relations, are not keeping up with the biological demands of the
community, since neither couple has produced children. The relationship of Ojiugo and Amarajeme works as a foil to the primary
couple. And because Ojiugo is also in a childless but loving relationship, Idu finds solace in her friendship. Like Adiewere,
Amarajeme was considered worthless until he married Ojiugo, and
then they became a happy, prosperous couple. In confronting their
unwitting inability to meet society's demands, Ojiugo denounces
the one-sidedness of a culture that ignores other types of women's
production when they cannot reproduce. All the good they do for
the community appears negated, while "women who misused themselves, who were prostitutes" have been blessed with children(38).

The sense of unbalance in Ojiugo's remarks pervade the atmosphere of this community and novel. Pressure from the community finally influences Idu to force Adiewere to take a second wife.
Adiewere marries again to conform to cultural conventions, but
the results are humorously disastrous for the polygamous household. Idu, rather than the "small wife," becomes pregnant, and
since this second wife no longer serves a purpose, she leaves after
being ignored by the both of them. However, if in Idu and
Adiewere's case, their attempt to conform leads to a comedic catas-
trophe, the result for Ojiugo and Amarajeme are tragic. In a more
desperate manner, Ojiugo gives into communal pressure as well
as her own desire to be proven adequate. Although she has known

for over two years that the man she loves is "not a man"(112), it takes her a long time to succumb to pressure. Finally, she leaves Amarajeme to live with a man who makes her pregnant. In a culture that structures personal success around continuing the group, adultery, in this case, is tolerated: "Although adultery in Igbo society can be punishable by death, it is condoned when the purposes is to have a child in a childless marriage"(Okonkwo 1971:146). However, when exposed that it is not—as assumed—the woman's fault, the community certainly has no room for a man who is impotent. When the malicious town gossip reveals to Amarajeme what everyone else knows, he realizes that "he is not like other men; that he would die without a baby girl or a boy to answer his name....What was he to do? So people had been laughing at him, so people knew all the time"(130). Soon after, Amarajeme hangs himself, and although it is an abomination to kill oneself, there is a tacit acceptance in the novel of his decision, since after Ojiugo leaves him, "in spirit he was dead"(127).

This acceptance of the possibility that death, at times, is a better alternative than life, prepares us for Idu's choice. Idu's choice to die to be with her recently deceased husband, rather than stay with her son and give life to her unborn child, is the critical incident of this novel. When Adiewere dies of an unnamed illness, Idu makes her decision to join him, disregarding her responsibility to her son Ijioma and the child within her. At the funeral, she tells her sympathizers that she need not weep because: "I am going with my husband. Both of us will go there, to the land of the dead. It will be even better there"(210-11). And her critical choice, mirroring both Amarajeme's suicide and the spiritual death of her friend Ojiugo after the suicide (216), is one that is not only shocking for the community but for the readers as well.[6]

Here is where Nwapa challenges the basic tenets of women's role in Igbo society by privileging the bond of husband-wife over mother-child.[7] However, Nwapa, grounding her novel in the stories passed down through family and community, states that the "bit about Idu is authentic. Something that happened to someone my mother knew." Only when pressed on her own agency in the novel, did Nwapa comment to me: "I was trying to make the statement that one can still love one's husband without giving him a child" (April 14, 1984). Theodora Ezeigbo questions this ending: "Is it aesthetically valid, given the importance attached to motherhood in Nwapa's recreation of traditional society....that [Idu] would will herself to die to follow her dead husband?"(159) If we counterpose

Nwapa's own remarks to Ezeigbo's, it is clear why Nwapa's own choice, in presenting the novel as she does, problematizes women's experience within the context of Igbo society. Idu's critical choice to will herself to die questions notions of motherhood and male/female relations in the world in which she writes, while at the same time, Idu's choice, like Efuru's, falls within what is acceptable in the community (at least in terms of the novel). Unlike Amarajeme who commits an abomination, Idu "wills" herself to die and is given a proper burial(218). Her friend Ojiugo is death in life, but Idu chooses a new life in death. This option for women is both historical and cross-cultural, despite the fact it is life-wasting rather than life-giving: the decision to "will" oneself to die only *looks* like a choice when the alternative in living is destructive to a sense of self. The response may not be configured as "feminist" but it is "Afr*a*centrist," since it exposes the complexities of women's experience within an African culture. And as Nwapa herself remarked, the novels are written to engage readers in the debate, in which we can make our own judgments.

One Is Enough (1981), written in Nwapa's middle period after she left Heinemann to start Tana Press, reflects many of the same concerns of both *Efuru* and *Idu*, but in an urban contemporary setting. The novel takes place after independence and the Biafran war, and it identifies many of the problems of postcolonial Nigeria. The main character, Amaka, like her predecessors, is a strong, competent woman, good at business, but is childless and in a marriage that restricts her. After being basically thrown out of her home in Onitsha, Eastern Nigeria, by her mother-in-law, Amaka decides to give up on marriage and move to Lagos. Her life in Lagos details her rise to power as she turns her skills in the "attack trade" during the war to gain business "contracts" in the capitalism-gone-wild world of postcolonial Nigeria, and in the process, finally becomes pregnant by seducing a priest. However, she decides that one husband is enough, and like Efuru, ends up alone and independent—although, in this case, much happier and less conflicted than her predecessor.

Some self-described feminist critics of African women's literature have seen this novel as more soundly in the feminist framework than Nwapa's earlier work, since Amaka happily succeeds in both personal and professional life, and is a powerful and independent woman. For example, Katherine Frank comments that the novel is a "painful, faltering, but ultimately successful movement of a woman from a traditional African world to a very different,

Westernized urban life"(18). Despite Frank's problems with the way in which Amaka uses sexual favors to succeed, she perceives the move as a positive model, a way out for women in the traditional (read patriarchal) society. In Florence Stratton's reading of the novel in *Contemporary African Literature and the Politics of Gender,* she finds fault with Frank, who sees Amaka's life in Lagos in terms of "moral laxity"(102), since Amaka gains favors in her business through sexual negotiations. Stratton resists this interpretation and poses Amaka as a counterreading to earlier novels of "prostitution" by male writers such as Ekwensi. She further comments: "In Nwapa's conception, then, it is men who are agents of social corruption. Prostitution, on the other hand, is for [Amaka] a strategy women have adopted for confronting male domination"(103). I have some problems with identifying what Amaka does as a form of prostitution, but accepting that she does trade sex for favors, we can discern that Stratton perceives this as a legitimate strategy for success within its cultural context. Stratton rightly notes that this strategy is part of Amaka's move to find her identity and position of power as a business woman. However, in both readings, whether Frank's Western-oriented one or Stratton's attempt to place the work within African women's context, neither critic questions what Amaka's success means to the growth and health of her nation!

Although many have analyzed the problems of the idea of the "nation state" in postcolonial West Africa (see, for example, Appiah's *In My Father's House*), the answer to this imposed national structure is problematic when the response is to totally "rip off" the society at large. Nnaemeka, in challenging Frank's notion of "Western is better," comments: "Feminists critics of African literature focus primarily on where rebellious women liberate themselves *from,* but it is equally, if not more, important to examine the politics of location that determine where they liberate themselves *into"*(92). Nnaemeka rightly questions the traditional versus modern binarism, but in this novel, there is another conflict that is equally significant and more culturally specific: To what extent can there be female liberation when it is gained at the cost of the community? Nwapa, in *One Is Enough,* clearly demonstrates that in a postcolonial setting, the politics of African women's cultural production is turned on its head, and questions of generational continuity and women's role in the society is transformed.

So what exactly does Amaka do? In her rise to power, she does what everyone else around her in Lagos is doing: she gets enormous kick-backs from government contracts. Like both men and women she meets in Lagos, she never questions the fact that the projects for the communities are done at exorbitant rates, or are paid for twice, or are paid for and not done at all! However, unlike Achebe's *A Man of the People,* there appears to be no authorial condemnation of these acts.

As one of her first conquests in Lagos, Amaka begins dealing with an Alhaji, whom she meets through her sister's lover, a Permanent Secretary. They happily use each other: she gives him sex; he gives her contracts. At this same time, she has met Father McLaid (Izu, also an Igbo from Eastern Nigeria) through her friends, Mike and Adaobi. Realizing his attraction to her, she aims to seduce him for both fun and profit. As part of their affair (although in this case, they actually have feelings for each other), he also uses his power to secure her military contracts. In one incident, before she has wised up and still thinks that people get paid to do work, Amaka is given a half-million-naira contract by Father McLaid to build a wall around some military barracks. When she arrives at the barracks, she sees the wall already erected. She asks one of the Brigadier's staff if she has made a mistake:

> He told her that she must be foolish to talk that way.
> She should just send her bill for payment. Was she
> the only one who did not know what was going on?
> If she had met the Brigadier, he would just take the
> contract from her and give it to another person. (82)

As she leaves, Amaka realizes she has just received 25% of a half a million naira for doing nothing (we wonder where the other 75% went), and we are told: "Amaka had finally arrived"(82).

In what is seen by Stratton and others as women helping women, Amaka lets her friend Adaobi in on the contracts. The only voice of resistance in this world of individual achievement at the expense of society is Mike, Adaobi's husband. He is a civil servant who is honest and doesn't take bribes; however, he is hardly perceived as a hero in the novel. First of all, he is seen as a chauvinist who doesn't want his wife to succeed the way that Amaka has. Second, as in the above quote by the unnamed military official, Mike is clearly identified as a "fool" for his attachment to values of honesty and integrity in his work. The omniscient narrator alerts us to this fact through Mike's internal questioning: "What could people like Mike do? They were civil servants trained by the colonial masters....Bribery and corruption was the order of the day....Wasn't it time for people like Mike to think seriously of retiring honorably"(65). Ironically, in this case, honesty and resisting corruption is tied to an earlier oppression, life under colonial rule. The only thing that is "honorable" for Mike to do in the turned-around world of postcolonial Nigeria is to get out of the way. This is evident when, under the ousting of Gowon (the general who took over after the war), Mike is fired from his job. He is furious, since he worked hard and is fired for "inefficiency and divided loyalty and drunkenness"(138). In short, he is fired for everything he hated in others, and he still doesn't "get it." Luckily, his wife, through contracts that she has garnered through Amaka, has bought them a bungalow to move into. But the transition for Mike, who believed in his own hard work and his country, is not so easy: "In the matter of some six hours he had aged six years"(139).

What I find especially ironic here is that, in the novel, individual advancement at the expense of the society in general is the complete reversal of what 'Zulu Sofola remarked about the basic tenets of an African community and one's responsibility to it: "One can object [follow one's individual notions] but not so far as to destroy the fiber of the society" (personal comment, 23 February 1984). Through the characters' actions, Nwapa demonstrates that the tenets that held together communities or clans do not seem to work in terms of an individual's commitment to the postcolonial nation. This is evident in what I interrogate as a critical incident in this novel—the clash of traditional values versus the contemporary lack of values (except in terms of individual achievement) through the conflict between Amaka and her mother. This conflict not only reflects a conflict of familial/community values but also problematizes the whole notion of generational continuity and women's cultural production within a contemporary West African society.

As in other Nwapa novels as well as in a traditional African setting, the elder women often function as "custodians of the 'custom'" (Arhin 1983:94). And in this novel, Amaka and Ayo's mother maintains this role in the family. Early in the novel, Amaka's thinking reflects her mother's important role in the community, as well as the changes in attitudes with increasing Westernization and Christianity. The Western "ambition" of getting a husband as the supreme success of female existence is an alien notion to her mother, despite what others of her "people" say:

But Amaka knew from the behavior of her *illiterate* aunt and mother that they did not share in this

belief of her people. Her mother brought them up
to be independent, but she did not emphasize marriage. She had several children no doubt, but her
emphasis was on self-determination and motherhood. She lacked the guts to ask her *formidable*
mother how she was going to have children without
being married. (26; emphasis added)

This passage not only demonstrates the transition from consanguinal
to conjugal relations in this postcolonial society, but also the contradictions in respect to the elder women's knowledge. Simultaneously,
Amaka perceives her mother and aunt as "illiterate" (a term that is
fairly loaded) while, at the same time, her mother appears to Amaka—
as fitting to her position in the community—as "formidable." Amaka,
as well as her sister Ayo, pay deference to their mother's power and
knowledge, yet at the same time, they see her as not comprehend-
ing the world of Lagos. However, through the fight between Amaka
and her mother over whether she should marry Izu after the birth of
her twins,[8] Nwapa examines the place of generational continuity and
women's role in contemporary society.

On a cursory level, one might say that her mother is not only
giving Amaka bad advice (to marry Izu even though Amaka is
through with husbands), but also going against her own earlier
advice—not to worry about a husband since the most important
thing is to have a child, no matter what. However, in interrogating
this section more carefully, we see the mother as not only defending traditional values of family alliances, but more importantly,
standing up for personal responsibility, a moral code lost in the
world in which Amaka lives. Amaka's mother scolds her: "Amaka,
did you not know he was a priest of God when you slept with him?
He was only good as a lover, as a man who arranged contracts for
you, and not good enough to be your husband?" (147). The mother, who grew up in a very different cultural setting, is frustrated in
trying to determine what values to pass on to the next generation.

Amaka, in the end, does not listen to her mother's advice, and
as in *Efuru*, "she slept very well"(146), happy with her decision. In
fact, the conclusion of the novel ties all the loose strings up, Izu
also returning to the priesthood. Nevertheless, the basic problem
in the novel remains: Can we really see Amaka as a fully realized
person if her self-determination as a woman is at the expense of
the greater good of the society? And, although it is clear that she
is no different from the other men and women who are sucking

the nation dry, still what does it say about considering this work one of Nwapa's most feminist? Nwapa herself does not answer these questions. After finishing the novel while living in Nigeria during the Buhari coup, Nwapa responded to my questions about the conflicts in the portrait of Amaka, characteristically: "Well, it is Amaka's story and it is her *own* story. There are many people who do this in our society. If the evils are relevant to the stories I am telling, I will include them; it does not mean I approve of it" (personal interview, July 1984). For those of us who are more polemical in our creative writings and criticism, Nwapa's remarks are disturbing while resonating a certain truth. She presents how women negotiate their positions in society; it is up to us as readers to make our own judgments. However, by her last novel, there seems to be a resolution of conflict, a more integrated sense of her self and her culture as she returns to Oguta to tell the story of *The Lake Goddess*.

In *The Lake Goddess,* to be published after the untimely death of its author, Nwapa presents her most resolved novel, as well as the one that most overtly confronts colonial and neo-colonial imposition of culture. The novel, according to Chikwenye Okonjo Ogunyemi, is a "sacred text, a voice beyond the grave, as it were, [which] is apocalyptic and oracular"(9). Here, Nwapa, in the circular narratives of women's cultural/biological production, returns to her beginnings and the "regenerative agenda"(Ogunyemi 1995:14) of Uhamiri/Ogbuide.[9] In the novel, the protagonist, Ona, born to a traditional healer, Mgbada, and his fanatically Christian wife, Akpe, has been chosen by Ogbuide to be a worshipper right from her birth (or even before, as the diviner Mgbada's visits intimate). The ensuing conflict of Ona's Christianized upbringing and her calling reflects contemporary debates on Western religion/medicine and traditional religion/healing. Moreover, the question of what values are to be passed on in the schizophrenic nature of postcolonial Nigerian society is posed, once again, through the mother-daughter conflict. However, as in other Nwapa novels, the strength of the community women as "mothers" and as part of a continuum are explored through the market women's relationship with Ona. And the ending is a blessing, a kind of metonym for her work as a whole.

In the novel, Ona, as a small child, is visited by the Lake Goddess, but her mother ignores the signs, and resists what she sees happening to Ona. Ona tells her mother that when she was at the lake with her grandmother, she saw Ogbuide: "She was dressed in velvet and coral beads. Her hair was very long. She could sit on it. Mother, I saw her"(73). Akpe, despite the fact that she is aware of such sightings, responds to her daughter by saying that she is "full of imaginations"; however, later she comments: "This child, this Ona is strange. God, please protect us"(75). Akpe decides that Ona must go to the Irish priest to help her become "normal," but when she tells this to her husband, he counters: "I am a medicine man. I can help our daughter. Father Millet cannot help us. The spirit of the water possesses Ona"(79). Akpe does not listen to her husband, and with the support of her own mother, she sends Ona to a convent. What is revealing here is the inability of the father to act as role model for the daughter. Although, Mgbada is a healer/diviner himself, he passively allows his wife to take control of Ona's education (as it is her traditional duty), despite the fact that it is clear that Ona needs to develop as a worshipper. Furthermore, Akpe's own mother is also uncomfortable about denying the Lake Goddess, even though she is a Christian: "Mama Theresa was confused. She had doubts about her own faith. She prayed asking for strength"(79). As Ona matures, marries, and has children, she is constantly plagued by her calling. From Ogbuide's spirit children who visit her to her own ability to divine the future, she is unable to resign herself to the role of mother and wifehood, defined for her by her rigidly Christian mother and later her husband, Mr. Sylvester, whose job, ironically, is to sell conventional medicines and pharmaceuticals (185). However, it is in the subplot of the novel, that we find the critical incident that exposes the Afracentric nature of the novel: the story of the two market women.

The narrative of the two market women, Mgbeke and Ekecha, runs parallel to the main plot of Ona's coming to terms with her role as worshipper of Ogbuide. At first, it appears that this storyline has little relation to the main plot. With the exception of certain remarks concerning Ona, most of the constant discussion of the two women as they seek out fish to sell, is about their own hopes and aspirations for their children, as well as general comments about the changes in their society. However, careful readers of Nwapa's works note that their voices represent the important voice of the community, especially the community women. At the beginning of Part II, Mgbeke and Ekecha question the position of women in this changing, postcolonial society, challenging the notion that Western education is positive for all women. In their discussion, they underscore the role of women in precolonial society, as well as illustrate the pitfalls of a supposedly "liberated" lifestyle. In speaking of one woman maltreated by her husband, one of the women (it is interesting to note that their voices blend so completely that it is sometimes impossible to follow which woman is speaking) comments: "You and I who have not gone to school could not take that. But here, our fellow women, who went to school, who have their so-called pieces of paper which can earn them money, behaving as if they have no pride—as if they have no homes to go back to"(104).

Her friend answers: "I know how to care for a man, but he must give me the respect that is due to me, otherwise *I paddle my own canoe*" (105; emphasis added). Here the market women use a metaphor linked to their activities and position of power as traders in the society, as well as to identify the strength from the lake that the women of Oguta receive. In many interviews, Nwapa has often referred to this strength of the Oguta women, and it is their voice which counterbalances the assimilative posture of Ona's mother, Akpe.

Through their parallel narrative, they position themselves as a voice supporting Ona—at times in person to her and always as an alternative voice for the reader. In relation to the above scene, we have already witnessed Ona leaving school to pursue her calling as a diviner. Furthermore, they quickly identify Ona as chosen by Ogbuide, who is of great importance to them since she supplies the fish they sell. They watch the spirit children who greet them in the market while Ona is buying fish. They are aware that Ona and the spirit children are two manifestations of Ogbuide (126), and thank the Goddess for selling their fish (127). Furthermore, it is one of their discussions of Ona and her relationship to her Christianized family and the community that I examine as a critical incident, reflecting the major conflicts of the novel.

The scene begins with the two women taking a rest from the sun and their paddling by beaching their canoe on the side of the lake. They evoke the Lake Goddess, chanting prayers: "Our Mother is Supreme" and "We are your children, and we are women like you"(193). They finish thanking her for the fish they sell and then begin to speak of their personal problems with their daughters, as well as about Ona. As many who have grown up with the Christian Church, yet are still attached to their traditional religion, they appear to accept the apparent contradictions in their belief system, evident in their conversation. In exploring ways in which to get their daughters married, they send them to convent school (these are two women who earlier decried education for the sake of learning traditional skills like trading), while at the same time, practice the ritual prescribed by the *dibias* for the same objective (195). Furthermore, when they speak of Ona, they call her both "blessed" and "mad"; and they also regard her as "a great embarrassment to her parents and grandparents"(194). Although they are apparently unaware of the inconsistencies in their perspective, we are not; as they talk, the sky darkens over the lake, filled with dangerous lightning and thunder. The two women are fearful, and beg Ogbuide to protect them, wondering what they have done wrong. Although it is not explicitly stated in the scene, there is a sense that their lack of support for Ona, as one of Ogbuide's worshippers, has brought this perilous weather upon them. As the weather shifts, they quickly return home, stating: "Our Mother is not happy with us today"(198). As in other of her novels, Nwapa problematizes the Western vs. traditional opposition by having the characters expose the inconsistencies in their culture within self, rather than having one character represent a particular position. Nevertheless, by the end of the novel, both women have not only accepted and honored Ona as a worshipper of the Lake Goddess, but have also asked Ona to divine for them.

Finally, Ona herself is resolved to her position as worshipper. Even if her own mother has not passed on the values of the culture to her, Ona is mothered by the market women, her grandmother, and her father, as well as by the Supreme Mother herself, Ogbuide/Uhamiri. As a worshipper, Ona is renamed "Ezemiri" (Queen of water), and she takes on her role as a priestess to Ogbuide. As her first incantation to the Supreme Mother clearly identifies her in her new position, it also reaffirms the notion of mothering in this matrilineal culture:

> Ogbuide, the lake goddess
> You chose me
> From my mother's womb....
> Ogbuide, the lake goddess
> I said to my mother:
>
> Mother,
> Accept me as I am
> Ogbuide has chosen me....
> Ogbuide,
> Our Mother. (231)

As we have seen throughout her writings, the role of mother, for

Nwapa, is broader than the biological—from community to spiritual mothers. And the introduction of these poetic orations in the conclusion of the novel also illustrate a change in the structure and tone of the novel itself: at this point, as Ogunyemi states, the novel takes on a strength of voice that comes from an ancient and immemorial source, as the utterance of Ogbuide takes over through Ona. As with Efuru and Amaka, Ona has found peace with herself by the end of the novel. However, unlike the others, her peace has not come with loss. Ona has fulfilled her one aspect of her duty to society through her husband and children, but she understands her role as worshipper and accepts it unconditionally. There is a sense of resolution in this novel that is not in the earlier ones. As she interprets the will of Ogbuide for the two market women, Mgbeke and Ekecha, Ona tells them that "Ogbuide wants all women to have voices. Women should not be voiceless"(241). Ona states prophetically, speaking of a woman who is now a worshipper of Ogbuide: "She had a voice"(241). Certainly, with the final praise-song to Ogbuide that ends this novel, we can say that most forcefully about Flora Nwapa as well.

In reading Flora Nwapa's literary production as part of African woman's cultural production, we can identify in Nwapa's writings what Susan Andrade (1990) calls African women writers' "self-inscription into history"(97). And although Andrade designates Emecheta's work as more specifically "located" than Nwapa's (for whom? the Western reader?), it is evident that Nwapa presents for us, despite the absence of dates (101), a "real" herstory of Igbo women's lives in all their complexities. In each of the critical incidents I have related in this essay, Nwapa exposes the contradictions in her culture, while passing on to us as readers the stories of the society she has known. It is in her role as recounter of culture, and women's particular place in it, that we understand Nwapa as an Afracentric writer. I am reminded of Toni Morrison's comments about her own writings. She notes, harkening back to the African presence in her own work, that she writes "village literature"(LeClair 1981:26). Morrison states further that her language and fiction "have holes and spaces so the reader can come into it...to make this book, to feel this experience"(Tate 1983:125). It is precisely Nwapa's own "village" literature that Morrison revisions, not necessarily as intertextual but as intracultural. Nwapa, as a foremother to writers and readers in Africa and the diaspora, has allowed the spaces and holes to let us in, to remember and recreate the world of the woman storyteller; in short, she had a voice.

NOTES

1. This message runs throughout Lorde's work; for example, see *Sister Outsider.*
2. As identified in the women writers, Afracentrism examines women's role in the transmission of culture as *complementary* to other (male) modes of production in the passing on of cultural traditions from one generation to the next throughout Africa and African-based groups in the Americas. For a fuller discussion of African women's transmission of culture in relationship to their art, see Wilentz, "Introduction," *Binding Cultures.*
3. For a critique of Afrocentrism and Pan-Africanism, see, for example, Anthony Kwame Appiah, *In My Father's House,* and V. I. Mudimbe's, *The Invention of Africa.*
4. Relevantly, at the Women in Africa and the African Diaspora Conference, held at Nsukka in July 1992, Nwapa does finally identify herself as a "feminist with a big 'f'" after hearing Ama Ata Aidoo speak on the issue (Nnaemeka 1995:83-84).
5. My concept of "Afracentrism" was first presented at the Africa 1995 Conference, Tel Aviv, Israel, June 1995, in a paper, "Afracentrism as Theory: The Discourse of Diaspora Literature."
6. Contemporary African critics, for the most part, found the novel's ending to be both fantastic and "morally" unacceptable. Response appeared to be based on moral outrage. See, for example, Adeola James's review of the novel in *African Literature Today.* A notable exception was Ernest Emenyonu's "Who Does Flora Nwapa Write For?" in a later issue of the same journal.
7. In *The Igbo of Southeast Nigeria,* Victor Uchendu discusses this opposition in the culture. See also Ifi Amadiume, *Male Daughters, Female Husbands,* for further examination of the consanguinal focuses of familial relations.
8. The focus of Amaka as mother of twins from a culture that once used to "throw them away" also reflects Nwapa's refusal to glorify either Western or traditional Igbo society.
9. According to Sabine Jell-Bahlsen, both the names Uhamiri and Ogbuide are identical in referring to the Lake Goddess (30). Why Nwapa chose to shift from Uhamiri to Ogbuide in this last novel, we will probably never know.

WORKS CITED

Achebe, Chinua. *A Man of the People.* London: Heinemann, 1966.

Amadiume, Ifi. *Male Daughters, Female Husbands: Gender and Sex in an African Society.* Atlantic Highlands, New Jersey: Zed, 1987.

Andrade, Susan. "Rewriting History, Motherhood, and Rebellion: Naming An African Woman's Literary Tradition." *Research in African Literatures* 21 (1990): 91–110.

Appiah, Kwame Anthony. *In My Father's House: Africa in The Philosophy of Culture.* New York: Oxford Univ. Press, 1992.

Arhin, Kwame. "The Political and Military Role of Akan Women." *Female and*

Male in West Africa. Ed. Christine Oppong. London: George Allen, 1983:92-98.

Brown, Lloyd. *Women Writers of Black Africa.* Westport, Conn: Greenwood, 1981.

Ezeigbo, Theodora Akachi. "Traditional Women's Institutions in Igbo Society: Implications for the Igbo Female Writer." *African Languages and Cultures* 3.2 (1990):149-65.

Emenyonu, Ernest N. "Who Does Flora Nwapa Write For?" *African Literature Today* 7 (1975):28-33.

Frank, Katherine. "Women Without Men: The Feminist Novel in Africa." *African Literature Today* 15 (1987):14-34.

Holloway, Karla. *Moorings and Metaphors: Figures of Culture and Gender in Black Women's Literature.* New Brunswick, N.J.: Rutgers Univ. Press, 1992.

James, Adeola. "Review of *Idu.*" *African Literature Today* 5 (1971):151- 53.

Jell-Bahlsen, Sabine. "The Concept of Mammywater in Flora Nwapa's Novels." *Research in African Literatures* 26.2 (Summer, 1995):30-41.

LeClair, Thomas. "The Language Must Not Sweat: An Interview with Toni Morrison." *The New Republic* 21 March 1981:25-32.

Lorde, Audre. *Sister Outsider.* Trumansburg, N.Y.: The Crossing Press, 1984.

Mere, Ada. "The Unique Role of Women in Nation Building." Unpublished paper. U of Nigeria, Nsukka, 1984.

Mudimbe, V. Y. *The Invention of Africa.* Bloomington: Indiana Univ. Press, 1988.

Nnaemeka, Obioma. "Feminism, Rebellious Women and Cultural Boundaries: Rereading Flora Nwapa and her Compatriots." *Research in African Literatures* 26.2 (Summer, 1995):80-113.

Nwankwo, Chimalum. "The Igbo Word in Flora Nwapa's Craft." *Research in African Literatures* 26.2 (Summer, 1995):42-52.

Nwapa, Flora. *Efuru.* London: Heinemann, 1966.

——. *Idu.* London: Heinemann, 1970.

——. *The Lake Goddess.* (Unpublished manuscript)

——. *One Is Enough.* Enugu, Nigeria: Tana, 1981.

——. Personal Interviews. Enugu and Oguta, Nigeria. March-July 1984.

Ogunyemi, Chikwenye Okonjo, and Marie Umeh, eds. "Special Issue: Flora Nwapa." *Research in African Literatures* 26.2 (1995).

——. "Introduction: The Invalid, Dea(r)th, and the Author: The Case of Flora Nwapa *aka* Professor (Mrs.) Flora Nwanzuruahu Nwakuche." *Research in African Literatures* 26.2 (Summer, 1995):1-16.

Okonkwo, Juliet. "Adam and Even: Igbo Marriage in The Nigerian Novel." *Conch 3.2* (1971):137-151.

Sofola, 'Zulu. Personal Interviews. 23 Feb. and 16 June 1984.

Stratton, Florence. *Contemporary African Literature and The Politics of Gender.* London: Routledge, 1994.

Tate, Claudia, ed. *Black Women Writers at Work.* New York: Continuum, 1983.

Uchendu, Victor. *The Igbo of Southeast Nigeria.* New York: Holt, 1965.

Wilentz, Gay. *Binding Cultures.* Bloomington: Indiana Univ. Press, 1992.

——. "Flora Nwapa." *Twentieth Century Caribbean and Black African Writers. Dictionary of Literary Biography,* ed. Bernth Lindfors and Reinhard Sanders. Columbia, SC: Bruccoli Clark Layman, 1993.

Efuru and *Idu:* Rejecting Women's Subjugation

Ada Uzoamaka Azodo

"...a girl is not a failure to have a boy; a woman need not be defined by marriage and, above all, motherhood; the woman who supports herself depends on men less."

—Susan Gardner

This essay contributes to the celebration of the legacy of Flora Nwapa, who was the first Nigerian woman-novelist, the first African woman to publish in English,[1] and to own a publishing company, Tana Press Limited. The pioneer of Nigerian African feminism, Nwapa raised for the first time the question of what should constitute feminism in Africa.

This study focuses on her attempts to present, as she saw it, the deplorable socio-economic situation of Igbo women in particular, and Nigerian women more generally, under colonial Nigeria.[2] More specifically, Nwapa paints a picture of life in the 1930s and 1940s, when untainted traditional life was beginning to disappear in the wake of the entrenchment of the British colonial government (Johnson-Odim and Strobel, 1992). Paradigmatic binary-oppositions were beginning to appear as those who had been to foreign schools lived among those who had not, and as British, Victorian capitalist economic interests of exploitation clashed with the indigenous people's will to pursue subsistence activities. Political independence came on October 1, 1960, after about fifty years of foreign rule. Nwapa cried out against what she saw as the relegation of women to the background. The men of the new nation in the movement for independence marched forward towards freedom, while the women were left behind in bondage to tradition, motherhood, and patriarchal subjugation.

At the time of the publication of *Efuru* in 1966, on January 15th of that same year, the first *coup d'etat* led by Major Chukwuma Kaduna Nzeogwu had toppled the First Republic. The spirit of nationalism was at its peak, and young patriots struck at nationals who, in their opinion, were leading the nation astray. In literary circles, the father of the African novel in English, Chinua Achebe, had published the classic, *Things Fall Apart*, in 1958 (Camara Laye had already published in French his equally very well-written novel *The African Child*, in 1953), in reaction to the racist writings of foreigners like Joyce Cary of *Mister Johnson* (1975), and Joseph Conrad of *The Heart of Darkness* (1983). Achebe presented traditional Igbo culture as a cohesive whole, which broke up only with the intrusion of Europeans. The by-product of Achebe's otherwise justified anger and laudable reaction was, however, disadvantageous to women's status in society. Recent feminist readings of Achebe are discovering that the author's attempts to put past history in proper perspective resulted in a masculinist literary creation in which the self is male and the other female (Stratton, 1994). In the extremely patriarchal Nigerian society of those days, women were neither heard nor respected in the community. They were accorded no human dignity. They were sold, whipped, shouted down, maimed, murdered, all at the whims of men, the power holders in society. The brutal colonial administration, run by male agents, compounded the problem. It appeared to confirm that the world was androcentric. Women under the patriarchal and racist colonial government had only a marginal place and role in the power structure. Under British administration, Nigerian women suffered a double blow, as they were now worse off than they had been under African patriarchy and its traditional institutional structures (Stratton, 7).

Efuru was to be followed soon after by Nwapa's second novel,

Idu, in 1970. After independence, both novels registered their author's disapproval of phallocentric images in society. Nwapa became the voice of thousands of voiceless women who were, in the real sense of the word, non-existent, and invisible, in the new Nigeria. *Efuru* and *Idu* will serve as extended examples of Flora Nwapa's feminist criticism. Both novels are similar in many ways. There are cross-references from *Efuru* to *Idu* and vice-versa. There is reference to Efuru's house in *Idu*. Both heroines, Efuru and Idu, are childless and each has a confidante, Ajanupu for Efuru and Nwasiobi for Idu. Both have a little girl ward living with them, Anamadi with Idu and Ogea with Efuru. Each novel has a village gossip from whom we learn all the negative aspects of the village community, Omirima in *Efuru* and Onyemuru in *Idu*. Both novels have Uhamiri, the goddess of the Oguta Lake, as a supernatural agent, directing the affairs of human beings (Azodo, 1993; Drewal, 1986). Nwapa's contribution to Nigerian feminism in both books is very significant for engendering criticism of African writing by an African woman writer. For the first time in Nigerian literary history, Nwapa posits a set of gynocentric stories, in contrast to writings by male novelists of the day.

Of course, traditional Nigerian women did not hear about feminism for the first time with Nwapa's publications. In pre-colonial times, thousands of Igbo women had always settled their score with offending males by laying siege on their households, until the erring husbands promised to turn over a new leaf. Until the wicked husbands promised to treat their wives more humanely, the women remained on their premises raining abuses on them in-between lampoons. Under colonial Nigeria, between 1929 and 1930, Igbo women dressed in men's war garbs, with black and white chalk painted on their faces to signify their readiness for the worst, marched against the colonial government in the then Eastern province. In what would later go down in Nigerian political history as either the Aba Women's War, or the Aba Women's Riot according to the whims of British derogatory reporting, women protested the taxation of unemployed women by the colonial office.[3] A similar scenario repeated itself in Abeokuta in 1946, when Yoruba women organized under the leadership of defunct Frances Funmilayo Ransome-Kuti to protest the British administration's confiscation of market women's goods to help the Second World War effort.[4] Feminism, indeed militant feminism, therefore, was absent neither in traditional nor in colonial Nigeria.[5]

One can also extend the observation and add that Africans did not learn about feminism from any other nationals, certainly not the women of the First World.[6] African feminism, as seen in *Efuru* and *Idu*, is an attempt by women of Nigeria to deal in the best way they could with issues—in their lives, in national history, and in culture—which make them playthings for men. This point is very important and worthy of particular note, especially given the present war raging between First World and Third World women about what should define feminism in their different cultures (Davies, 1986). Many African women believe that feminism in Europe and America is centered on fights for equality between the sexes and resentment of gender distinctions (Mitchelle, 1987). For Africans, however, these are merely the tip of the iceberg. On the other hand, African women believe it is more the very survival of women as humans that is in question. The harsh economic, social, and political situation of Africa in neo-colonial times make racism, imperialism, classism, and economic survival the very important factors of African feminism.[7]

The second purpose of this study, then, is to contribute to the definition of feminism in such a way that it has validity for all women all over the world. For Nwapa, African feminism can be described as "womanism" (Terborg-Penn et al., 1987; Nandakumar, 1971). Put differently, it is an effort to work for the communal good as an African woman. It is not a simple reaction to and a rejection of the First World feminism, with its radicalism, its Marxism, and its liberal gender protests against sexism.[8] Womanism for Nwapa, as for Alice Walker of *The Color Purple* fame, means being "committed to survival and wholeness of entire peoples, male and female."[9] It is working for the embodiment of harmony in human relations between the sexes and in the community at large. This involves tapping into the special intuitive talents women naturally possess for solving community problems. Nwapa falls in line also with such other black female writers as Maryse Condé, Filomena Chioma Steady, and Toni Morrison. As a well-educated, sensitive artist, Nwapa was not the least impressed by the position of women under colonialism and soon after independence in her country. She described their situation very succinctly, in discussing the theme of *Efuru*, when she said: "If I'm trying to prove something, it is that women are first and foremost human beings."[10] In this regard, it is pertinent to remind ourselves of Karl Marx, Sekou Toure, and Charles Fourier, according to whom the position of women in any given society is indicative of the stage of civilization or humanization in that society.[11] At Nwapa's death, Chinua

Achebe eulogized her for her interest in women's issues without being a feminist.[12] We understand this remark, quite unflattering as it appears for a woman who fought for Nigerian feminism all her literary life, as meaning that she was reformist in her tendencies. Nwapa's works espouse a feminist humanism which condemns submission, authoritarianism, and rigid, closed or absolute cultural value systems. On the other hand, this humanistic feminism (Terborg-Penn et al., 3-21), prescribes pluralism, criticism, skepticism, and deconstructionism in an atmosphere of love and harmony. It is a moral and political criticism.

It is surprising, therefore, that earlier critics of Nwapa did not detect Nwapa's politics through her feminist criticism. They have dwelt rather heavily on the author's apparent respect for Nigerian tradition and value systems. They argue that she worked hard to ensure their generational continuity. Few saw her disgust in opposition to, and protests against them.[13] With the exception of a few African critics, such as Ernest N. Emenyonu,[14] and the already mentioned Chinua Achebe, others, notably male critics like Eustace Palmer, Lloyd W. Brown, and Bernth Lindfors, destroyed the genius of Nwapa for several generations with their adverse criticisms of her writing. They compared her disadvantageously with her male contemporaries, as if all that is involved in literary criticism is comparison of male and female writers. They saw her as a pale and poor imitator of Chinua Achebe, Amos Tutuola, and Elechi Amadi.[15] They also attacked her plot, language, and characterization techniques, inferring that as a woman writer she had not lived up to the high standards of the male writers. Male authoritative language and rhetorics are replaced in Nwapa's writings, they observed, with women's gossip and dialogues in the backyard, at the stream, and in the market place.[16] For these masculinist writers, an omniscient stance is a precondition for writing.

It is only recently, with the innovations of new methods of criticism in the post-structuralist period, that Nwapa's more congenial, womanly and emotional tone is making a well-deserved return. A notable example among the new interpretations of Nwapa's works is Florence Stratton's *Contemporary African Literature and The Politics of Gender*, which has challenged both phallocentrism and phallogocentrism in contemporary African writing.

For our part, all of the women's gossips, dialogues, tales, songs, proverbs, riddles, and chants will become the poetics of feminist discourse. Whatever is experiential, personal for these women can also be seen as a public matter, given our belief that women who are not allowed freedom at home cannot hope to be accorded any in public (Gavison, 1994). When we examine the philosophy behind the subjugation of women—as well as the epistemological question of who is more effective in society, the domineering male or the nurturing female—we find that Nwapa's feminism refuses to privilege one gender over above the other. Even though most of Nwapa's male characters appear pale by comparison with an overwhelming number of women with very active "animus," yet the community of Oguta in *Efuru* still has the likes of the Western-trained doctor, Difu, and other males who are in every sense humanists and active members of the community. Difu even works during his vacation healing the sick!

A deconstructive reading (Elam, 1994) of Nwapa's *Efuru* and *Idu* is necessary for four reasons. Firstly, it permits a philosophical as well as a literary interpretation of the texts. The beaten Nigerian women, like the characters of Virginia Woolf's *A Room of One's Own*, will be seen as telling their men what they would like to hear, rather than pushing for generational continuity of tradition. Much like Virginia Woolf, Nwapa believes that social and economic emancipation of women will come only when women stop functioning as men's "looking glass."[17] In other words, Nigerian women need to have a consciousness of their oppression. Because oppression is woven into the fabric of societal structures, women's subjugation is hidden under the cover of tradition, making it difficult to detect or tackle. Secondly, deconstruction will allow a textual rereading of the texts, that is of events happening in the novels in social, economic, and political spheres, such that we can understand the present and argue for the future. Thirdly, we will be able to admit the views of the other without negating the self. Lastly, since a deconstructive reading has no closure, it is very well suited to our reading of Nwapa's *Efuru*, which equally has no end. A modern adaptation of a dilemma tale, *Efuru* ends with a question, thus leaving open-ended the discourse on Nigerian African feminism.

Nwapa explores in *Idu*, as a marriage counselor would do, the factors that make for a happy marriage, by presenting several cases/marriages for study. It will become evident, however, that Nwapa values equiarchy "under erasure"(*sous rature*), as Jacques Derrida would say, over matriarchy and patriarchy (Amoros et al., 1993). There is evidence of the author's overriding concern for human progress in society irrespective of sex and gender. Despite Derrida's ambivalence about whether feminism is compatible with deconstruction, given that the one wants to do away with a center while the other can be described as an organized movement,[18] one still finds enough ground to converge both terms in Nwapa (Elam, 1994). Their juxtaposition allows us not only to weigh the humanity of women, it also helps us to see the loss to mankind caused by the continued subjugation of women by men. Furthermore, we will understand the impact of past history on the present, as well as appreciate our duties towards future events.

This essay will argue, therefore, that men and women are equally responsible for the marginalization of women in colonial Nigeria. The fate of Nigeria today could have been different if the male population had been willing to extend recognition to the women as equal adults and contributing members of society, in the several generations leading up to the dawn of independence. The picture might have been better still if women themselves had not, consciously or unconsciously, always participated in handing down oppressive attitudes and habits to generations of future citizens. Women have yet to determine who they are as individuals and as a group, and what they can do. How many times have we heard feminist accounts of men's treatment of women like living organisms with no feelings or will? For how long shall we listen to complaints that men exploit women by treating them like sex objects and domestic slaves in patriarchal societies? How long shall we continue to wait for someone to tell women the bitter truth, namely, that it is about time they themselves started treating one another like subjects with feelings and emotions?

We will proceed, in the first instance, by presenting Nwapa's protests against men's attitudinal subordination of women, their oppression of women through rigid social structures, and unwarranted stifling of women's potentials with the burden of motherhood, including bearing and rearing of children. Next, we will show the added ill-effects to society of women fighting other women. Finally, we will conclude the paper by suggesting how much better everyone and society could be with males and females working together in partnership for the common good.

PROTEST AGAINST PATRIARCHY

Flora Nwapa protests the subjugation of women in patriarchal Nigeria. Not only does it hinder women from making their full contribution to society, she argues, it also involves a moral degradation of women. Women are taught early in life that they are a sensual class of human beings, created for the erotic pleasure of men (Amadiume, 1987). Women are to value their bodies, not their brains. Whereas men are free to carry on as they please, women are put under surveillance by their families and the entire community. Their every move is monitored under the guise of protecting them. A woman with a history, as the saying goes, runs a risk of never finding a husband. For a society which is pro-marriage and pro-natalist, it is a sign of being rejected and rejecting society, not to be married nor have children (Mbiti, 1990). Efuru, the heroine, was constantly harassed by her father and a male cousin while she was dating Adizua, her first husband. There is no doubt that both men drove her into elopement before the bride-price was paid on her head. Yet, Adizua, for his part, did not go through any travail with his mother, essentially because he was a man.[19] When later he deserts his family and disappears with a mother of two who had also left her husband, his misconduct is glibly dismissed as one of those things men do: "That is how men are." Equally in *Idu*, Anamadi, Idu's younger sister, was policed by the combined forces of her sister, the husband, and the entire community until she became rebellious, disappearing during the nights. One only has to listen to Idu to confirm this assertion: "Go to Anamadi's room and you will see it is empty. She has not slept in her room. Anamadi has got a man friend. Ewuu, what will I do about Anamadi? You said she was a child. Didn't I tell you that Anamadi is not a child, that we should hold Anamadi with an iron hand? Didn't I say that, Adiewere, didn't I?"(*Idu*, 19). Later, Anamadi hijacks a canoe with a group of other rebellious girls and disappears for days. A society which sets double moral standards for its adult population based essentially on sex and gender is setting up the society for perpetual condonement of discrimination against the weak.

Men's privileges do not stop with promiscuity. They are also allowed to practice polygyny, the marriage of one husband to more than one wife. In *Efuru* and *Idu*, polygyny is not only indicated in cases of infertility to provide offspring which ensure the man's personal immortality, it is also used to punish a woman with opinion, the woman who dares to ask questions, as is evident in *Efuru*. Such is the fate of Nkoyeni, the second wife of Enebiri, Efuru's second husband. Nkoyeni soon falls out of favor with everyone in the household when she demands to know why their husband deserted them for three months, and whether there is truth in the rumors that he had been imprisoned for theft for three months in Onitcha

(*Efuru*, 214). Nkoyeni is seen as bad and troublesome. It does not matter that in African ethics theft is one of the worst crimes a person could commit, second only to a woman's infidelity, with murder in the first position (Mbiti, 1990). A third and younger wife is quickly arranged for Enebiri, her husband, to compete with her and thus curb her excesses. A society where women's rationality and initiative are stifled in private already spells doom for their public emancipation (Gavison, 1994; Pateman, 1987). Such women will be very unlikely to contribute in public matters since all their energy is spent on seeking the marital security which is often denied them.

Nkoyeni's Western education is blamed for her stand on issues. So, society condemns altogether the new attitudes arising from the teachings of the foreign schools. The denial of education to women means that for a long time there will be a very small percentage of women who are in a position to contribute either rationally or skillfully to the growth of the nation. When there are scarce resources, boys rather than girls are sent to school to learn. Women's education is branded a waste of time, since they usually drop out of school to get married and raise a family. It does not matter that it is the men themselves who make a conscious effort to pluck them away early before they develop enough self-esteem to challenge their oppressive ideas. That is exactly why Nkoyeni's bright future is cut short, so that she can become the second wife of Enebiri. Such women as Nkoyeni always end up as housewives, "in the kitchen," as the saying goes. They lack the skills required to get ahead in the new Nigeria. Should they have many children, as most of them usually do, their destiny is forever blighted. There is practically no hope of getting themselves together again in life.[20]

By the same token, Nwapa disapproves of elitist education which produces lazy, idle, house-bound wives, in imitation of white women of the Victorian capitalist economy. Such education promotes the subservience of women and deprives the country of the contribution of part of its adult working population. The newly educated Nigerian women can no longer go back to the industry of their peers in farming and trade. Like wives of British colonialists, they sleep-in and engage maids and houseboys to do their household chores. They act condescendingly towards illiterate women, whom they see as harboring diseases which could infect them and their children (*Efuru*, 193-194). Married to rich husbands, these women remain prisoners in the houses of their husbands unable to help themselves. Should marital discord arise and they be thrown out by their husbands, as is often the case, they have no means of helping themselves. For fear of undesirable consequences to themselves, they remain with a man, taking untold insults and other abuses.

Nwapa is concerned lest the subjugation of fellow women continue in the new nation. She is aware that unless there are more than a few university-educated women like herself, who can see men's attitude towards the women in the country for what it is, there is no bright future for women, nor for the country as a whole. Humankind cannot make reasonable progress towards development if the women, who are an essential part of the society, are relegated to the background.

Protest Against Tradition

Patriarchy, as an attitude towards women, is embedded in the social structure. Tradition favors men over women in all spheres of life. The uniqueness of individual women is not taken into account, nor is their personal worth assured. And so, early in life, males and females are classified and assigned gender roles.

Men receive chauvinistic lessons on society's expectations of them in the public and in households, including being seen to be able to control their wives. Women, on the other hand, are tutored to be submissive to all males, including their husbands and brothers. They are to regard the primary foci of their lives as marriage and procreation, followed by the rearing of the offspring of those marriages. Anamadi is condemned by the entire community, including her family, for standing up for her beliefs and fighting in the market place. Even though she is being beaten up severely and her clothes nearly completely torn off her body, she refuses to yield. She is branded as bad, difficult, and as having no (common) sense. Had she been a boy, everyone would have seen the "sense" in her fighting to defend her valor. Her jumping out of the window of her room to escape her punishment of isolation would have been lauded as "manly." Yet, because she is a woman, she is condemned. Idu summarizes the thought behind all the condemnations of Anamadi's behavior by asking this barrage of questions: "Have you ever known such a thing in your life? What shall we do with a girl like that? Do you think she will ever get married?"(*Idu*, 7-8). The most important factors are that she is a girl and that she will still need to get married. Suitors might run away if they should hear that she is motivated enough to stand up for her rights and beliefs.

Ajanapu, the chief custodian of traditional values in *Efuru*, commands little Ogea, Efuru's maid, to "bend down and sweep like a woman" (*Efuru*, 44-45). The small girl, having finished her chore, sits down to a meal and Ajanupu shouts at her again: "You don't sit like that when you are eating. Put your legs together. You are a woman" (*Efuru*, 45). Even her mistress is not spared: "Remember," Ajanupu advises Efuru, "she is a girl and will marry one day. If you don't bring her up well, nobody will marry her. By the way, can she cook now?" (*Efuru*, 45). Good upbringing, according to patriarchal ideals, entails women learning the arts of submission, cookery, and service to a husband. Without these, they will inevitably face divorce (*Efuru*, 125-126). Good upbringing also involves kneeling down to give or receive things from one's husband(*Efuru*, 24). The wife must sit down by her husband while he eats, in order to pour his drinks. She may not eat with him. The meal over, she packs away his plates for washing while he dozes off to sleep. Early the following morning, she must wake up with the cock's crow to start the day's work, while her husband sleeps (*Efuru*, 173). When she delivers her baby, she must do it alone or with a female friend. She must be especially careful, if it happens at home, and at night, not to disturb her lord and master (*Efuru*, 32). With such excellent machinery for ideological indoctrination carried out by elderly women, it is no wonder that, by puberty, young women are already conditioned for a life of submission.

Surgical invasion of women's bodies, popularly known as female genital mutilation, takes over where ideological indoctrination stops, and the sensuality of women is placed in the control of men. No longer would these women have feelings for any man who is not their husband. Regarding sex with husbands, men merely pounce on wives when their sexual desires are aroused. There is no reciprocal longing for each other. Clitoridectomy ensures that sex becomes necessary for women only for procreation. Since many of these women have to go through abstinence for two to three years between pregnancies, a man does not have to contend with the burning desires of a wife for sex.[21] Furthermore, he has, perhaps, other wives to keep him satisfied until the new mother is again ready for another child. A man has sex with his wife when he needs it or when it is time to make a new baby. The operation of circumcision is seen as a bath, which all adult women need to take, in order to become clean and ready for marriage and childbearing. Evident here is the internalized myth according to which women are inferior and unclean. There is also therein an ambivalence about biological determinism. Is one a woman by birth, or does one become a woman by going through certain rites prescribed by tradition? Furthermore, the pain of infibulation is euphemized as the pain of hunger (*Efuru*, 13-14). Those girls and their mothers who hesitate or refuse to embrace this tradition are seen as foolish and stupid (*Efuru*, 13-14). Mythical stories are quickly supplied by mothers and grandmothers to show the dire consequences, including death of the newborn, which could result from going against this aspect of tradition.

The majority who succumb are sequestered at home or in a special house called a "fattening room." For durations of one month to six or twelve months, they are fed fat like animals for slaughter, as they are readied for life with their future husbands: "She was to eat the best food and she was to do no work. She was simply to eat and grow fat and above all was to look beautiful" (*Efuru*, 15). It is pertinent to mention here that Nigerian men nearly always prefer their women plumpy, since they say that when they hold a woman they like to hold flesh, not a bag of bones. The newly circumcised girls are branded "feasting girls." Their mothers, in turn, become "mothers of feasting girls." The young women are treated like erotic princesses, for whom the body rather than the brain is of primary concern. At the end of the fattening period, they visit the market where they put themselves on display for the admiration of the whole community. In some instances, depending on the means of the family, they go back and forth from the home to the market several times a day, and each time differently adorned. They are showered with gifts of money and other gifts in kind. Needless to say, Efuru, the one remarkable mind among these feasting girls, refuses to be part of this tradition beyond the minimum time stipulated. She refuses to listen to her mother-in-law, who wanted her to remain under sequestration longer. At the end of one month, Efuru returns to her trade (*Efuru*, 17).

Still, even Efuru is not always successful in escaping the clutches of tradition. She loses many nights of sleep worrying about her infertility. Childlessness is seen as a failure in her society and Efuru sees herself as a failure, despite all the other admirable qualities she has. When she finally has a baby girl, Ogonim, Nwapa achieves her first of three spectacular creations of women's freedom in the novel as the self arrives at an unqualified intensity of emotion and thought. Efuru, still not believing that her social stigma as a barren woman is finally over, exclaims: "Is it really true that I have heard a baby, that I am a woman after all?" (*Efuru*, 31). Unfortunately, motherhood has become a yardstick for recognition. When Ogonim dies at the age of three, Efuru believes her existence has run its course unless she can convince her deserting husband to return to her. Next, she leaves on a long journey for several months in search of the errant husband. Without a husband and a child, she may never have respect in the community. As her mother-in-law concurs, "Di bu mma Ogori," meaning that a husband is a woman's beauty (*Efuru*, 97). Unsuccessful in her quest, she returns to her father's home, since she has no meaningful existence in her matrimonial home without a husband. She later remarries, and even with her second husband, her equilibrium is still in question since she has no child. She achieves psychological balance only when compensatory dreams of Uhamiri, the Lake Goddess, who has no children herself, intervenes to restore her balance. As a human representative of this Goddess, she begins to see clearly into her situation for the first time. Nwapa creates her second moment of women's freedom when Efuru realizes that Uhamiri, the mentor she worships, cannot give her a child because she has none herself. Yet Uhamiri remains beautiful and happy. What then stops her, Efuru wonders, from doing the same? Uhamiri has no children and therefore cannot give her any children. One cannot give what one does not have. By the same token, one should also learn to accept a situation one cannot help (*Efuru*, 165; see also Brown, 1981:146).

Infertility seems to be psychological ammunition, conscious or unconscious, against the traditional belief that women cannot do much for themselves. Since the spiritual night that Efuru accepted her destiny, she ceases to dream of Uhamiri. She stops wasting energy longing for a child who is not about to come. Above all, she begins to prosper in her trade like all devotees of Uhamiri. Efuru adopts all the children of the clan and arranges story-telling sessions for them on moonlight nights. It is on one of such nights that the tale is told of the disobedient girl who went against tradition to choose her husband by herself and was punished by falling into the trap of a deceitful spirit-being. The narrative mirrors the life of millions of Nigerian Efurus, who are condemned for opting for personal sovereignty, even if they make mistakes in doing so. However, by the end of the novel, Efuru has arrived successfully at the end of her psychological voyage. She has achieved emancipation and is ready to divorce her second husband rather than sit there and wait one more time to be divorced. She also gives the indication that she may not get married again, a clear signal that Nwapa favors a mild rebellion in women if it will get them out of their quagmire.

The couple, Amarajeme and Ojiugo in *Idu*, prove that it is not always the woman's fault when there is infertility in the family. They also prove that infertility is not necessarily synonymous with the barrenness of a woman. After twelve years of pining away from not being able to conceive, and tired of covering up for her impotent husband, Ojiugo makes up her mind, without even mooting her plans to her good friend, Idu, to try to achieve conception from another man. She approaches a friend and age-mate of her husband, Obukodi, who already has eight wives to prove his virility. In no time at all, Ojiugo becomes pregnant, then she has to leave Amarajeme to join her new man. Amarajeme becomes inconsolable, refusing to eat or be happy, cooking a meal everyday in anticipation of Ojiugo's return. When Ojiugo's child arrives, Amarajeme can no longer stand the humiliation, because the whole community now knows that he is not really a man. He commits suicide by hanging himself. Idu expresses profound understanding of her friend's behavior, which would ordinarily be seen as adultery by saying: "About two years ago, people told her (that her husband was not really a man) but she did not believe it at first, but eventually she herself discovered. She wanted a child. Do you blame her when she went to the man who could give her one?" (*Idu*, 112).

Nwapa's approval of empowered women, women who circumvent tradition to prove themselves, is also seen in *Efuru*, through the author's treatment of Nwabata Nwosu, the woman who has so many children that she is forced to pawn them away for survival. Nwapa is, nonetheless, against parents who have more children than they can adequately take care of.

In *Idu*, the author disapproves of the couple, Ishiodu and Ogbenyanu, who have five children and not enough means to take care of the family. Both the man and woman are lazy and cannot survive without Idu and her husband. All Ogbenyanu does when she fights with her husband is trail her five children back to her mother who cannot support them either. There seems to be no happiness in bringing children into the world only to throw them on a relative, as Ishiodu and Ogbenyanu do on Adiewere and Idu (*Idu*, 70-71).

To dwell longer on Nwapa's approval of empowered women, which Ogbenyanu certainly is not, Nwabata, much like Efuru, goes through similar psychological growth and comes out the stronger for it. At the beginning of the novel, it is hard to predict the intelligence and initiative that she will later display. Nwabata has the intuition that robbers will visit their home to steal the loan of ten pounds received that day from Efuru. She hides away the money around her waist, where she believes the robbers will not suspect it to be. Their home is indeed burglarized, but unlike her husband who sleeps through the ordeal, Nwabata is awake. She outwits the robbers, deceiving them into believing that she too is asleep. In that way, she is able to save her neck. The robbers might have tried to harm her if they thought she was awake and capable of recognizing them the following morning. In a mocking voice, she later relates the incident to a neighbor, casting her husband as a good-for-nothing man:

> "Then Nwosu woke up and came out with his knife (after the thieves had gone). Yes, it was then that my lord and master came out with his knife. Kill me, I said to him. I am the thief. He fooled around and went and sat down outside the gate. That's the man who is my husband. Women are nothing. He, my husband, was asleep when thieves came to the house. But I am only a woman. What can a woman do?...In the end a woman does something and even then you still look down on women." (*Efuru*, 178 & 166)

The socialization process, the value system and the nature of economic opportunities open to women are some of the ways that society has put women in the subordinate position in which they find themselves. Nigerian women have no alternative but to choose from those alternatives already instituted by men for their governance. However, despite the firm grip of tradition on women, Flora Nwapa has consistently shown that when tradition is challenged it crumbles. She believes that unquestioning obedience to tradition is done out of habit. Women should, therefore, strive to cast off the yoke of tradition in order to free themselves. This is one of many reasons why motherhood, a universal issue for feminism (Andrade, 1990), is even more an issue for African feminism, where women's humanity hinges on it. The womb is that aspect of women's biology which, more than anything else, oppresses African women.

PROTEST AGAINST MOTHERHOOD

Women are unique human beings since they, not men, are capable of becoming pregnant and bearing children. Evidently, Nwapa is not against motherhood *per se*. Nevertheless, she disapproves of its limiting and defining women, who are then relegated to an inferior position. Women should feel free to engage in motherhood when it pleases them or withdraw from it anytime, through will or awareness. This implies that the self should be at the center and that personality and behavior should aim at satisfying the needs of the total self, including growth and maturity in love, self-worth, and fulfillment. This idea of maturation and autonomy, as manifested so abundantly in *Efuru*, implies that women should be allowed to follow their own system of values, without being subjected to values prescribed by society.

Yet, what we find is that marriage and procreation, the foci of existence especially for women, is the measure of a woman's humanity. A lot of time and energy is spent on getting married and having children. Nwapa protests "wife-selling" in the name of bride-prize or dowry: "Now a man who has four grown up daughters counts himself a very wealthy man. Each daughter will bring to the father at least a hundred pounds in cash, in raw cash" (*Efuru*, 191).

Nwapa introduces the concept of Androgyny (Treblicot, 1977; Elshtain, 1987; Braggin et al., 1977) to show traditional women's desire for cultural, economic, and intellectual emancipation, in deviation from traditional stipulations. Motherhood should open up the doors to women. But, when it does not, it should be cast aside so that women can proceed to realize their full potentials. The cult of Uhamiri, the Goddess of the Oguta lake, combines the interests of all infertile women oppressed by the tradition of motherhood to form an escape mechanism. The theme of the empowered woman who combines male aggressiveness in enterprise with her feminine nature, which is evident in the cult, shows amply that women can rise to lofty heights if they are given opportunity. As one medicine man puts it, the town of Oguta owes its development, in large part, to the devotees of Uhamiri: "Look round this town, nearly all the storey-buildings you find are built by women who one time or another have been worshippers of Uhamiri. All worshippers!"(*Efuru*, 153). In a later novel, *One Is Enough*, Nwapa elaborates on this theme by having a heroine who prospers in her occupation without feeling any form of emptiness due to the absence of a husband or a child in her life. Amaka, much like Efuru, chooses the life of a single woman after her husband, Obiora, takes a second wife because of infertility existing in the family.

By the same token, Efuru's defeatist attitude at the news that her husband wants to take a new wife meets with the author's disapproval. With irony the authorial voice mimics popular reaction:

> What is wrong in his marrying a second wife? It is
> only a bad woman who wants her husband all to herself. I don't object to his marrying a second wife, but
> I do object to being relegated to the background. I
> want to keep my position as the first wife, for it is
> my right....He is the lord and master, if he wants to
> marry her I cannot stop him. (*Efuru*, 53-55)

The fact that all Efuru fights for at this point in her relationship with her husband is the position of first wife, shows how low she has fallen. The full implications of Efuru's unnatural reaction to a deep insult on her dignity as a human being comes to light when the reader realizes that Efuru is only "her master's voice," merely parroting what she knows men would expect to hear from her. In almost the same words, her father advises her on community expectations of a woman in similar circumstances:

> Did he (Adizua) bring a woman into the house?....
> God forbid. Your life is not ruined my daughter. Our
> ancestors will not agree. It does not matter my
> daughter if Adizua wants to marry another woman.
> It is only a bad woman who wants to have a man all
> to herself....Meanwhile, go home and be a good
> mother to your daughter. (*Efuru*, 63)

Nwapa disapproves of the belief that women are all womb with no feelings or rationality (Christian, 1985). For this reason, she condemns Idu for begging her husband to choose another wife since the second, a mere sixteen-year old, ran away after trying unsuccessfully for months to find a place for herself in-between the happy marriage of Idu and her husband, Adiewere. Idu had been overly worried about villagers' gossip that she was the one stopping her husband from taking another wife. To her husband's question regarding her motivation for wanting him to marry again, Idu replies simply: "I only want you to marry another wife. I don't want to be called a bad woman any more. If you don't want to marry, then that is another matter, but I want you to marry. You never know what tomorrow will bring" (*Idu*, 90-91).

As long as women's primary role and responsibility continue to be wifehood and motherhood, men continue to take advantage of women's biology to oppress them. For the Nigeria of the 1930s, when disease killed many infants before the age of five, it is some-what understandable that motherhood was synonymous with human destiny. It was needed to perpetuate the human species and provide human hands for the farms, the primary sources of subsistence. Still, Nwapa seems to be saying that women should recognize their vulnerable situation and be ready to take care of themselves. Such indoctrinations as the following, neatly couched in traditional aphorisms, should be discouraged: "What we are all praying for is children. What else do we want if we have chil-dren?"(*Idu*, 150). If it is all right for women to give birth to chil-dren, they should also be equally prepared to give birth to ideas, to change that brings progress in society.

WOMEN AGAINST WOMEN

Perhaps the worst part is that women effect and perpetuate their own subservience and dehumanization. Part of the lesson of *Efuru* and *Idu* is that women should learn to eschew self-hatred and envy, in order to gain social, political, and economic reforms that would be of benefit to women. A culture of cooperation in utter sinceri-ty will help women achieve solidarity to overcome their oppres-sion.

We note particularly the insincerity of a so-called friend of Efuru who tries, in vain, to dissuade the heroine from taking advan-tage of child-care available to her, so that she can return early to her trade: "What is money? Can a bag of money go for an errand for you? Can a bag of money look after you in your old age? Can a bag of money mourn you when you are dead? A child is more valuable than money, so our fathers said" (*Efuru*, 37). By clinching her rhetoric with a proverb, this woman tries to lend authenticity to her deceitful advice. She wants to give the impression that her advice is a long-established folk-wisdom handed down by the ances-tors. She goes further to narrate the woes that many a maid has caused her mistress. One child that she knows ended up with blind-ness in one eye. Another child received a burnt face. But Efuru sees through the whole pretentious plan. She goes ahead, in spite of her friend, to bring Ogea in to help with child-care and house-hold chores.

Women are also in front in many attempts to displace infertile wives from their homes. It does not matter that the union of the couple is happy without children, no one seems to be able to escape societal prescriptions. Despite the birth of Ijeoma by Idu, the marriage between her and her husband is not seen as happy in their community, even though it is, all because a second child has not arrived. The couple is pressurized into bringing in a second wife. Onyemuru, as expected, leads the way in the attempt to get another wife for Adiewere, given that Idu is taking too long to get pregnant: "If Idu can't have a child, let her allow her husband to marry another wife, that's what our people do. There are many girls around"(*Idu*, 33). She then proceeds to enumerate examples of women like Uberife and Nwoji of Umuenu village, who did not stand in their husband's way to have children by other women in subsequent marriages. She emphasizes that nothing should come between a man and having children, not even love between the man and his wife. Later on, Adiewere's second wife absconds because Adiewere is just not respecting the conjugal arrangement whereby he is supposed to spend a minimum number of nights with each spouse, out of duty. He still prefers to spend every night with Idu. Following continued harassment from all corners to take yet another wife, Adiewere becomes totally depressed. He takes to drinking and keeping late nights. He contracts a terrible and, apparently, incurable disease and suddenly dies. Idu is heartbroken.

A month later, Idu is thoroughly disgusted by her brother-in-law's, Ishiodu's, overtures of marriage. She confides in Nwasobi: "Ishiodu came to put the thread round my neck this morning. Did you hear? I just looked at him, he put the thread on and left. He was not sure of himself. He is the only brother of Adiewere. He should marry me," and she laughed again, "That's what it is. Ogbenyanu and I will marry Ishiodu. Did you hear what I am saying?"(216). Idu concludes her narration by adding that she will not live to allow it to happen. She later wills herself to die so that she can join her husband. Before then, she prepares herself as one departing on a long journey. For the first time since her husband's death one month previously, she has good food prepared for her by her sister, Anamadi. She eats very well after cleansing herself as one indulging in some kind of ritual. She first feeds the ancestors and then herself. Then she has the floor swept clean. She spreads her mat on it, and then goes in to sleep the eternal sleep. In this way, this exemplary couple proves to their community that there is such a thing as soul love which can exist between two married people.

In the case of Efuru and her husband, Gilbert, backbiters go early to work:

"The important thing is that nothing has happened
since the happy marriage. We are not going to eat
happy marriage. Marriage must be fruitful. Of what
use is it if it is not fruitful? Of what use is it if your
husband licks your body, worships you and buys
everything in the market for you and you are not
productive?" (*Efuru*, 137).

Even wives spearhead the search for their replacement, in order
to enlist themselves in the good books of their husbands. The rationale of this act, according to another friend of Efuru's, is that he
will at least know that you (Efuru) want him to marry another wife
and have children"(*Efuru*, 164). The society, as well as the woman,
conclude from the beginning, without any scientific proof, that the
infertility has to be the woman's fault.

Women do not hesitate to blame other women for the faults of
the men close to them. It is in this way that Efuru chides Nwabata
for her and her husband's lapse in not showing gratitude early for
the farm loan she granted them: "And Nwabata, you have not done
well. You have the greater blame. You should correct your husband
for what do they know, after all" (*Efuru*, 168-169). When it is con-
venient, women are willing to accept that the same men who prescribe their every move in social, economic, and political matters
are ignorant beings, and irresponsible for their actions. The few
times men seem to be blamed for something it is, for example, for
not being in a hurry to take a new wife: "Foolish man. He (Enebiri)
sits down there and refuses to do anything. He doesn't see young
girls all over the place to take one as a wife"(*Efuru*, 175). He is further blamed for "letting a woman rule him"(*Efuru*, 175). Infertility
is mistaken for barrenness and Efuru is also blamed for "not allowing her husband to marry another wife when she is barren"(*Efuru*,
175). Efuru's mother-in-law receives her share of the blame for
complacency: "Did Efuru give you medicine that you have lost your
senses? You see your only son married to a woman who is barren,
this is the fourth year of the marriage, and you sit down and
hope?"(*Efuru*, 175). Other women advise her, using a very apt
proverb, to "begin early to look for a black goat," for at night it may
be difficult to find (*Efuru*, 139). Not only do women blame other
women for men's wrongdoing, individual women find a way to
blame themselves as well for their husband's crimes. In *Idu*, when
Amarajeme kills himself by hanging from the thatched roof of his
hut, Ojiugo is the first to convict herself of the crime: "I have killed

him. I am a murderer. I have killed him. He hanged himself because I left him"(*Idu*, 144). None of her friends could convince her to understand that Amarajeme hated the man he was and that was why he committed suicide.

Even when pregnancy occurs early in marriage (which should be a prophecy for happy days of conjugal living), as was the case with Nkoyeni, the second wife of Enebiri, women are still no less hurtful to other women. Omirima, the village gossip, not the least impressed, dismissed nonchalantly the news of Nkoyeni's pregnancy: "So what? Is she the only woman pregnant this season? It is expected that she should have a baby before the year is out. If no baby comes in the first year, it is our duty to probe her girlhood, and find out why"(*Efuru*, 197).

Then, again, women lead the way in showing preference for boy children as opposed to girls: "What is annoying," says Ajanupu, "is when some women have about six children and all of them are girls. What one will do with six girls I don't know"(*Efuru*, 184). Omirima advises Enebiri's mother on the issue of his son's illegitimate son: "But don't you think you should persuade your son to marry the mother of his son? A woman who gave birth to such a boy should be married. You don't know tomorrow. Nkoyeni won't be barren of course—she is pregnant already. But nobody knows whether she is going to have a girl or boy. She might take the footsteps of her mother who had four girls and a boy"(*Efuru*, 196). Efuru reportedly welcomed the boy with open arms. She was annoyed with Enebiri, not for keeping the existence of his illegitimate son a secret from her, but rather for not disclosing his existence until the week of his arrival. For that Efuru was seen as a good woman and wife. On the other hand, Nkoyeni, the educated one, who threatens to leave if the illegitimate child, a boy that he be, is not removed from their home and returned to his mother, is seen as a bad woman. Other women, who do not see eye to eye with her on many issues, condemn her action: "Why did you not let her go?" they ask her mother-in-law. "There are so many girls in town who will gladly be your son's wife. This is nothing but weakness" (*Efuru*, 197). They blame Nkoyeni's conduct on her foreign education, even going as far as to insinuate that she threw out the child due to a personal disappointment: "She thought she was going to have the first son for Enebiri. Serves her right"(*Efuru*, 196).

Indeed, it is a difficult situation for women. It is either that men are oppressing them with stringent rules, laws, and taboos, or they themselves are condoning men's acts by effecting and perpetuat-

ing them. Perhaps the worst tragedy lies in their apparent ignorance about the adverse repercussions to women of their tradition-
given role as custodians. Nwapa seems to infer that the new
Nigerian nation cannot advance to great heights with its women
relegated to the background. Not only will progress be hampered
with things being done only from men's perspectives, women will
be denied their natural right to advance to their fullest potentials
as human beings for a long time.

Nwapa's contribution to African feminism, through the writing
of *Efuru* and *Idu* certainly paved the way for advanced feminism
in the country. She encouraged women to think, read, and write
like women. She strove to ennoble and liberate women's latent
potentialities. In Nigeria, she saw the need to break long-prevail-
ing traditions, beliefs, and practices requiring women to have
wealth and happiness vicariously through their husbands. Her
hope for the future is that women should have a rejuvenated sense
of self-worth as human beings. On behalf of Efuru, who was sick
and weak at the time and, indeed, all the oppressed women of the
country, Ajanupu fought back, striking Enebiri when she was
struck. Gloatingly, Nwapa reports the incident: "Gilbert gave
Ajanupu a slap which made her fall. She got up quickly for she was
a strong woman, got hold of a mortar pestle and broke it on Gilbert's
head. Blood filled Gilbert's eyes" (*Efuru*, 217). The pestle for pound-
ing millet and corn has always been one of the symbols of African
womanhood and, on occasion, a weapon of war. The novel ends
with Efuru thanking Ajanupu for putting up a brave battle on her
behalf. For the third time, Nwapa achieves creation of women's
freedom. She portrays Efuru at peace with the world. She has left
her husband. She is dreaming of her mentor, Uhamiri. She has realized that she alone can make and unmake her own happiness. The
same goes for Idu who achieves her wishes through permanent
withdrawal from society. In doing so, she leaves us to discuss the
issue of feminism, rather than her refusal to become an inherited
widow, and what is worse, of a worthless brother-in-law. Evidently,
Nwapa is suggesting here that the way for women to earn equal
status and humane treatment in society is to challenge tradition,
age-old attitudes, and habits that keep them down. The focus, however, should not be on rivalry between the sexes, but rather on societal reform.

Nwapa's feminist criticism does not close with *Efuru* and *Idu*.
Her deep concern for women's progress continues in her later writings, most of which deal with women's issues. What we have done

here is explore the author's contribution as the pioneer of Nigerian feminist criticism under colonial Nigeria. One only has to be exposed to the works of Nwapa's followers to appreciate the quality of the plants that have grown out of the seeds that she, Nwapa, sowed. Such new Nigerian women writers include the poets Mabel Segun and Omolara Johnson; novelists like Ifeoma Okoye, Adaora Lily Ulasi, Rose Umelo, Buchi Emecheta whose novel with an ironic title, *The Joys of Motherhood*, developed out of the last line of *Efuru*; literary critic Omolara Ogundipe-Leslie, who is proving that women, too, much like men, are capable of making analysis and judgment with objectivity; historians Bolanle Awe and Nina Mba, who have spoken extensively on issues concerning women in education, politics, and history. A look at our bibliography will further confirm the contributions of sociologists like Chikwenye Okonjo Ogunyemi, Ifi Amadiume, and Filomena Chioma Steady.

Nwapa's findings seem to indicate that, in one sense, women of Nigeria have been handicapped because of the limiting effects of patriarchy, tradition, and motherhood. In another sense, the same women, perhaps unwittingly, have contributed to their situation by fostering traditions and attitudes which are inimical to their development and humanity. Progress in society today seems to demand that both factors be taken into account, and that women be reinstated in private and in public spheres as responsible and fully functional humans.

NOTES

1. Grace Ogot had published earlier, but in Luo. *The Promised Land* followed in English shortly after Nwapa's first publication. Grace Ogot, *The Promised Land* (Nairobi: East African Publishing House, 1966.)

2. Florence Stratton, in *Contemporary African Literature and The Politics of Gender*, states: "For, as feminist scholars in various fields have argued, colonialism is not neutral as to gender. Rather it is a patriarchal order, sexist as well as racist in its ideology and practices. What these studies indicate is that women's position relative to men deteriorated under colonialism. They also show that, while pre-colonial women had more freedom than their colonized descendants, male domination was nonetheless an integral part of the societies they lived in. Under colonialism, then, African women were subject to interlocking forms of oppression: to the racism of colonialism and to indigenous and foreign structures of male domination." (7)

3. Judith Van Allen, "Aba Riots or Women's War?" Nancy J. Hafkin and Edna G. Bay (eds.), *Women in Africa: Studies in Social and Economic Change*. Stanford Univ. Press, 1976:59-85.

4. Cheryl Johnson-Odim, "On Behalf of Women and the Nation: Funmilayo

Ransome-Kuti and the Struggles for Nigerian Independence and Women's Equality," in Cheryl Johnson-Odim and Margaret Strobel, (eds.). *Expanding The Boundaries of Women's History: Essays on Women in The Third World*, 144-157. See also Cheryl Johnson-Odim, "Common Themes, Different Contexts: Third World Women and Feminism," Mohanty, Russo, and Torres.(eds.) *Third World Women and The Politics of Feminism*, 314-327.

5. Ifi Amadiume, *Male Daughters, Female Husbands*, 36.

6. Nancy J. Hafkin and Edna G. Bay, *op. cit.*, 77-81.

7. See Buchi Emecheta, "Feminism with a Small 'f'!" *Criticism and Ideology*. Second African Writers Conference, Scandinavian Institute of African Studies, 1988: "Being a woman, and African born, I see things through an African woman's eyes....I did not know that by doing so I was going to be called a feminist. But if I am now a feminist then I am an African feminist with a small f....I write about women who try very hard to hold their family together until it becomes absolutely impossible." (175) "I find myself disagreeing on everything suggested by white women, even though I know that some of these suggestions could be quite relevant. I think that like the black boys in the school I taught, one simply becomes fed up with seeing oneself as a problem. I got up and shocked all those ladies, telling them to mind their own business and leave us Third World women alone. One could have heard a pin drop. I thought at one time I would be thrown out." (190) See also Ama Ata Aidoo, "To Be An African Woman-Writer—An Overview and A Detail," *Criticism and Ideology*: "Any time it is suggested that somehow one is important we hear that feminism is something that has been imported into Africa to ruin nice relationships between African women and African men. To try to remind ourselves and our brothers and lovers and husbands and colleagues that we also exist should not be taken as something foreign, as something bad. African women struggling both on behalf of themselves and on behalf of the wider community is very much a part of our heritage. It is not new and I really refuse to be told that I am learning feminism from abroad." (183)

8. Mohanty, Russo, and Torres, *Third World Women and The Politics of Feminism*, 315-316. See also Susan Gardner, "Culture Clashes," *Women's Review of Books*, XII.2 (November 1994):23. See also Alison Perry, "Meeting Flora Nwapa," *West Africa*, June 18, 1984.

9. Alice Walker, *In Search of Our Mother's Gardens*, (New York: Harcourt, Brace, Jovanovich, 1983):XI.

10. Interview with Gay Wilentz at Enugu, Nigeria, *Women's Review of Books*, XI. 6 (March 1994):8.

11. Juliet Mitchelle, "Women and Equality," 24. See also Sekou Toure, *Toward Full Re-Africanization* (Libreville, 1959):72.

12. Susan Gardner, "The World of Flora Nwapa," *Women's Review of Book*, XI. 6 (March 1994):9.

13. Gay Wilentz, *Binding Cultures: Black Women Writers in Africa and The Diaspora*, Indiana Univ. Press, 1992. See also Chinweizu, Jemie, and Madubuike, *Towards The Decolonization of African Literature* (Washington

D.C., Howard University , 1983):57.

14. Florence Stratton, *Contemporary African Literature and The Politics of Gender*, 102.
15. See Lloyd W. Brown, *Women Writers of Black Africa* (Westport, Conn, Greenwood, 1991):137-141. See also Eustace Palmer, "Elechi Amadi's *The Concubine* and Flora Nwapa's *Efuru*," in *African Literature Today*, 1 (1968):56-58.
16. See Bernth Lindfors, "Achebe's Followers," *Review de Litterature Comparee*, 48, (1974):569-578. See also Florence Stratton, op. cit., Ch. IV, 93 & 80-107.
17. Virginia Woolf, *A Room of One's Own*.
18. Raaman Selden and Peter Widdowson, *A Reader's Guide to Contemporary Literary Theory*, 3rd. (University Press of Kentucky):147.
19. Susan Gardner, "The World of Flora Nwapa," *The Women's Review of Books*, XI.6 (March 1994):10. "If you were to imagine a society without any restraints on men, that is the Nigerian society" (quotation of a journalist colleague of hers).
20. Juliet Mitchelle, *op. cit.*, 27. "In 1877, the first Trade Union Congress upheld the tradition that a woman's place was in the home whilst man's duty was to protect and provide for her." See also Ifi Amadiume, *op. cit.*, 136. See also Carol Boyce Davies, "Feminist Consciousness and African Literary Criticism": "The selection of males for formal education was fostered by the colonial institutions which made specific choices in educating male and female. Then, too, the sex role distinction common to many African societies supported the notion that western education was a barrier to a woman's role as wife and mother and an impediment to her success in these traditional modes of acquiring status....With few exceptions, girls were kept away from formal and especially higher education. The colonial administrations were therefore willing accomplices because they imported a view of the world in which women were of secondary importance." (2)
21. John S. Mbiti, *African Religions and Philosophy* (London, Heinemann, 1990), 22, passim. See also Efua Dorkenoo and Scilla Elworthy, "Female Genital Mutilation," Miranda Davies (ed.), *Women and Violence*, (London and New Jersey, Zed, 1994):137-148. See also Soraya Mire, "A Wrongful Rite," *Essence*, June 1994, 42, and "Her Say: Band Together to End a Painful Tradition," *Womanews*, Section 6, *Chicago Tribune*, Sunday, July 18, 1993, 10.

WORKS CITED

Achebe, Chinua. *Morning Yet on Creation Day*. London: Heinemann, 1975.
Amadiume, Ifi. *Male Daughters, Female Husbands: Gender and Sex in An African Society*. London: Zed, 1987.
Andrade, Susan "Rewriting History, Motherhood, and Rebellion: Naming An African Woman's Literary Tradition," *Research in African Literatures* 21.1(1990):91-110.
Amoros, Celia and Beth Buckingham. "The Matriarchal Myth," *The Literary*

Review 36.3(Spring 1993):415.

Awe, Bolanle. *Nigerian Women In Historical Perspectives.* Lagos/Ibadan: Sankore/Bookcraft, 1992.

Azodo, Ada Uzoamaka. "The Menage of Spirits and Humans in Africa:An Inquiry into Its Origins, Relevance, and Survival in Society." *Spirit Possession in Social Spaces,* ed. Ute Luig and Heike Behrend. Chicago: Chicago Univ. Press, 1995.

Brown, Lloyd W. *Women Writers in Black Africa.* Westport, Conn.: Greenwood, 1981.

Christian, Barbara. *Black Feminist Criticism.* New York: Pergamon, 1985.

Conrad, Joseph. *The Heart of Darkness.* New York: New American Library, 1983.

Davies, Carole Boyce. "Feminist Consciousness and African Literary Criticism." *Ngambika: Studies of Women in African Literature,* ed. Carole Boyce and Anne Adams Graves. Trenton, New Jersey: Africa World, 1986.

Drewal, Henri John. *Journal of Folklore Research* 25:101-134.

——. *African Arts.* 21.2 (1986b):38-45.

Duerden, Dennis and Cosmo Pieterse, eds. *African Writers Talking,* London: Heinemann, 1975.

Elam, Diane. *Feminism and Deconstruction.* New York: Routledge, 1994.

Elshtain, Jean Bethke. "Against Androgyny." *Feminism and Equality,* ed. Anne Phillips. New York: New York University, 1987.

Emecheta, Buchi. "Feminism with a Small 'f'!" *Criticism and Ideology.* Second African Writers Conference, Stockholm 1986. ed. Kirsten Holst Peterson. Uppsala, Sweden: Scandinavian Institute of African Studies, 1988: 173-85.

Emenyonu, Ernest N. "Who Does Flora Nwapa Write For?" *African Literature Today* 7(1975):28-33.

Ferguson, Ann. "Androgyny as an Ideal for Human Development." *Feminism and Philosophy.* Ed. Mary Vellering, Frederick A. Elliston and Jane English. Totowa, NJ: Little Field, Adam, 1977:45-69.

Gavison, Ruth. "Feminism and the Public/Private Distinction." *Stanford Law Review* 45.1(November 1994):1-45.

Johnson-Odim, Cheryl and Margaret Strobel. *Expanding the Boundaries of Women's History: Essays on Women in the Third World.* Bloomington: Indiana UP, 1992.

Mitchelle, Juliet. "Women and Equality." *Feminism and Equality.* Ed. Anne Philips, New York: New York University, 1987.

Mba, Nina. E. *Nigerian Women Mobilized: 1900-1965.* Institute of International Studies. Berkeley: Univ. of California, 1982.

Mbiti, John S. *African Religions and Philosophy.* London: Heinemann, 1990.

Mohanty, Chandra Talpade, Ann Russo and Lourdes Torres, eds. *Third World Women and the Politics of Feminism.* Bloomington: Indiana UP, 1991.

Nandakumar, Prema. "An Image of African Womanhood: A Study of Flora Nwapa's *Efuru.*" *African Quarterly* 11 (1971):136-146.

Nwapa, Flora. *Efuru.* London: Heinemann, 1966.

——. *Idu.* London: Heinemann, 1970.

——. *One Is Enough.* Trenton, NJ: Africa World, 1992.

Ogunyemi, Chikwenye Okonjo. "Womanism: The Dynamics of the Contemporary Black Female Novel in English." *Signs: Journal of Women in Culture and Society* 11.1 (1985):63-80.

Pateman, Carole. "Feminist Critiques of the Public/Private Dichotomy." *Feminism and Equality*, ed. Anne Philips. New York: New York University, 1987.

Steady, Filomena Chioma, ed. *The Black Woman Cross-Culturally.* Cambridge: Schenkam, 1981.

Stratton, Florence. *Contemporary African Literature and The Politics of Gender.* New York: Routledge, 1994.

Terborg-Penn, Rosalyn, Sharon Harley and Andrea Benton Rushing, eds. *Women in Africa and the African Diaspora.* Washington, D.C.: Howard University, 1987.

Trebilcot, Joyce. "Two Forms of Androgynism." *Feminism and Philosophy.* Totowa, NJ: Little Field, Adam, 1977.

Tutuola, Amos. *The Palmwine Drinkard.* London: Faber and Faber, 1952.

Walker, Alice. *The Color Purple.* New York: Washington Square, 1983.

Woolf, Virginia. *A Room of One's Own.* New York, Harcourt, Brace and World, 1929.

Breaking Through: The Publishing Enterprise of Flora Nwapa[1]

Ezenwa-Ohaeto

The role of Flora Nwapa as a pioneer female novelist placed her on a prominent pedestal to engage in other literary activities. Thus, the female writers who later emerged in Nigeria and Africa could be said to have not only defined themselves according to her achievements, but also gained confidence from her engagement in those other literary activities. In particular, Nwapa, like her contemporary and inspirer Chinua Achebe,[2] became involved in the publishing aspect of the Nigerian book industry. This literary activity made it possible for Flora Nwapa to not only create one of the most distinctive enterprises associated with African literature but also highlight the feminine perspective in the vision that led to the establishment of Tana Press Limited and Flora Nwapa Books and Company. These publishing enterprises were set up against all the constraints related to gender, economic difficulties, and even the ravages of the Nigerian civil war.

Considering the numerous societal restrictions and economic tribulations that beset the publishing industry in Nigeria, Flora

Nwapa certainly broke through the stranglehold of those difficulties with commendable determination. That success is best illustrated through the words of the famous novelist Chinua Achebe at a conference in 1973, when he insightfully described a publisher as: "a primary go-between from writer to reader in what I see as a dynamic social artistic relationship. I see him even in the role of an evangelist—publishing, broadcasting the good news, traversing time and space which the original lone voice unaided could never have hoped to reach, making the inner world of the artist available to the wider world of his community by using one of man's most profoundly important technologies—printing"(43). Achebe acknowledges the role of the foreign publisher in such a necessary dynamic publishing situation in Nigeria when he adds that "we must give due recognition to the brisk move in recent years by some foreign publishers in recruiting and training Africans in various aspects of the publishing trade. Indeed, like some high-minded and wrong-headed colonialists, they have worked assiduously for their own overthrow"(45). The implication of Achebe's observation is that the publishing industry in Nigeria benefited from such interactions between the foreign and local participants.

However, the argument of Chinua Achebe is that the indigenous Nigerian publisher should possess several qualities necessary for the growth of the industry. He states:

> [The African publisher] would not be exactly the same sort of person in every country or region of Africa. Perhaps his greatest quality will be a liveliness of imagination which is able to seize upon the peculiar characteristics of a place and make of them a strength for his task. He should not attempt to recreate the patterns of distribution sales and promotion from some foreign model. He will learn what he needs to learn from others but will make his own way in the world. This does not mean he will lack method and organization. On the contrary, his bookkeeping should be so scrupulous that writers will not hesitate to place their manuscripts in his care, or book sellers to do business with him. Literary gossip travels very far and very fast and a publisher who is a shoddy businessman will soon find no worthwhile manuscripts coming to him. (45)

That remark concerning the honesty and integrity of indigenous publishers is certainly important for the establishment of a beneficial rapport between writers and publishers in Nigeria. Later events confirmed the relevance of Achebe's remarks as some Nigerian publishers became notorious for several unscrupulous acts in their relationships with writers.

All the same, Flora Nwapa, who had no plans of establishing a publishing house in 1973 when Achebe made his comments, was lucky to have come from a tradition that emphasized on determination, hard work, and business astuteness in commercial activities. Those three factors are necessary for such a trailblazing role as that of a female publisher in Nigeria. Nwapa acknowledges those factors in an interview she granted to *Quality Weekly* magazine in which she described the women in Ugwuta society:

> We have had strong women who started trading with UAC[3] which came to Ugwuta in 1889. It was the women who first started trading with the company. The men were going to farm. It was later on that the men came on. So in Ugwuta in the 50s, there were already female personalities to look up to. They were wealthy in their own ways. Already they were very formidable. These women we are talking about invariably got married. Some had children, some didn't. So, because I grew up in this kind of society, I took it for granted that women were quite formidable; that they had their own rights and privileges. (Adeleke, 1990:34)

The fact that Nwapa became aware of formidable and enterprising women quite early in her life, obviously helped to mold her attitude to the publishing business. In addition, some of the prevailing circumstances in her society as well as tragic events like the Nigerian civil war added to the cultivation of such character traits and vision that made her embark on such significant enterprises. In confirmation, Nwapa reveals in another interview:

> During the war, that is the Nigerian civil war, women saw themselves playing roles that they never thought they would play. They saw themselves across enemy lines, trying to trade, trying to feed their children and caring for their husbands. At the end of the war, you could not restrict them anymore. They started enjoying their economic

independence. So what they tolerated before the
war, they could no longer tolerate. (Umeh 1995:26)

Nevertheless, the intolerance that Flora Nwapa perceived was not
the type of intolerance that generated discord and altercations.
Rather it was the intolerance of restrictions on self-fulfillment and
on the accomplishments of individual objectives in the social, political and economic spheres of life.

Actually, the type of activities involving women in Biafra
revealed their ability to be innovative, responsible, and imaginative. Nwapa explains that "during the war, we were blockaded and
it was hard for us to find food and useful materials. So the women
would dress as "Yorubas[4] and go into the enemy villages to trade.
It was us who found food for the men and kept the family going.
We were the backbone of the war. And for some women this was
the start of highly successful careers in trading"(Cavendish 1992:
4). To Flora Nwapa, the war only reaffirmed what she had already
observed in her native Ugwuta, but it added to her courage and
determination to tackle the special difficulties of living in a post-
war Nigeria. Furthermore, her appointment to the Executive
Council of the East Central State Government of Nigeria by the
Ajie Ukpabi Asika administration provided an opportunity for
Nwapa to fulfill some of her socio-political objectives in the society. Although she had known Asika when she was at the University
College Ibadan in the fifties, it was clearly her personal qualities
as a writer and administrator that convinced the authorities to offer
her the appointment. Nwapa served as the Commissioner for
Health and Social Welfare and later as Commissioner for Lands,
Survey and Urban Development between 1970 and 1975.
Interestingly, while she was a Commissioner, Nwapa was able to
write and publish a collection of short stories, *This is Lagos and
Other Stories* and a war novel, *Never Again,* through Nwamife
Publishers in Enugu. Before the civil war she had published *Efuru*
in 1966 through Heinemann Publishers; the second novel *Idu* was
accepted and released after the civil war in 1971. Such achievements were commendable for Nwapa confesses: "my days as a
commissioner were really very hectic days indeed. It was not the
type of atmosphere in which you could put down an idea as soon
as it came to you before you lost it to the endless meetings upon
meetings" (Ogan 1985:4).

All the same, political instability cut short Nwapa's career in
government in 1975. There was a military coup which removed

General Yakubu Gowon as the Head of State, and all those people like Flora Nwapa serving in government in the various states of Nigeria like Flora Nwapa were relieved of their posts. In an essay, "Writers, Printers and Publishers," Nwapa states: "we who were commissioners in various states of the Federation lost our jobs. Overnight I became once more an ordinary housewife. I had thought that with a good library and a spacious and comfortable study, it would be ideal for me to go on writing." Nwapa adds that she "found it difficult to write at home or to combine wifely duties with writing. I became restless. I diagnosed what was wrong with me. I needed a job. But I was over-qualified for any job, having risen to membership of the Executive Council of the State"(16). Thus, Nwapa felt that it was inevitable that she had to become self-employed and with the experience she had acquired in Ugwuta, it was not difficult for her to establish a viable business. Naturally, her ideas revolved around the book industry and she decided that she "wanted to be a publisher." But Nwapa comments: "As I was setting up my office, a friend told me not to publish alone, but to own a printing press. I could do commercial printing while doing publishing. So, I set up two limited liability companies, Tana Press Limited in 1977 and later Flora Nwapa Books and Company"(16). Nwapa further revealed in that essay, ten years later, that those two companies "printed and published children's books as well as novels and short stories. We have published some women writers as well as a few male writers. We are not going to discriminate against the men. But we are partial to the women. This is the story of a woman who is a writer as well as a publisher"(16).

Anyway, the publication list of Flora Nwapa was influenced by her objective to publish relevant books as well as her conviction that women could contribute positively to the creation of such books. Part of her decision to publish children's books is related to her view indicated in an interview that it is important for Nigerian children to read about their cultural heritage. She states:

> We have our own myths, folktales and all. We want them to know about these things. This is the era of the television. Children no longer have the time to sit around the fireside and listen to stories. If the writer can capture all these things in books, then children will read them and know about their own culture. Many publishers are now trying to publish books for children. This is a very positive develop-

ment indeed because previously, publishers were more interested in textbooks. We must make our children understand that they have a heritage and we can only do that through books. (Olayebi 1986:40)

While these objectives were at the back of her mind, Nwapa could not write for awhile even after she had established the publishing company. So, she was pleasantly surprised when her first book turned out to be a children's book. The inspiration for that book was dramatic, as she tells Amma Ogan: "It was a kind of dramatic thing. I had gone home to spend some time with Michael Crowder; I took him to Ugwuta Lake. We took a canoe; we went round. I was showing him the landmarks of the lake. Things like the Woman of the Lake and what we used the lake for when I was a child. Then I came back and wrote this Mammy Water story" (Ogan 1985:4). That story thus became one of the first books to be published by Tana Press.

By 1981, Tana Press had published about eight books and among them were *My Animal Colouring Book* and *The Miracle Kittens* written by Flora Nwapa. Tana Press also published *Eme Goes to School, The Adventures of Tulu, The Little Monkey* by Ifeoma Okoye and *How Plants Scattered* by Obiora Moneke. These books clearly added fresh dimensions to the number of story books available for Nigerian children. Nevertheless, the publishing and marketing of these books by Tana Press were not bereft of avoidable difficulties. Nwapa says in a 1993 interview:

> I am not very happy in terms of the marketing because in our set up today you have to lobby for the school system to accept your books despite the high level of its relevance or appropriateness. There are some of us who cannot lobby so we just sell our books. If you are interested, you buy them, but we do not have orders from the educational ministries. (Ezenwa-Ohaeto 1993:17)

This kind of encouragement from the educational ministries would have helped Tana Press to overcome its financial difficulties and recoup part of its financial investment. Nevertheless, that lack of support did not dampen the enthusiasm of the publisher, although Nwapa was aware of the unpredictable nature of publishing in Nigeria. In the essay on "Writers, Printers, and Publishers," Nwapa comments:

The publisher invests her money on the writer. Publishing is not a charitable organization. There is no sentiment about it. The publisher invests her money where she knows she is going to be rewarded financially. Publishing is capital intensive. It is not for those who wish to make money quick. I am saddened by university undergraduates who bring manuscripts to me and ask me how much money they were going to make, even before I accept the manuscript for publication. We have published about twenty children's books (in 1988) but out of these only two are selling fairly well. Not one has been accepted to be used in schools. Yet, money is not everything. A publisher must show the writer that she is fair. She must pay the writer whatever is due to her, as royalty and this must be prompt. The publisher must not give the writer the impression that the latter has been cheated of her royalty. (Nwapa 1988:16)

This comprehensive analysis, from all indications, illustrates Flora Nwapa's keen perception of the flaws and virtues of the publishing industry.

All the same, the relationship between the publisher and the writer is only one aspect of the problematic publishing scene in which Nwapa flourished. Perry in an interview with Nwapa comments that "Nwapa who wants to keep her publishing concern small" finds that the "three main publishers in Nigeria—Heinemann, Longman and Macmillan—which dominate the market, seem to resent her and feel threatened. She finds this difficult to understand and sees this as paranoia on the part of these publishers"(Perry 1984:1262). Alison Perry adds that what Nwapa "feels very strongly about is the lack of interest shown by the Nigerian government to indigenous publishers. She feels that she is left to struggle alone, alongside huge competition. If one of her titles, for example, was accepted in schools this would help her and encourage other Nigerians to set up their own publishing firms" (1262). This kind of support that would have helped Tana Press is confirmed by another Nigerian who has been involved in book publishing: Mokwugo Okoye. He insists that "publishing in Africa and much of the third world, one must concede, is a high-risk activity" and also that "the greater part of primary and secondary education is state controlled, which narrows the area of operation for most publishers to general and professional publishing." Okoye also stresses that "both these areas, however, require extensive state-sponsored support facilities—libraries, book shops, book clubs and trade schools—that are often woefully inadequate in developing countries"(Okoye 1982:4). It is not just the state, even individuals are reluctant to support local publishers like Tana Press. James Gibbs informs us that "Nwapa found to her dismay that many of her fellow Nigerians are distrustful of indigenous enterprises and their products. The demand for books published in Africa, which is heard at seminars, at symposia and at conferences of African writers and publishers is not heard, or not heard loudly it seems, in the offices of school principals and school proprietors"(Gibbs 1986:23). It is on record that Flora Nwapa challenged that unsavory reality through her publishing enterprise as well as through her insistent and quietly consistent manner of argument.

Another area which created difficulties for Tana Press is the difficulty in finding illustrators. Gibbs feels that Nwapa "apparently considered the rates charged by the more sensitive illustrators very high—but just what 'high' means in this context is open to debate. My own feeling is that illustrators, like authors, are generally underpaid"(Gibbs 1986:23). The importance of illustrators in the production of relevant and valuable childrens' books is well-known. Olajide Oshiga states that "the illustration presented for any publication has a character of its own, and ignorance of this character by the typographer or the layout artist can lead him to make mistakes of a technical kind or errors concerned with taste"(Oshiga 1982:20). Such mistakes do affect the children: Virginia Dike assures us that picture books in which illustrations are as important as the text perform dual functions: "they attract children to books by providing pleasure and visual excitement, and they stimulate the growth of perception and the development of visual literacy"(Dike 1981:15). This critic commends Flora Nwapa for bringing out coloring books and several picture books and also for relating an interesting story with illustrations by Obiora Udechukwu in *Mammywater*. But Virginia Dike adds that "in several books the physical quality seems to be comprised by financial constraints: the paper is opaque, the print is too heavy and bold for easy reading and the illustration lacks finish"(Dike 1981:17). Such technical flaws noted by Dike are part of the inadequacy of the machinery and not really the inadequacy of the publisher or illustrator. In spite of those flaws in some of the published picture books, there is no doubt that Flora Nwapa invited competent artists and made efforts to produce attractive books.

The high cost of production and the publication of Nwapa's personal works presented some technical problems for Tana Press. Nwapa acknowledged in an interview that "because of the high cost of production, publishers tend to publish those writers that are well-known—those that people will buy their books and they will get back their money." She explains further that when she "started [her] publishing business a ream of bond paper [was cheap]—not to talk about machines, spare parts, etc."(Afuba 1991:5). In effect, what Nwapa was saying in 1991 was that the cost of book production had affected the cost of books and that the publishing industry in Nigeria became threatened. However, because of her sense of industry and determination, Nwapa continued to publish books against such odds. In the process, she sought the advice of scholars, critics, and editors in order to make Tana Press publications attractive and relevant. It was in connection with that objective that Ifeoma Okoye, a Nigerian female novelist, advised her to give her personal manuscripts to some scholars and critics for editing.[5] It was advice that Nwapa accepted and a few posthumous publications, such as *The Lake Goddess,* as well as a play and a collection of short stories benefited.

It is clear that Flora Nwapa struggled to make Tana Press Limited and Flora Nwapa Books and Company relevant to the publishing and printing scene in Nigeria. She not only helped in the development of the book trade but also in the nurturing of talents like Ifeoma Okoye, whom she published early in her career.[6] As a woman who combined creative writing, motherhood, business acumen, government duties, and social work, Nwapa certainly made her presence felt in Nigeria. Through her publishing enterprise, she demonstrated to other female writers and entrepreneurs that it is possible for women to make such achievements in spite of the restrictions on her gender and the general socio-economic tribulations. Flora Nwapa may not have created a flawless publishing house or even flawless printing press but she contributed immensely to the development of the literary scene and the fulfillment of the educational requirements in her country. Thus, what her small publishing concern contributed to the Nigerian publishing industry was the demonstration that publishers can take certain risks—not just financial ones—in the interest of the society, which even the larger publishing concerns in Nigeria feel inhibited from taking on, in spite of their ambitions and wider responsibilities. In

effect, Tana Press was able to commence the publication of relevant children's books without considering the low level of demand
for such books then. Nwapa clearly utilized her limited resources
to the best of her ability while guided by the principle of fair play
and ethical standards in the multifaceted business of publishing.
For a woman and a writer who set out to create a small publishing
concern, the success of her enterprise exceeded expectations.

NOTES

1. The author is grateful to the Alexander Von Humboldt Foundation, Bonn
   and the University of Bayreuth, Germany.
2. Flora Nwapa acknowledges that Achebe not only gave her the encouragement to be a writer but also that he read, and edited her manuscript
   and sent it to his publishers in London. In addition, it was Achebe who
   suggested the title for that manuscript, *Efuru*. Such instances of help do
   not detract from the achievements of Flora Nwapa.
3. U.A.C. is the acronym of the United African Company, one of the early
   multi-national colonial companies that monopolized commercial activities in the Southern part of Nigeria.
4. The Yorubas are a major Nigerian ethnic community.
5. Ifeoma Okoye gave me this information in a conversation at Awka,
   Nigeria on August 24, 1995.
6. Ifeoma Okoye is at present a major award-winning Nigerian female writer. Her novels are *Behind the Clouds, Men Without Ears,* and *Chimere,* to
   name only three.

WORKS CITED

Achebe, Chinua. "Publishing in Africa: A Writer's View." In Edwin
Oluwasanmi, Eva Maclean and Hans Zell, eds, *Publishing in Africa in the
Seventies.* Ile-Ife: University of Ife, 1975:41-46.

Adeleke, Segun, Bose Adeogun and Ben Nwanne. "Quality Interview with
Flora Nwapa" *Quality Weekly* 6.8 (August 23, 1990):28- 35.

Afuba, Ifeanyi. "Conversation with Flora Nwapa" *The Outlook* (October, 19,
1991):5.

Cavendish, Luci. "Tract and Fiction: Interview with Flora Nwapa" *The
Guardian.* [London] (August 6, 1992):4.

Dike, Virginia. "Wanted: African Picture Books for Nigerian Children" *Pan
African Book World* 1.1 (August, 1981): 13-18.

Ezenwa-Ohaeto. "Interview with Flora Nwapa: April, 1993" *ALA Bulletin* 19.4
(Fall, 1993):12-17.

Gibbs, James. ed. *A Handbook for African Writers.* Oxford: Hans Zell, 1986.

Nwapa, Flora. "Writers, Printers and Publishers" *The Guardian* [Lagos] (August
17, 1988):16.

Ogan, Amma. "Flora Nwapa: The Stories of Women Come Naturally to Me"
The Guardian. [Lagos] (March 24, 1985):4.

Okoye, Mokwugo. "Books and National Development" *Pan-African Book World* 2.1 (February, 1982):4-5.

Olayebi, Bankole. "Conversation with Flora Nwapa" *The African Guardian* (September 18, 1986):40.

Oshiga, Olajide. "Design Concepts—Books Illustration" *Pan-African Book World* 2.1 (February, 1982):20-21.

Perry, Alison. "Meeting Flora Nwapa" *West Africa* No.3487 (June 18,1984):1262.

Umeh, Marie. "The Poetics of Economic Independence for Female Empowerment: An Interview with Flora Nwapa" *Research in African Literatures* 26.2 (Summer, 1995):22-29.

PART TWO
(RE)CASTING GENDER RELATIONS:
NIGERIAN WOMEN
IN THE COLONIAL AND
POST-COLONIAL EXPERIENCE

The works of Flora Nwapa lend themselves to considerations of gender relations, in spite of the fact that the tone in such works is neither strident nor shrill. Her quiet dignified style which is not in opposition to the current post-colonial consciousness, effectively encapsulates many of the issues that animate the fiction of women in most societies in Africa. The various critics whose seven essays are collected in this section grapple with Nwapa's (re)interpretations, (re)appraisals and (re)conceptualizations in a somewhat post-colonial context that situates the life of her fictive society.

Susan Arndt in the essay, "Applauding A 'Dangerous Luxury': Flora Nwapa's Womanist Re-Interpretation of The *Ifo* about The 'Handsome Stranger,'" suggests that Nwapa has adopted the *Ifo* about the girl who marries a man of her own choice in spite of parental objections. Arndt is convinced that *Ifo* have ambivalent meanings and therefore are to be decoded. It is the manner through which she decodes such *Ifo* in Nwapa's works that makes her ideas interesting as well as somewhat unsettling. Her thesis is persuasive when she argues that Nwapa subverts the notion of *Ifo* by applauding what the *Ifo* are trying to halt: women's strength, power and autonomy. However, *Ifo* possess depths that go beyond surface projections concerning the patriarchal elements, for some *Ifo* reinforce matriarchy while others either criticize or support patriarchy. Nevertheless, her study adds to the multi-dimensional possibilities for interpretation in Nwapa's works. This is a view that the other essays reaffirm.

Mary Modupe Kolawole, in "Space for The Subaltern: Flora Nwapa's Representation and Re-Presentation of Heroinism," insists that Nwapa re-presents women from a space of positive self-definition, revelation, rediscovery and relocation. This space she created is perceived as an expression of reality from the proximate setting of Igbo society as a paradigm. The heroines are symbols of social change and ideological voices that speak from the margins. They also transcend physical borders and individual predicaments in order to speak out against their oppressions. Kolawole adds that through their "dynamic actions they also show their resilience and refuse to accept the status of the muted voice of the woman as a subaltern." Similarly Ada Uzoamaka Azodo in her essay, "Nigerian Women in Search of Identity: Converging Feminism and Pragmatism," argues that Nwapa in her later writings advocates a synthesis of feminism and pragmatism. Azodo highlights some relevant contradictions in Nwapa's novels in terms of view points but she perceives them as qualities of flexibility. In effect, Azodo insists that her observations portray Nwapa as an experienced womanist and feminist as well as a pragmatist in the "face of overwhelming odds that women face daily in a complex society."

Undoubtedly, the multiplicity of Nwapa's narrative vision emerges through Emelia Oko's "Woman, the Self-Celebrating Heroine: The Novels of Flora Nwapa," where she insists that women through Nwapa's female characters embody specific qualities that are socially useful. Emelia Oko argues further that the novels indicate woman's self-celebration and the celebration of the agonies of loving. Her conclusion is that "the sustained position of Nwapa is that women are indeed different because of their genetic endowments. Therefore, they are faced with the challenge of maintaining their difference to redeem or change their world."

Oko's argument is convincing in many parts as focused on the latent qualities of Nwapa's works. It reinforces the feminine critical perspective that should be considered in order to present balanced views in the analysis of African Literature.

Adeline Apena's "Bearing the Burden of Change: The Colonial and Postcolonial Experience in Flora Nwapa's *Women Are Different*" contends that the challenges of modern women in Nigeria's colonial and postcolonial society demonstrate the contradictions of double patriarchal experiences of traditional and modern Nigeria. She argues that Nwapa's sense of pragmatism and determination to succeed is demonstrated in the women's strategies of adjustment, expressing the changing cultural environment. Apena stresses that while these strategies enabled modern women to gain economic power and social status, they did not necessarily ensure emotional stability.

Settling such issues requires multi-perspective insights which the views of the critics in this section reflect. In a very long essay, Osonye Tess Onwueme discusses "Shifting Paradigms of Profit and Loss: Men in Nwapa's Fiction" and attempts a projection of a possible interpretation of that fiction. Onwueme delineates how Nwapa portrays that while the images of men become diminutive as they go through experiences of tragic losses in trade, power and position, that the images of women become magnified as they celebrate "their new profit in trade, power and position." More importantly, Onwueme sees a double-entendre in Nwapa's fiction as she insists that the author also criticizes women who exploit and (ab)use the power of mothering/motherhood. Thus, she makes the observation that while Nwapa applauds the women for rediscovering, reaffirming and recentering themselves, she also undercuts, ridicules, satirizes and admonishes instances of reckless (ab)use and celebration of those powers by women. The unravelling of the layers of significations through the activities of the major characters signify that Onwueme has interestingly decoded several aspects of Nwapa's fiction but from a slightly different although equally insightful perspective similar to Chimalum Nwankwo.

Another part of that balanced view is what Chimalum Nwankwo projects in his essay, "*The Lake Goddess:* The Roots of Nwapa's Word," written with justifiable imitation at the misinterpretations, misapplications of theory and misinformation that currently stultify modern African Literature. Using Flora Nwapa's *The Lake Goddess,* he discusses narration, characters, naming, oral traditions and patterns of negotiation. The discussion of patterns of negotiation enables Nwankwo to argue that in a culture with a vatic capacity for marital tolerance which is part of the pragmatics of its policy, that it is not sensible to perceive altercations as familial adversarialism. Furthermore, this critic insists that in no work of Nwapa is she as politically acute as in *The Lake Goddess,* yet "the female characters who vehicle that politics conduct themselves with a cultural dignity which does not vitiate or repudiate male roles and spaces." The conclusion that Nwapa in *The Lake Goddess* rehabilitates certain political and spiritual principles which lacked eloquent articulation in other previous writings certainly reiterates the view that the roots of all her previous thoughts could be found in the novel. Chimalum Nwankwo accords recognition to Nwapa's themes and technique although he hints that the artistry is another "matter for all her critics to settle."

The significant ways that these seven critics emphasize the gender consciousness of Nwapa in her novels as well as the postcolonial context of her works exhibit major illuminations of the critical perspectives that have emerged in the criticism of this novelist. The readings may be somewhat contradictory but they are still readings motivated by the desire to delve into the inner recesses of meaning in Nwapa's novels and short fiction.

Ezenwa-Ohaeto

Applauding A "Dangerous Luxury": Flora Nwapa's Womanist Re-Interpretation of The *Ifo* About The "Handsome Stranger"

Susan Arndt

The Nigerian literary critic Obioma Nnaemeka reproaches African women writers for abandoning the strong women of African oral tales:

> If modern African women writers claim to be descendants of great female story-tellers in the oral tradition (in which, strong, radical women and symbols of womanhood are visibly recognized), why is there such a dearth of similar materials in the written literature? How can one explain the pervasive marginality and subalternity of radical women characters who symbolize change? Why do female pro-

tagonists consistently remain what I would call 'characters of reaffirmation' in the sense that they reaffirm the commonly accepted notions of women and women's 'reality.' (143)

In arguing that African oral tales are full of strong and radical women, Nnaemeka is right. But she neglects to add that these narratives would be unthinkable without passive and voiceless women figures who affirm official conceptions of womanhood. It is also true that many protagonists of contemporary African women's literature meet conventional social norms, while rebellious women, symbolizing change, are marginalized in these texts. The conclusions Nnaemeka draws from these observations, however, do justice neither to the nature of African oral narratives nor to the character of African womanist literature. This is due to the fact that she disregards the question of how these women figures are assessed in both types of literatures.

The powerful and autonomous women of African oral tales, who act on behalf of individual demands and wishes, are associated with maliciousness. Their activities are viewed with suspicion and result in disastrous consequences. Thus, women who want to achieve self-realization beyond motherhood and wifehood are ostracized. These women can expect only a woeful life in social isolation. The selfless and silent women who remain passive (or, at most, act according to the interest of their fathers, husbands, and patriarchal Igbo society) are (in contrast to self-empowered women) idealized. Their behavior is affirmed, praised, and even rewarded in the oral stories. This dichotomous appraisal stems from both the wishful thinking and fears concerning women in African patriarchal societies and those who are privileged by it—men.

Contemporary African women writers shape women figures from their own point of view. Their perspective, however, differs fundamentally from the one found in oral tales. Modern African women writers describe passive and selfless women with detachment and fear. They describe these women as victims who not only endure but also justify the discrimination leveled against them. Due to their submissive behavior, these figures are described by womanist authors as oppressed and tragic figures. The womanist authors sympathize instead with strong, self-determined, active, and autonomous women, who live up to their individual potential. This is evinced by the fact that such women find fulfillment and happiness in African womanist literature.

In short, the opposite of Nnaemeka's thesis proves to be true. Oral tales favor weak, voiceless, passive, and selfless women, who keep the gender-specific status quo of patriarchal societies alive. In contrast to this, the novels, dramas, and short stories of womanist authors emulate strong, self-confident, and autonomous women, who emancipate themselves from conventional notions of womanhood and pave the way for a new relationship between men and women. This oppositional—even—antagonistical— nature of both literatures' attitude to womanhood is illustrated here by comparing the handling of the motif of free choice of partners in *Ifo* (that is, the Igbo folktales[1]) and in Flora Nwapa's novel *Efuru*.

I. THE FREE CHOICE OF PARTNERS IS A "DANGEROUS LUXURY," OR THE *IFO* ABOUT THE DAUGHTER WHO INSISTS ON MARRYING A MAN OF HER CHOICE

Although mainly women and children come forward as narrators of *Ifo*, they are a "male authored" literature, in as far as *Ifo* both reflect and reproduce the official ethical and cultural values and norms of Igbo traditional patriarchal society. The conformist character of *Ifo* comes to light with their negotiation of the motif of free choice of a marriage partner. In traditional Igbo society, material and social interests of the families concerned dominate weddings. Arranged marriages are considered to be the norm(Uchendu 1965:50). Due to this notion, *Ifo* condemn any self-determined choice of a marriage partner as a "dangerous luxury."

One famous *Ifo* which exists in numerous versions is about a proud woman who rejects all of her suitors.[2] Thus, she not only acts against her father's will, but also violates the social norm of arranged marriages. Her motives are anything but noble—she longs for affluence and beauty:

> No man was good enough for her. This man was not tall enough; the other not handsome enough. This man was handsome, but not rich; that man was rich but not handsome. She rejected one suitor after another, and her parents became worried...."I must marry a very rich and handsome man," Adamma declared. "Anyone as beautiful as I am deserves a truly outstanding husband, so that we shall be admired by all." ("The Proud Girl and the Devil," Bordinat & Thomas, 88-89)

When finally a handsome stranger turns up, she immediately sets out to become his wife. Actually, her decision has nothing to do with love. Considering him a "complete gentleman," she falls for his supposed beauty and wealth. In order to have her way, she stubbornly counters both a parental prohibition against doing as she pleases and numerous warnings against marrying a stranger. In some versions of this *Ifo*, she even threatens to commit suicide. In the version called "Ogilisi Akwalaka" for instance, it is stressed:

> Her father, however, refused to give his consent, and her mother told her that she didn't believe that the suitor was a man; but the girl was firm. 'If you don't let me marry him,' she threatened, 'I'll hang myself.' (Ekwensi 1954:65)

After the bridewealth is paid, she finally leaves her parents' home with the handsome stranger. Her journey, however, ends not at the home of putative parents-in-law, but in a lonely, spooky place—in the forest, in the wilderness, or on a river, all places renowned for being the domain of evil spirits. This symbolizes the fact that from now on the young woman cannot count on the protection of her community. She has become an outsider. Only now does the woman realize her error. Her husband is a spirit, who had only temporarily taken on a human form. Now he reassumes his former, ghastly shape:

> He was very ugly, like all the other devils. He had one leg, one hand, one eye—and he was very short. He wore tattered clothes, and his skin was covered with craw-craw....He was a hideous devil...and ordered the girl to follow him, speaking in the harsh voice that devils speak. ("The Proud Girl and the Devil," Bordinat & Thomas, 89-90)

His frightening shape however gives but a mere inkling of the horrors that lie in store for her. The woman is maltreated and, as in most such *Ifo* she is finally killed by her husband.

It is exclusively the woman's pride, arrogance, opposition to an arranged marriage, autonomous decision-making, self-determination, and social disobedience that enable the spirit to get hold of her. The frightening spirit is therefore to be understood as a literary symbol of divine justice. In "A Proud Girl's Punishment," it is said:

> The spirits of the land...did not take the girl's uncouth behaviour lightly. They intervened on behalf of the humiliated bachelors. Particularly worried was the goddess of the sea. She sent one of her priests to go and beguile the proud girl and bring her to her senses. (Umeasiegbu 1977:94)

The fact that the handsome stranger "intervenes on behalf of the humiliated bachelors" and repeatedly suggests that the woman think better of her decision to marry him and return (that is to accept traditional marriage conventions) implies that he is really interested neither in marrying nor in harming her. His only concern is to teach her a lesson and to make an example of her in order to prevent other young women from suffering similar lapses. He stands for the idea that a woman who violates social norms by choosing a husband of her choice, and by claiming individual autonomy is bound to fall for a husband who will only bring her misfortune. Marriage is, in *Ifo*, understood to be the foundation stone in a woman's life. Thus, the protagonist's chastisement symbolizes not only an unhappy marriage but also a woeful life. This idea is stressed by the girl's death. It stands for the message that she will be isolated socially; that she no longer exists for her society. This consequence hints at the psychological tortures such a self-determined girl is likely to experience. As the woman has annoyed her family and community, there is nobody who is willing or able to intervene when she is maltreated by her husband. That such a woman is likely to suffer physically, too, is implied by the fact that the protagonist of the *Ifo* does not simply die off. She is rather brutally murdered by her husband. The painfulness of such a death comes to light in *The Beautiful Daughters*:

> Other spirits had seen the two beautiful things. They had begun to quarrel about sharing the girls when killed and cooked. Some said they wanted the breasts, others said they wanted the thighs and nothing else. One even wanted the ears. The girls realized that their lives were in danger and began to cry. But it was too late. The spirits pounced on them, killed them, and ate them. (Umeasiegbu 1977:96)

Nothing could be more frightening than the menace of a combination of both physical and psychological sufferings. Thus in "A Girl and A Python," it is moralized:

Sammi - o - oh
Samara
You were too choosy about a husband
Samara
You chose a python
Samara
And the python destroyed you
Samara. (Umeasiegbu 1977:102)

The stereotype of this disobedient and self-determined daughter stands for the fears of strong, outspoken, and independent women in traditional patriarchal Igbo society. It has its pendant in a woman figure, which represents the wishful thinking of a man's society. This stereotyped woman willingly accepts being married to a husband of her father's choosing.[3] While the father's motives and point of view are elaborated upon,[4] the young woman's thoughts and feelings remain unmentioned. She, who in some cases does not even enter the *Ifo*'s plot, is usually a silent and passive object. These attributes, which symbolize the women's social conformity, are not commented on and thus are affirmed by the *Ifo*.[5] That such behavior is worth striving for is also stressed in those *Ifo*, in which the once-rebellious daughter is able to escape from the marriage with the evil spirit. Turned into an obedient daughter, she goes back to her parents begging forgiveness and with the knowledge that disobedience always has fatal consequences. And so "the beautiful daughter had learnt her lesson. She married again, this time wisely, listening to her parents. She was also a good and humble wife"("The Proud Girl and The Devil," Bordinat & Thomas, 93) The way *Ifo* deal with daughters to be married indicates that *Ifo* offer their protagonists—and, hence, female listeners—merely the choice between happiness through obedience, passivity, and subservience on one hand, and suffering on the other, as a direct result of women's autonomy and self-determined activity.

II. The Quest for a Happy Married Life in Flora Nwapa's *Efuru*

Contemporary Igbo woman writers, however, have re-narrated the *Ifo* about the "handsome stranger," in order to question its notions of womanhood.[6] Even though the various re-interpretations do each have different meanings, what all of them have in common is the justification of women's free choice of partners and hence, symbolically, the claim of new social roles and possibilities for women. Thus, contemporary African women authors, to adopt Salman Rushdie's term, "write back" to *Ifo* (Ashcroft et al. 1989: 33). A discussion of Nwapa's re-writing of *Ifo* will exemplify this point.

Nwapa's first novel begins with the protagonist, Efuru, claiming that if she cannot marry Adizua, the man she loves, "she would drown herself in the lake"(7). As Adizua cannot pay the bridewealth, Efuru moves to Adizua's house in order to become his wife without either her father's agreement or knowledge. As is suggested by the proverb, quoted in the novel, "[H]owever good a suitor might be, he was never given a bride for nothing"(191). She therefore is flouting the marriage conventions of her society. Efuru and Adizua resemble the couple in the *Ifo*. Similar to the "handsome stranger" Adizua is handsome but a "nobody," who is—like his family—"not known"(7, 9, 18). Efuru, however, is beautiful, autonomous, and a descendent of a well-known family(7, 18-19, 23). Although, for the time being, her husband answers her love, the end of Efuru's self-chosen marriage has much in common with the protagonist's marriage in *Ifo*. Like the "complete gentleman" of *Ifo*, Adizua loses more and more of those attributes which made him lovable in Efuru's eyes. Exactly like the "handsome stranger," Adizua disappoints Efuru's trust in such a way that he becomes a stranger to her. He leaves Efuru and does not even attend the burial of their only daughter, thus causing Efuru to suffer. Just as the girl in the *Ifo* feels disgusted with the frightful spirit, Efuru finally loathes Adizua.

The plot in the first part of *Efuru* suggests that Nwapa has adopted the *Ifo* about the girl who marries a man of her own choice. Nwapa emphasizes the analogy between her novel and the *Ifo* about the "handsome stranger" by integrating a related *Ifo*.[7] This *Ifo* is about a girl who leaves the house against her mother's orders to pluck udara fruits. Her disobedience has fatal consequences: a frightening spirit decides to marry her, the girl's mother tries to save her daughter by offering him other young girls as a wife, but the spirit is not interested in any of the girls, only in the one who was disobedient. Thus the unpleasant husband embodies divine justice—a punishment for wanting autonomy. The girl's last hope are her sisters. Eke, Orie and Afo refuse to help, but Nkwo kills the evil spirit and thus rescues her rebellious sister.

Eustace Palmer is convinced that this *Ifo* has no reference to the novel and that—if Efuru went without this and other "unnecessary" expositions—"[t]he novel could quite conveniently have been half [its] length"(57).[8] By arguing this point he ignores the fact that in literature hardly anything happens without the intention of the author, and that integrated folktales are hardly ever without relevance to an African novel. They rather serve many intertextual functions.[9] Some of them are relevant for Nwapa's novel *Efuru*. Integrated folktales shape the atmosphere of the world depicted and thus contribute to its Africanization or—in Nwapa's case—Igboization. Furthermore, they may structure and enliven a novel's plot. The folktale in *Efuru* is told immediately after Efuru has left her first husband's home. Directly after the integrated *Ifo*, Efuru marries her second husband. Thus, with the *Ifo*, Nwapa offers her readers a diversity between the two major episodes of the plot. Moreover, she relates both marriages and the *Ifo* to each other and thus implies that more than a mere aesthetic interest is responsible for the integration of the *Ifo*. This leads to the folktale's importance as a means of supporting a literary message. Such integrated folktales may for one thing characterize a novel's figure, by stressing its peculiarities. As will be seen later, Efuru, other women figures, and the communal heroine in *Efuru*, are characterized by the tale's figure Nkwo. For another, there is often a direct relationship between both the *Ifo*'s and the novel's central idea. Either the integrated folktale harmonizes with and thus accentuates the literary message of the novel,[10] or the essence of both integrated folktale and novel is, as in *Efuru*,[11] diametrically opposed. In order to recognize this intertextual dialogue—whether supporting or opposing—the general means of understanding *Ifo* have to be learned.

Ifo have ambivalent meanings and therefore are to be decoded. There is a level on which the literal meaning is manifested. On this level the *Ifo* in Nwapa's novel argues that daughters who pluck udara fruits against their parents' wishes will be married to an ugly spirit. This obvious message hardly ever corresponds with the implied meaning of a folktale. What is literally said has to be interpreted in a wider social context in order to find what is to be understood as the actual message of the narration. For the integrated *Ifo*, this moral can be described as such: Daughters (or any women) who violate social norms by ignoring social prohibitions and by showing self-determination are punished with a life full of problems and suffering.

Ifo can also be used as an interpretatory key to the novel. By isolating the *Ifo* which is narrated within *Efuru* from its immediate context and by comparing it with the novel as a whole, its similarities with both Nwapa's novel and the adopted *Ifo* about the "handsome stranger" become striking. The three women figures

Susan Arndt

disobey social norms and experience a horrible marriage. Starting from the analogy between the handsome-stranger tale and *Efuru,* Gay Wilentz concludes:

> [Nwapa] is unresolved about the possibility of a successful marriage without the advice of the elders. The conflict between a Western-style approach to women's marital and family relations versus the security and stability of an African traditional system reflects a tension that runs throughout the novel. (Wilentz 1992:9)

Such an interpretation or, even more, a conclusion that the parallel between *Efuru* and the *Ifo* implies that Nwapa argues with the same tenor the *Ifo* do, doesn't do *Efuru* justice.

The difference between the integrated *Ifo* and *Efuru* emerges as soon as one compares the narrative perspective of the integrated *Ifo* with that of Nwapa's novel. While *Efuru* is narrated by Nwapa's mouthpieces (a chorus of women's voices) the *Ifo* is narrated by a man. By this shift in the narrator's perspective, Nwapa implies her critical detachment from the *Ifo.* She indicates that she, as a woman who writes in order to improve the situation of women in her society, has less in common with a man, who embodies a male authored literature standing for the reproduction and legitimization of the official patriarchal notion of both womanhood and the relationship between men and women. Hence, Nwapa primes her readers for the fact that related topics and motifs can be negotiated completely differently, even antagonistically, and that she is doing it in *Efuru.*

After her marriage has fallen apart, Efuru wonders: "If it has been..." she did not finish; "What is the use?" she asked (105). Efuru leaves the question not only uncompleted but also unanswered. However, the question can be answered by considering the intertextual closeness between Efuru's question, her remembrances of the beginning of her marriage with Adizua (105), and the performance of the *Ifo* that deals with a woman whose disobedience is punished with a terrible marriage (106-111). Efuru wonders whether the collapse of her marriage to Adizua was inevitable since she conducted herself like the disobedient daughters of the *Ifo.* Efuru herself answers this question when asked whether her marriage has fallen apart because she was too hasty in marrying Adizua:

"Well, maybe I was. But the success of a marriage
does not depend on that. Marriage is like picking a
parcel from numerous parcels. If you are lucky you
pick up a valuable one." (96)

With this Igbo proverb, Efuru denies the *Ifo*'s essence that the free
choice of a partner and a happy married life, or social disobedience and individual happiness are by necessity mutually exclusive. Realizing this critically informed intertextual dialogue, it is
difficult to ignore Nwapa's "writing back" to both the integrated *Ifo*
and the adapted *Ifo*. And it is this poetics which is the interpretatory key to Nwapa's literary message.

In order to challenge the *Ifo*'s moral, however, Nwapa does not
decide to let Efuru's self-determined love-marriage end happily.[12]
She approaches it from another side. She shows that the success
or failure of a marriage does not depend on the social suitability
of the wedding ceremony, but both on men's attitude towards
women and their behavior in marriage, and on society's attitude
to love. In this vein the proverb of the pile of parcels has at least
two implications for Nwapa's novel. They are embodied by Efuru's
two marriages.

Adizua's father married his wife "as a woman was married in
our day. He paid the dowry in full and performed all the customs
of our people"(59). Nevertheless, he leaves his wife all of a sudden
and causes her to suffer "as nobody has suffered before" (157, 159).
Adizua shows the same behavior towards his wife. A proverb stresses that Adizua's seemingly unmotivated abandonment of Efuru is
hereditary:

The son of a gorilla must dance like the father gorilla. Our elders were quite right when they said this.
Adizua is every inch like his father....(51) Adizua's
waywardness is in his blood and you can do nothing about it." (61)

Since Efuru's marriage falls apart due to her husband's wayward
and unfaithful character, which was assured by heritage, and she
meets the same fate as her mother-in-law, who married conven-
tionally, it becomes clear in Nwapa's novel that neither the non-
payment of bride price nor her father's disagreements are finally
responsible for the failure of Efuru's marriage. She just picked "the
wrong" parcel, which would have proved equally disastrous with
a different starting point.

Naturally, a rigid character like Adizua's cannot be considered a general cause for broken marriages. Normally, partnerships depend on more complex and flexible social interactions. A suggestion of a more universal reason for the failure of marriages is hidden behind the second implication of the proverb. It is exemplified by Efuru's second marriage with Eneberi. This wedding meets the social marriage conventions. Efuru "think[s] twice this time"(96) about it. The bridewealth is paid and her father confirms the union. As in the relationship to Adizua, Efuru finds love and happiness for the time being. This is symbolized in the novel when Efuru and Eneberi "went to the stream together, there they swam together, they came back together, and ate together"(137). This partnership measures up to Efuru's notions of marriage. It is, however, a social novelty (Jell-Bahlsen 1995:35). Thus Omirima, a representative mouthpiece of traditional conventions, asks:

> "Why must they go to these places together?... Are they companions? Don't they know that a man and a woman should not be seen together often whether they are married or not." (139)

As a social innovation in the relationship of husband and wife, Efuru's and Eneberi's intimacy breeds the envy and ill will of their kith and kin. This comes to light in a discussion among several villagers:

> "Husband and wife, they are swimming together," one woman began.
> "They come to the stream every day," another said.
> "Nonsense, why should they swim together? Are they the only happy couple in the town? I see them every time I come to the stream. It is disgusting. Can't anybody talk to them?"
> "They are simply showing off. I bet they are not as happy as they look. You give them two years, and we shall see what will happen." (137)

This jealous malevolence shows that Efuru's society is neither willing nor able to take a favorable view of her ideas of marriage. The villagers' reluctance stems from the fact that love does not find any social recognition. It is having children which decides whether a marriage is happy or not in the social context.

> Seeing them together is not the important thing...
> The important thing is that nothing has happened
> since the happy marriage. We are not going to eat
> happy marriage. Marriage must be fruitful. Of what
> use is it if it is not fruitful. Of what use is it if your
> husband licks your body, worships you and buys
> everything in the market for you and you are not
> productive? (137)

Because children have a higher social value than love and affection between husband and wife, there is no instance that can put a stop to Eneberi when he—similar to Adizua—denies Efuru increasing love, fairness, trust, and equality. He has a child with another woman without telling Efuru. And, which is even more grave for Efuru, he fails to attend her father's burial because he is imprisoned. At first he conceals that he was in jail. When Efuru finds it out herself, he does not tell her the reason for it. Eneberi's betrayal of Efuru's love reaches its climax, however, when Efuru falls seriously ill. Her disease is interpreted as a punishment from the goddess of the earth for adultery. It is argued that her only chance of survival is to confess her guilt. Although, in spite of all disappointments, Efuru nevertheless always showed Eneberi trust and solidarity, he now denies his wife any support. He does not defend her but asks her to admit her adultery. Efuru is deeply hurt. Eneberi's insinuation shows her that he is not the partner she has been looking for. Eneberi finally resembles the "handsome stranger" or the "complete gentleman" of the *Ifo*. He has both deceived and disappointed Efuru. Therefore, this marriage fails.

Efuru's relationship with Eneberi affirms the thesis, which was already suggested by Adizua's inherited behavior, that a conventional wedding is by no means a guarantee for the success of a marriage. It moreover illustrates a much more complex message, expressed by the proverbial metaphor of the pile of parcels. It implies that the choice of a man in the society in which Efuru picks up her husband is as restricted as the parcels in the heap. Regardless of the number of men from which Efuru can choose, the selection is nevertheless restricted as they are all from the same pile; that is, all men are products of the same social conditioning. In this sense the proverb that "the son of a gorilla must dance like the father gorilla" in Nwapa's novel not only argues that Adizua resembles his father in his incapability to cope with a marriage, but it also proves to be true. In a wider sense this proverb moreover stresses that all men who have been socialized in Efuru's society are—similar to both Adizua and Eneberi—alike in as far as they cannot cope with Efuru's expectations of marriage.

This dilemma is the real reason for the tragic ending of Efuru's free choice of a marriage partner. Moreover, she also points to the fact that Efuru, like any woman, has to assert herself and control her life. If the selection is so restricted, if one's chance of finding a happy (married) life is so limited, women should at least try to make the best of what is offered to them. Either they should be very fastidious about men and decide only for a husband that seems to be best suited to them. Or women should pluck up the courage to realize that, as Nwapa believes herself, "marriage is not the end of this world" and that "there are a hundred and one other things to make you happy apart from marriage and children"(James, 114, Umeh, 28). After Efuru has twice, but in vain, tried the first option, she finally decides to put an end to her life as a wife and woman in search of motherhood.

The second analogy with the integrated *Ifo* comes in here. The protagonist in Nwapa's *Ifo* is able to escape from her horrible marriage with the help of her sister Nkwo. Nkwo's support has a decisive metaphoric implication for the novel.[13] Nkwo is the name of an Igbo weekday. When Eneberi asks Efuru to visit her on Nkwo, Efuru herself points out the denotation of this day:

> "You talk as if you are a stranger in our town. How can a woman visit anybody on Nkwo day?...No woman in our town has time for any other thing except to buy and to sell on Nkwo day. After the market she goes to collect debts. Nkwo is not a day to make social calls, it is a day of business." (116)

This utterance identifies Nkwo as the market day of Ugwuta, the village which is the setting of Nwapa's novel and in whose context this *Ifo* is told. Moreover, it implies that the trade connected with this day is a domain of women. In this sense the *Ifo*'s figure Nkwo can be understood as a literary metaphor for market women, and the figures in Nwapa's novel as emblematic of the successful trading women of Ugwuta. Nkwo symbolizes characteristics which distinguish both Efuru herself and the novel's communal heroine: wealth, social recognition, power, and strength. In as far as this community of women is associated with co-operation and teamwork, Nkwo is also (and this is, in the novel's context, most important) to be understood as a metaphor for women's solidarity. In

Nwapa's second novel, *Idu,* this characteristic is stressed most explicitly.

> It was said that Nkwo was a woman of great kindness, who did not let her people starve....So on Nkwo days everybody assembled in the market. If one had no money, one went to Nkwo market; if one had money one was there also, so the saying went. (171)

Not only is the *Ifo* figure of Nkwo distinguished by this sense of solidarity. It is also a major attribute of Nwapa's main figures. Proverbs which are to be found in *Efuru* emphasize the significance of solidarity in their world: "It is true that a person who has people is better off than a person who has money"(131). This proverb implies that to experience solidarity is a crucial security in life. Another proverb argues that solidarity blesses the person who shows it with good luck, appreciation, and satisfaction in life: "*Ogworo azu ngwere eru ani*" (11) [translated into English as, "He who prepared fish for people has always a cheerful face."] Solidarity is an outstanding characteristic of Efuru. She lends money to the needy and pays for medical care for the poor. But Efuru herself experiences solidarity by other women. This solidarity reaches its climax when she is—similar to the *Ifo*'s protagonist—rescued from a fateful marriage by the solidarity of women associated with the market day. As marriage is of great importance in the life of a woman, this solidarity symbolizes essential help. When Eneberi leaves Efuru in the lurch after she is accused falsely, the trader Ajanupu shows understanding for Efuru's feelings and the trust Efuru would have expected from the man she loves. Ajanupu "said a few home truths to Eneberi" and beats him with a mortar—"the weapon of the women"(220). Furthermore, she takes Efuru to the hospital and thus enables her to recover and to prove her fidelity by undertaking a traditional ritual. Saved and strengthened by Ajanupu's solidarity, Efuru leaves Eneberi. Again the solidarity of a woman closely connected to Nkwo aids her. Uhamiri—the lake goddess, who supports many trading women—offers her a safe haven from a world which officially defines womanhood as wifehood and motherhood. She calls Efuru to become her priestess and guarantees her a successful life as a trader. By outlining the benefit of mutual help, Nwapa clearly affirms women's solidarity and sisterhood as a means of reforming patriarchal societies.[14]

As a married woman, Efuru has many wakeful nights. They were caused by Efuru's ruminations on her problems with her husbands and her barrenness; they symbolize her affliction. In so far as Efuru's sorrows stem from her endeavors to live up to her society's notion of womanhood, Nwapa protests against the moral of *Ifo* that women who obey social norms find bliss. By demonstrating the tragic consequences of reaffirming "the commonly accepted notion of women," Nwapa criticizes what *Ifo* are trying to reproduce. When Efuru leaves Eneberi, however, she challenges the commonly accepted notion of women. As a worshipper of Uhamiri, Efuru escapes from any direct male control. That Efuru now sleeps "soundly"(221) indicates that Efuru's sleepless nights belong to the past. This stands for the fact that she has finally found what she has been looking for all her life: happiness and fulfillment in herself. Additionally, she enjoys social recognition. By thus arguing, Nwapa "writes back" to the *Ifo* notion that strong women are the "source of all evil,"[15] who may only expect a woeful life. Moreover, she applauds what the *Ifo* are trying to undermine: women's strength, power, and autonomy.

NOTES

1. With the naming of this literary genre I refer to Emenjano (cited in Okoh, 75).
2. "Onwuero and the Three Fishes," Ekwensi, 1-2; "The Girl Named Lightning," ibid., 55-56; "Ogilisi Akwalaka," ibid., 65-68; "The Beautiful Daughters," Umeasiegbu 1977, 94-96; "A Proud Girl's Punishment," ibid., 94; "A Girl and A Python," ibid., 102; "Onwuelo and The Fish," ibid., 93-94; "The Proud Girl and The Devil," Bordinat & Thomas, 88-93.
3. "Sharing The Booty," Umeasiegbu 1981, 60-61; "The True Confession," ibid., 131-133; "Tortoise Gets A Wife with A Grain of Maize," Umeasiegbu 1977, 27-28; "Ndumihiri and The King," Chukwuma, 396-398; "The Tortoise Gets A Wife," Bordinat & Thomas, 62-64.
4. The father marries his daughter off to a friend as a proof of friendship, to ask this man's forgiveness, or in order to reward his help.
5. *Ifo* are a highly didactic literature. Their concern is to prevent Igbo people from violating social norms. Thus *Ifo* deal above all with figures who show non-conformist behavior. Figures who measure up to official social norms only act as protagonists when contrasted with a figure who disobeys the unwritten laws of Igbo society. Characters and actions which harmonize with official conceptions of conformist behavior are otherwise marginalized. As the assumed "normality," they remain uncommented upon.
6. See Emecheta, Osifo, Nwankwo, Ezeigbo, Okoye, and Nwapa.
7. This technique is also to be found in other novels: Osifo's *Dizzy Angel* is the story of a girl who by the community of an oracle, is to marry an old juju priest. In order to prevent her from rebelling against this fate, her

grandmother tells her the *Ifo* about the "handsome stranger"(81). During the same narrative event another *Ifo* is told, too. It is about a daughter who disobeys her parents and is therefore penalized with a horrible marriage to a frightening spirit(74-80). The protagonist in Okoye's "The Pay-Packet" marries a man of her own choice. Okoye hints at the intertextual analogy between her short story and the well-known *Ifo* explicitly. The protagonist refers to her husband as "You're a gentleman, B. A perfect gentleman...You're a fine gentleman"(23). Other figures title the man the same way (15, 21, 24).

8. It was little more than a decade later that Brown implicitly opposed Palmer's attitude by trying to relate the *Ifo* to Nwapa's novel. He points out that the "recital of a folk tale becomes a symbolic ritual within Nwapa's larger narrative"(Brown, 141). He understands Nkwo as a symbol of communal and family ties, which guarantee "continuity of life itself"(Brown, 141). This realization plays, however, all but an essential part in his interpretation of *Efuru*. Another decade later, Susan Andrade, Gay Wilentz and Florence Stratton discussed the relationship between the *Ifo* and Efuru anew. Andrade, however, bases her discussion on the false assumption that it is Nkwo who is married by the spirit and that she is rescued by Orie. That Andrade nevertheless finds a parallel between the integrated *Ifo* and Efuru once more outlines the danger that is inherent in an interpretation. It can carry away a critic from the actual text. Because of her incorrect starting point Andrade's assertion that Nkwo symbolizes Efuru's misery and that Orie stands for Efuru's rescue cannot be given any serious thought (Andrade, 99; also for a comment on Andrade's approach to the *Ifo*, see Stratton, 182). Wilentz misreads the *Ifo* too. She argues that Nwapa integrated the *Ifo* about the "beautiful, proud girl who will not marry anyone her parents pick for her but chooses a handsome stranger instead"(Wilentz, 8). Thus she concludes falsely that what Efuru and the *Ifo*'s heroine have in common is that both marry a man of their own choice, thus ignoring the fact that the protagonist of the tale was *forced* into her marriage (Wilentz, 9). Wilentz moreover stresses that unlike the heroine of the *Ifo*, Efuru has no mother (Wilentz, 9). In her interpretation of *Efuru*, however, Wilentz doesn't make much use of this interesting observation. Stratton uses the *Ifo* in order to illustrate that Nwapa attaches much importance to the mother-daughter relationship in her works "as a means of keeping alive the set of traditions Uhamiri embodies by passing them down through the female line from one generation to another" (Stratton, 93).

9. For a theoretical discussion of the functions of tales within novels by Igbo authors, see Acholonu, 236-250; Lindfors; Egudu.

10. In Okoye's novel *Chimere*, e.g., the protagonist, whose father left her mother before she was born, suffers from being labeled a "bastard" and a wayward woman because she accepts this stigma. A friend of hers criticizes this attitude. In his effort to change her behavior, he tells her an *Ifo* about a sick man, who is asked to sacrifice a goat for the gods(139). As he is much too poor to fulfill this demand, he passes a chick off for a goat, by putting it into a goat's skin. When the chick starts to cluck on

the way to the shrine, he says: "Little chick, whether you cry like a chick or not, what I know is that I'm carrying a goat to the gods and not a chick"(139). With this tale Chimere's friend wants to indicate to her that she has to overcome her inferiority complex. She should insist on being looked at as what she is—an honest and "virtuous" woman—and thus reject the stigma of being wayward and immoral by overcoming the official attitude of her society towards illegitimate daughters.

11. The same applies to Gracy Osifo's novel *Dizzy Angel,* which stresses that a woman should have the right to reject the suitor chosen by her family. Her protagonist practices this rebellion and will only therefore find a happy life. Thus the integrated *Ifo* (See note 7) embody a moral that is the antithesis of the literary message of *Dizzy Angel.*
12. In the novel there is, however, an example of a happy marriage which started with a free choice of partner (188).
13. Moreover, it has an important structural function. Nearly all events of importance in Efuru's life happen on Nkwo day.
14. That Eke, Afo and Orie refuse to help the *Ifo's* heroine because she herself was not cooperative at all when they were in need of salt, pepper or a pestle, respectively, implies the importance Nwapa attaches to women's solidarity (109).
15. From the title of Mineke Schipper's collection of African proverbs on women, *Source of All Evil.*

Works Cited

Acholonu, Catherine Obianuju. *Western and Indigenous Traditions in Modern Igbo Literature.* Ph.D. Dissertation University of Düsseldorf, 1985.

Adimora-Ezeigbo, Akachi. "Agarachaa Must Come Home." *Echoes in The Mind.* Ikeja: Foundation Publishers, 1994:2-23.

Andrade, Susan. "Rewriting History, Motherhood, and Rebellion: Naming an African Women's Literary Tradition." *Research in African Literatures* 21.2(1990):91-110.

Ashcroft, Bill, Gareth Griffiths, and Helen Tiffin. *The Empire Writes Back: Theory and Practice in Post-Colonial Literatures.* London: Routledge, 1989.

Bordinat, Phillip, and Peter Thomas. *Revealer of Secrets.* Ibadan: Africana, n.d.

Brown, Lloyd W. *Women Writers in Black Africa.* Westport, Conn: Greenwood, 1981.

Chukwuma, Helen. *The Oral Tradition of The Ibos.* Ph.D. Dissertation, University of Birmingham, 1974.

Egudu, Romanus. "Achebe and The Igbo Narrative Tradition." *Research in African Literatures* 12.1(1981):43-54.

Ekwensi, Cyprian. *Kolo The Wrestler and Other Ibo Tales.* London: Thomas Nelson, 1954.

Emecheta, Buchi. *The Bride Price.* New York: George Braziller, 1976.

James, Adeola. *In Their Own Voices: African Women Writers Talk.* London: James Currey [Portsmouth, NH: Heinemann], 1991.

Jell-Bahlsen, Sabine. "The Concept of Mammywater in Flora Nwapa's Novels." *Research in African Literatures* 26.2 (1995):30-41.

Lindfors, Bernth. "The Folktale as Paradigm in Chinua Achebe's *Arrow of God.*" *Folklore in Nigerian Literature.* Ed. Bernth Lindfors. New York: Africana, 1973:94-104.

Nwankwo, Bridget. *Drums of Destiny.* Ibadan: Heinemann Educational Books Nigeria PLC, 1991.

Nwapa, Flora. *Efuru.* London: Heinemann, 1966.

——. *Idu.* London: Heinemann, 1970.

Okoh, Nkem. "Categories of Igbo Oral Literature: The Enuani Example." *Journal of Asian and African Studies* 35(1988):73-83.

Okoye, Ifeoma. *Chimere.* London: Longman Nigeria PLC, 1992.

——. "The Pay-Packet." *African Women's Writing.* Ed. Charlotte H. Bruner. Ibadan: Heinemann, 1993:15-25.

Osifo, Gracy. *Dizzy Angel.* Ibadan: University Press, 1985.

Palmer, Eustace. "Elechi Amadi: *The Concubine* and Flora Nwapa: *Efuru.*" *African Literature Today* 1(1968):56-58.

Schipper, Mineke. *Source of All Evil: African Proverbs and Sayings on Women.* Nairobi: Phoenix, 1991.

Stratton, Florence. *Contemporary African Literature and The Politics of Gender.* London: Routledge, 1994.

Uchendu, Victor. *The Igbo of Southeast Nigeria.* New York: Holt, 1965.

Umeasiegbu, Rems Nna. *Words Are Sweet: Igbo Stories and Storytelling.* Leiden: E. J. Brill, 1977.

——. *The Way We Lived.* London: Heinemann, 1981.

Umeh, Marie. "The Poetics of Economic Independence For Female Empowerment: An Interview with Flora Nwapa." *Research in African Literatures* 26.2(1995):22-29.

Wilentz, Gay. *Binding Cultures: Black Women Writers in Africa and The Diaspora.* Bloomington: Indiana, 1992.

Space For The Subaltern: Flora Nwapa's Representation and Re-Presentation of Heroinism

Mary E. Modupe Kolawole

The position of African women is often centred on the problematics of voice, silence, and mimicry. Issues of self-assertion, mobilization, empowerment, and relocation of the African woman have become ubiquitous and important aspects of post-colonial, African, feminist, womanist, and Third World studies on women. Many Western scholars theorize about African women on the continent and some claim to represent their voice because, in their view, African women have been silent and invisible. Generally, African women are presented as a subaltern group in the margin of society, docile and accepting the multiple level of subjugation gracefully. The problem with this incorporation and condescension is that many see African women as "Other" and as a group that is inarticulate and thus should be represented by Western—women and men—writers. This attitude has its origin in colonialism, post-coloniality, and tradition, as well as the predominantly patriarchal structure of the African society. Yet, the specificity of African women's reality can best be represented by African women, since there is no substitute for lived experience. Toni Morrison's observation about Black women writers is significant to this discourse:

> We are...the subject of our own narrative,witnesses to and participants in our own experience, and, in no way coincidentally, in the experience, of those with whom we have come in contact....And to read imaginative literature by and about us is to choose to examine centres of the self and to have the opportunity to compare these centres with the 'raceless' one with which we are familiar.[1]

African women writers have therefore been confronted by the challenge to speak for their race and gender, revealing voices that emerge against diverse odds.

African women writers find themselves in a moment of flux and, consequently, they respond to the exigencies of a conflictual and changing society. They recognize their task as that of the ventriloquist who must represent and re-present the African woman and the collective self. They have disproved existing presentation of their reality as a tabula rasa for imposed ideologies. Literature has been a significant channel of African women's voicing, in accordance with women's role as dynamic voices of self-assertion in African orature (Kolawole, 1995). African women's literature is neither ahistorical nor acultural. Yet much existing theory about these women is divorced from cultural comprehension and apprehension. Since the historical emergence of African women's literature in 1966, African women have continued to establish an active tool of self-realisation through literature. The pioneers of African women's creative writing, Flora Nwapa and Ama Ata Aidoo, have revealed a progressive development of African women's voices. Various writers and critics from this region occupy different spaces in speaking out or speaking back. Most of these women writers recognize the African woman as the subaltern at varying degrees, however. The vital question raised by Gayatry Spivak ("Can the Subaltern Speak?"[2]) is relevant here. One needs to clarify this term as it relates to African women. The subaltern as the marginalized group denotes any oppressed or suppressed group whose voices have been muffled because there is no adequate space from which to freely speak for themselves. African women are products of several agencies that negate their self-enhancement, and often the social structures put in place are obstacles to their ability to speak out. Post-colonial conditions create socio-political agencies who do not take women on board while inscribing the group on their agenda. Attendant economic problems put African women in disadvantaged positions, as Gloria Emeagwali confirms:

> Women are directly affected by the perpetuation of this symptom of maldevelopment and have a formidable battle to fight as we prepare to move into the 21st century.[3]

The woman writer's challenges include the recognition that women's struggle needs to move beyond the acceptance of their subaltern space as a matter of fact. Because history is replete with instances of dynamic mobilization and rejection of oppression and marginalization by women in many parts of Africa, these writers create women that are true-to-life in resisting oppression and in creating new canons of relocation. Oral genres dominated by women reveal voices of self-assertion, as we see in Olele, Obitun, Gelede, Ekun Iyawo, and Egungun satirical songs by women in Yorubaland, Nwokoro and Nzema maiden songs in Ghana, Ipongo solo, Ila and Tonga of Zambia, Gala lampoons, and Kamba grinding songs. The deployment of oral literature to reveal women's voices is not limited to any part of Africa and in modern times, Ogori Ewere in Kogi state of Nigeria, Igbo birth songs, Fulani Bori songs, Zimbabwean women's club songs, and Buganda radio songs confirm the continuity of oral cultural production as avenues of enhancing women's voices and audibility. Irene D'Almeida problematizes the silence of African women:

> Silence represents the historical muting of women under the formidable institution known as patriarchy, that form of social organization in which males assume power and create for females an inferior status.... Reiterating the concern with women's silence acknowledges the fact that if women have made significant social advances by challenge and accommodation, by opposition, resistance, and subversion, their enforced silence—in particular the denial or limitation of their literate expression—remains nonetheless a common and painful reality.[4]

The formidable task of shifting from the tangential position allocated to women in early male literature in Africa is not a facile one. Indeed, silence itself could be a weapon of resistance against Western and "malestream" representation of their reality. But writers like Flora Nwapa recognize the need to emerge from silence and to break the "veil of silence" like many Francophone women writers who Irene D'Almeida has highlighted in her recent research. Irene D'Almeida has delineated various ways adopted by Francophone women writers in the process of speaking out.[5] This critic explores how women writers' voices are voices of self-discovery and self-revelation as writers identify the intersection between the personal, the familial and the collective communal voices. We see this in the works of Mariama Bâ, Naffissatou Diallo, and Ken Bugul. A significant issue raised by D'Almeida is "the full implications of being black, colonized and female"(44). Were Liking and Aminata Sow Fall have deployed literature as tools for revealing African women's space not as static but changing as they present the writer's voice as a social critic.

Like Francophone African women, other female writers have been very prolific in identifying women's voices as strong social and political voices as they attempt to speak from a subaltern space as well as transcend it. All of the works by Nawal El Saadawi present a radical view of Egyptian women as a subaltern category. Molara Ogundipe-Leslie's theory and actions have sustained a radical voice as she advocates searching for women's voices. In *Sew the Old Days,* this writer's Marxist commitment is highlighted as she touches on women's oppression and subjugation. In her recent critical work, *Recreating Ourselves: African Women and Critical Transformation,* she rejects the notion that African women as a subaltern group cannot speak. Although she identifies six mountains on the African woman's back that stifle her voice, she maintains that African women are speaking out, but many critics refuse to listen to their voices in appropriate places. She raises the fundamental question, "Are African women voiceless or do we fail to look for their voices where we may find them, in the sites and forms in which these voices are uttered?"[6] She herself provides an answer, "We neither look for their voices where they utter them nor do we think it worthwhile to listen to their voices."[7] This is the crux of the matter. Consequently, many African women writers have faced the challenge of speaking for these women to reveal the reality that African women are not as silent as we are made to believe.

African women occupy different spaces in dealing with the

problematic of women's voicing on the continent. According to Micere Mugo, ordinary African women (who constitute the majority) are neither invisible nor voiceless.[8] I do agree that history, orature, and reality confirm the theory that African women are not as silent as many scholars sustain and that they are speaking out and speaking back even from the subaltern space. In many places the subaltern status has changed considerably, and it is this representation of this awareness that makes Nwapa unique. Siga Jajne confirms the subaltern position of African women with specific reference to Senegalese women. She however identifies a traditional way of voice-throwing by women known as *Sani Baat*.[9] She observes that African women need to adopt this process of imposing their voices. African women are already forcing their voices onto existing discourse as subversive voices. This theory is validated by the category of iconoclastic rebellious heroines created by writers like Flora Nwapa in her later works. But African women's voice is heteroglossic, since the society is pluralistic. African beliefs present the world as a place of negotiation of values. Among the Yorubas, it is believed that there are always diverse routes to the market place, "*Ona kan ko wo oja*—there is no absolute single way to the market place." Ropo Sekoni has underscored this trope of the market place and its relevance to any theory of African women artist's approach. Significantly, the market place is the domain of women in many African societies, and women generally sustain this philosophy of the negotiation of space. African women's literature, consequently, has always revealed the gradual process of space negotiation through voicing while critics' voices continue to be equally multiple. Several critics from the region maintain that African women are not as silent as certain scholars suggest and that the problem involves locating their voice by searching and researching into areas of these women's audibility and visibility.

In the later fiction by the precursor of African women's literature, Flora Nwapa, she re-presents women from a space of positive self-definition, revelation, rediscovery, and relocation. She has created this space not as a romantic afterthought but as an expression of reality as she saw it and lived it in her proximate setting of Igbo society as a paradigm. Igbo women's resilience and resourcefulness is replicated in other parts of Africa, as seen in the economic independence of Yoruba women, the economic dominance and control of Ewe women of Togo and Ghana, and the myth of Nana Benz.[10] By re-creating themes such as displacement, migration, and redefinition of the self, Nwapa's heroines succeed in creating a new space and new voices for themselves. The journey motif is significant in this process of revelation, self-discovery, recovery of lost ground(s), and relocation. Often this involves subversion of existing voices and values and the imposition of a new set of values to create this space for women. The sheer diversity of heroines that exemplify this trend exposes Nwapa's genius and her keen sense of African womanism "from the horse's mouth."

Many are responding to Ogundipe-Leslie's call for the woman writer's dual levels of commitment: "the woman writer has two major responsibilities: first to tell about being a woman; secondly, to describe reality from a woman's view, a woman's perspective"(57). Early women writers had the odious task of plucking their work from an existing Eurocentric patriarchal portrait, and the difficulty of this task can be seen in the largely mimic nature of their literary production. The Swahili concept of kikusuku—parroting—could easily describe much of early women's literature from the continent. Women were presented by male writers as tragic heroines unable to speak from their subaltern position and this trend continued in some women's literature. Nwapa's early heroines, Efuru and Idu belong to the category of women who are iconoclasts, women who tower above others, above their husbands. These heroines, however, get destroyed in the process of trying to break the yoke of silence or strike a balance between traditional set limits and their own perceptions of their identity and self-actualization. Several other early writers like Ama Ata Aidoo fictionalize the dilemma of these ghosts caught between two realities.

Several younger or more recent writers have attempted to create a space for the subaltern woman right from their first creative works, as we see in the plays of Tess Onwueme and in Tsitsi Dangarembga's works. Onwueme's play, *Wazobia* presents strong revolutionary voices in a traditional setting while Dangarembga's women (and in particular her heroine Tambu in *Nervous Conditions*) represent women's rejection of a silent subaltern space. It is however remarkable that the older writers have progressively re-created and re-presented women to reflect the changing reality of African women's space. We see this trend in Aidoo's recent novel *Changes* and Nwapa's more recent works, *Never Again, Wives At War, One Is Enough* and *Women Are Different*. One needs to emphasize the fact that Nwapa has always presented women's voices. A traditional reading of her early works *Efuru* and *Idu* reveal women as silent, but a revisionist appraisal shows that Nwapa's women have always tried to speak, even from their subaltern space.

Nevertheless, the later heroines are more positively presented as voices of change and of dynamic relocation and empowerment.

Writers writing from a marginal space are better equipped to see issues from outside the hegemonic, conventional canons. Flora Nwapa's early heroines are created to show that the African woman as a subaltern has always tried to speak. The initial portraits of women underscore the tension within the individual who struggles against the grain. These women are consequently tragic heroines caught between tradition and the individual's quest for self-realization in traditional patriarchal settings. Their efforts are at best "stillborn" (in the words of Zaynab Alkali). This discourse however will focus on the later heroines as re-presentations of African womanhood from a more liberated authorial voice. One remarkable feature in Nwapa's later works is the composite or group heroism. By placing side by side the various actions of many women to reiterate and reinforce her preoccupations, Nwapa succeeds in recasting the collective nature of women's struggle for self-retrieval as opposed to individual lone voices in the society. Her early heroines Efuru and Idu fail to define a strong voice for many reasons, but each of these heroines is a lone voice. We have many more women speaking for women in later works like *Never Again, Wives At War, One Is Enough,* and *Women Are Different.* Nwapa's representation of women in these works transcends "speaking for them" as she re-presents a more dynamic versatile image of women—not only in domestic roles, but as professionals, radicals, iconoclasts, apostates, conformists, as well as wives and mothers. In doing so, she upholds "the right of formally un-or misrepresented human groups to speak for and represent themselves in domains defined, politically and intellectually, as normally excluding them, usurping their signifying and representing functions, over-riding their historical reality."[11]

In *Never Again,* we have an eyewitness account of women's dynamic involvement in the Nigerian Civil War through the authentic voice of a first-person narrator, Kate. Her main narration embodies her personal experience as well as the stories of the active involvement of several other women. The central motif is built around journeys and focuses more specifically on the mass exodus of Igbo people from one part of Biafra to relatively more secure places. The more significant exodus of Igbos from other parts of Nigeria remains in the background as well. As several women flee with their families, Nwapa presents women's active voices and dynamic actions at the core of the socio-political upheaval. The protagonist Kate is depicted as a woman who possesses tremendous strength, insight, and foresight. Her wise counsel is initially resisted by her husband and his friend, Kal. She perceives impending danger ahead of actual manifestation, as we see in her foresight about the direction of the war, which her husband Chudi and his impulsive friend cannot see: "I knew the game was up. I had told Chudi that Port Harcourt was in great danger, and would fall at any time. This was about two months before it actually fell. He was so upset that he threatened to hand me over to the Civil Defenders or the Militia"(1). His resistance to Kate's suggestion to move the family is typical of his attitude throughout the novella. Kate's sense of judgement undermines the more emotional reactions from her husband and she probes the war propaganda instead of being credulous like Kal. She is seen as a traitor to the cause but she is simply approaching the crisis in a more pragmatic way and she is always vindicated.

As Kate and her family move on for safety, she goes through a process of self-knowledge and self-awareness like several other women. Significantly, she mobilizes and organizes women at every opportunity: "I remained in Enugu helping to organise the women"(2). Many women and girls are actively involved in the socio-political situation and participate in the meetings as voices of encouragement and self-assertion for the Igbo people, in diverse ways: "On this occasion, a girl civil defender mounted the rostrum. She had good news to tell us and we were very eager to listen....She was at the S7T that morning, and one of the officers had told her that Lagos and Ibadan were being shelled...."(4) Although the authenticity of the story could not be verified, her intention is positive. At another level, Kate's mother-in-law is a voice of hope based on her faith in the power of the Woman of the Lake, Uhamiri, who is believed to be the protective power of victory for Ugwuta people. She draws on myth to sustain her belief, "They will not come. It is just empty talk. Never in the history of Ugwuta did an invader come by water and go home alive....She has never let her people down..."(5). In more practical ways, Kate and other women get themselves involved, refusing to be pushed aside and silenced. When a meeting of all able-bodied men is called, Kate insists on following Chudi there and some other women respond to the call of "able bodied men." Women politicians like Madam Agafa are particularly prominent at such meetings:

The women especially were very active, more active

than the men in fact. They made uniforms for the
soldiers, they cooked for the soldiers and gave
expensive presents to the officers. And they organised the women who prayed every Wednesday for
Biafra....It was one of the prominent women who
addressed the gathering of the men....she was so
powerful then that people were very careful not to
offend her. In short, she was highly respected,
feared and we had no alternative but to listen to her.
(7)

The women stopped at nothing in their struggle for survival during this war. We are told that men who did not join the war have
no work to keep them busy; in contrast, the women are involved
in the attack trade that involved crossing the enemy borders to
bring in essential goods.

Nwapa's portrait of women during the Biafran War is not an
abstract romanticization of women's role, but a true-to-life reflection of Igbo women's central role in the quest for survival and in
sustaining a cause they strongly believed in. Significantly, Kate and
other women succeed in striking the right balance between their
public commitment and their domestic duties as wives and mothers. This is reflected in Kate's sense of commitment to her children's
survival. Talking about her children, she shows a passionate commitment to her children's welfare and a deep sense of apprehension as well as a resolute desire to make sure they do not suffer:

"They had not started to suffer as grown ups were-
suffering. If the war lasted another year, they would-
begin to suffer, first from hunger, discomfort and
then ill health. I was determined not to see my children suffer. I would sell all I had to feed them if I
had to. They were not going to be hungry. They
would not suffer from Kwarshiorkor." (25)

When Kate's husband is invited to go out and join the combing, she
shows unusual boldness. She is seen as a protective voice and also
a voice of reason in curbing the excesses and wasteful destruction
of lives:

"My husband will not go with you," I said. I was-
surprised at my boldness. "Tell whoever sent you
that my husband will not come with you to the
combing.Why are you combing?....There are no

infiltrators here. Arrange evacuation of everybody
including the men and the youths. Yes, the youths.
Why should they die for nothing...." (33)

Nwapa also presents women as sacrificial figures in the death of
the pregnant woman who dies in labor along the escape road and
Ona who symbolizes resistance and gets shot by soldiers because
she refuses to move.

Wives At War fictionalizes the diversity of women's resistance,
self-definition, rejection of oppression and suppression, as well as
self-assertion. Women are not simply fighting a political war, they
are involved in ideological battle and a war between personal recla-
mation and imposed social values and norms. This collection of
short stories shows Nwapa at her best in recreating new images of
Igbo women as a paradigm of the changing status of women which
several critics of African women's reality fail to apprehend. From
the experiences of women during the Nigerian Civil War, we also
get a picture of the personal struggle of women who desire to cross
the Rubicon and transcend socially imposed borders. We see women
speaking from their subaltern space while others try to emerge from
liminal locations against several odds. The titular story, "Wives At
War," is a representative story of the harrowing experience of a
woman caught between loyalty to her family and the need to maintain her sanity. She struggles to come to terms with the war but her
agony takes root in the fact that her husband and she herself belong
to the two opposing factions, Nigeria and Biafra. She puts family
dedication above personal safety at first, moving from Lagos and
following her family to the war-ridden Eastern Nigeria. She asserts
a positive philosophy: "Bisi was a devoted wife, who was determined
to prove that inter-tribal marriages could and would work." Initially,
her husband, Ebo, is a responsible husband, but the war and its
dehumanizing effect change him and their circumstances. Unlike
her husband, she refuses to be dehumanized by the war and
becomes the voice of reason and positive moral correctness when
Ebo attempts to extort money from a personal friend: "The man
over there is your friend, Eze. You told me about him when you
were courting me. And you wanted him to pay you five Nigerian
pounds in coins. I ask you, will you eat money? You have a hundred pounds in coins. You gave it to me for safe-keeping"(6). Her
husband's corruption, selfish involvement in riotous life, and disregard for the family's safety results in her mental breakdown.

This is not simply a figment of the authorial imagination. It is a true-to-life depiction of a slice of reality as it affected Nigerian women during the war. This story elicits women's voices as positive voices in the reality of the civil war, which itself is a dehumanizing experience. It shows women's struggles and activities (as opposed to their being passive suffering victims): Bisi's mental breakdown and her husband's success in flying her out to Portugal on a relief plane brings out other salient issues and we see women as political voices. To get Bisi out of Biafra, he uses his influence to send a message that the First Lady is travelling out to London with a group of women to a meeting with the Queen of England. Ebo uses this pretext to fly his family out under confidential cover, but the news of the First Lady's trip to the queen spreads and other women react in a rather significant way. "At Umuahia, the following morning, women leaders got together for once to protest to the Foreign Secretary about their non-inclusion in the mission"(11). In the characteristic charged mood of the Igbo women's war of history, Nwapa depicts the women's fury and bold confrontation of the Foreign Secretary who faces an arduous task of convincing them that there was no such trip to the queen. These women assert themselves like the historical Aba women or Egba women of Abeokuta in resisting colonial administration. Biafran women's voices become voices of intimidation and terror to the Foreign secretary, "He had returned home to face Biafra and her women. The women. How could he cope with them? His wife could have helped him, but the women had [so] successfully turned his wife against him that he was obliged to send her and the children to the village"(11).

Although these women have their problems, such as rivalry among different women's clubs, the spontaneity of their mobilization and the boldness in presenting their case according to their convictions reveal women's voices as political voices. At a stage, one of the women insisted that women should be treated with courtesy, threatening, "You wait until the end of this war. There is going to be another war, the war of the women. You have fooled us enough. You have used us enough. You have exploited us enough. When this war is ended we will show you that we are a force to be reckoned with"(13). Nwapa shows the modern generation of women as ideological voices. This reveals women as symbols of social change and ideological voices. Here, Nwapa upholds a close link between literature and reality with an insider's eye-view. They eventually learn the truth of the matter, that no group of women travelled to the queen. Nonetheless the misinformation provides the women an opportunity for self-expression:

"Without the women, the Nigerian vandals would have overrun Biafra; without the women, our gallant Biafran soldiers would have died of hunger in the war fronts. Without the women, the Biafran Red Cross would have collapsed. It was my organisation that organised the kitchens and transport for the Biafran forces. You men went out to the office every-day doing nothing." (13)

This delineation of women's roles in the war touches on the central space occupied by women under the traumatic war conditions. One needs to reiterate that Nwapa's fiction is unapologetically close to reality. She did confess in her real life that her fictional women are portraits of women as she saw them in actuality. Women's role transcends the domestic and reproductive as one of the leaders of the women's associations affirm, "We have been poised for action ever since war was declared. We are independent....Right from the word go, we organised the women for a real fight. We asked for guns to fight the enemy. We asked to be taught how to shoot"(14). Women's reality during the Nigerian Civil War has been objectified by Nwapa to show that women's voices were not muffled voices of the subaltern but political and ideological voices.

Another story in the collection of short stories, *Wives At War,* underscores women speaking to positively shape the home and pilot the affairs of the family in a positive direction during the Nigerian Civil War in "Daddy Don't Strike the Match." Ndidi Okeke is a woman of deep insight, but her suggestions are usually rebuffed and rejected by her husband, Francis. When her husband refused to leave the North for the East in spite of the imminence of war, Ndidi took control and brought the children and their belongings home. She was financially independent and saved her family from an impending doom. Francis, who rejects her advice, escapes narrowly from Kano when the war eventually breaks out. As a mother, Ndidi is resourceful and ready to make sacrifices so that her children will not suffer. Her positive suggestions and advice are not accepted by her husband who sees her voice as typical women's unreasonable voice of fear, "Women, they are all the same. So unreasonable, so adamant..."(22). He resists Ndidi's advice to stop smoking with the same rebuff and this costs his life. He eventually strikes a match in an absent-minded moment as he is about to smoke in his laboratory which is full of materials for making explosives. His death could have been averted if he took to his wife's advice and caution. Significantly, their daughter, Ifeoma also had moments of premonition and apprehension before this tragedy. Here, Nwapa underscores African women's spirituality as exemplified in Ndidi and her daughter, Ifeoma.

"A Certain Death" reveals further the tragic dilemma of families in the war situation. The narrator, has lost all her three children as they were escaping from Lagos, and her brother has just lost a wife and two children in an air raid. Since her brother is the only living relation, she is determined to prevent his conscription into the army. She succeeds in placing a young man as a substitute after paying the man's family who needs the money for survival. Here, Nwapa is not interested in morality, but she presents the story of a woman who is ready to pay the price to secure her brother's safety. This is a common approach by this author and quite often, she creates women as tools of subversion of existing moral codes or social values to create a space for them to speak. Yet Nwapa presents the corollary of this in the short story, "Man Palaver." We have a documentation of subaltern women—abused women like Obiageli. Anyaga is a victim of a loveless marriage and a cruel husband. While Nwapa depicts women crushed by male cruelty—like Chiebonam who has a mental breakdown—she also presents others who have come to terms with their condition by taking comfort in their success and ability to take care of their children, like Anyaga. Chika, the narrator in "A Wife's Dilemma" feels neglected by a husband who is too busy and escapes into an affair in Lagos. This is short-lived, but it nonetheless shows how Nwapa has created women not only speaking out but also moving out of their subaltern space and finding fulfilment outside conventional moral codes.

The story, "Mission to Lagos," reiterates the way the women are breaking the yoke of silence and invisibility. Yetunde Smith represents the woman who crosses racial and intellectual frontiers in her relationship with a colonial officer, John Hammer. She is caught in the web between personal need and traditional attitude to Nigerian women who choose to marry White men. She is a strong symbol of self-definition and only agrees to marry John Hammer after a long ideological interrogation that touches on profound issues such as women's freedom, tradition, cross-cultural marriage, individual choice and the question of career and self-fulfilment. "The Chief's Daughter" presents similar themes concerning the individual's choice as opposed to parental imposition of tradition. Chief Onyeka's pompous traditional beliefs reiterate the forces of subjugation of women, "Just as her mother obeyed me in all things, so will Adaeze obey me in all things." But Adaeze's rebellion confirms that women are not only speaking out but are moving out of the subaltern space created by odious traditions. The chief learns this lesson the hard way when Adaeze returns to Europe to join her husband—despite her father's refusal to recognize her choice and his insistence that she will not marry but will stay with him and bear children outside marriage.

Women Are Different presents female characters who represent Nwapa's conviction that the subaltern can speak and are indeed already speaking from spaces created by them or for them. This novel is the most successful in this regard. The heroines are precocious right from their early school days and they have always proved to be determined young girls. "The three musketeers," as these three girls referred to themselves, are visionaries who at an early age have a clear picture of what they expect of marriage and family lives. They are idealists but recognize the limits set for girls in their society while being determined to transcend such ceilings. Rose, Agnes and Dora symbolize positive female bonding from their school days at the famous Archdeacon Crowther Memorial Girl's School, Elelenwa, the first girls' school started by the Anglican Church in Nigeria. The school principal, Miss Hill, is shocked at the precocious actions of these girls who defy regulations right from their first year. The three girls are idealists from their early school days and each of them desires to succeed and have a happy married life. Each later discovers that patriarchal society puts girls at a disadvantage as they each find themselves to be the "underdog" in different ways. Agnes is forced to marry early because of the pressure of a step-mother who is eager to send her away from home. She marries a man old enough to be her father. She is compelled to marry immediately after leaving the grammar school. But she refuses to be overwhelmed by her circumstances. While rearing children, she studies privately and secretly to enhance herself. She rises above her limitations by passing relevant examinations and strikes the right balance between her productive and reproductive roles, "Being a mother did not disturb her in anyway"(58). She insists on going to an evening school to improve her chance after having four children although her husband, Mr. Egamba, resists this initially. Her second marriage—to Ayo Dele—provides necessary motivation and encouragement in her quest for a university education. The death of her second husband exposes the callous treatment that widows receive in many parts of Africa.

Before Agnes could get herself together, Ayo Dele's children pounced on her "home" and seized everything she had worked for(62). She is not defeated by this problem. Significantly, her predicament is typical of women in the post-colonial set-up, but instead of accepting defeat, she rises above the woman's problems and takes comfort in a good job and the fact that she can adequately take care of herself. She is not a tragic heroine like the early Nwapa's female protagonists, Efuru and Idu. She responds to social change by facing the challenges.

Dora qualifies as a nurse and marries her childhood lover, Chris. Her husband is an unambitious man who lives above his means and is steeped in corruption. She later engages in business and becomes very prosperous. Chris begins to enjoy his wife's prosperity and travels to England on her savings after selling her house without her knowledge. The height of the shabby treatment is the discovery that Chris lives with a German woman when Dora eventually finds him and she has to return home in confusion. But she emerges from this low level and finds a new relationship with Tunde.

The third "musketeer," Rose, is a victim of a dubious man, Mark, and an impostor who proposes to her to trick her into sponsoring his admission to Harvard University. He deserts her shortly after their marriage *and* after spending her money to travel to Harvard. She subsequently gets involved with a man without knowing that he has a wife in New York. But she takes consolation in her work. Nwapa allows the three friends' experience to merge at the end of the novel. They are prepared to move the mountains on their backs, which become manifest in different ways as they share their experiences. Through this meeting point, we realize a change in the way the women perceive reality, women's space, and the new social order. Recognizing the way the Nigerian society of the seventies enhances women's subjugation at the hands of men who exploit them, they have adopted a more liberal attitude to marriage and family life. Their voices are more radical voices as they rise above their social problems. The next generation—as seen in the attitude of their children—has a different, more rebellious attitude to marriage. This is Nwapa's way of presenting more radical female voices.

Flora Nwapa, as the pioneer of women's literary creativity, has championed women's cause by presenting women's voices. Sometimes it is a voice of protest and rebellion; at other times it is a voice of compromise. But she has largely presented women speaking from their marginal position. Nwapa recognizes African women's multiple levels of oppression as she confirms to Adeola James, "You are oppressed at home, you are oppressed at work. Your husband oppresses you, your employer oppresses you and then your society piles upon you double, if not triple suffering."[12] Nwapa does not romanticize the African woman's reality, but in her later works they are no longer passive suffering victims: they transcend the borders and predicaments and speak out against their oppression. Through dynamic actions they also show their resilience and refuse to accept the status of the muted voice of the woman as a subaltern.

Flora Nwapa's later works have identified women's voices as voices of social transformation, self-recovery, self-definition and this she does through collective self-portraits. Centering gender issues has enabled her to raise new questions and to interrogate conventional images of African women's voicelessness. Reifying the role of women in diverse cultural settings and socio-political situations has foregrounded the changing nature of African women's consciousness as opposed to the prevalent Eurocentric and "malestream" depictions. She is not moralistic because breaking out of conventional norms facilitates the depiction of women's attempts to defy their marginal location. She highlights self-consciousness as opposed to self-effacement. This is in line with bell hooks' viewpoint, "Black women need to construct a model of feminist theorizing and scholarship that is inclusive, that widens our options, that enhances our understanding of Black experience and gender." Nwapa's contruction of heroinism is a departure from Eurocentric images by scholars steeped in their own particularity trying to create awareness about or for African women. Hers is a challenge to universalize attempts by non-Africans and what Spivak describes as "epistemic violence." Her success proves that fiction must necessarily pass through a cultural crucible as well as the author's ideological filter.

In problematizing the subaltern, the scholar has to take sides, willy-nilly, "the debate is a struggle between those who want to align themselves with the subaltern and those who insist that this attempt becomes at best only a refined version of the very discourse it seeks to displace."[13] As regards the problem of effecting agency for the subaltern raised by many scholars, Nwapa's loyalty is clearly with the downtrodden, which happens to be women. She overcomes the problem of the post-colonial intellectual as an informant or apprentice of the hegemonic group. As a colonized

African woman, Nwapa sees herself as a subaltern poised to re-create new positive images of African women as a group speaking out through heroines that are actively resisting subjugation. She belongs to the outer group, the "Other," but refuses to remain a shadow of the Self where the self represents the dominant group in terms of gender or race. She is speaking as a subaltern for the subaltern by making the African women visible and audible. This touches on another fundamental issue raised by Spivak, "How can we touch the consciousness of the people, even as we investigate their politics? With what voice-consciousness can the subaltern speak?"[14] Through literature, Nwapa tries to raise consciousness. Considering the level of literacy in Africa, a large proportion of the subaltern may not yet be reached in such a process of conscientization. But Nwapa has sown the seed of awareness-raising along with several other African women writers. She faced the challenge to speak as an African woman and she achieves her goal in the representation of women in the works discussed here. Finally, Spivak's basic question is directly linked to women as a subaltern category:

> It is well known that the notion of the feminine (rather than the subaltern of imperialism) has been used in a similar way within deconstructive criticism and within certain varieties of feminist criticism.... For the 'figure' of woman, the relationship between woman and silence can be plotted by women themselves; race and class differences are subsumed under the charge.[15]

Nwapa is not guilty of glossing over issues of class and race although African women writers generally need to focus more on the ordinary rural women that constitute the majority. They need to sing a song of the oppressed peasant women as Micere Mugo advocates. Nwapa has gone a considerable way to reveal African women's history and voice by narrating African women's reality in a laudable way and this is the hallmark of her creative achievement.

NOTES

1. Toni Morrison, quoted by Cheryl Walls in *Changing Our Own Words*. New Brunswick: Rutgers, 1991, 1.
2. Gayatry Spivak, "Can the Subaltern Speak?" ed. Cary Nelson and Lawrence Grossberg in *Marxism and the Interpretation of Culture*. London:

Macmillan, 1989. Quoted in Ashcroft et. al. *The Post-Colonial Studies Reader.* London: Routledge, 1995, 27.

3. Gloria Emeagwali, *Women Pay the Price.* Trenton, NJ: Africa World Press, 1995, 10.

4. Irene D'Almeida, *Francophone African Women Writers: Destroying the Emptiness of Silence.* Gainsville, Florida: Florida Univ. Press, 1994.

5. Ibid.

6. Molara Ogundipe-Leslie, *Recreating Ourselves: African Women and Critical Transformations.* Trenton, NJ: Africa World Press, 1994.

7. Ibid.

8. Micere Mugo at a Symposium on "Gender, Politics and African Cultural Production." Africana Studies and Research Center, Cornell University, Ithaca, New York, March 1992.

9. Siga Jajne. "African Women and the Category 'Woman' in Feminist Theory." Paper presented at the African Literature Association Conference, Columbus, Ohio, March, 1995.

10. "Nana Benz" is the term emerging from the domination of the economy by women who generally tower above men and are popular for acquiring luxurious Benz cars.

11. Said quoted by Ann Maxwell in "The Debate on Current Theories of Colonial Discourse," *Kunapipi* 13.3 (1991):71.

12. Adeola James, *In Their Own Voices: African Women Writers Talk.* London: Heinemann, 1990, 114.

13. Bill Ashcroft, Gareth Griffiths, and Helen Tiffin, *The Post-Colonial Studies Reader.* London: Routledge, 1995, 9.

14. Gayatri Spivak, 27.

15. Ibid., 28.

Nigerian Women in Search of Identity: Converging Feminism and Pragmatism

Ada Uzoamaka Azodo

"A woman who holds her husband
as a father, dies an orphan."
—Hausa proverb[1]

I

Ordinarily, feminism and pragmatism, being terms that advocate divergent viewpoints, should not meet. Pragmatism favors principled pluralism, and does not promote the artificial balancing of competing interest groups. Feminism, on the other hand, almost always promotes only women's points of view and exposes, in order to eliminate through reaction, the oppression and domination of women (Duran 1993:159). Feminism fosters women's values and recognizes their contributions. However, at a certain point, a converging effect occurs, when habit, closed systems, fixed principles and dogmatism embrace revolutionary ideals (Seigfried 1993:2). It is at this point that one can then speak of a feminist pragmatism.

The focus of this chapter is the textual analysis of Flora Nwapa's postwar writings. More specifically, in chronological order of publication, the study texts are her longer writings, including *This Is Lagos and Other Stories, Never Again, Wives At War and Other Stories, One Is Enough,* and *Women Are Different.*[2] All five books were written after the traumatic Nigerian civil war of 1966-1970, in which altogether some four million Nigerians and Biafrans lost their lives, not to mention human suffering in terms of dislocation of family, loss of abode, death of loved ones, and ravage and damage to infrastructure. This paper is further designed to go quite beyond the exploration of the social, economic, and political problems raised in the writings of Flora Nwapa to interrogate the status of women in modern Nigeria.

In particular, this paper follows the "new" Nigerian woman in search of her identity in the contemporary society, which is very different from the society of the past. While it is clear that she resents her suppression by the combined forces of tradition, patriarchy, and motherhood,[3] yet it is not yet clear that she has resolved her situation in modernity. To make matters worse, there is a consistent ambiguity and near confusion which is apparent in the thoughts of the author herself on issues concerning the identity of women in the new era. These matters need resolution.

If I frame my analyses in this mode, it is because Nwapa seems to recognize, since the writing of her first two novels, *Efuru* and *Idu,* that the Nigerian woman must, of necessity, now go beyond her narrow ethnic group to find herself in the context of a new Nigeria. Yoruba, Hausa, and Fulani names now rub shoulders with Igbo names in Nwapa's writings, a phenomenon that had not been seen before. Secondly, Nwapa recognizes that women in a new historical time period, in a postwar era, have new problems and desires. Nothing prepared Nigerians for the experience they were going to have during the civil war. Many had heard about wars from veterans of the Second World War, especially the war in Burma; still the ravages and trauma of the Nigerian-Biafra War were a unique experience. Nigerians were radically changed forever with the violence and brutal armed conflicts, which Providence had averted at the time of independence from Britain six years before. Citizens were shocked out of a state of innocence and idyllic living and thrown into a harsh one of want and hunger and disease. Non-medical populace was hearing for the first time of kwashiorkor, the nightmare of mothers who would do anything to save their children from being turned into skeletons before their eyes due to malnutrition. Many women went into "attack trade"[4] to ensure the survival of their family members. Pragmatic women learned to overlook the hostilities in order to make friends behind enemy lines in search of the necessities of life.[5] A by-product of this experience is that women also tasted, for the first time, the benefits of economic freedom. For the first time, many women realized they were as capable as men to do and dare. To regain control of their lives, therefore, women saw that they needed to have economic independence from men. Immediately, one recalls William James's assertion:

> A pragmatist turns his back resolutely and once for all upon a lot of inveterate habits dear to professional philosophers. He turns away from abstraction and insufficiency, from verbal solutions, from bad *a priori* reasons, from fixed principles, closed systems, pretended absolutes and origins. He turns toward concreteness and adequacy, towards facts, towards action and towards power. (1984:228)

Flora Nwapa's own pragmatism as a writer comes into play through her campaign for the restructuring of the environment, in order that social progress may be engendered. We see her as a progressive and a feminist pragmatist, though she was not a declared philosopher and so did not belong to any philosophical school. Through her unique insights and concerns as a womanist, she worked for human progress in her society across gender lines. Her writings bear testimony to the fact that she did not indulge in abstractions. Rather, we see her preoccupation with the concrete, the search for human meaning and value in postwar Nigeria. Nwapa grapples with inherent dualisms/pluralisms, disfunctions/deviances, dichotomies, opposing viewpoints and stands in the nation.[6] She tries to provide a method for responding to a situation in which no one point of view appears to be the absolute answer to the problems of society. In a recent article, Jane S. Upin (1993:38), describes a pragmatist as one who above all recognizes the need to modify his/her milieu in order to engender social progress. Such "progressives," she argues, advocate economic reform and social change. They challenge uncritical submission to the authority of the past, oppose laissez-faire political doctrines, support suffrage and labor movements while championing sweeping changes in education.

Nwapa qualifies, perhaps, as a "sage-philosopher,"[7] given her new insights into deviancy as a product/symptom of society's ills, and not as a failing on the part of the individual. Thanks to Flora Nwapa, a prostitute is no longer seen as a wanton, lascivious woman, but rather a hard-working, resourceful woman employed in the informal sector and trying to make the best she can of society (Nwapa 1981:130). The author challenges uncritical condemnation of women's behaviors, especially women in precarious living conditions, in-between survival and death. She champions changes in educational practices, and calls for economic independence of women, as a step towards ending the submission and subservience of women to men. It is her hope that well-educated women, economically independent of their husbands, will work alongside their men in nation building and take better care of their children, the future of the nation.

Because Nwapa bases her views and ideas on concrete facts as she observes them all around her with the keen eyes of a literary artist and an empiricist, we draw a parallel between her pragmatism and Nnamdi Azikiwe's methodology of "eclectic pragmatism." According to this Nigerian indigenous philosophy, the best of capitalism, welfarism and socialism are blended together—pragmatically—in response to the unique situation of Nigeria beleaguered with social, economic, religious, political and cultural problems. Eclectic pragmatism, according to Azikiwe, adopts the useful and practicable in opposing viewpoints and blends them together to achieve harmony in the nation. Apparently contradictory decisions lead to the achievement of set or desired goals. Agbafor Igwe adds that eclectic pragmatism "...accommodates the fact that man is selfish and altruistic; rational and irrational; individualistic and communalistic...."[8] To this end, pragmatism is a philosophy which is in keeping with an Igbo adage according to which "one does not watch a masquerade from one vantage point alone." American classical pragmatism emerged towards the end of the 1890s, in response to the development of the rationality of scientific knowledge, including specifically evolutionary biology. Resultant shifts in religious and intellectual thought brought about the necessity to account philosophically for human meaning and value (Seigfried 1993:17). In the case of post-1960 Nigeria, Flora Nwapa saw the need to give meaning and value to human life in the post-civil war era, deprived of the benefits of religion. The war had brought about a certain alienation. At the same time, it brought about a certain craving for tradition and for luxury products of a capitalist econo-my—houses, cars, jewelry—in a rather uncomfortable manner. In reading Nwapa's later works, we get a sense of evolving life and the efforts by a sensitive and committed artist to make sense of women's lives, attitudes, and behaviors in the difficult environ-ment they find themselves.

Ostensibly a criticism of the Nigerian post-civil war era, Nwapa's writings engage in a search for self with the modern Nigerian woman in complex times. In what follows, I will endeav-or to elaborate this feminist/pragmatist effort through the follow-ing four imperatives: mothercraft, education, justice, and peace. The questions that these four issues will allow us to pose will enable us to understand better the conditions of women in one country of West Africa.

Most immediately, the issue of mothercraft asserts itself the-matically in the relationships between mothers and daughters. Mothercraft involves nurturing, preservative love, and socializa-tion of the child (Ruddick 1989:15; see also Bailey 1993:189). In the first part of this work, which reviews mothering practices (the sec-ond part argues that maternal thinking provides a basis for femi-nist peace parties), Sara Ruddick pursues what she terms a "practicalist conception of truth"(PCT). According to this philoso-phy, truth does not reside in taking a transcendental stance. Rather the criteria for truth can be derived relatively and perspectively from practices in which they are made:

> [T]here is no truth by which all truths can be judged nor any foundation of truths nor any total and inclu-sive narrative of all true statements: instead dis-tinctive ways of knowing and criteria for truth arise out of practices. (Ruddick 1989:13)

According to Ruddick's PCT, only those actually engaged in any practice and/or those familiar with the practice have the moral jus-tification to criticize. Even then, they should engage only in self-criticism:

> It is sometimes said that only those who participate in a practice can criticize its thinking....There are moral grounds for critical restraint. People who have not engaged in a practice or who have not lived closely with practitioners have no right to criticize. (Ruddick 1989:26).

Flora Nwapa's feminist pragmatist approach enables her to explore what the role of the mother should be in nurturing, training and socializing her daughter in the peculiar Nigerian environment.

Specifically in Nigerian society, this further calls for girls to be taught the skills for survival, and the necessity to identify themselves as people, that is humans, without reference to any man. It also requires the mother to teach her daughter to feel adequately empowered to stand up for her beliefs and convictions. In *One Is Enough,* the mother of the heroine, Amaka, advises her to seek economic independence from her man, even while needing him emotionally: "Never depend on your husband. Never slave for him. Have your own business, no matter how small because you never can tell...never leave your husband, I did not leave mine, but I was independent of him"(Nwapa 1981:9). A little later she adds: "...the richer you are...the better your husband will be and he will really appreciate you. Your husband's relatives will appreciate you as well"(Nwapa 1981:10). The idea is that every able-bodied adult woman should strive to be employed in the formal or informal sector, according to her abilities and/or qualifications. After Amaka (who had been involved in attack trade during hostilities) suffers emotional and physical abuse from her husband, Obiora (who resents her success as a business woman) she decides to leave him and make a fresh start as a contractor in Lagos city. This separation from a husband is contrary to what tradition would have approved. Amaka's mother, however, completely approves of her decision. She urges her to stop at nothing in pursuit of success—to "make herself," as the saying goes. Amaka shows that she very well learned the lessons of her mother when she decides to lay her snares for a celibate Catholic priest, who was in a position to influence positively the award of a one million naira contract job to her from an agency of the government:

> For the first time she was going to put into practice what her mother had been teaching her. She was not going to wait, she was going for the kill. A priest was also a man capable of manly feelings. Father Mclaid was a man, not a god. Perhaps Father Mclaid had never been tempted. She, Amaka, was going to tempt. That was the task that must be done. (Nwapa 1981:130)[9]

As men and women scheme to take advantage of one another, Nwapa blames society for individuals' apparent moral depravity. She terms postwar Nigeria "a world that has gone mad. It is not their fault. It is the fault of their age, and the society. They cannot act differently..."(Nwapa 1986:130). Yet in "Man Palaver," one of the stories in *Wives At War*, Nwapa appeals to so many ungrateful, complaining women to learn to cherish God's "big mercies" in the form of benevolent, generous husbands, and to dwell less on the "small mercies," like the fate of living with a stingy, selfish, and abusive husband, who is nonetheless a husband. It appears that any woman, according to Nwapa, who is "lucky" to find a husband should be more than grateful for her find. She should then get down to making her own money. She believes that some women simply exaggerate the problems with in-laws, especially sisters-in-law. Many times, jealous sisters of one's husband try to turn away the affection that a man has for his wife by bringing in other women to entice him away from her. Nwapa condemns, however, hostile Nigerian men who resent their wives' success. She traces the origin of this hostility to the ideological indoctrination of the populace by agents of British, Victorian, capitalist economy in colonial times who turned erstwhile independent, and hard-working women into men's helpers and servants in the domestic realm:

> Times changed and men began to assert their masculinity over their industrious wives. Men made fun of husbands, at drinking places and functions, whose wives were well-to-do, saying: "Look at him, just take a good look at him. He is less than a man, depending on a woman to buy his shirts for him, to spread out the mat for him. One day, instead of him, forking [sic!] her, she will fork [sic!] him. And they spat to show their disgust." (Nwapa 1981:17)

II

The Nigeria/Biafra War banished erstwhile Christian morality brought in by British colonists and missionaries, which was based on hope and faith on a future better life. At the same time, it seemed to exaggerate and reconfigure traditional morality according to which paradise resides in the past, not in the present nor in the future. Nigerians seemingly interpreted their predicament as being one in which no moral stipulations or admonitions existed at the collective level. Individuals suddenly found themselves with the responsibility of choice among competing points of view, practices, and beliefs (Rotty 1982:202). What is truth? Who has the right to tell what is right and what is wrong, to stipulate the criteria for choice, to pontificate on what is admissible and what is not? All established methods of doing things suddenly evaporated into thin air. What is left is a wide array of options and alternatives to deliberate on and choose from. Nigerian women now found themselves able to make practical and meaningful decisions according to what obtains today, according to "present standards" (Bickford 1993:105).

Education appears to be that gateway to survival for the modern Nigerian woman. To be discarded is the age-old belief that a woman should measure her success through the achievements of her husband. Appropriate attention should be paid to early childhood education, for it is at this early beginning that girls are indoctrinated into submission.

In "Man Palaver," a short story we have already seen, Nwapa presents a series of women with opposing viewpoints on the natural abilities of boys and girls. She concludes that given equal opportunities, both sexes can hold their own, one against the other. Nwapa then adds that asking girls to turn "the other cheek was cowardly, it meant bringing about your own end. It meant you were soft. The world was tough. The women were taught to be humble"(Nwapa 1980:52). *Women Are Different* presents three case studies of high school friends, Agnes, Dora, and Rose, who are later able to pick up their pieces, thanks to a university degree and skills to fall back on. Agnes becomes an Education Officer when her marriage fails. Dora, a nurse, begins a bakery business of her own on the side. She becomes prosperous and is able to buy back her house which her husband pawned away to be able to travel overseas to study. As soon as Dora is able to confirm rumors to the effect that he is living with another woman, a German, who is keeping him, she goes back to the village and divorces him. Rose, for her part, does not succeed in finding a husband, but she has a good job as a Public Relations officer in a reputable firm. She is thus able to take care of herself in a respectable manner as a single woman.

Is a woman's education an investment? Does a woman become an investment merely because someone, usually a man, has paid her school fees? Is a woman bound by the dictates of the fee payer, reserving no rights whatsoever to order her life and destiny according to her choice? These are some of the pertinent questions that Nwapa raises in relating the predicament of Adaeze, the protagonist of "The Chief's Daughter" in the collection *Wives At War*. Her greedy father tries to extract the last service from her for the school fees he paid for her studies in London. The Chief refuses to listen to her pleas to be allowed to return to London to wed a young man whom she has met there. Adaeze, in turn, does not want to be treated as an investment, much as she is very grateful for the financial support she received as a student. So, at the earliest opportunity, she absconds. Back in London, she marries her sweetheart, Ezenwa. In no time at all, she is expecting their first child. Her father thus loses doubly. He is estranged from his daughter's new family and is denied the privilege of giving away his daughter, a privilege that every Nigerian man looks forward to.

The fact of a father "expecting" service from his daughter after graduation as a sign of gratitude for financial support she received in time of need appears to be a simple, understandable matter. It is even more comprehensible when viewed in the traditional African context in which the married woman is expected to leave her family of birth and completely integrate herself into the new adoptive family of her husband. In this light, therefore, a father appears justified to extract whatever service he can before he "loses" his daughter (financial investment, as it were) to a "stranger" who had nothing to do with the formation of the "money making machine." If poor Adaeze had been a mere source of production and nothing more, perhaps there would have been nothing wrong with her Chief father demanding what he paid for. Complication arises when we realize that the "object" presented as a material resource also doubles as a female of flesh and blood who has feelings and therefore emotional desires too. In order, therefore, to appreciate the full impact of this issue raised by Nwapa, we need to see the entire short story as a process of flow, that is, in the full context of modern day Nigeria.

In our view, Nwapa's chief concern in this matter is that the influence of tradition, the impact of modern education on women, and the dictates of modern day living all need to be taken into account in human relationships, be they filial or not. Present day living appears to be "a complex package of interconnecting items which often must be read together in order to comprehend a total message" (Caputi 1994:102-103). The message that is implicit is that women are refusing to be treated any longer as material objects. They are demanding their rights to be treated as humans. The covert criticism here is against the Nigerian society and governments which shirk their responsibility to render services to the needy citizens, thus forcing them to fall victims to rapacious blood relatives who hold them for ransom. Where professional politicians and established philosophers failed to admonish the Nigerian federal government for its lapses, Nwapa took on the responsibility to become the voice of the voiceless millions of Nigerian women taken advantage of by the combined forces of tradition, capitalism, and patriarchy.

III

Not only are women refusing then to be seen and treated as material objects, they are also demanding to be treated as equals in relationships with men. They are expressing their resentment of beating, sexual abuse, and sexual harassment. In addition, they are seeking custody of their children after separation, instead of automatically handing them over to the ex-husband and his family.

In this section of our study of Nwapa's later writings, we connect the neglect, indignity, poverty, abuse, violence and harassment that women suffer in Nigeria at the hands of men to the institutionalization of patriarchal culture in that country. The man is supreme and almighty whereas the woman is made to be submissive, meek, and weak. Men can marry as many wives as they like with societal support; women are not allowed to have relationships with more than one man, to talk of marrying him. Immediately, the collective conscience condemns that woman who dares to flaunt her liberty and freedom like a man. The implication is that a women is less than a human being, and therefore different standards are valid for the two different genders. Furthermore, it implies that in the search for truth and knowledge, different canons must be used to study both sexes of humans. Must there also be different laws for men and women in society? In criminal matters for instance, might we be pushed to excuse men from responsibility for their actions because after all, they are men, "animals" driven by "natural" impulses to either pet, sex, or batter their sexual partners? These are only some of the very serious issues which arise from the unequal treatment of men and women in Nigeria. Flora Nwapa was the foremost woman writer to see through the evils of the masculinist teachings of Christianity and to hail the dawn of a new era of pragmatism. To create peace in relationships between Nigerian men and women, Nwapa provides suggestions for an adequate feminist peace politics by treating from divergent—and sometimes dichotomous—angles the issues in man-woman relationships she observed daily.

Upon her graduation from high school, Comfort in *Women Are Different* shocks us with her radical scheme for the future:

Get a job, work two or three years, hook a man, get him to the altar. Have three or four children for him, and if he does not make it, leave him. I am not going to continue to count pennies all my life. And marriage for me will not be for better or for worse. Only fools think of that. I am going to live fully. If I am lucky and first land on the man who has already made it, all well and good. I could stay. But if not, I will be on the move...Love my foot. (Nwapa 1986:30)

Is Nwapa, as Ifeyinwa Iweriebor ponders, "a feminist, a womanist, a man-hater, simplistic, idealist, or obscurantist"?

Women are less willing to devote their lives to pleasing a wayward husband, by making up and dressing up, hoping thus to regain his attention which was captured prior to courtship. Men, like Eze, in "The Loss of Eze," a story with an ironic title, will allow themselves to be "captured" or "snatched" from their wives by another woman. When the urge captures him to abandon his older wife, nothing will keep a man like that at home. This is the bitter truth that many women have learnt, not only in Nigeria, but also in other regions of the world. Another story, "Mission to Lagos," from the same collection, *This Is Lagos and Other Stories,* recounts the story of Yetunde's personal experiences, with both a Nigerian and a British lover, John Hammer. She concludes that whereas Western men seek the personal and intellectual growth of their wives, the Nigerians only need their wives for sex (Nwapa 1980:68-80). Our view, however, is that this assertion must be taken with a grain of salt, since Yetunde seems to have generalized a personal experience, elevating it unnecessarily to a universal level. In "The Delinquent Adults," another story in the collection, *This Is Lagos and Other Stories,* Nwapa condemns wife beaters like the Chief Clerk, as well as in-laws who continue to maltreat the wife of their dead relative by fighting over property. Still, in spite of ill-treatment by husbands, many women go into a dilemma when they need to leave their husbands. In "A Wife's Dilemma," a woman confesses her difficulty in making up her mind to leave her abusive husband:

My husband kissed me only when we were courting. When we got married, he forgot all about kissing. If I tried to kiss him, he shut his mouth tight. I tried sometimes to take the initiative, but failed. So I gave up and turned my attention to taking care of my children. (Nwapa 1980:55)

Whereas some fail to take the bold step out of an unsatisfying marriage, other women have gotten out and felt good about the decision
they made. In "Man Palaver," Anyaga boasts of her achievements in
this fashion:

> My son is in a private school. I have a good job. To
> hell with husbands. I am a single girl with a son. I
> like my life. I don't go sleeping around, but I know
> whom I want to sleep with. The choice is mine. My
> job is full of challenges and I meet so many inter-
> esting people. I have had offers of marriage, but I
> turned all of them down. One marriage is enough
> in a life time. (Nwapa 1980:48-49)

In another glaring example, Chinwe, daughter of Dora, who is the
friend of Comfort and Rose, in *Women Are Different,* refuses to take
her husband's abuse and leaves him. She asks for no support
money or alimony, but rather asks for custody of her children
when she divorces her husband. She goes to a "Sugar Daddy"[10] who
is very generous with her. In nine years alone, the old man had
built her a house and later married her. Nwapa concedes, however, that the young girls miss out on a lot in life by giving their bodies to old cronies. She hopes that Chinwe will eventually return to
her husband whom she really loves. Her "irrational" decision is
blamed on youthful exuberance: "If she were older, she would have
stayed" (Nwapa 1986:117).

But Comfort has an opposing viewpoint, which she explains to
Rose in this manner: "If you come from a family that is so poor
that they cannot afford three square meals a day, then you will
understand what I mean" (Nwapa 1986:116). In a different view,
therefore, only extreme poverty, according to Nwapa, would seem
to justify such immorality.

Dora, however, is not impressed that all that young women like
her daughter need in life is material success:

> She was not impressed that her daughter could
> handle a one million naira job. She as well as Rose
> did not approve of a society where it is possible
> for young women of twenty-four or twenty-five to
> boast of possessing a Mercedez Benz, big business
> associates overseas, expense account paid, credit
> cards and so on, without a husband. At forty, what
> would this brand of women be? What was left to

achieve? (Nwapa 1986:118)

What is evident here is that young Nigerian women are becoming defiant and refusing to accept the kind of life their mothers had. As Rose chimes in: "What Chinwe is trying to say is this, 'Mother I cannot take what you have taken from father. I am going to have my own back. No man is going to hold me to ransom...'" (Nwapa 1986:117). Nwapa seems to concur when she adds:

> Chinwe had done the right thing. Her generation was doing better than her mother's own. Her generation was telling the men that there are different ways of living one's life fully and fruitfully. They are saying that women have options. Their lives cannot be ruined because of a bad marriage. They have a choice, a choice to set up a business of their own, a choice to marry and have children, a choice to marry or divorce their husbands. Marriage is not THE only way. (Nwapa 1986:118-119)

Bisi, in "The Traveler," is yet another story in the collection *This Is Lagos and Other Stories*, instructs fellow women that they do not have to accept sexual harassment as normal and natural. She deftly rebuffs Musa's advances. She condemns all Nigerian Musas who believe that women are there to hold, touch, and lure into bed at will. At the same time, she reaffirms her dignity and the dignity of all women as human beings with choices. "This Is Lagos" further criticizes men's promiscuity and abuse of women, two practices which are rampant in Nigerian villages and cities.

Indeed, it becomes imperative that in Nwapa's opinion, marriage and procreation should no longer be the foci of women's existence. What is more, children are not always a joy to their parents, especially when they mess themselves up with drugs as Ezeka does in "The Road to Benin," one of the short stories in *This Is Lagos and Other Stories*. "Marriage can make and unmake one," declares Amaka, in *One Is Enough*. She goes on to advise women to have an open mind about dating. The first option when a marriage is failing appears to be to sit tight and disregard the misbehaviors of one's husband. That should give a smart woman time as well to take stock of her life while caring for her children. So long as the children know and have their father, adds Nwapa, the mother's relationship with him appears secondary. As adults who are aware of the game the other is playing, man and woman can then continue to play a game of convenience and reality, while still cohabiting. No one pretends to be deceiving the other, Nwapa further explains.

However, the author frowns on women who engage in nudity and strip-tease, two aspects of pornography which have filtered down from the West to Nigerian cities. She seems to argue that there is an empirical connection between the abuse of a woman's body and the subordination of women which must be noted and taken seriously as a means of exposing yet another means by which men sexually violate women. Young women who are forced by economic circumstances to resort to belly dancing and strip-tease often complain of the assault to their bodies by men who watch, who often touch or grab the women as they dance. There is this constant threat to the woman dancer as the men violators attempt to establish or maintain control over the female performer on stage. In this sense, strip-tease and nudity are aspects of pornography which foster male sexual violence against women. There is reason for Nwapa to believe that Nigerian men, like their counterparts elsewhere, would attempt to duplicate the practices of abuse of women and entrench violence against women should pornography be allowed to take hold in the country. There have been reported cases of white men forcing their Nigerian girlfriends to have sex with their pet dogs. Even though such news items had been carried by newspapers in connection with white male foreigners exploiting young Nigerian women, who knows what Nigerian men might want to do in imitation of their foreign counterparts?(See Adams 1994:63-70).

Nwapa presents Rose in "Jide's Story," who accepts her errant husband with no questions asked when he, Jide, finally comes home, after gallivanting for days with prostitutes and several other nice-time girls. That attitude leaves her husband baffled about the new breed of Nigerian women:

> For days, I thought Rose would ask questions, but she did not. What was at the back of her mind? Was she being liberal? Did she love me at all? Why was she not jealous? Why did she behave in that superior manner? Was she making a fool of me? I was told that when a married woman had affairs with men, it was her husband who was the last to know. (Nwapa 1971:33)

Not only are women conducting themselves in a respectable manner by fighting less over men, they are trying different options that will enable them to grow as human beings. Women are also eager to free themselves from chores, husbands' daily demands and other conjugal duties so that they can pursue occupations that mean more to them. Some women even bring in younger women for their husbands, in order to free themselves to pursue their trades (Nwapa 1981:8-10). For those women who cannot have children of their own, adoption seems to be another viable option(Nwapa 1981:20-21). Some women throw themselves into vocations, in search of a full life (Nwapa 1981:22). Other alternatives for women, as we have already seen, are divorce and single motherhood. Even lesbianism is creeping in, but is not yet accepted as an alternative lifestyle. Women are not only seeking justice in relationships, they are also working towards the dismantling of societal structures which support the subjugation of women.

IV

When women are indirectly involved in peace politics,[11] they manipulate the men in their lives to get what they want. When they are directly involved, they exercise voting rights to put leaders in place. They also take part in decision-making processes as individuals or as members of a group. Women also can use their own money or power to get what they want. When they use money to buy themselves influence, such women are addressed as "Cash Madams." On the other hand, when they wield power, they are addressed as "Thick Madams."

In "Wives At War," Nwapa uses the dream motif to stress the need for the repoliticization of Nigerian women both at the local and national levels in the affairs of their country. In the first dream, Bisi learns from her experience as a bride abandoned at the altar, no longer to regard marriage as a means of upward mobility and success in society. In the second dream, a championship soccer match between two African soccer teams, Biafra versus Zambia or Tanzania (the author is not certain which), is suddenly interrupted by Egyptian migs. They drop bombs everywhere. When the casualties are counted, only all the spectators are found dead. Nothing happened to the players on both sides. Rather, they are subsequently airlifted to safety by huge Chinese planes. They are safely deposited on the Oguta Lake, which has become a symbol of a sanctuary for the author. In this very significant way,

Nwapa dramatizes her disapproval of political disengagement, while at the same time advocating participation in the politics of one's country.

Many Nigerian women already had full experiences during the civil war to fall back on. Some women were part of the militia and, in some cases, took part as combatant troops. Some worked in the background as patriotic citizens who poured hot water, hot soup or hot oil on the heads of enemy troops in order to scald them to death. Yet others ran food kitchens, making dry packs to feed the soldiers. Others acted as nurses treating the wounded soldiers. Some women kept the fighting troops' morale up in good company as nice-time girls. They gave succor to the refugees in camps (Nwapa 1980:13).

In "Wives At War," we find that women's clubs, which are political arms of the struggle, proliferate. Nwapa would later criticize this proliferation as being the reason for lack of solidarity among women. In Biafra, there is the National Women's Club (N.W.C.), an affiliate of the Nigerian Women's Club, which also has foreign headquarters in Europe and America. Then there is also the Women's Active Sense (W.A.S.), and the newest, born after the breakout of hostilities, the Busy Bee Club (B.B.C.). The latter is an extremely militant group under the leadership of a radical lawyer, who declares the stand of her group in the Biafran struggle as follows:

> We are the indigenous group. We have been poised for action ever since war was declared. We are independent. We are not affiliated to any redundant planless group. We are the creation of Biafra and our aim is to win the war for Biafra. Right from the word go, we organized the women for a real fight. We asked for guns to fight the enemy. We asked to be taught how to shoot. Did not women and girls fight in Vietnam? We asked to be taught how to take cover and how to evacuate women and children. But those who did not understand mounted strong propaganda against us. They said we were upsetting the women. But we were realistic. (Nwapa 1980:14)

It is in the absence of men that women discover that they can really accomplish any task they set their minds to do. In "Daddy Don't Strike the Match," in the collection *Wives At War and Other Stories,* Mrs. Okeke single-handedly plans and executes the movement of all their household property from the North to the South of Nigeria, in 1966, in order to escape the pogrom of the Hausa against the Igbo. Back in Biafra, Mrs. Okeke organizes a women's solidarity group. The aim is to request Colonel Chukwuemeka Odumegwu Ojukwu, the Head of State of Biafra, to allow women to carry guns for battle just like the men. Many women believe it is a just war, which calls for their full involvement in every way possible. Then Mrs. Okeke's little daughter, Ifeoma, dreams about becoming a fighter pilot in the future! Is this Flora Nwapa's premonition—that Nigerian women, sooner or later, are going to be in the forefront of defending the nation as worthy patriots? From all indications, things augur well for even greater political involvement of women in the future.

It is, however, in *Never Again* that Nwapa makes a political statement about the physical and spiritual disadvantages of wars and her disapproval of all forms of violent confrontations as means of settling political disputes:

> Why, we were all brothers, we were all colleagues, all friends, all contemporaries, then, without warning, they began to shoot, without warning, they began to plunder and to loot and to ravage and to desecrate and more, to lie, to lie against one another. What was secret was proclaimed on the housetops. What was holy was desecrated and abused, NEVER AGAIN. (70)

Armed robbery and military governments have since become in Nigeria legacies of the violence imbibed before the war, which worsened during the war. Nigerians learned to shell one another, to conduct air-raids against and bomb one another. They learned to comb out infiltrators from the bushes sheltering them in order to roast them alive. The civil service, which was relatively service oriented, has become an institution of corruption for and torture of the populace:

> Before the war, a government official rarely ever asked for a bribe. If you gave him a present when the job was accomplished satisfactorily and you were paid, he was grateful. Alas, it was no longer so. You had to give money first before you are even considered for a contract. (Nwapa 1981:45)[12]

Within a feminist framework, it can be said that the Nigerian civil war caused women to rebound into politics and prominence in society.

We have tried to study Nwapa's later writings, in order to show how she continues, in the manner of an elder, to advise women on what it takes to succeed in modern times. We have been able to highlight some contradictions in her viewpoints. These contradictions, however, must not be seen as signs of inconsistency or weakness in her writing. Rather, they must be viewed as qualities of flexibility on her part as an experienced womanist and feminist, as well as pragmatist in the face of the overwhelming odds that women face daily in a complex society. Nwapa advises Nigerian women to take life one day at a time, and to be prepared to change their tactics for survival according to the dictates of the particular circumstance in which they find themselves. The focus should always be on survival today. The Nigerian modern woman is showing a remarkable resilience in her will to achieve and succeed. Issues of morality and virtue appear to take secondary place to the desire to achieve and survive. The prevalent philosophy appears to be *carpe diem,* live today and let tomorrow worry about itself.

NOTES

1. This proverb was quoted by Flora Nwapa on the first page of *One Is Enough.* It is a timely warning to women to beware that their personal dignity be not trampled upon in marriage by a disrespectful husband.

2. Flora Nwapa's books studied include the following: *This Is Lagos and Other Stories* (Enugu: Tana, 1971), *Never Again,* (Enugu: Tana, 1975), *Wives At War and Other Stories* (Enugu: Tana, 1980), *One Is Enough,* (Enugu: Tana, 1981), *Women Are Different,* (Enugu: Tana, 1986). All five books have been republished in 1992 by Africa World Press, Trenton, New Jersey. All quotations from the books are from the Africa World Press editions.

3. See Ada U. Azodo, *"Efuru* and *Idu:* Rejecting Women's Subjugation" which appears in this volume.

4. "Attack Trade" was trading behind the enemy lines in the civil war. In spite of hostilities, ordinary people still took care of one another's needs.

5. Elechi Amadi, *Ethics in Nigerian Culture,* (Ibadan: Heinemann [Nigeria], 1982), 39.

6. See *Hypatia* 8.2 (Spring 1993). Special issue on Feminism and Pragmatism, ed., Charlene Haddock Seigfried.

7. D. A. Masolo, *African Philosophy in Search of Identity* (Bloomington: Indiana University, 1994).

8. See Agbafor Igwe's *Zik, The Philosopher of Our Time.* (Enugu, Fourth Dimension, 1982).

9. "To Keep Nigeria One is a Task That Must Be Done," was the war slogan of the Nigerians during hostilities. That gave justification for, first the police action, then the war, and finally the total war.

10. "Sugar Daddy" designates a "dirty old man," who delights in dating girls young enough to be his daughters.
11. *Hypatia*, 9.2 (Spring 1994), Special Issue on "Feminism and Peace," ed. Karen J. Warren and Duane L. Cady.
12. Elechi Amadi, *op. cit.*, Ch. 10:82-93.

WORKS CITED

Adams, J. Carol. "Bringing Peace Home: A Feminist Philosophical Perspective on the Abuse of Women, Children and Pet Animals." Karen J. Warren and Duane L. Cady, eds. *Hypatia* [Special Issue on Feminism and Peace] 9.2 (Spring 1994):63-70.

Amadi, Elechi. *Ethics in Nigerian Culture.* Ibadan: Heinemann, 1982.

Bailey, Alison. "Mothering, Diversity and Peace Politics." Karen J. Warren and Duane L. Cady ed. *Hypatia* [Special Issue on Feminism and Peace] 9.2 (Spring 1994):189.

Bickford, Susan. "Why We Listen to Lunatics: Antifoundational Theories and Feminist Politics." Charlene Haddock Seigfried ed. *Hypatia* [Special Issue on Feminism and Peace] 8.2 (Spring 1993):104-121.

Caputi, Jane. "Unthinkable Fathering: Connecting Incest and Nuclearism." Karen J. Warren and Duane L. Cady ed. *Hypatia* [Special Issue on Feminism and Peace] 9.2(Spring 1994): 102-103.

Duran, Jane. "The Intersection of Pragmatism and Feminism." Charlene Haddock Seigfried, eds. *Hypatia* [Special Issue on Feminism and Peace] 8.2 (Spring 1993):159.

Igwe, Agbafor. *Zik: The Philosopher of Our Time.* Enugu: Fourth Dimension, 1982.

Iweriebor, Ifeyinwa. "Remembering Our Flora: A Beautiful African Voice." *African News Weekly* 6.2(January 27, 1995):18 & 27.

Masolo, D. A. *African Philosophy in Search of Identity.* Bloomington: Indiana University, 1994.

Nwapa, Flora. *Never Again.* Enugu: Tana, 1975.

——.*One Is Enough.* Enugu: Tana, 1981.

——.*This Is Lagos and Other Stories.* Enugu: Tana, 1971.

——.*Wives At War and Other Stories.* Enugu: Tana, 1980.

——.*Women Are Different.* Enugu: Tana, 1986.

Rotty, Richard. *Consequences of Pragmatism.* Minneapolis: University of Minnesota, 1982.

Ruddick, Sara. *Maternal Thinking: Towards a Politics of Peace,* Boston: Beacon, 1989.

Seigfried, Charlene Haddock. "Shared Communities of Interest: Feminism and Pragmatism." Charlene Haddock Seigfried, ed. *Hypatia* [Special Issue on Feminism and Peace] 8.2 (Spring 1993):2.

Upin, Jane S. "Charlotte Perkins Gilman: Instrumentalism Beyond Dewey." Charlene Haddock Seigfried ed. *Hypatia* [Special Issue on Feminism and Peace] 8.2(Spring 1993):38.

Woman, The Self-Celebrating Heroine: The Novels of Flora Nwapa

Emelia C. Oko

Like Chinua Achebe, Flora Nwapa saw herself as a pioneer. Where Achebe fought by recreating authentic African life to remove "years of denigration and self abasement," Nwapa was consciously and sub-consciously inspired to tell woman's story. But she did this in a state of innocence but not ignorance, and therein lies the beauty and lyricism of her earlier novels *Efuru* and *Idu*.

It is an exercise in female heroinism using self as quarry. She had mothers, and grandmothers, whose stories were as poignant as Efuru's. The reader need not be over hasty in decoding biography and the novel. Such a reductionist method was deplored constantly by Engels in a letter to Margaret Harkness.[1]

There is loving heroinism in the name Nwanyi Efuru, which means "a woman is not a loss." Female power is charm, resilience, and benevolence. Woman's witchery is valued over male brute force in a trading community. Nwanyi Efuru, a woman of infinite beauty and delicacy, cannot be lost. She is valued for her intellect, unlike what happens in a farming community.

Contrast the dominant farming Igbos, where daughters are named to denote preference for the male child: Nwanyibuife for instance, means, "a woman is also something (or, better than no child)," implying that the desired is the male child. The second female child for the Igbos is named Chinwe—God's own, meaning that God knows why he inflicts daughters on African farming families. Efuru emerges from a community where the female is valued in trade, unlike the farming Igbos. Contrast the Oguta naming of daughters: names like Ogbenyeanu, which means, "a poor feckless man will not marry my daughter." Ajanupu is a valued daughter; the name means "will you be able to marry, cherish her?"—hence the insouciance and irascibility of that Ajanupu in Efuru, her foil. She sees clearly and does not equivocate in speech. Male is second-class to her superior intellect and clear-sightedness.

Female power resides in ideal constructs contrary to the Aristotelian denigration of woman. Efuru is a woman better than the average, beautiful, hardworking, successful, humane, and a philanthropist, but she demands only a little thing from the world: emotional satisfaction. It is narrow to define her ideals as love, unless of course we realize the many dimensions of loving Nwapa explores.

Her heroines are secure in parental and extended family love; social admiration insulates them in a sense of self-worth. They do not need husbands really to restate their obvious value which society affirms in everyday intercourse. Nwapa's symbolic naming is an index of the value structure surrounding her female characters and is a social index. Nwanyi Efuru meaning "woman is not a loss," means that a Nwanyi Efuru woman is valuable. Such a name arises from a culture cluster where woman's worth is not questioned as in the larger Igbo heritage.

Among the Oguta/Igbos, classified more accurately as Oru rather than Igbo, a woman's worth is valued higher than male worth. The value structure has a foundation in the economic and emotional fact of woman's nature and mode of existence. Her maternities, menstruations are times of vision. Woman among the riverine Orus is an astute and visionary trader as we see in *Efuru* and *Idu*. Her social status is as high as a man of achievement.

Among the Igbos, who surround the Orus geographically, woman's worth is devalued because she lacks economic strength. Cassava is a woman's crop while yam is a man's crop, the king of crops, as we are all now familiar with in Achebe's historical novels *Things Fall Apart* and *Arrow of God.* Yam is a crop that is difficult to grow, demanding in its nurture, but when harvested it is easier to store and lasts easily beyond one agricultural year. Such a crop naturally deserves the reverence it receives in the agricultural community. Did man deserve the symbolic equation of yam to maleness? Fact disputes this agricultural imaging. The male gender is the weaker biological sex and therefore there are less male children born than female children. But the scarcity of male children then naturally puts a higher economic value on maleness and therefore the justifiable reverencing of the male child as a king. However, most progressive societies through the ages recognized the fallacy of kingly power symbolized in maleness. Actual power resides in the Queen while the king is the ceremonial head. The priestess and the Queen, in rituals, religion, and politics, show in different cultures the encoded superiority of woman, masked by the ceremonial need of having a male king who is more than family head. He is the super; even if he fails, there are substitutes provided with political consent for the Queen, who is usually surrounded by eunuchs so that her affection is safely focused on the king for social security. Simone de Beauvoir[2] documents the details of this social masquerade of male ceremonial power on which Clark's tragedy *Song of a Goat* is based.

Nwapa, as we can see, comes from a social complex that values woman above men because it is women who are engaged in trade among the Riverine folks all along the Niger and the Atlantic coast from the Port Harcourt Igbos down to Onitsha to the Benin Kingdom of Oru. Ikenna Nzimiro traces the cultural link between the Oguta Ibos, the Bendel Igbos, right to the mythic Oru, which is not just mythic but a local government area near Benin.[3] The father of Emecheta's *Ogbanje* Ojebeta in *The Slave Girl* travelled to *Idu*na Oba to get the silver bells to scare Ogbanje's spirit companions and retain a loved daughter in the world of the living. *Idu* in Oguta is the rebirth of this female child of wonder. Both "Efuru" and "Idu" denote beloved naming of the girl child valued among trading people. Nwapa's feminism is a celebration of the female gender.

Efuru's value and Idu's value are located within a cultural complex that venerates woman as a symbol of the late goddess Uhamiri, who endows women with beauty, genius, intellect, and a humane nature, but unfortunately she does not give many children. The Igbos around Owerri venerate a woman's 12th maternity. This is really the center of agricultural Igbo and where farm labor is needed to wrest a precarious living from an unwilling soil as we saw in *Things Fall Apart*. Trade is a game not of numbers but of foresight and intelligent planning and the gentle duplicity of a beautiful woman canonized in Greek literature in Odysseus' entrapment by the sirens. Many children are a social drawback in an affluent society. The mythic pattern of Uhamiri can be clearly decoded in woman's economic power, industry, and charm. Myths always confuse whose attribute is the donna and which is social accretion. The myth of female beauty is a modern illusion. These attributes do not differ much from the mythic pattern of Uhamiri. *Efuru* and *Idu* celebrate female nature, female charm, female economic power, and her encounter with maleness. It is in keeping a male lover that her individualism is challenged to accommodate male otherness not as a transitory admirer but resident husband in *Efuru, Idu, One Is Enough* and *Women Are Different*.

The center of enquiry is humanistic and economic to decode the nature of female success and failure in self-definition. Why can she not keep this intransigent male otherness in tow? If she abdicates self to become a male plaything like Efuru's co-wife, she still loses a sense of self and does not retain male respect or love.

If she is fulfilled in her career, in trade and professions, then she retains male love, only in the limit of male capacity for generosity. Since man in patriarchy loves only himself and what advances him among his kindred, he must be allowed to love woman only within the narrow limits of his selfishness and fickleness. This means there is a tremendous recreation of romance but never enduring love of woman. The romantic Scottish poet Robert Burns writes about male hope of enduring love even when physical charms have been destroyed by age.[4] It is the difficulty of realizing this ideal situation where woman is valued for her selfhood, that Flora Nwapa's novels investigate. Her novels usually define being, the female sense of self before the encounter with male otherness which challenges the integrity of self. Efuru is a successful woman in trade, *beautiful* and *psychic*. She is well-endowed as a woman physically and intellectually. As a successful trader her "hand makes money."

She is a philanthropist. She has given Ogea's parents money for farming and paid for the hospitalization of Ona. The community sings her praises. And then she faces the challenge of finding completion in male/female relations. This is also the pattern in *Women Are Different* where three young women who go for entrance examinations pass into secondary school. It is after they define their ambitions as self that the author introduces the complication of their sexual encounter with maleness. What attracts a lover, as Shelley says, is the physical and intellectual "meeting of souls." The soul then casts an enduring light on the beloved as Shelley affirms in "Epipsychedion," a "hymn to love and psychic integrity."[5] The light of true love endures to light the vision of the beloved even when physical charms fade.

Men and women do reach this peak of romantic idealism in love. Marriage is supposed to domesticate passion as Elechi Amadi shows in *The Concubine*, but marriage as a social institution may domesticate dreams or, as in most cases, it is a breaker of illusions and dreams. This is why sociologists have documented that marriages survive more after the age of thirty in liberal societies like America when marriage is sustained not by illusions but by the need of two souls. In closed, culture-bound societies marriages endure, but it requires female self-effacement, female self-erasure to let such a union of unequal care and needs remain intact. Among the Igbos, adulteries are chastised in songs and penalties, such as a bottle of gin and a goat. The goat is the ever-recurring metaphor of female foolishness to show that marriages survive in patriarchy but it is by sacrificing the junior partner, woman. The couple that survives has been tempered by the fire of extended family interference, greed, and love. Nwapa's *One Is Enough* explores how woman contends with the soul destroying nature of marriage as an institution.

Amaka's mother-in-law is the image of economic greed. For Amaka's marriage to survive in *One Is Enough*, it needed her constant humiliation to her alter ego, her mother-in-law, and her constant self-effacement to her husband. She must pretend to be the junior in a union where she is obviously superior in intellect and in economic success. She buys a car for her husband and pretends to everybody that her husband bought the car. When she breaks the illusion of this masquerade, she clobbers him with a heavy instrument, symptomatic of her emotional fury.

As Simone de Beauvoir shows, woman is not allowed in patriarchy to realize herself in transcendental goals. She projects her ambitions into her children or husband. The wicked mother-in-law is the frustrated mother. Woman changes her financial status upwards by marrying a rich man, abdicates self-hood as a wife before she will be happy in her dominance as mother.

Ironically, the wicked mother-in-law is the converse image of the young wife. They are both aspects of one female self, but because they do not recognize their identical destiny they are locked in futile combat, while the husband is the eternal winner, complacent, venal, manipulating the women with a heart for their economic gains. Nwapa sustains the image of female existence as the weaver of spells/money. She is successful economically. In *One Is Enough* Amaka realizes the dead end of her marriage and her combat for her husband's favors. She breaks free to realize self.

This first break marks the ascent in economic and emotional self. The crossing of the Niger River becomes a metaphor of her passage from self-erasure to self-assertion. She was a successful teacher in the first part and bought a car (the metaphor of social mobility for her husband). A "good woman" even effaces the joyful fact that she can provide for her family, by hiding the fact that she bought the car. In Lagos she joins an a-moral society.

The concept of love and marriage in Flora Nwapa's novels *Efuru, Idu, One Is Enough,* and *Women Are Different* is conditioned by the liberal ethos of Oguta social life. This ethos of a trading community as we have seen must naturally be markedly different from that of her Igbo agricultural neighbors. Achebe shows in *Arrow of God* the moral ideal to which Ezeulu stood firm refusing prosperity by rejecting to be a warrant chief and an arm of the colonial exploiting machinery. The sacrifice of this economic adventurism of the Igbo is insignificant compared to the larger fundamental sacrifice of his own entire farmland and that of his clan. The yams of ritual he ate to mark the new moon are a symbol and a fact. Yam, celebrated as a mark of Igbo manly strength and achievement through Okonkwo, is sacrificed by Ezeulu to prove that these people who value group and personal achievement put an even greater premium on a moral ethos. This morality is central to the male principle that Ezeulu, chief priest and servant of Ulu, stood for.

The female principle in Nwapa resides around the mythic pattern of Uhamiri as we have seen. This female principle is humanistic and moral. The Oguta woman holds as a moral priority the integrity of her body. Romance becomes the motor of her siren quality, a gentle teasing of a beautiful woman that promises more than she delivers. Honesty and hard work win Amaka powerful friends among the military who use her as a cover for their national looting called business contracts. Contracts are awarded for jobs that have already been done so that military officers will share the contract money. Amaka joins in this lucrative charade and amasses a fortune; she builds her own house. She also meets Father Mclaid, the military pastor. Father Mclaid is a Nigerian orphan who was rejected by his family and adopted by the Irish and trained by them. He has always suffered the emotional instability of this first betrayal. He finds in Amaka emotional solace. Two battered psyches meet for a complement that yields twin sons. Father Mclaid is ready to leave the priesthood and take care of his sons and Amaka. But Amaka knew the perils of marriage as an institution for breaking dreams and the free flow of affection as Shelley demonstrates in "Epipsychedion." With Father Mclaid she finally becomes pregnant, and attains motherhood. But she refuses to marry him, declaring as the title says that one subjugation of female self in marriage is enough. Amaka means to keep men as friends and to bring up her twin sons without the social cover of marriage. The iconoclastic nature of *One Is Enough* is quite definite.

Marriage, as a social system, confines woman into accepted modes of obedience to man and subservience to all in-laws. The desired romantic reciprocity we saw in Efuru's romance with Gilbert is destroyed in the possessive structure of marriage. The Oguta woman taking her moral ideal from the Uhamiri cluster abhors casual sex. In fact, it is the central moral that sex is only in genuine love and genuine love presupposes the permanence of marriage. Marriage as an institution does not necessarily demand loving but the irony of human nature is that love demands commitment and man's nearest approximation to commitment is possession in marriage. In the 20th century, many African male writers have shown the deep agony and dilemma men and women face when marriage by its social demands destroys romance and finally even love. Marriage and romance are mutually exclusive. But marriage and love survive through sacrifice.

Nwapa's sensitive narration encodes the understanding of these humanistic verities which have taken male writers such a long time to understand. In fact, it is the Ghanaian poet, Kofi Awoonor in *This Earth My Brother!*, the Sierra Leonean novelist Lenrie Peters in *The Second Round,* and of course the ever sensitive iconoclastic Ghanaian historian novelist, Ayi Kwei Armah in *Two Thousand Seasons, The Healers,* and *Fragments,* who can decode woman's sensitive nature as part of maleness but man's sensitivity is devalued for social survival in a state of nature. Nigerian male writers are still in the dark, except for Elechi Amadi's *Estrangement,* and Bode Sowande's *Without A Home.* They seem to live perennially in a twilight world with women nurturing them, unable to identify sensitivity in male and female nature.

In Flora Nwapa's novels, following the truth of Oguta ethical norms, love is the pre-requisite for marriage, and love by its very nature is a very limiting human emotion but expands its light of benevolence to all. The lover is the most understanding, most benign of people but also the most jealous of beings. The archetype of the lover is Jehovah God, an understanding lover celebrated in the psalms of David in the most lyric poetry encoding emotion that in human terms translates easily into psychic and physical passion. But Jehovah God is immanent, immaterial. The God that spoke to Moses in the burning bush is disembodied. Human man was built in his image but man has never seen God, he cannot conceive of heaven. Therefore, human loving fraught with imperfection because, by his nature, man is imperfect and is perfectible only by love.

Love becomes the fulcrum that moves the world. It is love that generated the world. The same love redeems human relatedness and therefore the beauty and lyricism inherent in the recreation of love in all the genres: the novel, poetry, drama, sculpture, and painting. It is poetry that is the supreme genre for recreating the soul's response to the human need to relate meaningfully to each other. Can the reader blame Flora Nwapa for making love the center of all her novels? The lack of it will kill man. Too much of it makes the world of commerce and economy a meaningless ephemera as *Idu* shows. *Idu* the astute trader shortchanges herself when her mind is on her beloved husband and the prospect of the fruition of this love in childbirth. *Efuru* is the lyric of love and romance. It is interesting to see how long romance survives in this marriage. The early morning swim in the beautiful Oguta Lake, the male/female childish play in the lake draws the anger of the neighboring women whose lives are starved of romance. Efuru and Gilbert in their playful loving call up the pure lyric of erotic celebration of female self and male otherness. This idealistic love collapses because of Gilbert's intransigence but Efuru remains true to her quest for male/female meaningful relatedness as the meaning of life. Efuru goes back to her father's house without bitterness to rebuild herself.

It is as one gets older in chronological age that one appreciates the supreme genius of Flora Nwapa. The supreme feminist born in an egalitarian society, she explores how women contend with their existence before and after the central motor of loving. We say that Amaka is iconoclastic. She breaks the barriers that shackle her and even gets impregnated by the Catholic priest; he is a pauper, social pariah, reclaimed by the Catholic Church. In *Women Are Different,* the mothers in this novel break down barriers of hypocrisy and become successful business women, but then their children carry iconoclasm to the ultimate conclusion.

As the existentialists point out, if there is no God, if there is no certainty of God, then everything is possible. Everything becomes indeed possible for the undergraduate of 20th century Nigeria as we see in *Women.* If modern man is not sure of the divine, he is assured of the reality of sexually transmitted disease. Perhaps the advancement from Brecht's fear of congenital destruction of the human race through the pollution of the very source of life-sex and love was Brecht's artistic illusion. If syphilis is an illusion, Acquired Immune Deficiency Syndrome (AIDS), in which sexually transmitted disease ignores all the efforts of modern sciences, is not. In *Women* sexual foolery climaxes in a dance macabre. The literature of the absurd, not realism, is what will now encode modern sexual dilemma. Nwapa makes a tentative beginning in this great human enquiry of modern sexual dilemma, disease, and economic greed that mixes sex and drug abuse.

In *Women* there can be no conclusions. The novel is inconclusive. Structurally the story ends but the suggestion is that it will never end until man and woman meet each other in honesty and truth. The alternative is disaster for the human race. The pathos of Amaka's conversation with her mother-in-law breaks the reader's heart at the beginning. The language is that of the cry of an unrequited heart, recalling folk songs where the rejected child begs not to be thrown away and be allowed just mere existence. Amaka begs her mother-in-law not to reject her for her barrenness. The pathos of that confrontation is evocative, justifying her eventual rejection of a soul destroying union.

The lyricism in the language of *Efuru* evokes trading on the lakes that take her to places. She constructs the strategy of her trade as she rows on the lake. Beauty and guile are weapons of her trade. The language at the opening of *Efuru* sustains the illusion of a world full of potential and romance: "They saw each other fairly often and after a fortnight courting, she agreed to marry him"(1).

The illusion of ease is sustained by Nwapa's peculiar prose—a transliteration like (Achebe) of Oguta people's idioms: "the dust she trods on mean something to him"(1). The romantic nature of loving is Nwapa's belief in the power of love to survive human hardship. The illusion of love is communicated in her use of traditional idioms: "My brother, help me to talk. I am tired of talking"(94).

Talking—the active verb has been the mode of communion of lovers as they explore the wonder of their new found world of wonder, their body. To see male other in female eyes and female otherness in male eyes, that is the story of genesis. That Adam was asleep and lost his rib and is ever looking for, searching for, touching, and reaching out to woman, all verbs of active participation in the human rhythm of recreation of wonder in life and loving. This concept has been permanently enshrined in literature by the metaphysical poets who saw in woman a celebration of God, "My America, my New found land."

In *Efuru* the opening lyricism of the romantic celebration of love is sustained until Gilbert marries a second wife. This wife is quick to accuse him of being an ex-convict, sustaining Emecheta's mythic pattern of the hateful other woman as the obverse of the loving wife in *The Joys of Motherhood*. The loving wife is too busy with her love and trade, aspects of herself. She does not regenerate. Her other self—her co-wife—regenerates in hate because she in her self-hood is socially devalued into a reproducing machine. The language of romance becomes discordant in a house of two women, the obverse of oneself. The man introduces, with the connivance of the beloved, yet a third wife to even the equation. The triangular love becomes a quartet. Once the lyric of the duet is broken, the novel becomes a discordant medley of hate and distrust from which Efuru, the celebrating self, must extricate herself.

It now becomes clear that the novel *Efuru* is the lyric of woman's self-hood, a romantic celebration of woman as vital and central to the universe. The myth of Uhamiri the beautiful but non-generative rich woman of the lake is the prototype of Efuru. Why does she fail to regenerate is the sustained question that Emecheta's deified mother in *The Joys of Motherhood* answers. Why should she regenerate in a world that refuses to see the obvious—woman's genius, her visionary self? The world rejects her natural biologos that teaches her the sacredness of life in her maternities and informs the sensitive woman of her biological rhythms during her menstruation. (In *Idu,* Idu knows and smells her pregnancy through her urine.) It took all of the 17th, 18th and 19th centuries to convince the male gender that woman understands her body language better, and when she is at the height of her regenerative, creative self, the male gender, in fear and ignorance, describes her as mad, manic, predatory, a language that encodes the truth of her visionary self and her genius while ironically denying the validity of her vision.

In *One,* the naive pathos of Amaka's plea for social acceptance gives way to the aggressive language of business. But in the story of the man of courage, Okonkwo encodes also the poetry of his first and most wonderful lover, the audacious Ekwefi. The story of Amaka's success as a trader and business woman will be incomplete without its ancillary, the story of her loving.

Amaka means, "Nwa Amaka, a child is a beautiful thing." The business woman meets a social outcast, Rev. Fr. Mclaid, traumatized as a child, rescued by the Irish, he had lost caste and appreciates woman's social loss of caste in marriage, especially a woman brutalized in marriage like Amaka. Mclaid weeps tears as only angels weep. Macho male is not allowed the luxury of tears although Achebe suggests that they are allowed to weep on their wives' bosom but keep as a deep secret.

Elaine Showalter demonstrates in *The Female Malady* what the romantic poet Wordsworth argues in *Preface to Lyrical Ballad.*[6] They agree that sensitive men do weep human tears and their sensitivity qualifies them as poets and legislators of a moral universe. Mclaid's sensitivity is what awakens Amaka's dormant regenerative impulse. When her soul is at peace, she regenerates twice as much, a dangerous loving. To deliver one at a time for female physical safety is the Igbo man's prayer. Twins are dangerous as any gynecologist will affirm and Mary Slessor is a human legend for reclaiming twins in South East Nigeria. Nwa Amaka really comes into her own after rejecting marriage. She is the truth of her name, that a child is beautiful thing. Male gender may father but since they do not nurture life like woman for nine months, they never appreciate the joy and wonder and sacred duty of parenting. If men stay close to women they can share her world of wonder and the magical world of children's innocence—the miracle of human regeneration. This poetry of love is an ancillary code that runs through the second part of *One* as the harsher world of business considers contracts paid for work not expected to be done. The negotiations in the board room where there are no boards and classrooms without chalk or board is the irony redeemed by the sustained romantic illusion of man and woman surprised by love in Mclaid and Amaka.

Amaka's six years of marriage were not happy, as the narrator pretends to show. What she encodes is that Amaka lived in a shell of a marriage where her head told her love and her body told her hate. The contradictory impulse is sustained by that bizarre confrontation of Amaka with her other self, her mother-in-law, and the fact that her more sensitive body refused to regenerate in a union of conflicting impulses: hate and possession. The doppleganger effect of the *Bildungsroman* on the European novel of self-exploration is manifested in *One*. Multi- dimensions of self are seen in mother-in-laws and co-wives, aspects of unrealized self.

The lyrical language of woman's self-celebration and the celebration of the agonies of loving male other and his gendered intransigence is a common enough subject of love poetry from man and woman. This language of pure lyricism at the heart of love and the lament that love does die, indeed must die physically, is the core of *Efuru* and *Idu*.

With *One* the reader is astonished at Nwapa's radical turnaround latent in the very lyric of love. The remembering self we saw is the subject of *One*. Amaka talks of things past to help her define the present. It is the deadness of child/mother-in-law unnecessary charade, male refusal to mature and be weaned from the breast that is the subject of the first part of *One*—the latent satire on male gender is masked by the sweeping statement of the omniscient narrator; "six years of marriage" is too quick a summation of six years of innocence and guile to boost the ego of a dependent husband by a more industrious wife. But there is subtle guile in the metaphor of Nwapa's art; he is husband/son/lover. Such self-sacrifice needed to be destroyed to reinstate individual health for both male and female. Amaka then charts a new course with a sensitive male, Father Mclaid, while the former husband remains in his world of contrived warfare that benefits the superior partner of partriarchy—the male. The Onitsha Bridge marks a physical and metaphoric landmark and divide between negation of self and the full-flowering of self in the Lagos metropolis. In Lagos, colors are subtly mixed and matched in an ever-changing kaleidoscope where morality is also ever changing, ever shifting because all war is immoral. The engineers of war, the military, are the real profiteers, not the heavy female widows, outcasts in a benighted world where Amaka takes a doubtful place of honor in her right as beautiful, educated, moral, and hardworking, even though there is a selective morality in her world.

Women Are Different is astonishing in the complexity with which Nwapa interweaves the many voices of prose; the three women, Agnes, Dora, and Rose, are given cosmopolitan names in contrast to Oguta names in *Idu*, *Efuru*, and *One*. These women who meet at the entrance are all aspects of her self-image. They complement and oscillate in each other's life. Nwapa has indeed reached a new maturity, recognizing that maleness is otherness—fascinating, intransigent, but necessary in the generative act. Women do not need the male engineered war to win his favors, such as they are. *Women Are Different* follows an epic quest first for economic independence and then emotional attachment to male others, into the generation of daughters who love and are betrayed in love. If the mothers had a doubtful morality, the daughters are really Jagua Nana's daughters. They have learnt the subtle art of concealing what should be exposed and exposing what should be concealed, a paradox on male sexual fantasy which Nwapa replicates in doubling images of mothers and daughters, lovers/earth mothers, a very strange medley in which it is difficult finally to disentangle who sleeps with whom. But as men would say, "Who is talking of sleep, only the dead sleep?" The language of this epic has the grandeur of the mock heroic: it is celebrating and diminishing. Agnes faces her arranged marriage with a stoicism and intelligence that command respect. She refuses to be denied the chance of getting her basic "O" Level from the Secondary School. In her marriage, love and sex are social duties. Sleeping with her husband is nothing special. She submits herself like a lamb ready for slaughter (57).

But she falls in love all the same with Ayo Dele, who encourages her to further her studies until she gets a university degree and can then negotiate with her husband from a position of economic strength. Agnes lives for herself but she knows social responsibility. She experiences passion and love but the narrator kills off this romantic ideal when Ayo Dele dies so that Agnes must be content with life as self without the support of a romantic lover. Her daughter Zizi has no romantic illusions. It seems, as her society becomes more economically buoyant with oil wealth, Zizi's generation loses the enchantment of love and romance.

Zizi's marriage is a farce; Zizi and Theo agree to live separate lives after marriage. Theo uses Zizi to get economic wealth, houses in New York and London from his parents. Parents of spoiled sons, Theo's parents are deceived, although the novel does not explore this male failure and its parental antecedents. Intransigent husbands are reclaimed to give family stability. Too much egoistic love leads to a mania we feared in Elechi Amadi's *Concubine.* These women are nobody's concubines, only adept at female witchery or witchcraft, made wholesome by exposure into the light of realistic recreation of how women sustain their economic independence and then reach out to a plenitude of self-loving.

Dora laments her daughter's immorality after her own self-sac-rificing motherhood and wifehood. But Dora is a reluctant wife to a husband who abandons her for a German girl. Chris has always been inadequate and fraudulent, both in his government work and in his relations with the two women, Nigerian and German. If Dora accepts mediocre maleness, her daughter is entitled to define her own morality. Chinwe and Zizi define a new Nigerian morality like Amaka in *One Is Enough*. In a world where the rich do not become rich by hard work as their traditional forebears, but by petroleum bunkering and cocaine trafficking, how do you teach the young? If the parents are teaching wrongly through practice what does Nigeria expect of the young? These are in fact the alarming ques-tions Zizi and Chinwe pose for a moralizing, immoral older gen-eration:

> God help Elizabeth and Theo. They are part of the world that has gone mad. It is not their fault. It is the fault of their age and the society. They cannot act *differently.* (130, emphasis mine)

But the sustained position of Nwapa is that women are indeed dif-ferent because of their genetic endowments. Therefore, they are faced with the challenge of maintaining their difference to redeem or change their world. It seems Zizi and Dora do not have their mother's strengths in a tenacity of purpose that is also morality. J. P. Sartre, the existentialist philosopher, argues that man must choose and in creating himself create meaning in his world.

Dora, Agnes and Rose were defined by choosing their world:

> Her generation was doing better than her mother's own. Her generation was telling the men, that there are *different* ways of living one's life fully and fruit-fully. Their lives cannot be ruined because of a bad marriage. They have a *Choice,* a choice to set up a business of their own, a *choice* to marry and have children, a choice to marry or divorce their hus-bands, marriage is not the only way. (119; empha-sis mine)

The reiteration of choice challenges the individual. Dora, Agnes and Rose have chosen in their ways. Zizi and Chinwe are choos-ing too—only their ideals seem too microscopic, ideals of self-sat-isfaction without social responsibility.

NOTES

1. See Georg Lukacs' book, *Writer and Critic and Other Essays,* p.83.
2. See Simone de Beauvoir's brilliant study on the female estate, *The Second Sex.*
3. Ikenna Nzimiro gives an in-depth analysis in his book, *Studies in Ibo Political Systems: Chieftaincy and Politics in Four Niger States.*
4. Robert Burns explores this idea in his book, *Poems and Songs.*
5. See Percy B. Shelley's creative work, "Epipsychedion."
6. See Elaine Showalter's *The Female Malady: Women, Madness and English Culture: 1830-1980.*

WORKS CITED

Achebe, Chinua. *Things Fall Apart.* London: Heinemann, 1958.

___.*Arrow of God.* London: Heinemann, 1964.

Amadi, Elechi.*The Concubine.* London: Heinemann, 1966.

___.*Estrangement.* London: Heinemann, 1986.

Armah, Ayi Kwei. *Fragments.* London: Heinemann, 1974.

___.*The Healers.* London: Heinemann, 1979.

___.*Two Thousand Seasons.* Nairobi: East African Publishing, 1973.

Awoonor, Kofi. *This Earth My Brother!* London: Heinemann, 1970.

Beauvoir, Simone de. *The Second Sex.* Trans. Parshley. Middlesex: Penguin, 1973.

Burns, Robert. *Poems and Songs.* Ed. James Kingsley. London: Oxford Univ. Press, 1973.

Emecheta, Buchi. *The Joys of Motherhood.* New York: George Braziller, 1979.

Lukacs, Georg. *Writer and Critic and Other Essays.* Ed. and Trans. Arthur Kahn. London: Merlin, 1970.

Nwapa, Flora. *Efuru.* London: Heinemann, 1966.

___.*Idu.* London: Heinemann, 1970.

___.*One Is Enough.* Enugu: Tana, 1981.

___.*Women Are Different.* Enugu: Tana, 1986.

Nzimiro, Ikenna. *Studies in Ibo Political Systems: Chieftaincy and Politics in Four Niger States.* London: Cass, 1972.

Peters, Lenrie. *The Second Round.* London: Heinemann, 1965.

Shelley, P. B. "Epipsychedion." Ed. Kermode and Hollander. *Oxford Anthology of English Literature.* Vol. II. London: Oxford Univ. Press, 1973.

Showalter, Elaine. *The Female Malady: Women, Madness and English Culture: 1830-1980.* London: Virago, 1987.

Bearing the Burden of Change: Colonial and Post-Colonial Experiences in Flora Nwapa's *Women Are Different*

Adeline Apena

This article is based on Flora Nwapa's perspective in her novel, *Women Are Different* (1992) which provides the background and information for the analysis in this study.[1] The analysis raises new issues in the study of women in contemporary Nigeria. The interpretations are drawn largely from the originality of empirical observation, personal stories, experiences and accounts from the writer herself, a product of independent Nigeria.[2] The study discusses the responses and reactions of two generations of Nigerian women to the challenges of modernity in traditional society and modern Nigeria. It focuses on women's capacity to adapt and adjust to change. Its objective is to demonstrate that in spite of serious patriarchal constraints, modern Nigerian women have been taking significant initiatives and making remarkable achievements.

The term "modern women" in this work, used interchangeably with "new women," refers to women who obtain substantial levels of Western education qualifying them for positions in modern professions and businesses. Largely urban dwellers, modern women differ in many ways from the majority of women who are rural peasant farmers, still bound by customs and traditions.

The study is in three parts. The first which is the introduction, highlights the social, economic and political dynamics which have been affecting change in Nigerian society. The second focuses on reactions to change, emphasizing achievements and failures, while the third, also the conclusion, analyzes the strategies of adjustment and adaptation.

PART I: INTRODUCTION

Flora Nwapa is a pioneer first generation modern woman, being one of the first few Nigerian women to receive university education. As an educator, novelist, administrator and publisher, she served as a role model for women, blazing the trail in both private and public sectors. Her struggles, trials and accomplishments seem to reflect the experiences of the women in the novel, *Women Are Different.*

Nwapa narrates the experiences of four friends: Rose, Agnes, Dora and Comfort, the musketeers as they prefer to refer to themselves, from their youthful days as classmates in a Christian missionary girls high school (Archdeacon Crowther Memorial Girls' School) to adult life experiences as modern women in different professions and businesses (3). The story emphasizes the reactions of new Nigerian women to the challenges of changing values of modernity, highlighting their difficulties, successes and failures. Focusing on the influence of Christianity and Western education, the novel provides insights into the nature of Nigeria's colonial and post-colonial society. Specifically, it identifies nationalist struggles for independence, urbanization, Nigerian civil war, the militarization of politics and the oil boom economy, as the major driving forces in Nigeria's socio-political and economic development in the period of study.

THE DEVELOPMENT OF CULTURAL CONTRADICTIONS

In general, two major factors, a British-type political economy and a Christian education, provided the dynamics which determined the nature of modern Nigerian society. Growing materialism, self-

interest and increasing significance of "cash" for survival depreci-
ated communal values as power began to revolve around the individual. Western education and material wealth became the main
determinants of power. Based on patriarchal values, these factors
propelled men and marginalized women. Also, it is important to
add that as centers of administration, commerce and education,
urban centers founded on male values and interests, favored men
rather than women.

IMPLICATIONS OF CHANGE

The cultural environment of the modern society affirmed the superiority of Western values, the values of the colonizing master, Great
Britain, and the inferiority of indigenous traditions. Lured by the
material comfort of Western lifestyle, people in the modern sector
struggled to be part of the superior culture, jettisoning their traditions.

The institution of a capitalist economy in a predominantly peasant non-technological and industrial society created profound dis-
articulation in all sectors of society. Cultural confusion criminalizing
the moral order of traditional Nigerian society, reflects the problems of a disarticulate society. In spite of this, there has been expansion of opportunities and a greater scope for individual initiatives
and choice. It is under these circumstances that the modern
Nigerian women discussed in this article have been functioning.

PART II: REACTIONS AND RESPONSES OF MODERN WOMEN TO CHANGE

Christian missionary enterprise, in particular, and Western education, as important aspects of the colonizing mission, were important strategies of socialization and in propagating the ideals of the
colonizer.[3] The major goal of female education was the production
of a new type of woman, equipped for modern careers when the
British handed over independence.[4] The missionaries were more
eager to groom these women to be law-abiding, humble and faithful citizens, to guarantee the Christian moral code against all other
influences. As Rose in *Women are Different* claims, "we are taught
to be humble and gentle and humane, we are not taught to do violence to no man" (33). Although Christianity and Western education liberated women from some of the constraints of traditional
society, modern women had no desire of giving up all their cultural values especially those pertaining to wifehood and mother-

hood. They did not intend to constitute a "bunch of spinsters" like their principals and teachers (42).

The new women had a choice between early marriages limited to mothering, cooking, cleaning and sewing and higher education with the potential of lucrative professions, power and privilege. They encountered different pressures from school and traditional society and family. School principals and teachers expected them not to be distracted by teenage love affairs in order to ensure high academic performance and pure Christian life. On the other hand, the women were pressurized by traditional society to marry early. For example, Agnes is pressurized by her parents to drop out of school and marry a man as old as her father. While it is depressing and distracting to Agnes, it is frustrating to the school principal who contends, "What was the use of spending so much time and energy teaching a child algebra, geometry and all other subjects if she was not going to make use of them?" (22).

It is evident that the "patronizing" British missionary educators were strange to the customs of the people they were converting. Their concern was educating modern women for careers not necessarily as wives. "Having shunned the world to do the will of God," these missionaries, themselves not wives, wanted to empower these women in a new way. This influenced Agnes's decision to complete her high school education and to gain the Cambridge School Certificate before marriage (22).

Western education also affected the perspectives of women. It changed women's images of women as well as their attitudes to marriage. It inspired women to take advantage of opportunities in modern society, to actively embark on self-improvement and to participate in national development. The ideal physical figure of women (the heavy and fleshy figure) symbolizing true womanhood became less attractive to modern women. Slimmer and smaller figures increasingly came to represent the ideal physical figure of women. Dora stated that "she was not going to be like other women, who, when they had their first babies, stopped taking care of themselves, added weight, and forgot to watch their figures...as soon as she had her baby, she would do exercise to keep her tummy down" (28). In addition "she would mind what she ate so that she did not bulge in the wrong places" (28).

The focus on marriage shifted from community and the extended family to the individual's values. Emotional and sexual love between men and women became the deciding factor rather than family approval.[5] Marriage qualifications became centered on good looks, high academic qualifications and lucrative professions rather than family approval. For example, Dora and Rose married Ernest and Chris, their high school boy friends who attended neighboring male high schools, after completing their high school education.

Modern marriages are risky affairs, unsupportive of women's stability and security. The lack of strong family involvement and community support contribute to the vulnerability of women in modern marital relationships. Foreign values are not seriously rooted to bind modern husbands and wives, in particular as they do not live in the cultural environment of the marital laws they are operating within. When crises arise in modern marriages, partners seek to protect their own interests even at the expense of the children. This explains why Chris abandons Dora and his children, to advance his education overseas in order to qualify for a lucrative profession (law) to place him in a position of power and privilege.

Rose has no children but she shares the experiences of Dora. Self-interest was the dominant factor in the decisions of their husbands. Going abroad in search of higher education for self-improvement and ultimately national development was a positive trend. What is dysfunctional about it is the sacrificing of emotional welfare and the material needs of wives and children. Male self-interest in the modern society contradicted male responsibilities, obliging husbands to protect wives and family.

One of the consequences of the flight of husbands was the emergence of single families headed by women as evident in the experiences of Dora and Rose. Increase in economic power enables them (modern women) to cope with family responsibility financially. As in the case of Rose, it enables single women (without children and husbands) not only to be self-supporting but also to be confident and contribute effectively to societal growth. What modern women lack is emotional security, being exploited by the physical and moral flight of men from marital obligation. Thus, in reality, women bear the burden of change and ultimately pay the price of human and national development.

WESTERN EDUCATION AS THE BANE OF MODERN MARRIAGES

Christian missionaries educated men and women differently, producing inherent contradictions affecting relations between men and women, especially marital relationships. The emphasis in women's education was good behavior including Godliness, cleanliness, obedience and hard work. While they were instructed to shun greed and selfishness, they were not prepared for the vagaries, uncertainties, and competition in the real world outside of the schools.

Men were educated with the objective of becoming professionals and leaders in the modern society; they were not trained to be responsible fathers and caring husbands. Modern men acquired the arrogance of a male-dominated western society. The combination of both traditional and modern patriarchal values seem to intimidate modern women.

The majority of the schools were single sex institutions until after the civil war in 1970, limiting interaction between young men and women. Physical separation fostered distrust, suspicion and a lack of understanding. Consequently, relationships whether in or out of marriage were not founded on mutual understanding and respect. The urban environment, where the modern women live, lacks the traditional support system, in particular dependable emotional support systems, capable of sustaining marriages in times of crisis. Friendship groups contracted at school seem to be the only reliable support for modern women in urban environments. But with all of them in the same "boat," sharing similar experiences, they are not able to solve each other's problems effectively.

CHALLENGES OF CHRISTIAN MORALITY

How are modern women trained in Christian morality able to cope with the challenges posed by greed, distrust and competition in a fast-developing society? Is it possible to maintain Christian morality when confronted by the uncertainties of self-interest and inordinate ambition for wealth? Are they going to be greedy and undignified? Will they remember Christian ethics and fulfill the goals of missionary education, countering dysfunctional influences of society? Will they shun bribes and corruption or will they be strong enough to pressurize their husbands not to be corrupt? (58) Separation during youthful days estranged young men and women and did not facilitate confidence and trust making it difficult to sustain mutually dependable relationships.

ACHIEVEMENTS OF MODERN WOMEN

Why are modern women successful in professions and businesses despite serious challenges? Basically the new women are ambitious and dynamic. They focus on the advantages of the new professions (power, privilege and influence). Agnes pursued higher education after marriage and having children despite the opposition of her husband (68). After gaining a university degree in mathematics, she became a teacher and later an education officer. When confronted by the choice of remaining a wife and obtaining a higher education, she opted for the latter because of the lure of power and privilege in the new professions. However, the cost of her success was a broken marriage, single parenthood and emotional insecurity.

Dora's career exemplifies the struggle and achievement of modern women in independent businesses. Unsatisfied with her nursing career, she established a catering business and later expanded to urban housing (61). But her husband Chris has an unsuccessful working experience; he is accused of bribery and corruption. Dora is unable to influence him because of the lack of mutual respect and trust. In spite of her invitation that he joins her business, he abandons her to go overseas to further his studies after selling the house built by both of them, without her knowledge, to pay his way (67). Once overseas, he reneged on his marital obligations and contracted another marital relationship. The actions of Chris depicted the instability of modern marriages and its effects on women. Self-centeredness, patriarchal arrogance, and lack of respect, characterize the attitude of husbands to wives and families.

Rose earns a higher university degree in Education, becomes a teacher-education officer, and finally an executive in a private public relations firm (69). Thrice she was deceived by three husbands in circumstances similar to that of Dora (70). But unlike Dora, she has no children making her emotional abuse much more difficult to bear in Nigerian society. The women are not daunted by emotional abuse and psychological exploitation. Rather, they face the challenges with determination to succeed in their professions. They draw strength from their achievements while they protect themselves by the power and privilege of their positions.

THE EFFECTS OF THE NIGERIAN CIVIL WAR ON MODERN WOMEN

The Nigerian civil war transformed social values and almost revolutionized gender relations and attitudes towards sexuality. This change produced a new breed of Nigerian women described here as second generation modern women. The war introduced a new mentality and moral order based on immense desire for material wealth. Communal bonds suffered serious erosion as individual ambitions clashed with communal interests. The trend coincided with the growing militarization of society. Military culture dominated civil society and undermined social values. The culture emphasized ostentatious lifestyles based on easy wealth acquired irregularly. A new class of rich men emerged, deriving enormous wealth from military patronage. These men wielded considerable power and influence.

Societal responses to sexuality was dramatically affected by an unprecedented emphasis on wealth. Women's reactions to sexual acts were transformed, adopting a more flexible attitude towards sexual advances. This study contends that the need for protection from the dangers of war and the threat to life and personal welfare compelled women to make shifts in their sexual attitudes. The sexual act became a source of power, referred to as "bottom power."[6] Apart from ensuring personal protection, it is used for advancing financial security and emotional comfort. The traditional values constraining sexuality no longer seemed to be effective as the commoditization of sexual acts became increasingly part of the military-dominated life style. Luxurious and ostentatious living (expensive homes, clothing, cars and social ceremonies) is part of this trend.

National wealth in the post civil-war period was derived from oil exports. The oil boom from the early 1970s to the early 1980s advanced economic growth, including urbanization and industrialization. Also, there was expansion in education, medical care, roads and other infrastructure. These developments fostered greater mobility and flexibility in social attitudes, including sexual responses and marriage. In addition, desertion and divorce became common experiences.

Power derived from wealth came to supercede that of education. Wealth became the greatest attraction for marriage. Marriages contracted on account of wealth known as "money marriages" became popular among second generation women.[7] The attraction of money marriages was due largely to the absence of governmental provision of social welfare and security for the people. The need for social security was urgent in the face of the growing erosion of communal support. Thus, the protection and security derived from "money marriages" constituted the only reliable source of social security and insurance against poverty and despair.

In addition, it is plausible to assert that the reaction of the new generation of women can be regarded as revenge against the "insensitivity" of their fathers to their mothers. Chinwe, the daughter of Chris and Dora states, "Mother I cannot take what you have taken from my father. I am going to have my own back . . . ." (117). Gradually, some of the first generation new women began to accept the new sexual attitude and money marriages because of the failure of their marriages based on the insensitivity of the men. Also, they were influenced by growing individualism and survival needs in the absence of public support. Comfort, one of the musketeers and a first generation new woman remarked, "I will encourage my daughter to go where the security money is." She argued that "these men are filthy and they do not know what to do with their filthy money" (82). Concluding she stated that "the entire society was rotten and she was not going to be the angel among the devils" (84). Referring to Chinwe's husband she asserts, "This man will look after Chinwe" (112). Money marriages became a strategy of adapting to the economic and socio-political changes, as well as for coping with the challenges of insecurity. In addition to the lack of traditional authenticity, money marriages are affected by frequency of marital failures jeopardizing family cohesion and the welfare of children.

BEER PARLOR BUSINESS AS A STRATEGY OF ADAPTION

Beer parlors are eateries dominated by the consumption of beer. From the early 1970s, the "beer parlour" business became an important aspect of the informal economy.[8] Controlled by women, they served as recreation centers, promoted formation of social relations, including marital relationships. Beer parlors promoted money marriages, but at the same time they encouraged marital infidelity, divorce and desertion..

Chinwe met her second husband in a beer parlor. Commenting on Chinwe's husband, Comfort states that "the man is wealthy . . . he throws his money about . . . already bought a house for Chinwe" (116). Rose remarks "A rich man meets a young mother at a beer-parlour, who recently left her husband and after nine months builds her a house. That is tempting I must say" (116).

Beer parlors appear to be replacing the traditional compounds of extended families. Christian ethics which facilitated the growth of first generation new women do not support this type of lifestyle. But, the majority of these women are giving up the Christian morality to be able to cope with the difficulties of a changing society. Comfort exclaimed to Rose representing a small percentage of the first generation women still faithful to the Christian moral code that "your missionary ways have not left you . . . you are daft" (116).

It has become apparent that Christian morality as presented to the first generation women has become obsolete and it is no longer able to sustain modern women's expectations in the new society.

Rose observed that "a society in which young women of twenty-four and twenty-five boast of possessing Mercedez Benz, big overseas business associates, credit cards and expense accounts paid, without a husband . . . at forty what will this brand of women be? What was left to be achieved?" (118). It should be noted that in some ways, women's responses to the challenges of change is contradictory. While women divorced their husbands because of infidelity or marrying other wives, they are prepared to marry men who have other wives.

This situation seemingly continues to empower men. It could be perceived as exploitation of women by men supporting Comfort's claim that "the young women are giving more in these new forms of relationships because they give their bodies, youth and vitality to the filthy sugar daddies" (116). Or was it the women exploiting the men in view of the material gains accruing to them? It may be contended that if women are empowered by the wealth of money marriages, they are exploiting the men, who are exploiting their society. Comfort asserts that Chinwe's second husband, a contractor for the Nigerian army received mobilization fees which he squandered without laying the foundation of the building" (117).

Unlike their mothers, the second generation women taking advantage of the increasing moral flexibility in society, are quick to seek divorce as a way out of unpleasant marriages. Divorce became a remedy for emotional stress, shifting marital situations, self-liberation and survival. They see no need to endure emotional depression especially as several options had become apparent. Chinwe claimed that "a man who could neglect his children was not worth taking seriously" (118). Not only did she seek divorce after eighteen months of marriage, but she also wanted no maintenance money for her children because she was industrious and desired personal liberty (118). Thus, the second generation Nigerian women are independent and more concerned about promoting personal freedom and self-importance.

The actions of the second generation women have serious implications for women's quest for freedom and equality. They indicate that women have choices to marry, establish business, have children or change unhappy marital situations and that women's lives should not be ruined because of unpleasant marital relationships.

Part III: Conclusion

The analysis showed the various stages of Nigerian history from the colonial period to contemporary times through the periods of the Nigerian civil war and the oil boom economy. It examined women's changing responses to changing historical conditions. It analyzed the strategies of adaptation and adjustment including Christian education, money marriages and divorce. The study also noted the effectiveness of Christian education as a strategy of adaptation up to the civil war and oil boom periods. Thereafter as the economy expanded and society turned more militarized, Christian morality became inadequate and the second generation of modern women emerging after the civil war adopted different and more adjustable methods including money marriages and divorce to cope with the problems of change.

This discourse also emphasized the impact of education on the rise of new women to positions of power and privilege. The work identified the role of education as a factor for upward social mobilization. A small percentage of the entire women's population, first generation modern women, laid the foundation for the growth of a second generation of new women. As agents of change, the second generation of women are more aggressive and effective in seeking change to patriarchal constraints on women's personal freedom and effective participation in national development. Their strategies emphasize the growing flexibility of society and the increasing decline of traditional constraints on women.

Furthermore, it stresses the influence of boarding house education and urban life experiences on gender relations. Boarding houses, relatively an unfamiliar environment, constitute a different "world" based on the British "spartan" cultural system. It deprives young women of the opportunity of receiving traditional education on womanhood and wifehood and more so, intimacy with mothers. Thus, while not being socialized as traditional women, they are not totally socialized as housewives in British households. Boarding houses deprive men of basic education on traditional manhood especially as husbands. Like the women, they are not socialized as husbands in British households. These incoherences produce dysfunctionalities in modern families and gender relations.

Modern women found it difficult to solve their emotional problems because in the contemporary society of Nigeria, the emotional problems of women seem to be the strength of men in the face of competition for positions and increasingly scarce resources. Contradictions between the new generation and older women complicate the problems of the new women. The difference in world view of these women and their mothers make it difficult for modern women to benefit from the generational experiences of their mothers. Traditional mothers are less materialistic and highly communal in their orientation while the modern women have developed a materialistic and individualistic approach to life.

Finally, the determination of modern Nigerian women to take advantage of expanding opportunities play a significant role in their achievements. Modern women seem to draw on the strength of traditional women, in their ability to cope with stifling patriarchal constraints while establishing control of some of the most vital sectors of society including marketing. More than any other sector of the population, modern women serve as agents of change, bridging the gaps between modernity and traditional society. But still reflecting the contradictions of changing cultural values, they bear the "double yoke" of modernity and tradition. The struggle of modern women for gender equity and independence continues in spite of the obstacle posed by growing militarization of society which is an extension of patriarchal tendencies.

NOTES

1. See the 1992 edition Flora Nwapa's novel, *Women Are Different*, published by Africa World Press, Inc.
2. As a product of independent Nigeria, I was born in, nurtured, educated in Nigeria and worked as a professor in the Nigerian university system for more than one decade.
3. See E. A. Ayandele's excellent book, *The Missionary Impact: A Political and Social Analysis*, p. 286.
4. Nigeria obtained independence from British colonial rule in 1960. See Ayandele, op. cit., p. 294.
5. Traditional conditions of marriage include good character, family reputations, compatibility and mutual respect, a sense of responsibility and the ability of husbands to provide material and emotional support, as well as the ability to procreate.
6. "Bottom power" became a descriptive term for women's sexual power and influence.
7. "Money Marriages" are new forms of marriage based on the desire for material comfort.
8. The beer parlor business boomed from the early 1970s after the Nigerian civil war and raised women's visibility in urban economies.

WORKS CITED

Ayandele, E. A. *The Missionary Impact: A Political and Social Analysis.* London: Longman, Green & Company, 1966.

Nwapa, Flora. *Women Are Different.* Trenton, New Jersey: Africa World Press, 1992.

Shifting Paradigms of Profit and Loss:
Men in Flora Nwapa's Fiction

Osonye Tess Onwueme

Post-modern culture is more and more a market
culture dominated by gangster mentalities and
self-destructive wantonness. This culture engulfs
us all—yet its impact on the disadvantaged is devastating, resulting in extreme violence in everyday life.
    —Cornel West, *Race Matters* (10)

I am quite concerned about the fate of the black
woman, whether she is in Africa, North America
or the Caribbean. She faces many problems.
I think the crux of these problems is economic.
If the black woman is economically independent
she and her children will suffer less.
    —Flora Nwapa[1]

The World is a market place.
If you bargain well, you profit.
If you do not, you lose.
        —Aniocha-Igbo Proverb[2]

The above (con)texts take us to the cutting edge of trade and commerce, which continue to shape the contemporary body politic. In
the (new) world order, where profit is the soul of business, the
body, mind, and heart, like everything else, become as infected as
they are commodified in the spirit of capital. It is within this cultural matrix that Flora Nwapa positions her fiction as a means of
(re)locating, (re)constructing, and (re)writing the margins and
paradigms of profit and loss, and thereby producing a new balance-
sheet of power for the reading of gender politics in African
Literature. In this regard, we have chosen to invoke the powerful
insights of Cornel West in his critique of the post-modern culture
as a relevant text to the understanding and analysis of the multi-
dimensional signs of the market-culture that Nwapa inscribes in
her fiction. In my specific readings, what Cornel West states in his
critique of contemporary American culture in *Race Matters* (1993)
is a (re)configuration of the Aniocha-Igbo mythic vision of the
world as a market-place. Although Cornel West particularizes the
violence done to women and other disadvantaged constituencies,
he also problematizes this condition by his general conclusion that
"[T]his culture engulfs us all" (10). A lot is implicated here. At one
level, it is implied that the virus of "gangster mentalities and self-
destructive wantonness" (10) are by no means discriminatory, nor
gender-specific. Therefore, women are not excluded. Like their
male counterparts, women cannot be acquitted from the guilty verdict that is often assigned to carriers for (en)gendering, with ease,
the disease of violence.

   Also implicated in this politics of the market-place/culture is
the knowledge that both men and women are driven by personal
self-interest to buy and sell for profit. And since fierce competition
defines the unstable spirits of the market-place, traders experience
and suffer from the "anxiety of affluence," with the attendant "anxiety of loss." For in keeping with the roaming spirits of the free
market economy, nothing is taken as a "given." Values remain in
a state of flux, changing dynamically and fluctuating, as the huge
winner today may be the huge loser tomorrow. In view of this tur-
bulence, the individual's capacity for ingenuity or skillful bargaining is consequently challenged, since every participant is

charged and violently engaged in the trade for personal profit. And why not? Profit imbues power, authority, and control. With the spaces and margins of power/profit so sparse and limited, ferocious conflicts endanger the climate of fairness and goodwill as each one tries to undo/outdo the other by resorting or degenerating to violence and scavenging as a means of facilitating the appropriation and monopoly of the state/center of power. Thus, on these precipitous roads to power, individual stakes are high, community interests are marginalized, and the spirits of partnership and compromise lose currency because charity is out of fashion in this violently free market where men and women trade and traffic in power.

Furthermore, special interests define the norming modalities of the market place. The winner takes all, thus engendering a contentious hierarchy/dichotomy that ensures polarities between the winner on the one hand, and the loser on the other. In this dispensation, the winner takes center-stage, writing and rewriting the norms, and (re)mythologizing the world from this nexus and praxis. The loser is cast aside as (an)other[3] at the margin. The loser becomes simply an outsider, powerless or disempowered, voiceless and/or silenced. And the trauma is aggravated where the winner appropriates the other as fodder and converts "it" as (an)other object only worthy to be exploited for the optimal empowerment and security of the [winning] subject. This is the precariously undulating landscape of the market-place which the Aniocha-Igbo people euphemize and which Cornel West renames, and reaffirms for what it is—"a market culture dominated by gangster-mentalities and self-destructive wantonness"(10). Thus my intention in this discourse is to show how in the exhibition of their spoils and exploits, the (wo)men in Nwapa's fiction become the living cast who give testimony to West's prophetic visions of violence in the contemporary market-place.

As part of this preamble, I would like to remark that West's (re)configured ideology of a market-culture has triggered my own questioning of some relevant essential motifs and beliefs about the Igbo cosmic exegesis which Chimalum Nwankwo recalls as follows: "[W]hen one thing stands another stands by its side"(48). Nwankwo cites this proverb to validate his argument about the inherent dualism in the Igbo world view. According to the critic:

> [T]he Igbo world is a world of fluid dualities, and the
> pragmatics of existence derive from those dualities.

> The flexible relationships between man and
> woman, between human beings and their *chi or*
> guardian spirits, between society in general and
> other metaphysical entities and realms indicate a
> world that is never closed in terms of meaning and
> possibilities. To understand what Nwapa has done
> with that specific reality, we cannot afford to read
> the *word* in isolation and see the lake goddess mere-
> ly as 'a goddess of contradiction.' Contradiction
> denies the Igbo sense of wholeness, the product of
> the spirit of negotiation, the spirit of the healthy
> paradox.(46)

While I understand how Nwankwo derives this conclusion about the
principle of dualism which hinges on recognizing "the healthy paradox," I also wonder if this perspective does not negate or idealize the
"possibilities" of tensions and conflicts that define and underlie the
unique existence and separateness of the entities. After all, even
now, in spite of the new accords and treaties that are stated/signed,
the ominous realities of the Cold War still remain visible, at least in
the mind, if not in the signatures of the statesmen and their subjects. On the surface, there may be no apparent signal calls for war.
But this too may not exclude an internal violence or injury generated by internal struggles in warfare. It is this concern that leads one
to wonder about the true relationship that exists between each one
that stands from the other. Does this standing beside the other necessarily imply mutuality, neutrality, or the absence of tension
between one and the other? Could being beside the other not inten-
sify the fact /act of "otherness"?[4] Would it not be valid also that, contrary to outward posturings, an absence may mark and heighten a
presence? Indeed, is it not possible that each one standing by the
side of the other may be positioned as such, to exist outside the other
in order to bargain the silent spaces and margins between the self
and the other? My own thinking is that the dialectics of "negotiations" do not necessarily foreclose "the spirit of healthy paradox,"
but more than likely may inspire the spirit of tragic paradox. And
perhaps in our attempts to construct harmony out of difference, we
run the risk of being liquidated by the high, hidden costs of our utopian mythopoeic visions. From this skeptical background of questioning, I stand to engage, in particular, my feminist sisters, who
tend to read the bodies of (wo)men as the exploited losers, and to
urge them instead to (re)read these signs of (wo)men as the winning

exploiters in the trading posts of profit and loss in Nwapa's fiction.

Reconceptualizing our myth-making about women is necessary, for I am convinced that whichever side of the polarities one stands/positions the self, winning or losing generates consequent tensions. Existing beside the other also implies that the struggle may continue dynamically in spatio-temporal dimensions, since the loser, on the one hand, may continue to bargain and to negotiate the self beyond the limiting margins while the winner, on the other hand, may continue to (re)invigorate action to ensure self-security at the center. From this perspective also, being in the center could become a tenuous business that one may lose or hold onto, as the primary producing agent of power is bound by the dialectics of market gravity to shift places with the other. After all, as Nigerians say, "No condition is permanent."

Flora Nwapa is much too familiar with that world, where no condition is permanent, and she recreates the dynamic character of the fluctuating values in this market-place, with her fiction. Indeed, as she persistently depicts in her fiction, if there has been anything permanent about the Nigerian (and by extension, the African) world, it has been change. Change is the only constant in this world and she takes us through the uncertain course of that change, shifting from the rural settings in *Efuru*[5] and *Idu*[6] to the urban metropolis, and as she tells us—*This Is Lagos* (1971). Following this course of change, Nwapa constructs her fictional texts from the privileged insights of a participant-witness, who is testifying, signifying, and revising the visions of power and gender politics concerning men and women in that social milieu. My understanding is that as Nwapa writes and (re)writes the changes from one text to another, she discloses that, contrary to the popular opinions that have gained currency among the feminists *and* in the circles of the "rabid sexists of the society"[7] as Ogunyemi calls them, women have been the prime movers, the lead actors, the significant producing agents and shareholders who drive the market forces and values of change and power. It is from this perspective that I note the problematics of Helen Chukwuma's reading that:

> Nwapa's literary domain is women. In her novel [sic] and short stories, Nwapa bared the soul of woman, she showed her as flesh and blood nursing her own dreams and aspirations, being hurt and giving hurt and breaking with tradition when it chokes her. This has never been done before. (115)

The critic is right in stating that "this has never been done before," and we join Chukwuma in commending the author for making "women her domain." However, we note that Chukwuma, like most other feminist critics, fails to see beyond Nwapa's preeminent positioning of women, and thus, to decode the author's critique and interrogation of women for their contradictory roles in power: in one position as objects and victims of power, and from another position as subjects and producing agencies who (ab)use power.

My argument therefore is that—contrary to what [feminist] critics are willing or able to acknowledge—the targets of Nwapa's critique of power are women; women who have bargained or fought their way to power and then subsequently turned their power into a vendetta against men, who then become violently marginalized by the women. From this position, I wish to propose that Nwapa's dominant constructions of women in her fiction has not just created a gynocentric[8] tradition and a new market-place for myth-making about women, which she achieves by countering and altering the dominant "phallic criticism"[9] and writing of African Literature, but more significantly, one that dialogically critiques, interrogates, and rewrites the positions and assumptions of women's power promoted under the auspices of Feminist Studies of African Culture and Literature. Nwapa's voice, particularly in *One Is Enough,* where she excels in double-entendre, should be read as multi-dimensional and polyrhythmic, for in this (con)text, she is as much affirmative and celebrative of women's power as she is signifying and satirizing (wo)men's excessive and oppressive use of power to "give hurt"[10] to men. My reading of Nwapa's text will therefore seek to identify these multi-dimensional signs which should help us locate the shifting paradigms of profit and loss relating to men in Nwapa's fiction. In this regard, it will be appropriate to begin with Nwapa's resonant notes, through which she affirms and celebrates women's power.

Until Nwapa entered the scene with the publication of *Efuru* in 1966, the dominant legacy, voice, and primary producing agency in African literature was essentially male. It was indeed the man's world. As "God made man in His own image," the primary protagonists and products too were forged, fashioned, and framed in the image of man. And even after Nwapa dared to enter this market-place, and was soon accompanied by Efua Sutherland and Ama Ata Aidoo, female critics of the literature were still at large. And so, the field remained predominantly in the powerful hands of the male critics, who in their divine wisdom either ignored, ridiculed, scorned or descended and scavenged on the very soul of the women's literature(s). Nwapa herself bemoaned this fact in her interview with Adeola James:

> The writers? If you talk about the Nigerian male writers, we have a very good relationship. They don't feel threatened at all. But the critics? I won't say that they have been too kind. A friend of mine, Ama Ata Aidoo, brought this out in a paper she presented at one time. She said that some male critics don't even acknowledge female writers. Every writer thrives on controversy, so you are killing the writer if you don't even talk about her. Being ignored is worse than when you are even writing trash about her. (114)[11]

Nwapa has every reason to cry out against the injustices of "phallic criticism,"[12] which Stratton has already magnified in her own study. It was this practice of gendered-criticism that once prompted Femi Ojo-Ade (1983), a lone male voice in the critical paternalistic hegemony to remark this absurdity in his essay, "Female Writers, Male Critics":

> The female counterparts of the Soyinkas are a rare breed. They do exist, however. Nwapa, Aidoo, Njau, Emecheta and Head are names that come to mind. For that rare breed, practicing the art of writing poses problems at once similar to, and greater than, those faced by their male 'masters'....The male writer, like the male social animal, is more fortunate than the female. His presence is taken for granted. The publisher seeks him out, unlike the woman whose silence is also taken for granted.
>
> Cultural misconceptions and taboos abound. It is believed that women must keep quiet when men are talking....Fortunately, taboos die—though slowly—with the times. With the new, warped, so-called luminaries emerging from the colonial roots, even the female deaf and dumb have acquired a voice. It is a voice of confusion, of confrontation, of commitment. The men have had their say; they continue to have it. It is the woman's turn. (159)

Nwapa's emergence as Nigeria's first woman novelist has not just opened a new stage where women break and cross the boundaries of silence to recreate their own destinies, she has also opened new frontiers for other women to produce and market their own fictional wares. It is thus significant that Nwapa has charted a new direction and has mobilized a choir of women writers as new winners whose voices are heard as the chorus in this new, alternative market-place for women's concerns. Perhaps we might regard this as the first stage of Nwapa's commitment: to bargain, to (re)construct, and to shift the paradigms for the value and product of African literature from the monopoly of the male center, to one where women too can and do engage in it as share-holders and producing agencies.

But Nwapa does not end her mission at this level. Indeed, she sustains and grows from it. In time, she develops more challenging and complex shifts in the (en)gendering and constructions of power relationships in African Literature. And it is this other dimension, the more complex aspects of the transforming paradigms in Nwapa's literature that will constitute the subject of this discourse. Specifically, I intend to focus on two aspects of Nwapa's creation of alternative traditions and shifts in paradigm, which I hereby respectively designate as: (1) Nwapa's Emancipatory Economic Theology/Ideology for Women, and (2) Nwapa's Gynocentric Vision: Signifying/Rewriting *The Joys of Motherhood,* Patriarchy, and the Tragedy of Men.

I. Flora Nwapa's Emancipatory Economic Theology for Women

Through her writing, Nwapa leads us to appreciate the ethics and politics of the market-place. She uses both her male and female characters to reveal to us that the ascribed values are subject to the vagaries of market instability and fluctuation. And, above all, that value is determined by the magnitude of the bargain negotiated by the prevailing producing agency for the product. This means that there is a dynamic mechanism built into the system to allow for shifts and variables that may be determined by the agency of power, and which may in turn impact the values/prices to sustain the defiance of statis. Consequently, change and transformation remain constant realities. Women, in the contexts that Nwapa writes about, remain conscious of this shifting character of the market place, and have traditionally taken advantage of change to benefit and to profit. As a female writer myself, I share these values to which Nwapa has subscribed. Indeed, the critic Afam Ebeogu, in his analysis of my drama, *The Reign of Wazobia* remarked that:

> Onwueme argues that traditional political and socio-cultural institutions, essentially erected by men as mythic structures of domination, have built-in mechanisms for subverting their very effectiveness, and that the time has come for women to identify these traditions and institutional structures and use them to liberate themselves. Theirs is a fight not for the reversal of roles, but for the balancing of such roles; not for the denial of natural endowments, but for the utilization of such endowments in achieving self-fulfillment, equality and respect.(99)

In some respects, Nwapa's dynamic female characters share these attributes that Ebeogu notes about Onwueme's female protagonists, as well as other related female protagonists in African women's writing. However, this is as far as we can go with this comparison. For Nwapa's female protagonists are uniquely constructed as *sheroes*[13] with significant differences from other female characters in the literature. As we shall see, a close reading of Nwapa's female protagonists, particularly in *One Is Enough,* shows that whereas their contemporaries may be fighting for "self-fulfillment, equality and respect"(99) according to Ebeogu, Nwapa's sheroes engage their own iconoclastic mission of change with an irrevocable spirit of vengeance and conquest that the author affirms, undercuts, and ridicules at the same time. Her female protagonists in their own (con)texts consciously make it their business, not just to subvert the so-called mythic structures erected to dominate women, but actively penetrate and transcend these thresholds, and then use the system to marginalize and erode itself. What is more interesting is that while consciously dislocating the system, their male counterparts who thrive on the false security of patriarchal norms and institutions of social control, remain largely ignorant, blind, and unaware of women's schemes to overhaul such systems. In fact, the ignorance of men who think they control these social institutions leads them to approve and legitimize the alternative routes that women construct and navigate to displace and to disempower men.

One good example of these alternative routes for women's empowerment is trade. Traditionally, the market-place was the exclusive preserve of women. According to Kamene Okonjo (1976):

[In] traditional Igbo society marketing was the woman's domain....The *omu,* in her capacity as the head of the female side of the community, with her cabinet determined the rules and reg- ulations under which the market—the domain of women—was to function. (48-49)

Nwapa's women also thrive, rule, and control the market-place of their world like these Western-Niger Igbo women that Okonjo writes about. Like anyone engaged in the trading business, they strive for good bargain as the means to enhance their status, and thus achieve economic independence. For this reason, the women engage in dynamic activities of trade; buying, selling, producing, and contracting businesses, both legal and illegal, as exemplified in the "attack trade" that has created new stages of power for the elevation and promotion of the (in)famous "Cash Madam Club."

Another route that Nwapa's women take to advance social mobility is Western Education. Nwapa creates this alternative paradigm through the activities of the "Three Musketeers" (the Agnes-Dora-Rose triangle of sisterhood), who become highly suc- cessful by acquiring Western Education, and by committing to the mutual assistance of one another to show that *Women Are Different* (1986). Nwapa reaffirms this paradigm in the strong sisterhood of Amaka-Adaobi-Ayo, with the "Cash Madam Club." Western Education is only the stepping-stone to power and economic inde- pendence. As Agnes's experience reveals, this kind of education is not as swift in ensuring the desired economic freedom and inde- pendence, which the trading activities can readily facilitate for women. And to demonstrate this fact, the author presents her *shero,* Amaka, who after acquiring the education and becoming a teach- er, makes a professional shift into the trading and contracting busi- ness:

> Amaka went on with her business in Onitsha, sup- plying timber, sand and food. She was a contractor, one of the numerous female contractors who had sprung up during and at the end of the war. Before the war, she had been a teacher. At the end of the war, because she took part in the 'attack trade,' she rediscovered herself. (4)

It is not surprising also that Amaka's friend, Adaobi, who is appar- ently comfortable in her profession as a nurse, soon becomes Amaka's new convert into the sisterhood of women who bargain their way into outstanding profit and power by trading. With this rise in economic power, Adaobi is not just able to build a house for her family in the coveted area of Ikoyi, but is quite expedient in redeeming her economically disempowered husband from ridicule, homelessness, and shame. The narrator approvingly records Adaobi's rise to economic power through Amaka:

> She was putting up a bungalow in Ikeja on the plot of land she bought without saying a word to her husband. Amaka had helped her do one or two contract jobs she got herself and they shared the profit. (80)

In this way, Adaobi becomes like Agnes and Dora (in *Women Are Different*), who earn their meteoric rise to power by trading. Much like their rural counterparts—Efuru (*Efuru*), Ekecha, and Mgbeke (in *The Lake Goddess*),[14] these modern educated women are conscious of their potential. They know that the market-place belongs to them, and that they can transcend the thresholds of competitiveness by bargaining men completely out of place and power. For this reason, the women engage in trade, including the "attack trade," as the more viable and more rapid routes for making profit and empowering themselves economically so that they will ultimately gain independence from men. As Nwapa reaffirms this position in her interview with Adeola James (1990), it is imperative for women to own and to control their own business in order to gain economic independence. Nwapa is quite convinced that: "[I]f the black woman is economically independent she and her children will suffer less" (112). One is, therefore, not surprised that Nwapa deliberately positions this subject with such prominence and power in her novels, to the extent that trade crosses the boundaries of business to become a religion—an "Emancipatory Economic Theology for Women"!

Strengthened by this new faith, old gods (wearing the image of men like Obiora and Mike) are cast aside or displaced to make room for the new goddesses of the market-place. In the new arena, women mount the high altars of power to divine their life possibilities from the present to the future, by locating themselves in the central positions in the market-place. Amaka is the new priestess of this new religion. Once she has become initiated, she commits herself to winning new converts, and Adaobi is one of Amaka's early converts to this new ideology/theology. In this new stage, the tables are turned over—and it is no longer men who control the power and value of the new market-place, but women. The paradigms have shifted. While the images of men become diminutive as they experience tragic losses in trade, power, and position, the images of women become magnified as they celebrate their new profit in trade, power, and position.

Nwapa persistently returns to themes concerning these *sheroes* who chart their individual course to progress and fulfillment by engaging in trade. The women and their author celebrate trade as the new place of worship of profit and power. Efuru's success in trade empowers her to give her husband, Adizua, the money required to pay her own bride-price. In *One Is Enough*, Amaka is the "Born-Again-Efuru." As a new Efuru, Amaka transcends the illiteracy which in the past, Efuru had lamented as her handicap. In her "second-coming," therefore, [Efuru-]Amaka supplies the missing link [in Efuru's life] by strengthening her knowledge and power with Western Education. The narrator states approvingly that:

> Amaka went on with her business in Onitsha, sup-
> plying timber, sand and food. She was a controller,
> one of the numerous female contractors who had
> sprung up during and at the end of the war. Before
> the war, she had been a teacher. At the end of the
> war, because she took part in the 'attack trade' she
> *rediscovered* herself. (49; my emphasis)

Apart from other issues revealed here to unveil Amaka's past undertakings and struggles, the narrator consciously signifies the important, transforming impacts of the "[attack] trade" in Amaka's development, for it is through this medium that "she *rediscovered* herself." The implication here is that it is through trade that Amaka is born-again to become the new woman that is educationally aware and economically astute, strong and powerful. Furthermore, if Amaka received her "baptism" and "first sacrament" through Western Education, it is by trading that she earns the requisite "confirmation" to become economically and socially empowered for her independence and fulfillment. And in this manner, even more than acquiring Western education, Nwapa's women gain economic and social prominence by engaging in trade as channels for promoting and advancing self-esteem, fulfillment, and absolute independence.

Added to this retinue of successful female traders in Nwapa's novels are the two fish sellers Mgbeke and Ekecha in *The Lake Goddess*.[15] These women continue the legacy of self-assertive and economically independent nation of women that inhabit and dominate Nwapa's fiction. From text to text, Flora Nwapa constructs and connects her nation of women, who, in spite of the constraining psychological, social, and cultural impediments, still transcend the limitations by recentering and "rediscovering" (49) themselves through the trade and commerce that Amaka exemplifies. By trading, the *sheroes* gain control, pride, authority, place, voice, and admiration in the community. Therefore, in the novels, the recurrent issues of trade and commerce should not be regarded as merely coincidental, but as carefully textured markers for signifying the liberating theology which Nwapa has created to promote her ideology for women's empowerment.

Amaka's aunt is one such woman that is imbued with the exemplary qualities that the author ascribes to her women. In this context, the aunt serves as Amaka's symbolic, surrogate mother. In addition, Amaka's aunt easily plays the role of the high priestess of the new religion [the liberating economic theology for women], that Nwapa has founded and inscribed in her fictional characters. Like any clergy, the role of Amaka's aunt is to proselytize and to win new converts. Amaka is a most eligible candidate and her aunt excels in her mission of strengthening Amaka's fledgling faith by historicizing her own experience, and as a means of educating and mobilizing Amaka into the new [economic] creed as she sermonizes:

> 'I got a sixteen year old girl for him. Yes I married her for him. I said to her, this is our husband, take care of him. I am going to take care of our children. I must see that they [children] all have a good education. Good education means money. *So I am concentrating on my children and my business....Never slave for him [your husband]. Have your own business no matter how small, because you never can tell.'* (9; my emphasis)

Soon after this encounter, Amaka herself passes her period of apprenticeship and becomes the new spokeswoman for promoting the new ideology/theology. Amaka acts out her new role in her encounter with the new bride who was beaten up by her husband, reminding Amaka of a similar experience she had at the hands of Obiora during the very early stages of their marriage. As the narrator indicates, "Amaka learnt one thing from that incident and it was that she would never helplessly watch a man, least of all her husband, beat her. She must defend herself"(27). Having grown from this experience, Amaka fulfills her new mission of teaching and instructing the new bride about the politics and psychology of rebellion:

> 'He will do it again. Men are beasts, so watch out. Think of what it was that led to the beating and make sure that this trouble does not repeat itself....But the trouble with men is their ego. *They refuse to appreciate their wives. Mind you, they do appreciate their mothers and sisters, but never their wives.* Your husband will always show you that he is a man. Your rightful place is under his thumb. Your rightful place is not in the kitchen as we erroneously think, but right under his thumb. *He would like to control your every movement, and it is worse if you depend on him financially.'* (27; my emphasis)

The above should not just be seen as the ideology informing and shaping Amaka's vision of men, particularly husbands, but should also be viewed as the nucleus of Nwapa's iconoclastic interrogation, critique, and (re)vision of women's power which she achieves through a progressive disclosure of the matriarchal foundations of patriarchy that she develops in the plot of this novel. Besides, the author employs Amaka in this context as the effective spokeswoman and agency to disclose and promote her liberating economic theology for women that is underlined in Amaka's statement to the bride that "[H]e would like to control your every movement, and it is worse if you depend on him financially"(27). There is no understating the facts here—Amaka is the author's persona, and is hereby giving the new commandments to guide and ensure woman's passage into freedom and independence, while at the same time providing the critical arrow for confronting and deflating the excessive dimensions of that power and freedom.

What is most remarkable however is that, using this trade metaphor in her fiction, Flora Nwapa discloses her ideology of the market place to reconstruct, reinterpret, and rewrite the place of women and power, especially in the Igbo market-place/society. Furthermore, by the time Nwapa writes *One Is Enough* (1986), the author shows how women have bargained themselves out of the margins of power to position and assert themselves at the center of power, with the larger consequence that the male contenders for power lose tragically, as they become marginalized and silenced by women. In *One Is Enough*, Nwapa not only pulls together the connecting threads between Efuru, Idu, Dora, Rose, and Agnes, and other *sheroes* who struggle to (re)negotiate power from men, she repositions and shifts the agency/center of power from men to the women. Amaka, her mother-in-law, her mother, her aunt, her sister and sisterhood of the "Cash Madam Club" become the lead-actors creating, performing, celebrating and exhibiting the acquired power with which they assault, silence and marginalize men. The author speaks as woman and names her subject—*One Is Enough!* Here there are no tentative assumptions, for she has gone beyond signifying to naming, the sole authority, agency, and controller(s) of power—women. Here, women have gone beyond negotiating power with men. They have totally appropriated power which they wield at center-stage. In *One Is Enough*, therefore, the speaking subject is woman. Amaka is the archetype of the new-breed-woman who bargains her position beyond the margins of power into the center, where she now speaks and exerts her monopoly of power. Through this *shero* in particular, the author simultaneously begins to indicate her concerns about the appropriation of power by women. Nwapa progressively questions and deflates the extreme positioning and exertion of power by women, and it is this concern that brings us to the underlying questions about the (dis)place/me[a]nt and (dis)location of men in Nwapa's fiction.

In *One Is Enough*, in particular, the actively listening ear cannot miss the echo of the author's voice in the following questions. In this alternative ethos where power is solely invested in woman, what becomes of the man? And what is Obiora's or Izu's place in this unequal, undulating, and shifting landscape of power? Through the *sheroes*, the author consistently signifies that in this world where women rule and control, every man becomes (an)other, sharing the common tragic destiny of being the outsider that is silenced, victimized, and marginalized. Perhaps, more than anything else, Nwapa creates the alternative images of men who are dislocated and disempowered by women as a way to caution women to engage in self-examination concerning their quest for power, and she stresses the troubling consequences of power that is abusive, exploitative, and oppressive. In this process, the author reverses the trend of "herstory," to tell and to highlight "history." In other words, the focus shifts from images of the oppressed woman as seen in the first episodes of the novel, where Amaka is beaten and rejected in her matrimonial home, to (re)enact the "history" of the oppressed man who becomes the victim of woman's new power and success. Repeatedly, the author insinuates that just as men are acutely implicated in this (en)gendering/(en)danger-ing of power, so are the women. Specifically, what Flora Nwapa seems to portray in this novel is that women (as mothers, mothers-in-law, aunts, and sisters), who at one time or the other experience victimization and oppression from men, ironically tend to resort to violent, oppressive, and abusive use of power and thus continue to perpetuate the unjust systems that they claim to abhor. Nwapa thus indicts women who exploit and enforce power not just for their own benefit and personal fulfillment, but ironically for the destruction of men, and even for their own self-destruction. This appears to be the tragic consequence of power; that in the contentious conflict and struggle to win/profit and place the self in power, women too, turn into oppressors, who tyrannize, terrorize, silence, and maim others—particularly men—with their power. In this reconstruction of power, the tragic ethos is reversed and shifted. And having lost power to women, men become the other losers at the margin. It is this tragic destiny that Obiora shares with Reverend Fr. Mclaid/Izu that Flora Nwapa consciously "signifies" in *One Is Enough.*

Thus from this platform, Nwapa targets the powers of mothers/motherhood for interrogation and critical evaluation. In this regard, she positions mothers as the prime movers and sole agencies who exploit their traditional role and power to rule, control, and marginalize their children, especially their sons. At the superficial level, therefore, Nwapa advocates and affirms the need for women's empowerment in *One Is Enough.* At the symbolic level, however, Nwapa resorts to the use of double-entendre to expose, critique, and chastise the so-called powerful successful women for their reckless use of power. More importantly, she exposes the hidden, violent exploitation of power by women against men, perhaps as a means of cautioning women against the reproduction of excess and injustice which destroy harmony in society. Thus in one breath, Nwapa promotes women's power and, in the next, she derides women's (ab)use of that power. Using this device, Nwapa interrogates and problematizes women's power, and it is this dimension of the author's vision that we must turn to now.

II. Flora Nwapa's Gynocentric Vision: Signifying/Rewriting *The Joys of Motherhood,* Patriarchy and Men's Tragedy in *One is Enough*

One Is Enough is framed as a provocative, subversive text that is aimed at revealing, reconstructing, reinterpreting and rewriting the conventional notions of women's (dis)empowerment. In this process, the writer relocates and uncovers the shocking and disturbing truths about women's conditions of power and disempowerment. Instead of the conventional weight of culpamania heaped upon men for oppressing helpless women, Nwapa reverses the direction and constructs an alternative ethos where men are seen as the victims of women who are their oppressors. Reading the tragic and tormented faces of men in this novel, it becomes obvious what Nwapa aims to accomplish—to debunk and recreate certain myths about the oppressive powers of male patriarchy, and to underwrite and relocate those powers on the oppressive matriarchal foundations of patriarchy itself. It is no wonder then that feminists, who have been used to romanticizing and framing the tortured images of women as the oppressed, silent victims of male power and patriarchy, have tended not to see or appreciate this other ugly face of women's predatory powers of life and death over men.

Perhaps, too, Nwapa undertakes this mission of role reversal in the spirit of bargain and [fair] play. She discloses that women exploit and (ab)use the powers of mothering/motherhood invested in them by tradition. In this manner, women too alienate, marginalize, silence, disempower, and dislocate others from power, particularly their sons. Revising the politics and interpretations of motherhood, Nwapa shifts the (con)text and repositions the negative visions of "motherhood" as "otherhood"[16] in order to construct her alternative visions of "mothering" and motherhood. Nwapa employs the oppressively controlling mothers in the text to suggest that contrary to the popular beliefs held by feminists especially, women/mothers wield the absolute power conferred on them by tradition for their own empowerment, and for the disempowerment of others; the worst victims being their sons. As the women/mothers grow in their power, they push the men to the margins as outsiders and losers. With women's maternal endowments of power, men are naturally handicapped and cannot contest or renegotiate the powers/joys of motherhood.

This is one unique and special endowment that nature has

absolutely denied men from birth. Like a trickster, Nwapa leads us into the hidden powers/"joys of motherhood" in *One Is Enough*. While she is applauding the women for (re)discovering, (re)affirming and (re)centering themselves as the productive agencies of power, she is simultaneously undercutting, ridiculing, satirizing, and admonishing the reckless (ab)use and celebration of these powers by women. Herein lies Flora Nwapa's double-entendre in *One Is Enough*.

It is in this novel that the author particularly targets and challenges the feminist/womanist sisterhood of writers and critics to disclose the hidden cost and the unspoken paradox of *The Joys of Motherhood* to men, who suffer the silence and the "othering" imposed on them through invocations of the powers/"joys of motherhood." Indeed, *One Is Enough* provides us with enough evidence to argue that Nwapa employs the text as a new market-place to renegotiate and rewrite the contemporary discourse and understanding of the politics of African womanhood/motherhood. Using this medium also, she provokes an inter-textual dialogue to enable her to reposition and reconstruct her own previous visions and ideologies of the politics of womanhood/motherhood as enunciated in *Efuru* and *Idu*.

More significantly, however, *One Is Enough* sets the stage for the author to interrogate and rewrite Buchi Emecheta's ironic *Joys Of Motherhood*. Flora Nwapa signifies in many ways who and what her targets are, and I sometimes wonder why or how feminist/womanist critics should continue to (mis)read her target in this (con)text. In *One Is Enough*, Nwapa crosses the borderlands in celebration of women's power and goes "the extra mile" to reconstruct, critique, and problematize both the produced and prospective powers of women in society. The text provides Flora Nwapa with a new market-place and the fictional matrix not just to promote her gyno-centric vision, but to interrogate, challenge, revision, and question the negative (mis)representations and lop-sided visions of both patriarchy and motherhood that Buchi Emecheta champions as a new creed for feminists in the ironic *Joys of Motherhood*. In my opinion, Florence Stratton misreads Nwapa's vision when she makes the assertion that "[I]n *One Is Enough*, then, Nwapa interrogates such male texts as Ekwensi's *Jagua Nana* and Ngũgĩ's *Petals of Blood*, the main ideological function of the novel being to refute the prostitute topos."[17] On the contrary, my position is that Nwapa's target in *One Is Enough* is not so much to these male texts, as it is Emecheta's and the related feminist texts. In *One Is Enough* especially, Nwapa consciously engages Emecheta in an intertextual dialogue. She effects this by creating alternative characters that she positions to respond to Nnu Ego's lament about her dystopic "joys of motherhood." If Buchi Emecheta constructs Nnu Ego to embody the female who is tragically oppressed, disempowered, suffering, silenced, and weighed down by the demonic powers of patriarchy, Nwapa creates the tragic male alternative in Obiora and Fr. Mclaid as mirrors of the "other," whose cry of pain is muted or muffled by the heavy cloaks of mothering/motherhood. Thus reversed, patriarchy is framed externally to mask and cover-up the tortured faces of the sons in the hands of mothers. With this alternative construct, Nwapa shifts the paradigms of profit and loss to produce a new balance-sheet of power. If in the past, women have been portrayed as victims of patriarchal traditions, she shows us the other losers—men. Men too have been victims as the author reveals in the experiences of Obiora and Fr. Mclaid/Izu in *One Is Enough,* much in the same manner that "Adizua becomes feminized"[18] by Efuru, Amaka's alter ego. While exposing the demonic powers of tradition in *One Is Enough,* Nwapa unveils the matriarchal roots of the tradition which have continued to impede and marginalize (wo)men, especially in Igbo society. I also wish to recall here what Kamene Okonjo stated in another relevant context. According to this scholar:

> [T]he existence of dual-sex systems in West Africa is particularly interesting because most West Africa societies are patrilineal and patrifocal. As elsewhere, men rule and dominate. Seeing this *outwardly patriarchal framework,* many observers concluded that the position of women in these societies was totally subordinate; as a result of their misconceptions, they produced a distorted picture of the "oppressive" African men and the "deprived" African woman. This situation has been noted by Afigbo (1974), and it is now the task of other writers, particularly African writers, to help correct it. (45; my emphasis)

Gay Wilentz made similar observations in her own study where she stated the following:

> The child is central to this matrifocal culture that Nwapa writes about, and the woman—mother, grandmother, aunt, older sister—is the link between

the child and community. Nwapa's novel as woman-centered oraliterature, focuses not merely on woman's position and mother, but on her role in child-socialization and its implications for the community as a whole. (9-10)

I agree with Wilentz up to this point. However, what I find rather curious is that instead of insisting that women must accept the responsibility for upholding the customs (patriarchal and all) that endanger the ultimate fulfillment of both men and women in society, feminists and womanists resort to scape-goatism and witch-hunting of men for the misfortunes of women. Is it not the same critic [Wilentz] who admits openly that:

[A]lthough the women are in charge of the education and initiation of young girls into adulthood, they are not free to teach what they like. As upholders of the traditions, they are compelled to act 'in line with past practice' (Abrahams 4) whether they agree with the custom or not. In a dual sex role culture, this aspect of women's responsibility is one which demands great respect from the society, yet the contradictions inherent in what *they* teach reflect the perceived oppositions in Nwapa's discourse. (10)

One cannot but wonder why critics like Wilentz persist in exonerating women from the negative roles they have assigned themselves, ostensibly to ensure the preservation of the cultural norms and values. What does Wilentz mean when she states that "they [women] are compelled to act 'in line with past practice...' whether they agree with the custom or not"? Compelled by whom? And what essentially is this past practice? And who invents the practice? There are a number of contradictions in Gay Wilentz's propositions here. How can these women that she describes as the "upholders of the traditions" be presented also as such passive actors who must simply toe the party line like zombies? The argument of this critic distorts the issues, and it is precisely this kind of warped vision of patriarchy and motherhood that Nwapa questions and revisions in *One Is Enough*.

As the following examples will show, Nwapa uncovers and unveils the mask that women/mothers put on patriarchy as peripheral signs to camouflage the tyranny of matriarchy. Our first example is Obiora's mother in *One Is Enough*. Obiora is pushed over and silenced by his own mother who wields her power indiscriminately against him as her son and against Amaka as her daughter-in-law. Obiora's mother brandishes the power conferred upon her by the tradition. One obviously hears the satirical and critical voice of the author intruding in the narrative during Obiora's encounter with his mother: "Will you shut up and let your wife talk"! shouted Obiora's mother. "Who asked you to come here anyway? Please remove yourself from my room" (2). Following this mother's command, the intrusive narrator quickly informs us that without any rebuttal, "Obiora left the room and the two women were alone"(2). As implied, Obiora is displaced in this contest for power between the women. He progressively becomes the outsider, as his mother takes center-stage and consolidates her power. Until Obiora rebels after losing everything to his mother, the narrator consistently drops the hint to us about who is in control and in power. It is not Obiora, but his own mother.

Similarly, Amaka's aunt, who functions as her surrogate mother is also dictatorial. She stifles Amaka's voice and choice in the selection of a marriage partner. She instructs Amaka: "What is important is not marriage as such, but children, being able to have children, being a mother"(8). And not long after this, the narrator remarks about Amaka's anguish: "Only her fate played pranks. God had deprived her of the greatest blessing bestowed on a woman, *the joy of being a mother*"(20). Nwapa may be asking some underlying questions here. Is it not apparent that women know and relish the hidden powers/joys of motherhood? Why do the Igbos name their children Nneka—"Mother is Greater/Supreme"? Greater than who or what? Does "Nneka" not imply that an uncertain, unspoken hierarchy exists in the positioning of mother as supreme? Is mother therefore not positioned on a higher scale than the silent other—the father? On the other hand, if Nneka—Mother Is Greater, who is the other? Who is mother greater than? The culture is most silent on this issue of male marginality in society. And perhaps this provides the rationale for the erasure and diminutive imaging of men in the society which Nwapa appears to be signifying in this novel. With the exceptions of Mgbada and Sylvester, who are portrayed as devoted, caring, loving, and responsible fathers and husbands in *The Lake Goddess,* men are generally presented as what Chikwenye Ogunyemi has referred to as "feminized men"(148).

In *One Is Enough,* for example, there is a remarkable absence of strong father-figures. And indeed, where fathers are present or mentioned, they are presented as useless, reckless, and irresponsible men who are so powerless that they are virtually of no consequence to their family and society. Amaka's mother frequently reminds her of her "own imbecile father's family"(32). The narrator provides this special information about Amaka's father and family:

> They had been brought up to despise their father because he was of no consequence. He had too many wives and he was a drunkard. He was of no use to them when they needed him most, and when he died, it was more or less a relief for them. It was good riddance, and they almost celebrated his passing away. Their mother was at her best. She so demonstrated her grief that she nearly convinced the people that she cared for her late husband. But she did not care. All her children knew that she did not care. The boys especially knew, and it hurt them so much that they put up feeble fights in support of their father, but they did not succeed. In the end, their mother won them over. (23)

Amaka's aunt magnifies this image of worthless men as she tells Amaka: "I have neither respect nor regard for my husband. You can see that, can't you?"(8) Thus with the erasure and closure of powerful male-figures, the women/mothers gain extreme prominence and power in the society. It is these extremities in women's positioning and appropriation of power that Nwapa highlights, interrogates, and critiques in the novel. The subject is encoded in the portrayal of the experiences of Obiora and Amaka under the heavy hand of the mother who positions and promotes herself as a deitess. Amaka obliquely reminds us about this truth as she instructs the new bride who has been beaten by her husband: "Mind you, they [men] do appreciate their mothers and sisters, but never their wives"(27). But then, as the events develop, the underlying truth is also revealed that the men are merely pawns in the hands of their mothers. As the author signifies, Amaka's problem goes beyond that of fulfilling the traditional expectations and social pressures of motherhood. Hers is a spiritual quest to attain the divine and traditional powers and attributes of motherhood in the society. Like all Igbo women, she yearns for that absolute power of motherhood.

With this concern, therefore, Amaka's problem is not so much with her husband [Obiora], or with her *chi*-[personal god] for that matter, but more significantly, Amaka's problem hinges on the unspoken struggle for power with her mother-in-law and other mothers who inscribe and regulate the norms that everyone else must conform to in society. In this manner, the problem shifts from the superficial inter-gender conflicts to the intra-gender conflicts between Amaka and her powerful mother-in-law. The narrator suggests this shift as pivotal to Obiora's loss of voice and power and, consequently, his wife's loss in the concluding paragraph of Chapter I: "Six years later, there was no child. Obiora's mother was tired of waiting and so she had come to a final solution. Obiora must have an heir, because his brothers and sisters all had heirs"(12).[12] And then she continues:

> 'Whether you hear or not, it will end today. Everything will end today when I finish with you. The hold you have *on my son* will end today. Do you hear me? I have waited for six years, and I cannot wait for even one day more. Didn't you see how *I hushed up Obiora when he came in to interfere?* He is a stupid son. Sometimes I wonder whether he is my son. But I know he takes after his useless father. . .' I hushed up my son this morning? I hushed him up too last night, and stopped him from striking you because he wanted to strike you when he flared up.' (13 -14; my emphasis.)

It is interesting that Obiora's mother states that he [Obiora] "came to interfere." What is clearly revealed here is that Obiora is an outsider. As his mother displaces him and speaks for him, Obiora shifts and drifts into nothingness, and he instinctively feels his powerlessness as he says to Amaka "[W]hat mother told you will happen"(19). From this point onwards, Obiora becomes progressively weakened, silenced, and stifled by the omnipresent and omniscient mother. It is the mother's voice (not Obiora's voice) that Amaka and the reader hear in these announcements: "Obiora has two sons by another woman"(14). It is also Obiora's mother that gives Amaka this important piece of information:

> 'My son's wife and mother of his sons wanted you thrown out of this house. But I told her I would have none of it, that you will not be thrown out because you are the first wife. I too am the first wife of my husband. So I told her categorically that you would not be thrown out.' (15)

As age and motherhood confer authority and power on women in the tradition, Obiora is portrayed as the outsider, (an)other intruder and loser in this market place of power. With the emphatic "I" as the doing subject and the primary agency, the mother asserts her place at the center of power. As the woman takes center-stage, contesting or negotiating power, Obiora is forced into the margins, disempowered and emasculated by his mother. Worse still, one gets the impression that Obiora is emotionally underdeveloped and acts like a juvenile delinquent as he goes about asking Amaka: "[H]ave you seen my mother"? In fact, he exposes himself to more ridicule and doubt in this warning to Amaka:

> But let me warn you that if you step out of this house in protest when my wife and my two sons arrive, you stay out forever. You must not come back. I am not the type of man to go begging with cap in hand for you to come back. *I have outgrown that.* (20)

Has he really? A "Freudian slip," maybe? But the fact remains that Obiora fails to show that he has "outgrown that," even his mother's grip on him. The narrator hints at the man's progressive feeling of helplessness and powerlessness by noting that "Obiora said weakly"(19) in his rebuttal to Amaka's listing of her accomplishments and chivalrous operations to salvage and to promote him. Obiora feels his loss and it is this realization of his emasculation and impotence that aggravates him, and so "[h]e rushed at her"(26). It is clear that Obiora's rash action is not premeditated and that he resorts to this brutal force as a final defensive act to regain power and affirm his manhood.

The narrator's empathy appears to be with him in his progressive experience of marginality. Obiora suffocates in this traditional trap where he is ostensibly placed to reign as "king" and head of household, but where, ironically, the mother rules him out of power and place. Unlike the wife who is temporarily displaced, Obiora is permanently silenced and displaced in his home. As he loses, and chokes, his level of stress builds up accordingly, and we know that it is only a matter of time before he will explode. Obiora's anguish reaches a tragic climax as his voice ruptures in this moment of epiphany as he confronts his mother:

> 'Why shouldn't I be upset. You have driven away Amaka and the mother of my boys. Are you now going to be my wife'?'

How dare you talk to your mother in that way? How dare you?' As she lifted her hand to strike him, he dodged but she held her outstretched hand in mid-air.'

'Now listen to me, Mother, listen very carefully to me. I hear Amaka is coming home and I am going to see her. I am going to beg her to come back, and you are not going to interfere in anyway, because if you do, I am going to kill you. It is not an empty threat. I am going to kill you.' (82)

The moment of truth has come for Obiora. He will no longer be silent and evasive before this predator who eats up his manhood, and so he names it categorically—Mother! Once he has located this agency of oppression, he threatens to "kill" her metaphorically as a step towards regaining his manhood, dignity, and power which he has tragically lost to the tyranny of motherhood.

Thus like a trickster stage-managing this episode, Nwapa leads us away from woman's suffering and takes us right into the bleeding heart of man in this tragic ethos. As Obiora's mother invokes and asserts the powers of motherhood—for bringing him to life, she expects total respect, obedience, and veneration from him, and he never seems to have the ability to question that authority until she transgresses with that power.

Men like Obiora are not alone in the agonizing experience of the powers/"joys of motherhood." Women too suffer victimization and dislocation from mothers/motherhood. Amaka's experience with her mother and her aunt exemplifies this phenomenon. Amaka's aunt evokes the powers of age and the umbilical ties of motherhood to rule and choke her. This is highlighted in the episode where Amaka's aunt dictates to her: "No, my daughter, *I don't like him.* I know his family well. He *is not* going to be a good husband... And that's the man who wants to marry you. No, my sister's daughter..."(8). No doubt, Amaka's aunt makes it clear that her power and decision on the subject is not negotiable. And she asserts that power with uninhibited authority: "And that's the man who wants to marry you? No, my sister's daughter"(8). It is clear that Amaka is in no place to bargain any other position since the aunt has appropriated the center and pushed Amaka's voice to the margins. The point is not whether Amaka likes or loves Bob. "No, my daughter, *I don't like him*"(8), her aunt states, and with a note of finality, she declares: "I don't like him. I know his family well. *He is not going to be a good husband*"(8). The all-knowing, all-seeing aunt, as surrogate mother, has spoken and her word/power is not negotiable. Amaka will not, cannot, have any voice in this matter as far as the older woman is concerned, and there is no contesting that power.

And so, knowing that she has a ready platform upon which to build her power, Amaka's aunt proceeds to instruct her, perhaps being mindful that Amaka will soon become a representative of the next generation of mothers who must continue to promote the creed of motherhood as she states: "What is important is not marriage as such, but children, being able to have children, *being* a mother.... Have your children, be able to look after them, and you will be respected" (8; my emphasis). What Amaka's aunt is doing here goes beyond the customary roles of parenting and socializing the young. It begins to sound like the catechism of some religious missionary, inscribing and handing down the articles of faith, creed, and commandment that are designed to ensure the institution of motherhood from generation to generation. In some ways, this sounds like a parody of Buchi Emecheta, who, in *Our Own Freedom,* envisions the future as that of "[M]others handing down the future to their daughters"(47).

Perhaps it is this legacy that women/mothers are determined to preserve and to promote for themselves and for their daughters at the exclusion of men that Nwapa questions and derides in *One Is Enough.* The author's laughter gains momentum as Amaka's aunt proceeds to give her own testimony of how she renegotiated and redeployed the powers of motherhood to rule and control her husband without his being conscious of it:

> 'I married a man I did not like. And in spite of the fact that I have my children, I have neither respect nor regard for my husband... When I had seven children in seven years of marriage, I knew I had enough, and stopped sleeping with my husband. Of course he protested. He reported it to my mother, but I would not budge. No more... He had given me seven children... What else did I want from him? When he began making too much fuss about this, I got a sixteen year old girl for him... Yes I married her for him. I said to her, this is our husband, take care of him... So I am concentrating on my children and my business. That was how I turned my back on my husband....'(9)

The author leaves no one in doubt in signifying that Amaka's aunt

uses her powers for dictatorship. Amaka's aunt is made to impli-
cate herself in this highly suggestive monologue in order to height-
en our consciousness of the hidden and dangerous powers of
motherhood. In the process, Nwapa underscores for us the apparent powerlessness of men. Beneath it all is the indication that men
wear the mask of power; women rule men behind the mask.
Amaka's aunt has located and positioned herself at the center and
therefore "[she] will not budge" to anyone, least of all her husband.
It is obvious that her husband (the man) has been pushed to the
margin by the woman. For one thing, she has already admitted that
she has no respect for him. Therefore, the man is objectified, rei-
fied, and marginalized to the point of silence. The woman knows
it and celebrates her absolute power and denies him any power of
choice or voice by marrying a younger wife, (an)other victim, "for
him." In this way, she extends her own empire of subjects to rule
and control. And to ensure that the point is not missed, she
reasserts her power: "[Y]es I married her for him"(9). Neither the
man, nor the younger woman/wife, as the symbolic initiate (who
is now taking her place in the lower scale of the female contestants
for power), has any voice or choice in the matter.

Amaka's aunt consciously stresses the point that it is through
motherhood that she has earned this authority and power, and that
it has been conferred upon her by custom/tradition. But on a much
more sinister note, the portrayal of Amaka's aunt suggests that men
like her husband are totally powerless and indeed are used, exploited, and manipulated consciously by women. In this regard, the
man becomes the means to an end: a sperm-donor. And as Amaka's
aunt affirms her own ability to bargain the man out of power, she
celebrates her accomplishments and victory with this rhetorical
question: "[H]e had given me seven children. What else did I
want?"(9). Here, Nwapa excels in the signifying and satirizing
devices through which she exposes the woman's excessive uses
and abuses of power, in order to debunk it. And in a prophetic manner, Amaka's aunt gives this injunction to Amaka: "Marriage or no
marriage, have children.... As a mother, you are fulfilled"(11).

Like Efuru and Idu before her, Amaka is charged to go in quest
of motherhood and to achieve it by any means possible. Why not?
After all, her teachers, her foremothers command it. They expect
her and other succeeding generations to preserve and promote that
legacy [of motherhood]. Frustrated about her current impotence to
fulfill the word and prophetic mission, Amaka's anxiety grows and
she begins to lament her fate and failure: "God had deprived her of

the greatest blessing bestowed on a woman, the *joy of being a mother*. . . was that really the end of the world? Was she useless to society if she were not a mother?"(20). In this passage, Amaka's naming of her quest—"the joy of being a mother"—also ritualizes Nwapa's crossing beyond the borderlands of creativity in her own text, to engage in critical inquiry and inter-textual dialogue with Buchi Emecheta on *The Joys Of Motherhood*. Moreover, Amaka's rhetorical questions about the expectations of "society" provoke thoughts about the primal interest and constituency that Amaka must serve. Indeed, one would like to know to whose agency, and whose "society" does Amaka owe this debt that she must pay in the high currency of motherhood, if not the allegorical "woman's world"? What is being suggested in this context is that women need to look inwards and re-examine their own norms and values, which may be responsible for certain impediments and traps that they face in society, instead of blaming all their failures and sufferings on men and patriarchy. Nwapa uses *One Is Enough* not just to disclose the other dimension of motherhood, that is, the powers of motherhood which Emecheta and the sisterhood of feminists quite often distort or derogate, in order to stimulate new thoughts and dialogue on the politics of motherhood in African culture and literature.

The author presents the mother as power personified. In the legal as well as the divine configuration, the mother positions herself as the police and priestess of the tradition. Obiora's mother in particular is employed by Nwapa to dramatize the extremities to which mothers can push and promote their power to the point of hurting and silencing others, and especially men. Thus in the mother-figure, Nwapa effectively critiques the profound powers of motherhood which women like Obiora's mother and Amaka's aunt deploy in the tradition to empower themselves and marginalize or disempower others. We could summarize Nwapa's critique of the hidden powers/joys of motherhood with these questions: (1)Do women, especially successful and powerful women, reproduce the oppressive norms which they often decry about patriarchy? (2)Are women really the victims of men, or is it really men who are the victims of women? (3)And therefore, are women simply targeting the wrong enemies of their progress? (4)In view of the mother-in-law's tyranny against her daughter-in law, who really is Amaka's oppressor—the man or his mother? (5)What about women who oppress men, including innocent men? It is this last question, in particular, that will lead us to examine Nwapa's portrayal of the unequal relationship between Amaka and Rev. Fr. Mclaid (Izu). As the author discloses, Amaka transforms Izu into her toy-object in her aggressive and manipulative hands, just as she terrorizes him with her new powers of motherhood.

Once Nwapa shifts the narrative to the Lagos metropolis, we are faced with women who violently turn the market-place of trade and power into a theater of war. In this economic and social battlefield, where women who are economically successful rule and control power, men experience marginality and tragic loss. Here, the territory is no longer under the control of mothers who choke their sons with the umbilical cords of motherhood. It is a land conquered, and under the siege of another brigade of sisterhood with "gangster mentalities"(10). The author presents Amaka stepping into the parade as she earns her place in the (in)famous "Cash Madam Club." This is the new league of women who grow into power by sapping and lynching men, metaphorically: "They were the new generation of women contractors. There were about ten of them. Six were widows and the other four had left their husbands to start life again. They were all involved in the "attack trade" during the war (49). Having survived the Nigerian/Biafran war, these women have become psychologically initiated, trained and conditioned like an army. And like an army also, they do not simply fight for survival, but for conquest and triumph against the target enemy—man. Similar to the settings of the "post-modern culture" that Cornel West critiques, this city is characterized by anomie. And its flexible morality offers a ready cesspool for the economically strong brigade of women to terrorize and conquer the weak. These women are war veterans, and are steeped in the strategies and psychology of war. They know, for example, that in the war zone, there is no place for sentiment and compassion—the winner takes all. In a war zone, what does anyone care about the enemy or the wounded? They are simply dismissed as casualties and as injuries and casualties are the constant features of the battlefield. The new legion of sisterhood is fighting not simply for survival, but for gaining total territorial control from men. In this contest even innocent men like Izu become (generic) enemies and easy prey to these female gangsters. And to fight in this new battlefield, the new breed of women do not come empty handed. They come armed with vengeance. Some enter the battle field as new ideologues and "matroits" with their dogmatic agenda. In Amaka's case, her mission is clearly stated: "The erroneous belief that without a husband a woman was nothing must be disproved" (24).

But other women enter the force, ready to use their bodies as bullets and ammunition which is reinforced with the ready-reserves of sisterhood to conquer men. For this army, what counts is not "how" one gets there, but *that one gets* there—to win. For them, the end justifies the means. Madam Onyei, as the "general" of the new power, typifies the spirit of this new breed of women. The narrator is unsparing in her critique and social commentary against this woman:

> Madam Onyei. All she did was send for her eldest daughter who was in school and left her to the mercy of men. Now she does not know what to do with the numerous contracts she has.... [So] Madam Onyei went on making money. The joy of having grows by having....The way she threw about her money made people start to gossip. She did not bother, but went on making money and spending it. (48-49)

As exemplified by Madam Onyei, the new breed of women become predators of men. They have transcended their position from being petty traders and fish-sellers, to become the fishers and sellers of men. More than anywhere else in the novel, Nwapa is at her best in presenting this farcical, absurd drama of women in power. From this point onwards, the author is out to undercut and ridicule the women's (ab)uses of power. In fact, the author's intrusive voice takes on an unmistakable resonance and it is here that her criticism of women's power is most acerbic. Madam Onyei is presented to us as a lead character in the new cast of women who are trafficking in power—body and soul—over anyone around them. This is why Madam Onyei can leave her daughter to the mercy of men. But it is suggested that the mother gains money and power by exploiting and throwing her innocent daughter as an object to injure men's hearts and wallets, and therefore, both the men and her daughter lose place and power to her.

The next in this rank of lead actors is Amaka's sister, Ayo. The narrator/author tells us early that "Ayo was a schemer" (48). In a sense, Ayo is presented by the author as the marks-(wo)man and strategist, who trains and prepares Amaka on how to shoot down her male targets. Once she has trained Amaka, the new recruit joins the squad and goes into the battlefield to gun down the enemy, men. As it turns out, her first casualty is the Alhaji who helps her to organize and execute her first contract. In the battlefield, there is no place for the weak hearted—no place for any conscience, and so like Madam Onyei and the "Cash Madam Club,"

Ayo too is celebrating her new power. Her mother already boasted about that power much earlier, during the strategic episode to motivate, mobilize, and intimidate Amaka into joining the woman's force against men:

> Ayo is the only one, among you who is like me. She took no nonsense from any man. When her husband came up with his pranks, she left him and got herself 'kept' by a permanent secretary whose wife went to the land of the white people to read books. Foolish women, to leave her husband for that length of time to read books. Ayo moved in. In four years, she had four children. In four years, her husband had sent her to school to improve. She is cleverer than all of you....She qualified as a teacher. In the fifth year *she was able to make* her 'husband' buy her a house in Surulere, and that year the wife returned without anything, *and my daughter moved out gracefully with her children, into her own home. In her position, what does she want from a man?* (33; my emphasis)

More implicated than anyone is this episode is the so-called "husband" that Ayo exploits as a material object to achieve her goals. And yet, the man remains ignorant and blind to her schemes. The man lacks consciousness and fails to realize that he is just being used, exploited, and tragically manipulated by the scheming woman, Ayo. Again, underneath this is Nwapa's muted question: So where is the legendary power of the so-called "husband"?

Additionally, Amaka's mother skillfully innoculates and immunizes Amaka against any form of identification and empathy for men, so that, like a good soldier, she can go into the battlefield to conquer and to win, without ever losing. After all, didn't Amaka become baptized and initiated in the metaphoric "attack trade"/war? The mother celebrates her new *shero*, Ayo, who her mother proudly claims as "my [her own] daughter"(33) because of her conquest of men. Thus echoing Amaka's aunt, the mother rhetorically asks this question: "In her [Ayo's] position, what else does she want from a man?" History repeats itself. Ayo is really her mother's daughter. It is implied that she has reproduced her mother's use and exploitation of her husband, a man that was only worthy enough to give her children, and therefore was used as a means of fulfilling the goals and needs of empowering the women. It is understood that Ayo's power is symbolized and inscribed in the four children, as well as the new home that Ayo has got her illegitimate "husband" to buy for her. And in this ascending order of importance in the roll-call, Amaka is presented to us. Amaka grows in her new position to play the lead role in this new league of female economic gangsters. Amaka becomes a better student once she has graduated from the "military school" of her mother and her sister, Ayo, and the "Cash Madam Club" who are now serving as her surrogate mothers in the city. To prove her new knowledge/power, Amaka goes on the offensive. Her first target and casualty is the Alhaji, who is exploited and eliminated with military precision and speed. Again the author is not neutral. From this point, the author becomes more intrusive in the narrative as evidenced in this disclosure:

> It was the Alhaji who *had helped her* get that first-contract. The Alhaji liked Amaka very much and had made advances. But Amaka again did not feel up to it. The Alhaji had helped her secure more contracts and *had helped* her in the execution of them. But he had *never* asked Amaka what profit she made. Amaka had told him about the three thousand naira she made on the toilet roll deal, but all he said was, 'Thank God, it is all yours.' (66; my emphasis)

It is suggested that the expectation in the jungle is for the one who has won power, to consume the other, or to marginalize their profit. But the Alhaji is different. The narrator also repeatedly suggests that the Alhaji's personality is imbued with an admirable quality of humility and dignity. He is not just like any other player who makes it a business to win and profit in every trade that they engage in. And to stress this point, the narrator reaffirms that Alhaji actually, "liked Amaka very much," and "never asked Amaka what profit she made"(66). What is remarkable, therefore, is that while the Alhaji is driven to Amaka for romantic love, Amaka, on her part, is driven completely by her narcissism, and the inordinate ambition for money and power. Amaka is out to play her own game of profit with a vengeance as she manipulates and exploits the man to execute her contracts, and buy the site for her own prospective home. Thus in ignorance, the Alhaji becomes trapped by Amaka to pay the price of her growing power. But what is even more significant is that after becoming suspicious of Amaka's scheme, and that she will not yield any ground to reciprocate his affection, the Alhaji still drifts towards her as if she had cast a spell on him. In the end, Amaka proves herself a parasite—she siphons all for her own needs of empowerment and profit, and dumps the Alhaji as waste. The Alhaji's response to Amaka's profit in the toilet roll deal heightens the tragic irony of the man's condition: "Thank God, *it is all yours*"(66). What is this "it"? On the surface, "it" refers to the profit in the business of the toilet roll. But at the metaphorical level, "it" is the Alhaji, and indeed, all men. "It" is all "yours," women, and especially so for women like Amaka. And just as Amaka makes her total profit from the toilet roll, the new sisterhood of women completely bargain men out of the game of profit and power. Once the women have made their profit, they dump the men as Amaka does with the Alhaji as if he were mere toilet roll or waste paper. Amaka sets out from the beginning to use the Alhaji just as she would use and dispense with toilet paper:

> She wanted a man in her life. All women should have men in their lives....The men could be husbands or lovers...she did not want him as a lover but she could not yet make a clean break with him. *He was precious to her just then.* She needed him more than he needed her. (66; my emphasis)

It is signified that like the toilet paper "it" (the Alhaji) is of temporary "precious" value to Amaka "just then." But once "it" has served her purpose, "it" is thrown away like any other article that is of no value to her. This illustrates how the men are turned into "things" and material objects in the manipulative hands of the women. And what is most tragic is that the men remain grossly unaware, lost, drifting, lacking the necessary control and (in)sight that could help to redeem them from the women. To a large extent, therefore, even though the men may be engaged in sexual relationships with the women, they are presented as juveniles who exhibit profound ignorance and naivety that renders them vulnerable to and impotent around women.

Having (ab)used and exploited the Alhaji, Amaka dumps him and moves to another object. Her new target is Fr. Mclaid/Izu. Amaka's (man)handling of the Rev. Fr. Mclaid is, perhaps, the most classic and pathetic example of men who become trapped in this new tragic ethos. What is particularly significant about Fr. Mclaid's tragic experience in Amaka's hands is that, unlike the Alhaji who at some point became suspicious of Amaka's insincerity, The Man of God is acutely ignorant of the treacherous "ways of the world" and, more so, of the sinful snares of women like Amaka. Like the post-modern Ananse-spider, Amaka spins her webs to entrap the priest. By the time he realizes it, he is maimed, disempowered, and lost. He loses his identity. Izu's crisis of consciousness is echoed in this dramatic irony: "I know what I am doing, but I cannot help myself"(76). In the above statement, he subconsciously reveals his vulnerability and powerlessness before his predator, Amaka. In the final analysis, the priest suffers tragically for his sinful "fatal attraction" to Amaka. Nwapa replicates the biblical myth of Eve, the temptess who seduces Adam out of the utopian garden of Eden. Like Adam, Izu loses his grace and power. He loses his own blood in the accident, his twin boys, and his identity to Amaka, who continues to take and gain all. In spite of the priest's legendary spiritual powers, the post-modern Ananse has spun her webs to control and maim him and in the end he gains nothing while she profits from all his actions. What is revealed is that behind the sacred cloak/mask is nothing more than a man, and a blind man for that matter. His metaphorical blindness leads him into painful losses.

In some ways, I see Fr. Mclaid/Izu as Obiora reborn and reconfigured in the metropolis. Like Obiora, who is outwardly powerful, Izu is a man of God who should be imbued with spiritual powers. Both men are choked and humiliated by the same woman they loved. They are used as ritual sacrifices on the violent and bloody altars of Amaka, the powerful goddess of the post-modern market culture. It is obvious that Izu is still searching, as suggested in the sad story that he tells about his experiences in the past—rejection, alienation, split-identity, and homelessness, which resulted in the absurdity of his adoption by a white Irish priest, Father Mclaid. But more importantly, Izu's painful and nostalgic account of his own mother's tragic circumstance and unfortunate death (which consequently denied him a mother's love and care) underlines his inner longing and anguish for both a mother-figure and a family to serve as symbolic home and emotional center. Figuratively speaking, Izu, as a "motherless-child" who is still searching for mother's nurture, subconsciously identifies Amaka, the "babyless-mother" who is still longing for "the joy of being a mother" as a prospective mother. Amaka is that mother-figure that Izu is yearning and searching for. From this perspective, Izu's attraction to Amaka transcends the physical search and thirst for her body, to become a spiritual, mythological quest for roots, identity, home, and anchor. With the birth of the twin boys, Izu must have thought that his quest for center(ing) had become fulfilled. This is suggested in the degree of passion, love, care, and devotion that he exhibit-

ed as he held the twins in his arms. Moreover, his determination
to give up his name and place in the church now that he had found
his rightful place and home in family reinforces the rationale
behind his passion. In this regard, Izu's ultimate quest is spiritual.

For her own part, however, Amaka's primary quest is to gain
material, economic, and social power. Her incidental desire is to
become a mother, and then to attain the legendary powers of motherhood for herself. Reading between the lines of Amaka's attitude
and behavior after the twins' arrival, one gets the sinister feeling
that Amaka's attitude is one of emotional satisfaction for adding
yet another feather to her already colorful cap. Thus the twins may
yet be other sets of objects which she will now possess and add to
her worldly acquisitions and possessions, as a means of validating
and extending her territorial and material space and power. Izu,
on the contrary, exudes passion and affection that is deeply rooted in his heart and soul for these children that will now provide
him with the needed spiritual anchor, belonging, and home. The
narrator suggestively confides this in us:

> It was only a father who could carry a baby like
> that....It was simply too much for Father Mclaid. The
> secret was out. 'Amaka, you had twins. I am a twin
> myself. Oh my God, my twins, both boys. I cannot
> believe it.' He embraced Amaka. (124)

In this manner, Izu regards the twins not simply as the mere extensions of family, but as his symbolic, spiritual, ritualized home and
place. The twins do not, therefore, have the same meaning and
value for Amaka and Izu. Both of them are traveling in different
directions, metaphorically. Although they have met, had intercourse, and crossed paths in the mythical modern crossroads of
Lagos, it is clear (at least on Amaka's part) that she will not allow
him to cross the imagined boundaries that she has constructed
between her and every other entering that space.

Amaka has given up love and loving since reaching this market-place of Lagos. If in the past she loved, and failed, she has built
a solid wall around her heart in the present, and she does not
appear to truly love anyone—not even her legendary mother and
sister, Ayo. On the other hand, Izu, who is carrying his load of true
spiritual and romantic love and passion is thus in many ways an
intruder who is only worthy to be tolerated for as long as he can
be used for Amaka's purposes. In this sense, Amaka is just as emotionally sterile and handicapped as Obiora, who is emotionally

underdeveloped. And compared to Izu, who is fully emotionally developed, one suspects that his love for Amaka is destined to be a "fatal attraction" and it is only a matter of time before he will crash. The narrator builds up empathy for Izu by constantly suggesting that he is innocent and ignorant of her ways/schemes. To heighten that empathy, the narrator drops the hint about Izu's progressive loss of power because he is blind and lacks knowledge: "Father Mclaid had been around at this period but he suspected nothing and Amaka told him nothing either"(101).

Perhaps more than any male character in Nwapa's fiction, Father Mclaid/Izu is the most tragic. He lives out the tragic paradox of his existence as a priest, whose spiritual power and identity is stripped by Amaka in the very same manner that she strips him of his clothes in bed. In spite of his unrelenting struggles to retain Amaka and the twins as his symbols of place, power, and center, Amaka denies him all and instead takes all for herself. Thus Izu is symbolically stripped, naked and maimed at the end of the day. And in the final moment of truth, knowing that he has lost everything, he returns to the foster-care and orphanage of the church—the alien make-believe place that is not at all home.

Izu also loses his name and identity before Amaka. Although the public reveres and respects his identity as Fr. Mclaid, a priest of God and an apostle of the church, he is simply "Izu," (an)other man before Amaka. Referring to him as Izu, he becomes devalued to just another man and object that Amaka uses for her own empowerment without his being aware of her schemes. The narrator constantly employs innuendoes to reveal Izu's progressive loss of power and victimization. But ironically, while he is genuinely interested in "giving," Amaka is obsessed with "taking" and exploiting his very gentle, affable and caring nature. Acting as Nwapa's persona, the narrator reveals Izu's tragic crisis and loss with great empathy and compassion. As soon as Izu is introduced, we are told that "[H]e was an easy sort of person, who endeared himself to his parishioners"(66). The narrator appears to dwell on this theme of the man's goodness which crystallizes in this valedictory remarks about his compelling and outstanding love for his chosen one, Amaka: "He had loved her with a great passion which was mature and considerate"(101). This is important, for it clearly demonstrates that Izu is not the regular kind of man, who might be mean, chauvinistic and brutish. And even in his moment of crisis, denial, and rejection by Amaka, he continues to demonstrate remarkable affection and consideration for her, to the extent that he unilaterally undertakes to give up the priesthood for Amaka's sake, and to facilitate the process of his proposed marriage to her. But Amaka remains insensitive, uncompromising, and relentless in her self-centered visions and pursuits of power and progress. Happiness for Amaka is for her personal profit, and profit alone motivates all her actions as implicated in her thoughts:

> She was going to exploit the situation. What drove her to see Father Mclaid was just the contract and nothing else.Now other things were working in her mind. She would *play it* cool. For the first time, she was going to put into practice what her mother had been teaching her. She was not going to wait, *she was going for the kill.* A priest was also a man capable of many feelings. Father Mclaid was a man, not a god. Perhaps Father Mclaid had never been tempted. She, Amaka, was going to tempt. That was the task that must be done. (54; my emphasis)

It is no longer the voice of the narrator that one hears above, but the critical authorial commentary of Nwapa who is lampooning, criticizing this so-called transformed, successful woman, who, instead of using her new power positively, uses it negatively to destroy others, and most of all the innocent. As Amaka prepares to spin her webs and consequently "kill" her victim, the author prepares us for her final act of transformation: from being the victim to becoming the victimizer.

The narrator/author takes us through the evolution of the new woman (Amaka) as she shifts the tragic ethos (from living at the margins of her world as the exploited), and instead, pushes the man (Izu) into that tragic ethos as her exploited victim. Here, the man, Izu becomes the "other," the loser, and the new object within the tragic margins and ethos of the woman's world. In this new dispensation, there is no sentiment, or sympathy. And like any trader, Amaka (who is determined to earn personal profits, even out of the losses of others) shows in her actions that she lacks clemency. For this reason, she proceeds with business as usual, once she has bargained her way into huge profit. In this regard, as usual, the author's intrusive voice persistently provides innuendoes and critical commentaries that expose Amaka's insensitivity. The author's disapproving remarks about Amaka culminates as she describes Amaka's sexual encounter with Izu, in juxtaposition with Izu's sad narrative. In this scene, as Izu desperately attempts to reveal himself and the circumstances surrounding his birth and life to Amaka, his efforts yield neither compassion nor humane understanding from her. At this point, the narrative voice has completely lost its neutrality and remains poised to expose Amaka's hidden, savage, insensitive (ab)use of the man. With painstaking details, this narrator discloses the woman's treacherous, venomous, and predatory drives as she prepares to devour her prey:

> 'I haven't told you about myself, Amaka. I want to tell you tonight. You remember your question, when I was introduced to you as Rev. Fr. Mclaid?' 'I remember very well,' said Amaka. *She had changed into her nightie and was lying on the bed. Father Mclaid went and lay beside her.* 'But I want you now, Izu. You can tell me afterwards, please.' (69; my emphasis)

In this very scene, the author shows exceptional skills at signifying. On the one hand, as the man is instinctively driven by humane concerns to bare out his soul, unveil his past, and then share deep emotional bonding and affection with her, Amaka, on the other hand, is mechanically driven by a special instinct to conquer the man. The scene presents a sharp contrast between the strong, aggressive, and mercenary woman, and the weak, gentle, and considerate man. It is highly remarkable that as the man draws her attention to his mother's very special tragic human experience, or rather, "(wo)man experience," which should have been of special interest to Amaka as a woman, she rebuffs and shocks everyone, including Izu, with the studied disinterest, impatience, and petulance marking her words: "But I want you now, Izu. You can tell me afterwards, please." Amaka clearly shows that her goal is to silence the man. In this context, Amaka betrays her narcissism and egocentrism to the point that she loses the reader's empathy for her previous tragic experience and marginal condition. The narrator elevates Amaka, only to ridicule her by remarking immediately that "(S)he had changed into her nightie and was lying on the bed. Father Mclaid went and laid beside her"(69). Thus, contrary to the expectation of activities surrounding a scene of romance and love, one gets the sinister impression that Amaka is preparing herself for battle, and that her target enemy lacks the essential knowledge and insight to fight her back. In this regard, Amaka turns her bed into a theater of war, as she lies in ambush with her body and nightie as part of her armament, ready to strike her prey. By reading the signs of danger ahead, the reader's fear, anxiety, and sympathy for the male victim grows until it reaches a point of aggravation when Amaka gives her command: "But I want you now, Izu..."(69). Perhaps, subconsciously anticipating the brashness and brutality of her impending action, she quickly adds "please" as mere perfunctory schema and afterthought. From this point onwards, it is clear that the narrative voice has lost neutrality and is determined to undercut Amaka's inordinate ambition for conquest and power. The disapproving voice culminates as Amaka shifts the roles, not just by proposing sex to the man, but also by dislocating the man from the center of power and turning him into a mere object of desire. As such, she does not request or plead: she demands and commands, just like any boss would command a subordinate. Through such innuendoes Amaka is unmasked and condemned, as it becomes evident that she cannot and will not permit spiritual bonding since this might arouse in her some understanding, compassion, and love for her prey. And maybe, in fact, Amaka thinks that showing love and affection to the man might be a sign of weakness and, therefore, she makes a conscious effort to be firm, impervious, and resolute to secure her power. Therefore one is bound to identify with Izu as he pleads: "Just listen to my story, darling," he said, caressing her lovingly and gently"(70). And not long after, Izu gains total control of the reader's empathy with the highly emotive statement: "So as far as my community was concerned, I was a non-person"(75).

But Amaka's narcissism and inordinate quest for power stand as a huge wall between him and her. Izu's struggle for empathy yields nothing, and he loses himself to her:

> Izu was crying. She said nothing. She went to the toilet, got a clean towel and wiped his tears away. *She caressed him, she touched him, in forbidden place, and he was aroused. They made love again.* It was the first time in her life that she had planned the total annihilation of a man, using all that her mother taught her, which she had sadly neglected because of what the spinster missionaries had taught otherwise. (74; my emphasis)

Izu's tragic revelation to her that he had always been an outsider, a victim of the rigid customs, has no emotional impact on Amaka. One would think that Amaka, who herself in the past had been a victim of the rigid traditions, would have empathized with another innocent victim of tradition. Of course, it is not at all surprising that his plea or cry for mercy falls on the concrete walls that Amaka has constructed around herself. Like a predator now lying in wait in ambush, Amaka is poised to strike and devour her victim. His life, his fate is totally in her hands and as she holds him, he is transformed into a "thing," (an)other object that she has the power to (man)handle.

On his own part, Izu is like a child who desires nurture, love, and understanding from Amaka as a mother-figure. But Amaka defies that role and chooses instead to objectify, reify, and seduce him as a mere sperm-donor for her twins. Again the author's intrusive voice delivers the final blow to Amaka's excessive and reckless instinct for conquest and power in the following: "Amaka had succeeded in tempting him as she said she would. She was going to play her card very well"(77). Amaka's oppressive and exploitative actions provokes more sympathy for the man, especially as she stifles and mutilates his identity. Izu's tragedy is felt even more in his final days of anguish as he struggles and pleads desperately for her to accept his hand in marriage. Worse still, we are aware that while Amaka has lost nothing and has continued to gain everything from him, Izu has progressively lost everything in the relationship with Amaka. What is most remarkable is that in spite of his humiliation by her, he continues to plead his course to the extent that he actually takes the unconventional steps of giving up the highly powerful government position, as well as the church that has been his foster-home, as the ultimate price that he is willing to pay for home, place, identity, and belonging with Amaka and the twins. And yet, Amaka remains determined to prove her point that "One Is Enough"! Thus rejected by Amaka, Izu drifts, wandering and dislocated in the streets of Lagos with their yawning, bloody mouth. It is in this context that he drifts into the mouth of another woman, a prostitute, who chews up whatever was left of him and his identity. It is no wonder then that Izu ends up in the physical accident which the author uses to signify the man's total tragic psychological maiming, disfiguration, disempowerment, and loss.

Amaka neither appreciates nor shares Izu's condition and crisis of identity. By disclosing the distance between Amaka and Izu, Nwapa shows the "other" tormented crying face behind the mirror—the man. Besides critiquing the powers of motherhood, Nwapa uses the same double-entendre to criticize the immodest, reckless use of power by women who are successful. Therefore, my argument is that the underlying question that Nwapa seems to pose in *One Is Enough* concerns how women use power. Underneath it all,

the author seems to pose these troubling questions: Granted that women suffer marginality, could Amaka be justified in the context that she chooses to prove that "One Is Enough"? Could Amaka's position be valid at all times? Is it possible that she is being irrational in this context that she chooses to punish the wrong victim, Izu? Are all men really evil? Or rather, to what extent are men like Izu and Obiora oppressors? These appear to be the underlying critical questions through which the author cautions women against the destructive (ab)use of power. This is the food for thought that Nwapa proposes in *One Is Enough.* "One Is Enough" for whom? And if one is enough for one, is it enough for the other? If one is enough, the implication is that while one gains, the other loses. Therefore, in *One Is Enough,* Nwapa stages a contest between the prospective winners and losers: between men and women, and it is abundantly signified that men are the huge losers while the women are the huge winners in this race for power. In many ways, therefore, *One Is Enough* is Nwapa's most allegorical critical reappraisal of the powers/joys of motherhood in both their extreme positive and negative dimensions.

NOTES

1. Flora Nwapa strongly expressed her position on the necessity for women to gain economic independence in her interview with Adeola James in 1985.
2. The Aniocha Igbo are separated from their Ugwuta neighbors in the east by the Ugwuta Lake, which also serves as the life-essence of the strong female characters in Nwapa's fiction.
3. See Frantz Fanon's psycho-analytical study of the traumatic effects of marginality and "otherness" as these relate to the black race and identity in *Black Skin White Mask: The Experiences of A Black Man In a White World.* Explaining this phenomenon, Fanon states that "[a] slow composition of my self as a body in the middle of a spatial and temporal world-such seems to be the schema. It does not impose itself on me; it is, rather, a definitive structuring of the self and the world-definitive because it creates a real dialectic between my body and the world" (111).
4. For further understanding and analysis of the concept of "otherness," see V. Y. Mudimbe's "Discourse Of Power And Knowledge of Otherness."
5. See Chikwenye Ogunyemi's insightful figurative readings of this text in her "Introduction: The Invalid, Dea(r)th, and the Author: The Case of Flora Nwapa, aka Professor (Mrs.) Flora Nwanzuruahu Nwakuche," *RAL* [Special Issue On Flora Nwapa] 26.2(1995): 1-16.
6. See also Ogunyemi's symbolic association of the tragic events in the setting of this novel with the numerous adversities in the Nigerian condition.

7. Ogunyemi used this terminology in the above analysis.
8. This is my own formulation for the central position that women occupy in Nwapa's fiction.
9. This was first used by Mary Ellman in *Thinking About Women.*
10. This is Helen Chukwuma's main thesis in "Floral Nwapa Is Different."
11. Nwapa's interview with Adeola James in 1985.
12. Tess Onwueme originated the term "shero" as part of her presentation following the international premiere of her award-winning play, *The Desert Encroachers,* during the first international Women's Playwrights' Conference in SUNY Buffalo, New York, 1988.
13. As reported in *The Exchange* 11. 6 (May 1988), "The Nigerian author of this article [The *'Shero' Present To Future*] Tess Onwueme has coined the word for her title from the male oriented word 'her,' as a pun on the concept of heroism" (1).
14. This is Nwapa's novel which will be published posthumously by Africa World Press.
15. See Nwapa's unpublished novel, *The Lake Goddess.*
16. See the rest of Cynthia Ward's provocative argument concerning "What They Told Buchi Emecheta: Oral Subjectivity and the Joys of "Otherhood," *PMLA* 105. 1 (January 1990): 83-97.
17. See Florence Stratton's imagined reconstruction of the "prostitute topos" as Nwapa's primary target in *One Is Enough.*
18. This refers to Ogunyemi's analysis of Adizua's marginal condition in *Africa Wo/Man Palava: The Nigerian Novel by Women.* Chicago: Chicago U P, 1996.

WORKS CITED

Chukwuma, Helen. ed. "Flora Nwapa is Different." *Feminism in African Literature: Essays on Criticism.* Enugu: New Generation Books, 1994, 115-130.
Ebeogu, Afam. "Feminism and the Mediation of the Mythic in Three Plays by Tess Onwueme." *The Literary Griot* 3.1 (1991): 97-111.
Ellman, Mary. *Thinking About Women.* New York: Harcourt, Brace & World, 1968.
Emecheta, Buchi. *The Joys of Motherhood.* New York: Braziller, 1979.
Emecheta, Buchi and Maggie Murray.*Our Own Freedom.* London: Sheba Feminist, 1981.
Fanon, Frantz. *Black Skin White Masks.* New York: Grove Press, 1967.
James, Adeola, ed. *In Their Own Voices: African Women Writers Talk.* London: Heinemann, 1990.
Mudimbe, V. Y. *The Invention of Africa: Gnosis, Philosophy And The Order Of Knowledge.* Bloomington: Indiana University Press, 1988.
Nwankwo, Chimalum. "The Igbo Word in Flora Nwapa's Craft." *Research in African Literatures* 26.2 (1995): 42-52.
Nwapa, Flora. *Efuru.* London: Heinemann, 1966.
——. *Idu .* London: Heinemann, 1970.
——. *Never Again.*(1975). Trenton, NJ: Africa World Press, 1992.

——. *One is Enough* (1981). Trenton, NJ: Africa World Press, 1992.

——. *This Is Lagos and Other Stories* (1971). Trenton, New Jersey: Africa World Press, 1992.

——. *Women Are Different* (1986). Trenton, New Jersey: Africa World Press, 1992.

Ogunyemi, Chikwenye Okonjo. "Introduction: The Invalid, Dea(r)th, and the Author: The Case of Flora Nwapa, *aka* Professor (Mrs.) Flora Nwanzuruahu Nwakuche. *Research in African Literatures* 26. 2 (1995): 1-16.

——. *Africa Wo/Man Palava: The Nigerian Novel by Women.* Chicago: Chicago Univ. Press, 1996.

Ojo-Ade, Femi. "Female Writers, Male Critics." *African Literature Today* 13 (1983): 158-179.

Okonjo, Kamene. "Dual-Sex Political System In Operation: Igbo Women and Community Politics in Midwestern Nigeria." *Women in Africa: Studies in Social and Economic Change.* ed., Nancy J. Hafkin and Edna G. Bay. Stanford, California: Stanford University Press, 1976, 45-58.

Onwueme, Tess. "The 'Shero' Present To Future." *The Exchange* 11. 6 (1988): 1-2.

Stratton, Florence. *Contemporary African Literature And The Politics Of Gender.* London: Routledge, 1994.

West, Cornell. *Race Matters.* New York: Vintage, 1993.

Wilentz, Gay. *Binding Cultures: Black Women Writers in Africa and The Diaspora.* Bloomington: Indiana University Press, 1992.

The Lake Goddess:
The Roots of Nwapa's Word

Chimalum Nwankwo

I. THE FIFTH FIRE

Every tragic item in African historiography has been explained away. Slavery is explained away by the West in these terms: Africans also had slaves. Africa was complicitous in the instituting of Western slavery and therefore cannot exculpate itself. In the unraveling of African institutions and culture by the contumelies of colonialism, we are also compelled to not only believe but to accept that Africans were also complicitous. In the big power machinations which perpetrated the octopal ravages of the Cold War, from economic cannibalism to the mind boggling internecine feuds and strifes, Africa is also perceived as complicitous. In the spread of AIDS, the most deadly plague of modern history, Africa is this time not merely complicitous, the experts have divined and decreed that Africans and the monkeys from their infernal jungle provided aboriginal habitat for the virus and therefore must accept the scientific evidential role of vicarious vector or direct vehicle.

The historical trend is not as frightful and as loaded with torment as what I call the fifth fire. The fifth fire is that chaotic echo chamber of literary criticism in which, fortuitously or by design, there is a conflagration of African value systems lit by all forms of invidious and deracinating stances. Action without fear of consequence is privileged over action with circumspection. Drunk and consumed with a logos more potent than any viral invasion, a multitude of African scholars in sometimes tragic (because it is sober and deliberate) participation, opine directly that it is impossible for the African consciousness to till its own earth, cultivate with its own tools, and farm what it desires for its nourishment; that it is impossible to sustain by supplementation and careful borrowing; that it is Romantic to look back carefully and see whether lessons do not exist in the deeper layers of the historical pattern outlined above. Unfortunately, in all this, because of the slightly unwary critical stances of Chinweizu and his collaborators who rushed over the inescapable questions of recursiveness and heteroglossia, the realm of vigorous critical possibilities exists in another firmament.

To continue to speculate about this is as time-wasting as the astral and useless modes of interrogation conducted by a celebrated coterie of Africans in the Western Academy, lost saviors or scholars whose motives are as questionable as they are dangerous, inimical, and traitorous. To dare to speculate is as frustrating as contemplating the psychological condition of another breed of scholars whose status of exile and migrant laborer are guarded with mouthing the "myriad subtleties" picked from under the great table of their treasured and monistic logos. Dazzled by the neon of technology, they cannot even begin to imagine how mournfully irrelevant the wonders of their subtleties and their phony lambent fires are in the cold dark realities of devastated African cities and countrysides; how pathetic, hollow, and tragic their voices of eureka ring as they, figuratively, run naked throughout the streets of the Western Academy. And to dare to speculate is to marvel, then recognize, how, from the disarray and poverty of nations, exiles live in another country and blindly see it as their country. They are wise about their lost homeland but in crass ignorance of the mocking laughter of their hosts who worry about a wisdom potent in the halls of Western Academy but is ineffectual ash in the burning homeland. So rather than speculate and interrogate like the lost saviors and other exiled cohorts, let us simply call the fifth fire, the present intellectual condition with regard to rarified modes of literary criticism and the consequences of the African human condition, not just now but in the future. For whether one likes it or not, after the fifth fire, the productions of today will flow back into African libraries of the future, and they will be consumed by future Africans whose cultural bolsters have been burnt and blown away. Permitting such sentiments to shape one's critical directions is probably called writing "combatively," but reneging from such directions could also be a denial of space for all those things which are associated with the Lake Goddess and Flora Nwapa's legacy.

II. THE WATER OF THE GODDESS

The Water of the Goddess cannot flow into the African hearts or minds from Nietsczche, from Paul de Man, from Derrida, from Foucault, from Heiddeger, from Cixous or Irigaray. These and other favorite sources proudly referenced for legitimation and intellectual certification are tainted and blighted. They are tainted and blighted by their inexplicable connections in their own homeland by either fascism, nihilism, or a sordid and questionable attitude to the human body or mind, and of course, to the idea of God! Whether those attitudes are valid or invalid in the West, an African critic need not be compelled to divine or determine. And if a tiny moony group of African scholars and pseudo-salvationists pass positive verdict about those Western thinkers, such verdict is of course in a so-called democratic universe invalid, as is unfortunately (or fortunately in this case) all court minority verdicts! Those Western thinkers are incriminated and convicted in the African ontological court by the prevailing African climate and attitude with regard to what the Western philosopher calls "foundations." It is the implicit subversion of foundations celebrated by those thinkers which completely stultifies and dishonors and insults the passing and memory of that great spirit, daughter of the Goddess.

No one wants to live in a capsule of time, sanitized from the cosmic's propelled tide of change or flux. But no one also wants to live inside the eye of a storm, for even virginity in humans (or is it shemans?) must be violated for procreation. No one wants to live in a sacred grove talking to the trees without the periodic tonic of laughter, as no one wants to live in a wild universe of eternal revelry and no sweat. The mathematics of cosmic laws derive always from medians for equilibrium, and the multiplex dynamics of give and take. This is what one reads in the articulation of the Igbo way by Igbo minds and scholars: D.I. Nwoga, Obiechina, Njaka,

Onwuejeogwu, Uchendu, Afigbo, from the powerful sociological contributions of Ifi Amadiume, and the ceaseless historical finds of Elizabeth Isichei (Igbo by marriage), authenticated by what we are now recognizing as Igbo pragmatism. The Igbo way will dismiss tragedy with laughter for blunting events and the purchase of time for contemplation. The Igbo way will accept tragedy and will insist according to one popular profound Igbo name, *Mbagwu,* a nation never dies. Not even from the conflagration of the fifth fire! To surrender to the fifth fire is to ignore the wisdom of the Igbo mythical star, the tortoise, who before submitting to certain death in an over-matched fight with a jungle king, proceeded to scratch the earth furiously. The reason of the wise tortoise? It is a testament after death that, at least, some action was attempted before destruction. Action, as we also see in Nwapa has to be read in that same vein or Ona, in *The Lake Goddess,* will pass as a lunatic, as indeed she does before some around her.

Those who study Nwapa, in dedication, must refrain from the present on-going posthumous appropriation of Flora Nwapa into the turbulent, chaotic, anti-foundationalist, and apocalyptic water of Western feminism. Conscious of the nature of the elements of that storm, Nwapa repeatedly disavowed, disassociated, pleaded almost, one would say on her knees to be excluded from that water. In Igbo culture, to give or call one a name which one does not like and accept is another ultimate insult. Flora Nwapa has been read as subversive, rebellious, deconstructive, and so forth, given all imaginable labels to qualify her as the slayer of the Gorgonic African male. All such readings constitute some of the greatest tongues of flame in the fifth fire. And when those who understand the Igbo way critique her craft, as familial counsel of improvement, those champions immediately assume their imported clownish calico of mourning and proceed to wail louder than the bereaved.

The great trick of such Western critics and their lost allies is Kubla-khanism, the erection of stately domes of pleasure and/or agony far away from the territory of cultural combat. Egged by the worst aspects of Western feminism, drifting without moors, fleeing far away from the vigorous intellectual cannonades from the compendious and magnificent minds of some of their own veritable rebels like Camille Paglia, these feminists pitch tent and preposterous phalanxes in Africa. As some of us are afflicted by exile woes too dreadful to contemplate, these feminists regale in triumph in the imaginary African male-female battle fields. The difference, of course, lies in the fact that where the African critic in equiva-

lent Quixotic engagement is at best tolerated, the alien female critic is celebrated and sometimes rewarded with symbolic human
heads. Like all mercenaries, they are worthy collaborators far away
from where their African counter-parts are soaking in scorn. How
can an Igbo scholar or critic begin to understand or rationalize
much of what one encounters in the cultural ostriching and imperialistic leger-demain of travelogues such as *Motherlands or
Contemporary African Literature and the Politics of Gender?*[1] Whose
mothers live in those lands? And what is the immediate relationship between subjects and prefects of the politics?

African writing on Nwapa can rise above all trickery and con-
centrate on the deep cultural lineaments which will lend appropriate mystical aura to not just the world of her writing but the
ultimate purpose which Nwapa's candor and humility constrained
her from claiming. The water of the Goddess is good water for dous-
ing all the insidious flames of the fifth fire which, left alone, will
establish new heaps of rubble for future generations of those questing for all those kinds of verities upon which the guiding epistemologies of all nations devolve. The fifth fire is the kind of
conflagration which leaves in its eddies of ash quests for the glories, real or imagined, of Timbuktu, ancient Egypt, Songhai, Ife,
Nok, Cush, Zimbabwe, Igbo Ukwu, and so forth. To write now with
responsibility, without fear or dishonest sentiment, and without
the peripatetic angst of penury or the survivalism of exile and material want, is to deepen the spiritual canals of the Lake Goddess.

III. *The Lake Goddess*

No one knows the origin of the para-normal impulses which generate farewell paeans such as Flora Nwapa's *The Lake Goddess, a*
technically speaking, hasty anticipation or prophecy of passing.
The Lake Goddess as part of her *oeuvre* is the coda of a trajectory
that was already being impressed in *Cassava Song* and *Rice Song.*[2]
The Lake Goddess is a work which looks back and edits, like a rewritten autobiography the crucial mis-steps in her creation of the
many worlds which remake the Igbo material of her works. In a
previous piece on Nwapa's writing, I concentrated on the word and
the craft, with a faulting of the craft.[3] This paper concentrates on
the word, especially its intricate strands and roots in the present
and the past.[4]

In the core of my position in African literary criticism is a posture which insists that the responsibility of the artist and indeed

the critic is to provide a vision or criticism which advances the course and cause of the given context in line and consonance with what exists in the complex of peculiar foundations available to that context. What happens in other foundations or to other foundations should not bother the African writer or critic. If God has died in Berlin, and history has ended in Tokyo, and the author died in Paris, London, or Washington, it should not matter as long as the traditions of other spaces retain their integrity. Emenyonu observes in his *The Rise of the Igbo Novel* that "the modern Igbo writer, with his legacies from the past, cannot be oblivious of a deep sense of inviolability and commitment which characterized his predecessor, whether narrator, carver, sculptor, spokesman, orator, chief, priest or drummer" (2).

It is not a very wild speculation to imagine that *The Lake Goddess* has benefitted from some so-called masculinized criticisms which demand, foremost, in the Igbo landscape of Nwapa's creations the pragmatic or balances deeply entrenched in the Igbo polity, accretions from Igbo cosmogony, and realities and total history. One must emplace a caveat here for those critics who in ill-concealed ignorance and academic haste enclose under the usual African umbrella, their sweeping concepts of nation: universal notions of the lineaments and *modus operandi* of African religious institutions: generalizations of historical experience and modes of acculturation in continental Africa's contact with the west and so forth.[4]

The world of *The Lake Goddess* is an Igbo world which duplicates some of the physical and metaphysical ambience of *Efuru* and *Idu*. If there is a difference here, it is that the ethereality of presences acquires substantiation from an area of character action and cultural identification that in previous works appeared dichotomized. The power of *The Lake Goddess*, Uhamiri, occupies some orbit with other autochthonous forces. Here, the role of all such forces is to act as superiors of incursive forces, not only rejecting but countering, with clear recognition, the inimical and disruptive character of those alien forces. What we have here is a strong authorial assumption of political responsibility. There is neither equivocation nor favor nor the old egregious gender-privileging that marked the previous works. It is the story of a stressed world responding with comprehensive vigor to external threat. Cultural implosions are guarded because destructive criticism is eschewed in favor of the more cautious and constructive; the matter-of-fact is clearly specific and politically strategic. On the whole, the word is anchored in the following.

Narration: Structurally, the story moves from threat of fragmentation, through leit-motif of fragmentation to restitution and victorious anchor in the autochthonous. The authorial voice is partisan, and much critical irony is in favor of the underdog: women and the people and their culture in general through the ubiquitous Lake Goddess. The resilience of Igbo culture is consistently a concomitant vehicle.

Characters and the Word: The territory of action is as divided as we find in Igbo history and the overall socio-political frame. The world of men is different from the world of women, as we find in the activity of the age grades, from their good-natured lewd jokes to the role of visible support system that unites people during critical moments such as Mgbada's traditional wedding. Women on their part rise to the same level of bonding and cooperation in the activities of the Umuada. Even so, this is still a woman's story, the Lake Goddess bonds all, as she is mentioned in conversations by women and men, and summoned in reverence or through sightings in the form of representations through children, and the dreams of the tutelary rituals of Ona toward the end of the story. This novel presents a male-female world in which conflicts arise from the non-factitive and natural, and man and woman *negotiate*[5] themselves in and out of advantage or disadvantage.

Patterns of Negotiation: The cultural legacy in *The Lake Goddess* is taken more seriously than in any other Nwapa text. This goes beyond character action into authorial ventriloquisms. Mgbada, a foundational character in the novel, sits astride two worlds. He accepts the inevitable, and that is, that Christianity and the West cannot be wished away. He never fights that fact frontally even when such would be considered justifiable, nor would he accept the kind of defeat we find expressed in Nwoye in *Things Fall Apart*. Mgbada is at ease in the new world and gently insists and affirms that "two religions can be practiced by one person, that one complemented the other"(LGMS 51).[6] He cannot "leave the ancestors alone"(52), as his mother-in-law would wish. His position is synchronic with the partisan, cynical and skeptical authorial voice which quite early in the novel adopts a nationalist prong: "When Mgbada was ten years old, strange people came to the town with strange ideas. They talked of a God who was born by a woman and who died for the sins of the world. They criticized the religion of the people calling them pagans and heathens"(2-3). The people in their turn, through Mgbada's father are prepared to resist:

> What kind of religion preaches that one should abandon the worship of one's ancestors? Why should these people who are foreigners for that matter, be concerned about where one goes when one dies? We know that there is life after death and so when we die we join our ancestors and continue to live. (2)

Even though the foreign religion thrives and the people are divided, continuity is maintained because of other people such as Mgbada's father and his resolution, for "before he died, he charged Mgbada: Whatever you do, whatever you become, don't forget the worship of our ancestors. You have the *ofo* of our family. You are the first son. Much is expected of you. You must carry on after me"(2). Mgbada carries on pragmatically, as we find in his patience with the fanatical Christianity of his wife and his mother-in-law, and his vocal and active identification with the ways of tradition, his daughter, and the Lake Goddess, throughout the novel.

In the first generation of characters in *The Lake Goddess*, the marital situation of Mgbada and Akpe indicate that Mgbada is the more active agent. In the second generation, the marriage of Mgbada's daughter, Ona, to Sylvester reverses that situation in the story. For Mgbada, a modern Igbo man, the response to changing circumstances for wholeness of being is to accommodate by what one would call *positive equivocation*. Ona, a "modern" woman answers or is made to answer in a firm rejection of the West. She is a symbol of another kind of wholeness, an integrity of spirit beyond the confines of family, education, and religious indoctrination. It is a belief and submission to a cosmos centered in the largesse of the Lake Goddess, in a romantic harmony which endures like the lake itself with its transcendental serenity, beyond the mundane desires for children or wealth in its various forms. If Mgbada's condition is the expression of a schizoid identity, product of the manicheanism of culture conflict, Ona is the resolution of that conflict, the restitution of the lost felicity. But all that is not without price, for the Igbo believe that the whole business of life is like that of a real market in which everything is negotiable.

A careful look at the relationship of those major characters reveals something interesting with regard to the switch in comportment. Mgbada's wife, Akpe, accommodates her husband's propensities with somewhat of the familial faithfulness with which Sylvester accepts Ona's afflictions and her eventual reclusive devotion to the Lake Goddess.

With these characters in view, it is easier to see how the rest of the characters fit into Nwapa's word. What becomes clear is that the degree of attraction or repulsion of each character is proportional to a capability to negotiate, to what in Igbo eyes would be considered humane. Nwafor, Mgbada's mother, is as careful in her relationships with people as Akpe's mother is almost reckless, especially with her dubious fanaticism. Our reactions are affected or indeed mollified, such as in the agreement of the two women who connive and pretend that a circumcision had been performed to please cultural conservatives. Other characters in subsidiary roles such as Ekecha and Mgbeke start off cantankerous and socially intransigent but after sloughing off their stigma conclude redeemable. Madam Ogbuefi Ugo's arrogant and domineering style cannot escape our condemnation because the culture does not welcome such abrasions. Ojuru's humility finds reward in Mgbada's traditional healing practice, while Mbeokworo who does circumcisions is silenced in the benevolent conspiracy to shield Akpe from pain. In a previous paper, I observed that Ogbuide, Uhamiri, or the Lake Goddess is a paradoxical deity. Nwapa's final work fleshes out that paradox in significant bivocalities which arch back deep into Igbo cosmogony and history. Note, for instance, that the same Goddess who endorses, through Ona, *voices* and education for women emphatically endorses marriage despite its *well-known* attendant problems. That Ona is not in marriage at the end of the novel is not even an issue. Above all that, rather than talks of divorce throughout this novel, though Sylvester's people suggest, in the face of his predicaments, to "sever all connection"(207), the people of the Goddess endorse accommodation of one sort or the other, responding to S.E.N. Anyanwu's finds that "etymologically, among the traditional Igbo society, the word divorce does not exist"(187) and, that "divorce is a marital tragedy"(201).

In an interview with Sabine Jell-Bahlsen[7] and Francis Ebiri, Nwapa makes a point which we have encountered in *Efuru;* that is, that Okita, spirit of Urashi River, is always quarreling with his wife. In the course of that explanation, Nwapa laughs about the fact that it is so in real life: husbands and wives are always quarreling. Such conduct is with humans as it is with gods. One wishes that this theme were more carefully explored creatively and then carefully integrated into the Igbo world picture of Nwapa's works. From that point, it would then be easier to understand and appreciate the role of all those numerous quarrels between husbands and wives, men and women in Nwapa's novels. In the absence of that critical item or aspect of culture, we lose the implication of numistic fallibility and its obvious relationship with the fallibility of humans, and the replication of such as artistic imperative. Upon all that, we also miss the wider implication of that in the Igbo cultural stipulation for negotiation in reference to the thing called divorce. Consequently, we then mis-read quarrels as endorsement of rifts, as "voicing" or an aspect of empowerment in the ignorant manner numerous foreign feminists and their collaborators celebrate their discovery of Igbo male brutality in Okonkwo beating his wife in *Things Fall Apart*. So, for a culture with a vatic capacity for marital tolerance, part of the pragmatic of its polity, we substitute insensate projections of familial adversarialism. Generally, having said all this, one could still say that the relationships in *The Lake Goddess* affirm that this is a culture in which the norm is represented by patterns of negotiation.

Naming: "Ogbuide wants all women to have voices. Women should not be voiceless. Ogbuide hates voiceless women,"(240) declared Ona in the final scene pre-figuring the rehabilitation of Ekecha and Mgbeke. The greatest irony of Nwapa's career is that in no other work is she as politically acute as in *The Lake Goddess,* yet the female characters who vehicle that politics conduct themselves with a cultural dignity which does not vitiate or repudiate male roles and spaces.

The integrity of spaces appears to be fortified through naming in this manner. All the characters in *The Lake Goddess* have names that are homonymously constructed into the cultural frame, somewhat in keeping with the continental African tradition of naming as circumstantial signification with social and spiritual implications. For example, *Mgbada* is an animal considered swift and graceful by Igbo people, possibly implying in *The Lake Goddess* Mgbada's capability to negotiate effectively for safety culturally. Ona is named a jewel possibly because of her absolute victory and steadfastness with the forces of tradition. Ogbuefi Ugo is a woman with a name that is rarely given to females. Ogbuefi is a titular male name, "killer of cows." Because Ogbuefi is rarely unisexly given, it is obvious that the author intends more. A woman with a man's name, her brashness appears in contrast to the idealized Igbo woman's nature represented by Efuru in *Efuru,* who is at a point invoked iconically in *The Lake Goddess.* Characters whose conducts are tainted by foreign ways bear their Igbo names with foreign names to signify blight and we see our confused hypocrite, Mama

Theresa, battling with that condition quite early in the novel. All these are accented by the presence of a Father Millet and an alienated Madam Margaret, who "prayed more than six times a day" (87). To make sure that we do not miss the political and cultural implications of Madam Margaret's choices, the absurdity of her status and station as a woman is named thus: "she ritually washed the statue of the Mother of Christ and her son which was in front of the church....As she did this, she said loudly, omitting *Maria the mother of Christ* (my emphasis): Jesus Christ, my son/They have dirtied your face/Never mind, my son/Margaret has come/To scrub it clean"(87). Clearly, the authorial voice comically underscores Madam Margaret's hypocrisy and fanaticism and aligns sympathy with the victims of cultural imperialism and a peculiar invidious male chauvinism, both affirmed by Father Millet. The name, Mgbeokworo, has a deliberately rural ring, to harmonize with her traditional responsibility.

Oral Tradition: *The Lake Goddess* draws immensely from the relationship between folk ways and oral tradition, extending resource from the real to the unreal, through one dimension and by chiasmus. This is of course in keeping with the Igbo notion of dualities, especially the idea of perpetual traffic between humans and spirits in a cosmos of zero spatial zones with assorted dependencies and complementarities. Several times in *The Lake Goddess,* children are seen or regarded as messengers or impersonations of Ogbuide. Ogbuide, Uhamiri, or the Lake Goddess herself is sighted under various circumstances by different people, men and women. In this novel, tradition suffers incursion but survives by a clear authorial privileging that surpasses the political engagement of previous works. The Lake Goddess remains in the center of all this; and that Ona triumphs is an expression which calls for and affirms a continuity of traditional values stretching from Mgbada's father, through Mgbada to the fictional present. The situation responds to the widely accepted perception of tradition as a continuum in that well-known essay by T. S. Eliot, "Tradition and the Individual Talent."

IV. GENERAL IMPLICATIONS

Approaching the roots of Nwapa's word, *The Lake Goddess* fruitfully begins with a studied rejection of the major signposts of thought in the Western world, from ideas of reality to the icons and iconography of those ideas. We must eschew a world, historically and logocentrically fabricated, with its linear order and hier-achization. That Nwapa's world intersects or is subsumed in the history of the colonial or post-colonial program does not mean that one must accept what has already been pointed out in Masolo's cri-tique of the practice of certain so- called African philosophers who permitted "the written African philosophy" to emerge "as part of Western discourse rather than as an African discourse on itself"(42). When we read Nwapa as an Igbo woman writing about Igbo women and men under certain specified historical and cultural circum-stances, it becomes easier to discard inevitably associated notions of nihilism and the pseudo-apocalyptic conclusion of events and human order which go as principal baggage with the so-called post-modernist impulse. This point is pre-eminent in my contention because what now reigns as feminism is socio-politically a benefi-ciary of the unraveling of a specific Western traditional cumulus.

What must follow this understanding is an acceptance of the general African systems of thought as the product of a legitimate parochial and alternative reality. That alternative reality derives from something in the kernel of Mbiti's half-playful contention that Africans are a notoriously religious people(1). One might question the existence of a congruence of epistemologies, but such a ques-tion, despite the ravage of colonialism, will hardly make a dent on a general African ontological frame. For if, as Mbiti suggested, we accept that Africans are notoriously religious, one must articulate efforts of understanding Africans and African artifacts through such a pervasive trait. In accepting this, it might be useful to hold in view an old assumption of Melford Spiro that "every religion con-sists of a cognitive system, a set of explicit and implicit proposi-tions regarding the super human world and man's relation to it, which it claims to be true"(96). Once again, if such no longer holds true in the West because of the modernist and post-modernist con-clusions of certain ideas and events, such distant interpretations of reality must be left where they belong. The world of Nwapa, especially the Igbo frame, must be approached based on the assumption that a cognitive system is still in place and that Nwapa, being African, derives from that cognitive system.

A celebrated Nigerian journalist, Peter Enahoro, in contem-plating religions, facetiously suggested that the problem of the black God is that it suffers from *literatus ignoramus;* and because of that affliction, the spirit-human codes of relationships have been undocumented and uncodified. If, in more serious terms, such is accepted to be the case in written terms, the African artist has more than provided enough materials for reading the ethos of the Nigerian world. Let us quickly read this world in general African terms before reducing it to the more specific Igbo frame.

We may follow the Western engagement in history and thought from an antiquity of dual founding mythological ancestors toward the so-called end of history and modernity, and in Christian and even Communist terms toward a utopian restoration or recreation of a lost felicity. Though the African universe in many cultures also entertains such mythological foundations, its projections include no such restorations. What African art reveals through its practitioners and some of its diligent scholars is a different paradigm that is worth watching and noting. For Herbert M. Cole, who has done some extensive work with distinguished Igbo artists and art scholars such as Chike Aniakor, his observations should be quite instructive:

> The human couple is a paradoxical unity in life and art. United as one, it is the elemental force igniting sexuality and reproduction to maintain the species. As two it is the equally elemental force of opposition and reciprocity. Matched pairs of sculptures carved or cast at the same time for a unitary purpose are simultaneously two and one...Dualities are often articulated in African belief systems and are implied in art, where they serve as explanatory models. Maleness is understood in relation to femaleness, hot to cold, night to day...Dualistic systems such as these compress practical and philosophical notions associated with gender, morality, time, space, cosmology and transformation. (68)

The dualism in African art has been found to be pervasive in Igbo art and clearly takes its source from Igbo cosmogony, history, and politics. We have seen this expressed and defined, where duality translates into all kinds of complementarities rather than bellicose opposition. What we could find in all this is the kind of dialectical continuum articulated again by other art scholars. The necessity for male-female harmony, almost, one could say, *at all costs* (to the chagrin, I am sure, of the Western feminist), is a reverential recognition of that fact. Certainly, the Nwapa reader cannot miss the level of sacrifice imposed on female characters in the name of peace. That imposition is part of an Igbo female reality which must be carefully studied vis-a-vis whatever constitute the male reality and the contiguity of that reality to the workings of the total polity. According to Roy Seiber and Roslyn Adele Walker in *African Art in the Cycle of Life*:

> One of the most pervasive concerns of African societies is continuity. The future of the family and the group depends on the ability of the present generation to sire and bear children. Additionally, an individual's sense of social and biological completeness lies in his or her ability to become a parent, for one must depend upon one's children for the proper respect and consideration that is due age. Children not only guarantee the well-being of the individual in life, they will also provide a proper burial and ensure the transition of the spirit of the parent to the after world to take its place as an ancestor and possibly to reincarnate as a member of the family. Thus it is not surprising to find in African thought and art aubiquitous emphasis on human fertility. (28)

From the above, it should be clear that for the African, the Igbo inclusive, the issue of fertility is an issue with implications of transcendental gravity, an issue directly linked to an ontology predicated on a clearly expansive eschatology. That this is an issue which one might consider irrational or mundane elsewhere does not mean that no one else can regard it as grave religious or spiritual necessity, or that one should simply and dismissively ascribe the issue to the atavistic realm. Hypothetically, for continuity, Igbo men and women must get married. For continuity, they must have children. And for continuity, if a child cannot come from one woman, the Igbo permit polygamy. The delicacy with which this issue is regarded sometimes, in the past, necessitates in certain African societies very discreet surrogate siring, a condoning of what would in other cultures be regarded as unfaithfulness or worse. Here, among the Igbo and many Africans, all kinds of imperatives conflate in practical issues of life and death.

Reading Nwapa must, in the light of the above, involve in a very serious sense some of the attitudes we bring to our reading of religion. The Igbo world which frames Nwapa's art is a religious world. The Igbo cosmogonic myths and mysteries, realities and paradoxes, which constitute the Igbo polity and its pragmatic are all clearly posited in Nwapa's word. That Nwapa is from Ugwuta, far away from the ritualistic foundations of the Nri axis, does not mean that she is not Igbo. It only means and calls for the recognition of the following when we, for instance, want to compare the pragmatic of her own world with that of say, Chinua Achebe, who has worked with Idemili, the tutelary equivalent of Uhamiri.

1. The discussion of gender in Nwapa deserves to be handled with the same delicacy and caution with which Nwapa handled the issue of her loyalty to the notion of feminism. Obviously she constantly, at best, equivocated because she was aware of what Jell-Bahlsen later found out in her research and discussions with Nwapa that "MammyWater and the local goddess, Ogbuide, or Uhamiri, are identical (and that) the term MammyWater transcends gender. It is equally applied to male and female water deities, and also to divine pairs."

2. Nwapa writes as a woman "steeped in Igbo tradition" and the viewpoints she presents must be the viewpoints of Igbo women. This is a legitimate viewpoint not from the so-called perspective which claims a subversion and deconstruction of patriarchal realities and power but from the creative construction of a womanist reality which Igbo women would understand and appreciate from their cultural and historical vantage point.

3. While looking at this new female reality one must juxtapose it with the prevailing cultural male reality in complementarity, rather than in supplantive opposition, in keeping with the ruling pragmatic of Igbo foundations encoded in Igbo art and socio-political history.

4. When we consider the character of the Lake Goddess and the political and spiritual implications of her pervasive invocations in Nwapa's work or world, we must make some effort to see and note how the Lake Goddess relates to the general Igbo frame. Certainly, what is obvious in all Igbo country with regard to all religious worship of female deities is that there is an unmistakable consistency in social attitude. Agency, functions, and followership are flexible in terms of what constitute the roles of men and women. When a deity or the principle guiding relationships between the people and the deity translocate or relocate, the new locus becomes peculiar in character or characteristics. Each habitat designs a new symbiosis between the people and their needs in keeping with Igbo pragmatism.

5. Unless we understand all these, critics, especially the armchair feminist gurus, international monitors of universal genderization, will keep mixing issues of social justice with troubling and unresolved questions of where cultural imperialism begins and

where it ends. The logical and informed answer to that question should satisfy those readers of Nwapa who cannot see the relationships between Achebe's Idemili and Nwapa's Lake Goddess and indeed other female deities not just in Igbo country but all over Southern Nigeria, especially the delta and riverine cultures.

A man writing as a man about a woman is always afflicted with limitations which sometimes craft might be able to mitigate and sometimes are impossible to mitigate. A woman has similar problems in dealing with the male world, but what we find in *The Lake Goddess* is an honest effort to mediate such problems and rehabilitate certain political and spiritual principles which lacked eloquent articulation in other, previous writings.

We are looking here at a sense of tradition with political implications, far and apart from the attitude which we find defined in Paul de Man's injunctions to one of his students who was curious about the relationship between theory and quotidian realities: "Don't confuse any of this literary theory with your lives"(*French Lessons* 166). If the Lake Goddess is to Ugwuta people merely conceptual, like theory, the Igbo demand stands in contradistinction to Paul de Man's. Art is part of reality to the Igbo, hence the further insistence of Emenyonu that "the traditional artist had a clear conception of his immediate society, its problems, projecting through the ethical formulas in his tale a direction for his society, and the individuals caught in the dilemmas of society"(3). To the Igbo, concepts do not generate or provoke distances between ideals and practicalities, neither are they supposed to merely disguise for meaningless reification or the perfunctory. *The Lake Goddess* is like all enduring myths or reliable foundations which, according to Raphael Madu, are supposed to "rescue feeling and fear from silence and confusion"(272). Hence, culturally "the insertion of counter-pole figures...seeks to exculpate humanity from its fate, is also a call for human kind to accommodate its fate"(271).

What has happened in *The Lake Goddess* in its Ugwuta environment is what has happened to most cultural items in Igbo country. There are questions about her authochtonous origin in Igbo country which Nwapa tried to answer with her interviewers. The issue of the origin is not as important as the fact that the Lake Goddess undergoes a transformation that is still functional in the new syncretic shape. That new syncretic shape does not alter the original cultural function, for all devotees, to rescue feeling and fear from silence and confusion; neither does it deter from the prevailing dualistic patterns which mark the Igbo polity where male and female function in complementarity. *The Lake Goddess* attempts to gather all these into one enduring word, probably beyond all previous efforts, so much so that one could find in all that the roots of all previous thought. Whether this has been done successfully in an artistic sense is another matter for all her critics to settle.

NOTES

1   While regarding books of this nature with silence, my expectation or hope is that African women who really understand the political implications of such productions will eventually attend to them with appropriate critical lasers from African culture. Obioma Nnaemeka has touched on some of the troubling issues but deeper and wider incisions are still needed!

2.   It is not quite certain why Nwapa chose the medium of poetry to make her most committed cultural and political statements.

3.   This paper, which was mostly an evaluation of the manner in which Igbo culture and history were used successfully or otherwise by Nwapa, has been found annoying by certain critics who on their own part have been trying to record the number of hits scored by Nwapa, on their behalf, against African men. Whether the whole thing is the consequence of an ostriching of sorts, cultural imperialism, plain inanity mixed with ignorance, or the unfortunate intervention\intrusion of Ekwensu (a highly regarded evil divinity in the Igbo pantheon), it is hard to say. See *RAL* 26:2 (1995).

4.   Please see the unfortunate article by Elleke Boehmer in *Motherlands.*

5.   To appreciate the issue of negotiation, there is no text better than Njaka's Igbo Political Culture. It discusses Igbo culture clearly in the days before jargon became wisdom in cultural studies. It should give ballast to Obioma Nnaemeka's idea of nego-feminism especially while discussing Igbo women in literature.

6.   LGMS is an abbreviation for *The Lake Goddess* in manuscript form. All page references which follow are to the same unpublished manuscript.

7.   This is a very useful interviewer whose efforts should be quite revealing, especially to the non-Igbo or rootless Igbo. One should be very careful though about the kind of problems which we all encounter while studying unfamiliar cultures. Jell-Bahlsen indicates that "Uhamiri" is spelt rather than "Uhammiri" by Nwapa as a deconstructive act. Not so. The Igbo language has numerous orthographical discrepancies imposed by provincialism, and no one, educated or illiterate regards such as construction or deconstruction. See Jell-Bahlsen's note 10 in *RAL* 26:2 (1995).

WORKS CITED

Anyanwu, S. E. N. *The Igbo Family Life and Cultural Change.* Marburg: Marburg Univ. Press, 1976.

Cole, Herbert. *Ideals and Power in the Art of Africa.* Washington: Smithsonian Institution Press, 1989.

Emenyonu, Ernest N. *The Rise of the Igbo Novel.* Ibadan: Oxford University Press, 1978.

Kaplan, Alice. *French Lessons.* Chicago: University of Chicago Press, 1993.

Nwapa, Flora. *The Lake Goddess.* (Unpublished Manuscript)

Madu, Raphael Okechukwu. *Myth and Meaning.* New York: Peter Lang, 1992.

Masolo, D. A. *African Philosophy in Search of Identity.* Bloomington: Indiana University Press, 1994.

Mbiti, John. *African Religions and Philosophy.* New York: Praeger, 1969.

Sieber, Roy and Rosalyn Adele Walker. *African Art in the Cycle of Life.* Washington: Smithsonian Institute Press, 1987.

Spiro, Melford. "Religion: Problems of Definition and Explanation" Michael Banton, ed. *Anthropological Approaches to the Study of Religion.* London: Tavistock, 1966.

Photo courtesy of Nina E. Mba

Flora Nwapa and her colleagues at Priscilla Memorial Grammer School, Oguta, 1953.

Photo courtesy of Nina E. Mba

Flora Nwapa with her students at Priscilla Memorial Grammer School, Oguta, 1953.

Photo courtesy of Nina E. Mba

Commissioner Flora Nwapa Nwakuche in her office, Enugu, East Central State, Nigeria, 1971.

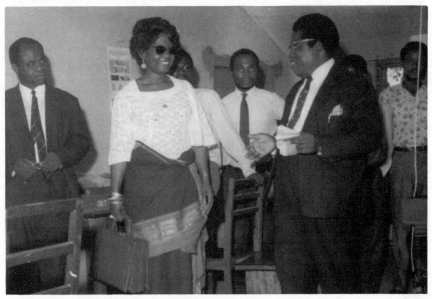

Photo courtesy of Nina E. Mba

Commissioner Flora Nwapa Nwakuche, 1971.

b

Photo courtesy of Nina E. Mba

Flora Nwapa Nwakuche, Commissioner of Lands and Survey, East Central State, Nigeria, 1971.

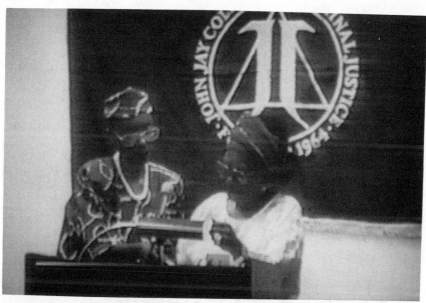

Marie Umeh and Flora Nwapa at John Jay College of Criminal Justice, CUNY, during a Black History Month celebration on February 18, 1982.

Photo courtesy of Sabine Jell-Bahlsen

Obioma Nnaemeka and Flora Nwapa at the 1992 WAAD (Women in Africa and the African Diaspora) Conference at UNN (University of Nigeria, Nsukka).

Sikiru Babalola and Flora Nwapa on February 18, 1992 during John Jay College of Criminal Justice, CUNY's Black History Month celebration.

d

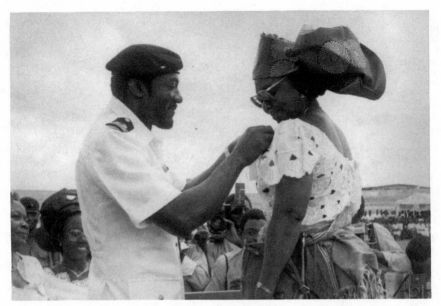

Photo courtesy of Nina E. Mba

Former Governor Amadi Ikweche of Imo State honoring Flora Nwapa in Owerri, Nigeria, 1992.

Photo courtesy of Ejine Nzeribe

Chief and Mrs. Gogo and Flora Nwakuche at Port Harcourt, Nigeria, during the Nigerian Civil War in 1967.

Photo courtesy of Ejine Nzeribe

Flora Nwapa Nwakuche with her three children (left to right) Uzoma, Ejine, and Amede in 1981.

Photo courtesy of Ejine Nzeribe

Flora Nwapa Nwakuche with her daughters, Amede (left) and Ejine (right) and Flora's sister, Mrs. Emeni, at Ejine's graduation from the University of Jos in Nigeria, 1982.

Photo courtesy of Chief Gogo Nwakuche

Chief Gogo Nwakuche and his wife Flora Nwapa Nwakuche at a traditional ceremony in Oguta around 1983.

h

Photo courtesy of Ejine Nzeribe

With her husband, Chief Gogo Nwakuche (center), Flora Nwapa Nwakuche (right) and their children, celebrating Uzoma's Call to the Bar at the Nigerian Law School in Lagos, 1991.

Flora Nwapa Nwakuche (second from right) at her niece's wedding in 1993, Owerri, Nigeria. (Left to right) Flora's niece Nkechi, Flora's mother, Mrs. Martha Nwapa, Flora's sister-in-law Mrs. Evelyn Nwapa, the bride and groom, Flora's brother, C. Fine Nwapa, Flora Nwapa, and Flora's niece Uzoechi.

Marie Umeh and Flora Nwapa's son Uzoma Nwakuche, Esq., in New York, 1994.

PART THREE
AESTHETICS AND POETICS
IN NWAPA'S CANON

The appreciation of the beauty in both artistry and themes, as well as the principles that guides Flora Nwapa's writings is essential to the understanding of her literary contributions. Such aspects of critical appreciation deepen insight on the motives, the notions and the ideas that coalesced to inspire Nwapa. Thus, critics who sincerely hope to interpret and discuss such inspired works read them closely while granting the writer the right to use the subject matter in ways that best encapsulate his or her thoughts and views on life, society and the human inhabitants of such societies. The critics whose works are collected in this section reflect such dimensions of critical appreciation in their varied but significant essays.

In the opening essay of this section, Ernest Emenyonu in "Portrait of Flora Nwapa As A Dramatist" presents an elaborate study of Nwapa's three major plays: *The First Lady*, *Two Women in Conversation,* and *The Sychophants,* which will be reaching the literary world posthumously. In terms of social relevance and artistic finesse, Ernest Emenyonu is of the view that these plays portray the best of Flora Nwapa as an imaginative artist. He declares that it is possible that Flora Nwapa's posthumous publications may achieve for her the fame and critical acceptance which she never seemed to enjoy in her life-time. Her plays will probably be in the fore-front for that honor. Emenyonu uses textual analysis of the plays to show Nwapa as an artist confident in herself, in full control of her subject and unapologetically fierce and uncompromising in tone.

Jane Bryce in her essay, "Evading Canonical Constraints: The 'Popular' As An Alternative Mode—The Case of Flora Nwapa," examines Nwapa's fiction by recording the criticism of her works and uses that critical recording as a launching pad to re-emphasize the literary achievements of Nwapa. Bryce argues that in her later works, Nwapa moves away from the traditional which is solid, secure, value-laden and conservative, in order to envisage a subject that is contemporary, urban, self-determining and open-ended. She employs the categorization of "elite" and "popular" rather than "good" and "bad" to contend that Nwapa's novels side-step the values of the establishment through ambiguity and lack of closure. The conclusion that "what the texts celebrate is what most ordinary people struggle for—survival—an outcome which cannot be taken for granted" interestingly problematizes this reading of Nwapa.

Obododimma Oha turns to another aspect of Nwapa's creativity that has not received much critical attention. He agrees with Ezenwa-Ohaeto, in his essay also in this section, that Nwapa's poetry has been unfairly ignored. Oha in the essay, "'To the root/of dear Cassava': Rhetorical Style in Flora Nwapa's Poetry," considers the female poet as a singer of her work. Thus, he discusses the rhetorical style in the work by relating it to the constraints of genre and register; he identifies a rhetoric of praise and condemnation in terms of "aesthetic ideology," cultural contexts and rhetorical purpose. Oha depends extensively on the views of feminist critics and theorists for his analysis as well as devoting much attention to the section "Cassava Song." His conclusion that he has narrowed the discussion to the major aspects that suggest the text as a mode of interaction between poet and society is relevant to the understanding of the poetics of Nwapa's canon. This study, indeed, complements Ezenwa-Ohaeto's "Orality and the Metaphoric Dichotomy of Subject in the Poetry of Flora Nwapa's *Cassava Song and Rice Song.*" However, the second essay by Oha, entitled "Never A Gain? A Critical Reading of Flora Nwapa's *Never Again,*" examines Nwapa's war novel. Oha analyzes the representations of pro-war propaganda in the context of *Never Again* and the significance of gaps and silences in the narrative. Reasonably prescriptive is Oha's comments on aspects of the narrative that the "novelist could have gainfully developed" especially on the fact that the narrative mode of rhetoric does not transcend the emotional weakness of the pro-war propaganda that it seeks to undermine and demystify. His view is that the narrative undermines itself at such narrative points, as well as reinforces a tyranny of Nwapa's ideological position. The identification of the connotation of words, such as "vandal" and "saboteur" and the submission that it is not necessarily Biafran propaganda that is misleading and destructive (to the individuals in Biafra), but the masculine (re)interpretation of warfare as an expression of masculine superiority and nobility appears quite interesting, although the war call up other associations that are unsettling both mentally and physically.

Such mental and physical associations identifiable in some of Nwapa's short stories offer an avenue for Naana Banyiwa Horne in her essay, "Flora Nwapa's *This is Lagos:* Valorizing The Female Through Narrative Agency," to discuss other aspects of aesthetics and poetics relevant to the literary works of Nwapa. Horne is convinced that the stories in *This is Lagos* valorize female voices and also that the women are imbued with voices that ring out to be heard. This feat, the critic feels, is accomplished through the agency of narration or the way the elements of the narration are constructed. Those figures she considers character-bound-narrators are used by Nwapa to emphasize the interrogation of patriarchy. Horne insists that motherhood is a dominant influence in the lives of the women in the stories and the reaction to situations emanating from that role constitute the urge for the raising of the female voice. This study positively affirms that Nwapa's women in the stories "prove themselves capable of dealing with the complexities of city existence. They are movers and shakers who struggle to either combat patriarchy whenever possible and often pay a high price for attempting to retain control of their lives." This insistent interpretation of patriarchy in Nwapa's writing particularizes her consciousness in creating her characters. It is this consciousness in vision and creativity that Theodora Akachi Ezeigbo stresses in her essay entitled, "Vision and Revision: Flora Nwapa and The Fiction of War." Ezeigbo argues that Nwapa explores the war from a position that is unambiguously feminine. The critic also discusses the uses of autobiography in this war novel and she states that the novel, *Never Again,* is limited in aspects of time and space because "there is every reason to believe that the limitation of Nwapa's account is consequent upon her limited experience of the war." Interestingly enough, Ezeigbo perceives that limitation as empowering Nwapa's vision in a different important direction by making her write with a focus on Ugwuta society, thereby providing insight on the effects of war and propaganda. It is this propaganda rhetoric that Obododimma Oha has analyzed and Ezeigbo proceeds to explore it as an "idiom of violence" associated with it. Thus, she explains that "the women in the works are assertive and economically active as Nwapa focuses on women determined to survive the war" but Ezeigbo also agrees with Oha that there are instances of technical flaws. The outcome of the study, nevertheless, is that women appear to be a vital source of hope for the regeneration of the society as a result of the human ideals they embody in Nwapa's writings.

The discussion of some of those human ideals and inter-gender relationships is at the heart of Ezenwa-Ohaeto's essay, "Orality and the Metaphoric Dichotomy of Subject in The Poetry of Flora Nwapa's *Cassava Song and Rice Song*," which closes the section. Ezenwa-Ohaeto points out that even feminist critics have ignored the poetry of Nwapa but that the subject matter of *Cassava Song and Rice Song* is important for she uses poetry to highlight various aspects of her society. Thus, the study portrays the poetic indicators in Nwapa's poetry where cassava symbolizes social and political issues, as well as motherhood and the feminine essence. On the other hand, the critic demonstrates that rice symbolizes alien, disruptive ways that constantly challenge the values of cassava. Ezenwa-Ohaeto stresses the consciousness of Nwapa in appropriating artistically the oral traditions in her society, while at the same time making poetic use of irony for political commentaries. This study appropriately concludes with the view that Nwapa "may not have produced technically arresting poetry" but her developments of elements of orality, ironic and satiric symbolizations, as well as "the major metaphoric focus of her work transform it into a worthy piece of literature that interrogates reality, subtly revising prevalent notions and humanistically reflecting life and society."

These multi-dimensional appraisals of the works of Flora Nwapa erase some of the misconceptions associated with them and they open new interesting perspectives to the understanding of such literary works. For a writer whose aesthetics and poetics have not been adequately discussed these essays are illuminating studies.

Ezenwa-Ohaeto

Portrait of Flora Nwapa
As A Dramatist

Ernest N. Emenyonu

Flora Nwapa concealed her talents as a playwright until the very end of her life, and then left three highly remarkable plays—*The First Lady* (1993), *Two Women in Conversation* (1993) and *The Sychophants* (1993)—to reach the literary world posthumously. Although the three plays bear 1993 as their publication date (the same year Flora Nwapa died), they are yet to be released to the public. The plays are said to have been edited by Emeka Nwabueze and Nelson Okonkwo but there is nothing whatsoever to suggest that Flora Nwapa had a helping hand in writing the plays. The work of the editors seems, in the absence of the author, to have been limited to technical matters of processing the completed manuscripts for publication.[1] Flora Nwapa's prolificity in her year of death is something both interesting and enigmatic to scholars and critics. Awaiting publication when she died on October 16, 1993 were three plays and a novel, *The Lake Goddess*, and most probably one or two more as yet undisclosed works. In these works Nwapa makes her most potent and poignant social and political comments on Nigeria and Nigerian life. Nwapa's death, so untimely, so suddenly, has introduced death as a metaphor not only in her life, but also in her writings. What did she know that she didn't want to tell in her life-time? Her three plays reveal a lot.

In her excellent and highly evocative essay, "The Invalid, Dea(r)th, and the Author: The Case of Flora Nwapa, *aka* Professor (Mrs.) Flora Nwanzuruah Nwakuche" (easily the most insightfully authoritative criticism of Flora Nwapa's fictional art to date) Chikwenye Okonjo Ogunyemi examined intrinsically death as metaphor in the life and fiction of Nwapa as an artist, and came out with a model analysis which, by its intensity and profundity, com-pels the attention of every scholar of African literature.

Ogunyemi calls Nwapa the *"Ogbanje* artist"* who tantalized her human audience and ancestral cohorts with her mesmeric appear-ance and disappearance.[2] It is very possible that Flora Nwapa's posthumous publications may achieve for her the fame and criti-cal acceptance which she never seemed to enjoy in her lifetime. Her plays will probably be in the forefront for that honor. A close study of the three plays reveals an artist confident in herself, in control of her subject and fiercely aggressive in tone. The themes reflect thoughts and ideas which seem coterminous with specific phases in Nwapa's professional career and social life. Traces of auto-biographical references cannot, in fact, be totally ruled out.

After her relative success with the novel, Nwapa was very cau-tious with releasing too soon the products of her experiments with other art forms. For instance, although *Cassava Song and Rice Song* (poetry) was published in 1986, it had been written more than a decade earlier, and before its eventual publication, Nwapa had passed it on to some colleagues for private critical reactions. The three plays have 1993 publication dates, but some of the events which they touch upon either by direct references or allusions, would seem to have been documented at significant points in the author's political or public service career. *Two Women in Conversation* is dedicated to "Zaynab Alkali (author of *The Stillborn*), Kujara, Anthonia and others who made my stay in the University of Maiduguri a most memorable one," which suggests that the play was written after Nwapa's visiting professorial appointment at the university in the 1990/91 academic session. The play, by linguis-tic and other stylistic references, bears testimony of that trans-cul-tural experience.

In her prose fiction, Nwapa's primary and ever-consuming motif is the element of the supernatural in human domains—the journey of life and the journey beyond death with the Lake

Goddess (Woman of the Lake, Mammywater) as the vehicle as well as the catalyst. It is significant that Nwapa in her "last journey" devoted a whole novel to the Lake Goddess. From *Efuru* through *Idu* on to *Never Again*, Nwapa has been systematically revealing the facets and personality of the Lake Goddess. In the plays, the presence of the goddess is not physical, but her aura and impact are clearly evident. The female characters are full grown women who have matured into taking their destiny into their own hands. They determine and construct their agenda of action and existence. They act in concert with other women to assert their freedom and independence. Nwapa shows them as resilient in the face of irrational oppositions from men, and unflinching in the face of complex odds. In these plays, especially *The First Lady* and *Two Women in Conversation*, Nwapa firmly asserts that women-empowerment is the key to gender equilibrium by which universal harmony can be assured. The new woman of Nwapa's plays (she had been in gestation in Nwapa's prose fiction) resembles in essence, the mystical *Uhamiri*—a paradoxical mixture of motherly tenderness and masculine fierceness. One critic observed that "Uhamiri furnishes her with a deeply sustaining and insightfully spiritual model" (Ogunyemi 1995:5). Ogunyemi further declares quite succinctly and accurately that:

> Her [Nwapa's] novels—*Efuru, Idu, Never Again, One is Enough, Women are Different*, and the unpublished manuscript, *The Lake Goddess*—short stories, poetry and children's books, are, in miniature, representations of Nigeria in different (post) colonial phases.... Each of the novels diagnoses the signs and symptoms of the characters of the community/country, all suffering from maladies whose prognoses she articulates. (6)

The plays diagnose other maladies and proffer other solutions. These solutions are bitter pills for men already trapped within the sexist ideas of the familiar status quo. Through these plays, Nwapa has projected for her society, the inevitable final stage of the chaos which had become the order of the day in Nigeria. In *The Sycophants* the mind has vegetated beyond remedy but the physical man seems to enjoy and relish in the stupor. In *Two Women in Conversation* the society has reached a state of anarchy but the men who run its affairs seem oblivious of the ever ticking time bomb. The marital war is at its zenith. Unlike in Nwapa's novels, the women combatants in the plays are poised for a fight which no longer seems uneven or predictably conclusive. Women are no longer content to wait for their oppressors to bequeath freedom on their own (men's) terms. The women are poised to wrench it from their grips. In *The First Lady*, female assertions of responsibility and leadership are uncompromisably evident. In sum, the plays are the windows through which the reader views Flora Nwapa's new women and their world. The decay in the old order is dramatized in all its putrefaction but there are persistent dreams of a new dawn. Nwapa shows flickers of hope, of restoration and of rebirth. Each play reflects a particular social dimension and a particular dream. Ogunyemi states that "Her (Nwapa's) call is for action among the rulers and the ruled and among men and women to heal the sick society; ... she wants the treacherous old ways buried to make possible the rebirth of the country and its peoples" (6).

I. *THE FIRST LADY*

This extremely elaborate drama is also the most multi-faceted in thematic variety and stylistic diversity. It is a play about illusions, appearances, greed, vaulting ambitions, torn hopes and metaphorical cleansing. It is also about identity and conflicts in values and standards. It is about corruption—corruption of the state of womanhood and the battle for regeneration and ascendancy. It seeks to set straight the records of the status of women in the traditional society vis-a-vis the misconceptions of contemporary society. Nwapa addresses eloquently such issues as women leadership and political activism, education and universal suffrage, accountability and individual freedom, patriotism and individual honor. The author's techniques in communicating these are as robust and diverse as the thematic ramifications. There are vivid and elaborate stage directions. Nwapa employs also the techniques of flashbacks, ironic twists, linguistic code-switching and use of native idioms at tense moments. Nwapa is consistent in her stylistic devices in the three plays but each has its unique features as closer studies of each will show.

Nwapa seems to have anticipated Ifi Amadiume's incisive essay, "Beyond Cultural Performance: Women, Culture and the State in Contemporary Nigerian Politics,"[3] when she wrote *The First Lady*. Amadiume had argued convincingly that the "instrumental use of culture by the political classes, power groups or elite groups" to advance chosen and pre-determined political actions and positions was contrary to the purpose and functions of traditional women's groups (42). She explains that "political mobilization in itself is not the same as social movements, which have a structural impact or aim for structural change" (41). In the traditional society women's organizations, existed independent of men's groups–cult or social; the women had their councils which regulated actions and imposed sanctions on both men and women in the community. They were never intimidated or coerced into taking actions against their will, by men's groups. In a cultural sense, women were revered and their power-in-council was formidable. According to Amadiume, "Collectively, the women also held power of opposition through strikes, non-cooperation or mass exodus.... We have a great deal of data on traditional African women applying traditional strategies of rebellion to protest patriarchal systems imposed by either Islam, or Christianity" (Amadiume, 43-44). In the Nigerian situation, governments in power in contemporary times systematically embarked on a program of neutralizing the solidarity of women's groups by bringing them under government surveillance and control. This was achieved by making all women's organizations come under the umbrella of a pan-national organization under the leadership of the wife of the Head of State who then manipulated the women to act as active supporters of government policies even where those policies were detrimental to the general welfare of the women and their families. The wife of the Head of State is, of course, appropriately, designated the "First Lady."

In the play, *The First Lady,* Nwapa not only ridicules and satirizes this unelected leadership among the women, but uses the play to expose the government's strategy of muzzling women's voices in local and national issues. Where women appeared too stubborn to "fall in line" with the mandates of the State, force, coercion, intimidation and humiliation were among the tactics used to subdue them. In the play, Nwapa questions through the women characters some fundamental tenets of patriarchal authority unheard of before, such as when they argue with a local traditional chief on the role of women in politics and society:

Chief Akuma: In our tradition, the wife should stay
                      home, welcome the husband when he
                      returns home, and....
Barrister Egenti: No. That is not our tradition. If by "our
(a woman)   tradition" you mean Igbo tradition, then I

should put it to you that you have got it all wrong. The traditional Igbo woman is not a stay-at-home. She goes to the farm. She trades. She does many other things.

Chief Akuma (lamely): Yes. But things have changed.

Barrister Egenti: And human beings have not changed?

Chief Akuma: There was no politics in Igbo traditional society. I mean in our traditional past, our traditional way of life

Barrister Egenti: Did women not set up and run their own councils? Did those councils not oversee both women's affairs and men's affairs?

Chief Akuma: I don't mean that kind of thing. Listen to me. I am running for a high political office. Our ancestors did not have this type of thing. Now, I return home, and what do I find? I find that my wife is away campaigning. I go to bed hungry. Before I am up the next morning, my wife is already out. I don't see her for days. She does not know whether I eat or not....

Barrister Egenti: And you have no cook?

Mrs. Nkana (a woman leader): Easy, Barrister. Chief, what you have just said can cost you dearly if we should reveal it to the women. We have come a long way in this country and in this age, Chief. Whether you like it or not, neither you nor any other man can set back the hands of the clock. You can no longer do without us. We shall not allow you to do without us.... (100-101)

The climax of this encounter comes when Nwapa reports a gesture and body language which has serious cultural implications. In the middle of their verbal exchanges the Chief and the two women pause but in the interval, the women show unmitigating acts of contempt towards male authority ("He pauses, looks at them. They look back at him, looking him straight in the eye. He is the one who looks down first. Then he raises his face and begins....") (101).

The focus of *The First Lady* is change, the inevitability of change, and no one can turn back the hands of the clock. Ironically, when the chief asserts that things have changed, he does not seem to understand the full import of his statement; change also affects female perception of women's rights and roles in the changing world. Expressions such as "listen to me," and "this is my home" are all sexist, used for coercion and intimidation, while the invocation of "our ancestors" was a decisive way of asserting the age-long patriarchal authority and silencing female agitations which, obeying the status quo, should have been voiceless. Also, the expression that a man (married) "went to bed hungry" was an implacable way of drawing attention to the absolute failure of a wife in the home. But those assertions and expressions no longer shocked the women in the play. Instead, in the highest cultural manifestation of feminine revolt (through innuendo), "the women look him (Chief) straight in the eye," signifying not just an assertion of equality, but much more. In some cultures, looking at someone "eyeball to eyeball" may be an uncompromising way of eliciting the truth but in the cultural setting of the play women cannot look men, let alone the chief of the community, "straight in the eye," for it signals the flouting of constituted authority. Nwapa shows the women "interrupting" the chief while he is talking and further arrogated to themselves the power to "put things to the Chief," a judicial expression used to correct someone in error, or educate someone who has ignorantly taken a wrong position. The response of the women folk to male aggression, oppression and subjugation, is not shown in emotional acts such as weeping, hysteria and resignation. The new women leaders in the play are mature, educated and self-assured. They stand their ground and deflate the male ego with superior knowledge. The women have organized themselves for action. They are in politics for a purpose: "We are going to bring sanity to politics.... We are going to have a female president in Nigeria in my lifetime" (112). To achieve this the women must dismantle all structures and forums set up for women by men, supposedly for the women's welfare but indeed with the veiled motive of weakening women's groups and subverting their solidarity. Nwapa lambasts and ridicules women who act as stooges in the male design to "keep women in their place." As the play opens, the wife of a gubernatorial candidate is presumptuously, at her insistence, being addressed by everybody as "Her Excellency, The First Lady" (lending the play its effusive title) even before the election is held. Nwapa makes her an object of ridicule and satirizes her to no end. Her actions are sheepish and she is a mockery of a successful woman at home and in public life. Her pursuits and goals are puerile; her vision is demented. Her husband is not sure of winning the election but in her juvenile fancy, she is already making concrete plans for the swearing-in ceremony.

> Ada Akuma:I have already ordered the outfit for the occasion. It is being sewn in Lagos. I am going to swing into action. Whatever the Deputy Governor's wife wishes to be called, will be her own business.... I am going to have selected women with no party loyalty in my entourage. About twenty or so.... I must make it a point to go to church every Sunday with the women. And of course, the cars we are going to use. Nanny, I am thinking of an English car. Margaret Thatcher of Britain uses a Jaguar. I once saw her emerge from a white Jaguar in front of Harrods. Ever since then, I have been in love with Jaguar.... The women in my entourage will use Peugeot 505 cars. The cars will be the property of the government, but I am going to be personally in charge of them. (115-116)

In the end, Nwapa cuts her drastically and mercilessly down to size by having the honor of 'the first lady' go to the first wife—a law student in Britain—(emphasizing the author's priorities for women) who sees nothing special or spectacular about the honorific title and returns to her legal studies in England after her husband had been sworn in as governor. Humbled and deflated, the second wife, alias "Her Excellency," follows her footsteps and returns to school to get the formal education she never even had. This represents a process of growth because at the end of the play, she had attained significant awareness and maturity from what she was at the beginning of the play. She had learnt to differentiate between appearance and reality. She informs her Nanny:

> And when this is all over, I am going back to school. And, Nanny, please none of this Excellency rubbish again.... I am going to be a lawyer.... It is good to be called Mama Lawyer. But it is even better to become a lawyer. If Mama Umeh can do it, I too can. (198-199)

Throughout this play, Nwapa is at pains to draw a clear distinction between a leader of women in the cultural sense and a leader of women's organization in the conventional sense. The former is a product of a democratic concept in its true sense. Mrs. Nkana (a leader of women) explains this idea to Chief Akuma, a worldly politician with all the corrupt ramifications:

> You see, I am in politics to see whether I can bring discipline to the party. The same applies to Barrister here. I am called the leader of the women or the women's organization. But there was really no election. We have no secretary, no chairman, but we have a treasurer whom we all trust. Every kobo is accounted for. (104)

A leader chosen by the women themselves has their confidence and their mandate. She is accountable to them. On the other hand, a leader hand-picked by the government is responsible to the government and can, (and often does) betray the women at critical moments. Nwapa's emphasis in *The First Lady* is on good leadership as an essential ingredient for true women's liberation struggles.

Next to the true meaning and nature of leadership in women's affairs, is the theme of marriage in the play. *The First Lady* and *Two Women in Conversation* both show Nwapa in extreme involvement in the destiny of womanhood in the African situation. Never before had Nwapa shown such deep-rooted concern in the complexities of modern marriages. Marriage, essentially a male created institution, has been used to destabilize and complicate the lives of ordinary women in the society. Many of the women in these plays have marital problems but in general, marriage as an institution seems to have been so structured that it invariably always leaves the female partner at a disadvantage. In *The First Lady,* the most successful women are independent, often unmarried or if married, are living semi-independent lives. It seems that Nwapa is suggesting that for women to fully realize themselves, they may have to step outside marriage for other options of self-development and fulfilment. Talented women must liberate themselves from the shackles of unequal unions and get involved in the more challenging task of nation-building: "No nation can advance without the support of her women folk. For it is said that the hand that rocks the cradle rules the world." (25) Nwapa uses Chief Akuma and his second wife to illustrate the way marriage can under-develop womanhood. Chief Akuma is a politician, a chief, a businessman, a polygamist and a community leader who firmly believes that the woman's "place is really at home looking after our children and us. I have been thinking. Women should not be involved in politics." (98) Yet he could not have won the gubernatorial election if some highly talented, intelligent and pragmatic women had not stood by him. But women must always remain in the shadows of men even if the women have better things to offer. Men marry girls (as second, third or fourth wives) who are young enough to be their daughters for the selfish reason that a man as he advances in age, "needs a young girl to convince you that you are not getting old." (187) Is this then a good reason for marriage? Is this the ultimate purpose of marriage? Nwapa seems to ask these and other vital questions. Why do women marry? Why do men take on multiple wives? Why do women accept consciously the positions of second wives only to wage wars against each other inside the household? Is marriage as conceived and executed in patriarchal communities not a post-colonial tool of enslavement? Has marriage as an institution in Christian and Islamic African cultures not become a veritable prison yard for women? Unlike in her novels where women essentially complain, nag and sometimes, die in silence unfulfilled, Nwapa, in the plays, sets out an agenda first for the mobilization of women for self-liberation and secondly, for political action.

Chief Akuma's second wife, Ada Akuma, is an example of an exploited woman. In a dexterously handled artistic device, Nwapa shows through a flashback how Ada became Chief Akuma's second wife when she was still a very young teenager. Chief Akuma saw her in a flash and desired her. He lied that he wanted her as a maid, then lied that he wanted to take her away from her foster parent in order to give her an education. He even lied that he had the full support of his first wife. It was all a scheme, a conspiracy in which Ada's future was compromised by all parties involved. To bring out the full pathetic import and sad consequences of child-marriages, Nwapa allows the now full-grown Ada to re-live her childhood and tell her story:

> I was only a child. I was at school at Aba. My parents were very poor. I was smaller than my age mates and my dresses were shabby. Our school was an all girls school, and we girls used to talk and dream of what we wanted for ourselves. I still remember the day my mate Nko married a man who already had three wives. The man Nko married was old enough to be her father, or even her grandfather, but he was very rich and he was a chief.... (177)

Like her mate Nko, Ada in her circumstances wanted to end the hardships in her young life by marrying a rich man no matter the overwhelming odds. It was entirely a marriage of convenience in which emotions, love or any intrinsic values played no role. Ada confirms:

> That was how I came to marry Papa Umeh. I did not do any scheming. I could see he had money, and my girlish wish was to marry a man who had money. I had known poverty and I would embrace any opportunity to escape from it forever. So I went along with their plans, but I was not a party to it. (190)

Ada's marriage and subsequent life allow Nwapa the opportunity to raise moral issues on such marriages. She creates a situation in which Ada and her childhood mates at school discuss and argue over the issue of generation gaps and age differences in marriage. Nwapa seems to caution all youth to look carefully before they leap into marriages especially where the age differences are so glaring and the only reason for the marriage is material consideration. In all such cases, the marriage soon loses its glitter after the scales have fallen off everybody's eyes once the initial euphoria is exhausted. As if unsatisfied with the level of rational thoughts in the teenager's perception of marriage, Nwapa was yet to devote another full length play in which two mature women from broken marriages argue eloquently on the pros and cons of marriages of unequal unions and their views allow Nwapa the opportunity to make some stringent pronouncements on the universal issue of marriage incompatibility.

Although authorial comments are heavy and varied in *The First Lady*, they did not preclude Nwapa from equally pouring her venom on other issues especially those that have contributed to the contemporary national malaise in African countries in general and the Nigerian nation in particular. At the center of these sub-themes is politics and government agencies versus public welfare. There is no doubt that Flora Nwapa was totally disenchanted, if not disgusted with the political situation in Nigeria of the 1970s and the years after. She is concerned about the extent to which the governments of the day had failed in their responsibilities to address the issues of social welfare for the generality of the masses, especially in the rural areas:

> What a country! Here in my village, in this mansion
> with its wall-to-wall carpet, with furniture straight
> from Italy, there is no regular water supply; no elec-
> tricity and no telephone facility. I have to send my
> secretary to Enugu to make a call to London. (60)

From the lack of these basic amenities, to fraud and mismanagement of the few available facilities, the general consequence is the deterioration of the quality of life of the Nigerian citizenry. Nwapa condemns the politicians and governments for this unpardonable criminal negligence which is largely responsible for all sorts of crimes including juvenile delinquency and armed robbery in the country.

For Nwapa, although to all intents and purposes the institution of marriage has all but collapsed as much as the political institutions, yet the former has better prospects of salvageability than the latter. With Nigerian politics and political institutions, Nwapa is disillusioned, disenchanted and sorely pessimistic about the future. Several characters, male and female voice the author's sense of exasperation with Nigerian politics:

Madam Arum:	This is Nigeria. Anything can happen here.... Seemingly impossible things are possible in politics. The false can become the true and the true false. That's politics for you. (119)
Ada Akuma:	The party is full of so-called people of timber and calibre. Everybody is a law unto himself. Imagine the National Chairman already running for elections which are still four years away while we are yet to get through the present elections. Isn't he selfish? (111)
Mrs. Nkana:	What is an election if this woman could be in possession of ten thousand voters cards, a woman who sells food in the market and whose husband is a cook? What I have learnt in this election is that rigging begins right with the registration of voters.... Before you are nominated by your party, you proceed to get ghost voters into the voters register.... So before I began to organize the women, the party in govern-

	ment had bribed everybody. The electoral officers are mainly teachers in the employ of the state government. So they do what they are asked to do. (130-132)
Mrs. Nkana:	I have lost faith in democracy and the ballot box. No wonder the soldiers keep coming back.... Yes, our party has won, but we have destroyed our country.... Auntie, we are not ready for the rule of law and democracy. (173-174)

This is Flora Nwapa's final assessment and pronouncement on Nigeria before she took her eternal exit. It is borne out of years of fruitless patience and the agony of unfulfilled hopes. Nwapa can talk authoritatively about government, politics and the public service because she was for years a faithful insider who labored to make the system work and seems to have come to the sad conclusion that the Nigerian political mess defies any solutions which the 20th Century could offer.

Flora Nwapa is effective in her medium as she is in the message of the play. The play is replete with a variety of literary techniques and devices not seen before to this degree in Nwapa's previous works. They are the experimentations of an artist who has finally found her voice and her medium and is fully in control of both. Displaying the artistry of an experienced professional director and producer, Nwapa has an abundance of vivid stage directions including movements on the stage and precise lighting instructions. The scenes of flashback in the play are facilitated by skillful stage directions and technical guides.

Nwapa has the right eye and ear for details. People's dresses are described down to their make and color. The feelings and moods of the characters are presented both from external perceptions and inward psychological musings. In this way, the reader is able to share in the characters' motives for actions while the visual portraits make the presentations more real and easy to relate to. The men in the National Chairman's entourage "are all large men with immense bellies, all extravagantly dressed to show their wealth.... Fulsome salutations are exchanged, with the men saluting one another by their full string of titles and nicknames as Chief Akuma leads them into the house" (64). The National Chairman himself "is bedecked with the fake coral beads which have become the trademark of the newly-rich who have bought chieftaincy titles.

He has a very large multi-colored leather fan with which he occasionally fans himself ceremoniously. His speech and general comportment are correspondingly pompous." (65) When Friday, Chief Akuma's driver uncharacteristically bursts into his employer's bedroom the reader senses trouble and Nwapa makes this conclusive with aside-information in italics and in brackets. ("Tense pause, during which Friday and Chief Akuma look each other in the eye") (138). The act of both men of different statuses in life looking each other in the eye, signals the impending violence. Nwapa uses songs to introduce moments of heightened emotional tension. The songs range from children's refrains to church choruses with adaptations in the wordings to suit the occasions. These songs wherever they occur, are rendered in Igbo and Nwapa does not endeavor to interpret them, leaving clear indications of her target audience. A good example is on page 88 where Nwapa adapts a popular church chorus used to extol the majesty of Jesus, (with a prayer by the singer to be counted among the friends of Jesus), into a song of praise for Chief Akuma. It probably also is an ironic way of portraying the image that Chief Akuma wants the people to perceive of him. The song remains the same in both arrangement and phraseology except that where the original song has "Jesus," Nwapa substituted with "Akuma." Cultural greetings are also rendered in Igbo without any attempt at interpretations or transliteration. When the women visit Chief Akuma in his home, he greets them in the popular formula in which women groups are greeted in the Igbo culture. A politician must be familiar with it. Any public speaker must be versed in it. Anyone seeking the attention and support of women in any undertaking must ingratiate himself in their hearts with this structured pattern of greeting which extols and praises the women for their productivity, creativity, and solidarity:

Umu nwayi kwenu! Tuanu!
Muanu! Zuanu! Kwenu! Kwezuenu! (85)

This greeting is indeed a public declaration that she who rocks the cradle rules the world. The women are complimented for child bearing, nurturing and tender loving care. Nwapa does not try to interpret the Igbo expression but it is so apt and appropriate to the situation. Nwapa indeed knows her women, what gets their attention and what touches their emotions. She also knows her Nigerian politicians and all their theatrical maneuvers to draw attention to themselves. Nwapa portrays Chief Akuma's verbal acceptance of the women against his internally entrenched philosophy that women must be seen not heard in political arenas. Nwapa's use of irony is infallible and devastating yet subtle and subdued. Nwapa's feminist streak manifests itself in the way in which she juxtaposes men's failures with women's successes without appearing to be setting women up consciously against men for the purpose of comparison. The reader witnesses the different attitudes of women and men's groups to fraud, fiscal principles, accountability and moral integrity. Women in their quiet but resolute ways repair what men have destroyed or mishandled. The issue muted in all such scenes throughout the play is, when will Nigeria and Nigerians realize that the mess which politicians have thrown Nigeria into since independence in 1960, indeed calls for a change of guards and a handing over of the batons and mantles of leadership to women?

Nwapa was often criticized for poor quality of dialogues especially in her first two novels. The plays are totally different. The dialogues flow smoothly, naturally and authentically. The characters speak as one would hear them speak in real life situations. Entrenched in the dialogues are a variety of linguistic features which make for great imaginative communication - humor (often sardonic), anecdotes, proverbs, sayings (witty and cryptic) each appropriate to the situation, the setting, the age and status of the speakers. Here is Nwapa's rendering of a brief conversation between the highly placed Chief Akuma and his cook, John. Such occasions are rare but indeed some prominent people do have some of their domestic servants as their greatest confidants. They unburden their minds to them when there is no third party. Nwapa shows the status differentiation in the manner of speaking of the chief and his servant:

Chief Akuma:	John. You remember I promised you something?
John:	Oga, I remember. But I no want am, Oga.
Chief Akuma:	You don't want another wife?
John:	No, Oga. My Mama bring one for me last year. I send am back to my Mama. I dey all right, Oga. My wife and the pickens dey home with my Mama. No wife, Oga. Women trouble too much.
Chief Akuma:	Wise man. What do you want me to do for you then?
John:	Nothing, Oga. No worry, when you become governor you go do plenty for

> me. The fact that I dey cook for governor
> enough for me. Na prayer I dey pray for
> you make you become governor. (61)

Nwapa effectively uses Nigerian pidgin English here to depict this as well as her other less educated characters and this device adds humor to the plot as it lessens tension. John as a character is credible - a middle-aged (possibly older) Igbo man with little or no education but who is sufficiently self-confident to stand up to a future governor and decline the offer of a second wife, free of charge. Yet, with all his down-to-earth pragmatism and disarming humility, John could not fool a woman's seasoned intuition. Ada Akuma (the chief's second wife) had seen through his hypocrisy and facetiousness much earlier and had characterized him as treacherous and potentially dangerous but her husband would not believe her until too late.

Sprightly dialogues, effusive humor and subtle irony are not Nwapa's only seasoning spices. There is a generous sprinkling of Igbo expressions, adages and proverbs. It is a familiar feature of writers in a second language situation to take to code-mixing, transfer of semantic patterns of indigenous languages into English and the translation (transliteration) of local idioms also into English. The traces of these techniques are highly limited if not non-existent in the play. Nwapa instead, freely uses Igbo expressions in their regular forms, offers no translations but allows the reader to arrive at their contextual meanings through imaginative associations and inferences. The significance of the insertions of these Igbo expressions is that they lend additional authenticity to the play as they establish Nwapa as someone who knows her environment quite intimately.

In the same vein, Nwapa interjects allusions and references to actual events in Nigerian life and history. There is allusion to the first military *coup d'état* in Nigeria in 1966. There is also the mention of the military take-over in December, 1983, barely three months after a civilian government had been re-elected into power. There is the mention of the Nigerian civil war with sad undertones.

| Mrs. Nkana: | Yes, Dr. Ijele, we shall pay for the rape of democracy. |
| Dr. Ijele: | But we have paid. We paid in 1966. Then we fought the civil war. We have shed blood. |

| Mrs. Nkana: | But we shall pay again. Pray that there will be no bloodshed this time around. (173-174) |

The First Lady is a highly successful play. The language is poetic and flows effortlessly. Nwapa writes as a seasoned playwright. Her handling of the "play within the play" (the flashback to Ada's childhood and early marriage) is a major artistic feat. The issues around which the actions in the play center, are current, topical and inherently problematic to the continued existence and stability of the Nigerian nation. The things that detract from the excellence of the medium and message in *The First Lady* are not of Flora Nwapa's making. One notices with great dismay that the play is littered with embarrassing misprints, unpardonable spelling errors (not excluding Igbo words), grammatical infelicities and confusion of tenses among others. These occur too frequently, too glaringly and too wantonly to be ignored. They are unconscionable. They are a disservice to Flora Nwapa. One can't help but ask, "so what did the two respectable editors do?" One hopes that future editions will rid the play of these minor setbacks and allow it to remain in its purest form as Flora Nwapa's major legacy to women's liberation and democracy.

II. *Two Women in Conversation*

In *Two Women in Conversation,* Flora Nwapa returns to the issue of marital incompatibility, failed marriages and the effects on the female victims. The discourse is at a highly serious and mature plane. The conversation is between two women - sophisticated, educated, well traveled and embattled by domestic violence. The format of a conversation at home, in the kitchen, over food and drinks, allows the women to talk freely and unrestrained, in response to their innermost feelings and emotions. Unlike *The First Lady,* Nwapa pin-points the precise details of her setting, from the specific dates of events, actual physical and historic situations, down to the ages of the characters. The stage direction at the opening of the play states:

> The play is set in 1990, a few months after the unsuccessful military *coup d'état* of April 22nd of that year. The conversation takes place between Niki and Juma, two modern, educated, articulate, highly independent Nigerian women whose mar-

riages have failed. Juma is fortyish, Niki past her
mid-forties. (72)

Nwapa had written about failed marriages before in her novels. The
setting had more often than not been in the rural areas and the
female characters, distinguished by their virtues of dynamism,
astuteness and constant fidelity to their erring husbands. What is
strikingly new in *Two Women in Conversation* is that the women are
excessively independent, mentally mature, sophisticated in metropolitan terms, and they have traveled extensively including sojourns
abroad where they received higher education. They have acquired
the type of international and across-the-boundaries exposure which
their counterparts in the novels lacked. Because of their contacts
and exposure they can analyze their situations from personal as
well as intellectual viewpoints. They come up with ideas and conclusions which are self edifying as well as instructive to other
women caught in similar predicaments.

Two Women in Conversation x-rays the hidden closets of marriage as a modern social institution. The author's role is like that
of a psychologist who prepares a conducive and appropriate atmosphere and environment, induces her disoriented clients to talk
uninhibitedly, and then listens to them bare their charged souls.
When it is necessary to prompt the clients, the relevant catalyst is
administered. Sometimes the discussion is at a purely personal
level, at other times it rises to dimensions of a highly tuned philosophical discourse. The play begins with a questioning of the conventional assumption of *love* as the basis of marriage.

Niki:	Love.
Juma:	Love? What is love?
Niki:	What you feel for someone whom you like.
Juma:	What a definition?
Niki:	A kind of warm feeling. A kind of affection towards someone at a point in time.
Juma:	Good. I like that qualification: at a point in time. Because at a point in time again you feel like murdering that someone.
Niki:	Do you sometimes feel like murdering your husband?
Juma:	(Without hesitation) Yes.

Undoubtedly the stage is set for unpleasant revelations and unorthodox propositions for the resolution of unpalatable acrimonious relationships. Upon further prodding, Juma reveals that hers was a marriage of circumstance, not of emotional love.

> I was young.
> A war was on, school was shut.
> And I was bored stiff.
> Then I met him.
> For lack of something to do,
> I stuck to him.
> And he took me to America.
> If there was love,
> I never experienced it. (96)

All the wrong reasons for marriage, and yet the problem is almost universal as Niki is told "You are not the only one, my child." The prognosis on failed marriages unearths a variety of reasons ranging from lust for "the other woman," to indiscreet permissiveness of some religious faiths and cultures. Domestic abuse and violence shift from physical assault to the politician's strategy of hiring thugs to cause harm to offensive opponents. Juma's husband who walked out on his young wife and children returns three years later with hired thugs to force the wife and children to follow him to an undisclosed destination. Juma was only sixteen years old when she ran away from home to live with the soldier in the Biafran army whom she was to formally marry at the age of eighteen in far away New York. The marriage existed virtually in name when they returned to Nigeria nearly two decades later. Juma had picked up a few enlightened habits in America. She had developed into a woman with a distinct personality. She had graduated from a university, had her own bank account and had bought a gun which was always loaded and ready to be used. The gun came in handy at that inauspicious moment of her estranged husband's intrusion:

> I pointed the gun at my husband, I said: 'Now if you don't get out, I am going to shoot all of you dead....This is no film show,' I said, 'If you don't move immediately, I am going to shoot. Now move, shameless bastards. Get out of my house!' (94)

The personality and language of Juma were no longer those of a local Nigerian lady. They are obviously those of a woman of the world, familiar with Hollywood films and dramatic heroic encounters. Nwapa beautifully juxtaposes two violent backgrounds—the war going on in Nigeria at the time, and the "war" going on in the home of two people who had seen and participated in the war at home before escaping abroad. Nwapa attributes much of the chaos in the families to the evil consequences of the war. Juma admits that the Biafran war ruined many people psychologically, especially young girls.

> It made us bolder and worldly. Overnight we lost
> our childhood. And became women. Dangerous
> women. (84)

The women are "dangerous" in the sense that they can now stand up to their men folk and fight for their rights. They had stood side by side with the men in the rough theaters of war (Nigeria-Biafra War) and had known male strengths and weaknesses, watched them fight, lose and win battles, and seen them at their weakest moments when they had to drop their guns and flee in the face of superior enemy gun-power. That which had made the men weak had emboldened the women who survived the war. They are *new* in all respects, mentally, socially, morally and intellectually. They have learnt how to be on the offensive when a situation calls for it. The men caught unawares by this new revolutionary phenomenon react much the same way as Juma's husband:

> My husband was the first to run.
> And his thugs jostled one another.
> To overtake him.
> I fired a shot into the air.
> Just to show I meant business. (94)

Nwapa advocates women fighting for their rights in marriage, but not necessarily through violence:

> What you should do
> Is be positive in your thinking
> Be generous in dealing with him
> And as I said, no bitterness
> If you want to leave him
> Do it quietly, and with dignity
> But remember that
> You shoulder great responsibility.
> The children you have to take care of,
> Single-handed. (137)

Nwapa's women have come a long way in their mental and emotional development, and have gone the full circle of marriage under unbending patriarchal authority. Having stepped out of the circle they can now offer to others the wisdom gained from experience. What *Two Women in Conversation* offers is a kind of testament of survival to emancipated women distressed by chequerred marriages. Frequent authorial comments take the semblance of "the sermon on the mount" where women are admonished on the do's and don'ts in marital relationships in a manner that only Flora Nwapa, herself knowledgeable in these intricacies, could have written. A central idea that seems to be emphasized is that when women fully and properly understand who they are and equip and fortify themselves adequately, they can fight and win marital battles on their own terms. But even more significant is the beneficial opportunity offered men to understand their essential roles and contributions towards the success or failure of contracted marriages.

For, although young girls have learnt wild ways during the Nigerian Civil War and have grown into womanhood with them, Nwapa seems to say that the major factors that lead to failed marriages are to be found in men, the worlds created by them, and the cultures that sustain such worlds. Nwapa writes with a deep sense of concern and involvement on how to make the best out of a situation seemingly evil from the cradle. Displaying the type of philosophical vision that often grows from injured merit, Nwapa offers seven "practical commandments" to married women in search of marital stability or harmony in the home:

1) (You) must make a man feel that he is a man.
2) You must not make women out of our men.
3) You must not give a man inferiority complex.
4) You must not make a man feel that you know anything.
5) Don't ever make him rich.
6) Don't ever help him in his business.
7) Leave him to struggle, if he makes it, you share in the wealth.
    Whereas if you helped him
    And he became really rich
    He would have a complex
    And would always remember
    What you did for him
    And resent it.
    And even go as far as getting another wife.

Who would wallow in his wealth
And think him the greatest creation of God.
He would enjoy lavishing his money on her.
As for you.... (111)

All forms of marriage come within the range of Nwapa's bitterly charged firing missiles. Each facet, which dehumanizes as it humiliates the woman, is brought to the forefront of condemnation. The customary marriage system is attacked because in the event of a break-down in marriage the customary laws of divorce are overtly in favor of the man and clearly grant no rights to the woman who must return her dowry no matter who is at fault and regardless of the length and fruits of the marriage:

It is a terrible custom
This return of dowry
No consideration is taken
For the years of marriage
And for my children.
Niki, I slaved for that man,
I brought up our children single-handed
Now I am required to return the dowry
It is most unfair. (123)

Nwapa goes from the local to the universal. In the Nigerian situation in general, the courts seem more sympathetic to the man than the woman in divorce proceedings; perhaps understandably so because the men constitute the legislature and the judiciary. It is the sense of total loss which the entire system leaves the disenfranchised female victims, that leads the playwright to exhale a diatribe on human injustices.

Whoever told you
That the world is fair?
We live in an unjust world
Sometimes I think
That God himself is unjust
God, please forgive me
I don't mean to blaspheme
But you do understand
Don't you, God. (124)

Despite the humor, the venom sticks. Nwapa equally comments on women and property rights where the woman is disadvantaged even unto death:

In our custom
The wife's property when she dies
Belongs to the husband.
If she has no children
The husband inherits everything
It is that bad, Juma. (125)

The aspect of distaste here lies in the fact that this custom applies
even in situations of estranged couples. Marriages under both
Islamic and Christian religions are not spared either. With their
condemnation, Nwapa has virtually discredited all forms of marriage known and practiced in the Nigerian situation - marriage by
native laws and customs, polygamy in the traditional and Moslem
situations, purdah, and monogamy in the Christian frame of reference:

...think of this
One man one wife thing
In Christianity
And think of the
One man four
wives thing
In Islam
And you wonder....
I don't mind the
One man four wives thing
What I mind is not being
Sincere.
If you want many wives
By all means go ahead
Have them, and say
To hell with Christianity
Why have a church wedding
And then go for more wives
Later, just because you have grown rich? (133)

For the first time in her fiction, Nwapa takes a very serious swipe
at the purdah practiced in the Moslem North of Nigeria, due no
doubt, to the insights gained during her one year visiting professorial appointment at the University of Maiduguri during which
she met and socialized with educated Hausa-Fulani women:

Niki:	But what you do not know is that in Islam, you can have concubines.
Juma:	Concubines? What about purdah?
Niki:	You have them as well.
Juma:	Oh my God! What do these men take us to be? Their chattels? (133)

Nwapa thus becomes the voice of all disenfranchised women suffering deprivations and abuse under Islamic hegemony, be they in India, Pakistan, Afghanistan, Iran, Iraq or Nigeria. She goes further to satirize the Moslem law that permits marriage of four wives by one man who supposedly is to love the four wives equally:

> Love again!
> Equal love
> For four wives
> Who are by no means equal?
> Let's stop deceiving ourselves. (134)

After traversing the marriage circuit through native, Christian and Moslem habitats, Nwapa's perception of failed marriages is that men-monsters see their female victims as chattels and in her pessimism envisions eternal obliteration of any institution of marriage that allows marital injustices to exist and prevail:

Juma:	As I see it, Niki, the institution of marriage will soon disappear from the face of the earth.
Niki:	Well, if you say so.... But I don't think so. It is fashionable to marry and as long as it remains so, the institution will not die. Look at the fuss made when people are getting married. What young couple wouldn't want to experience it?
Juma:	Then the separation. Soon after that fuss. Everything is a farce. Pure photo-trick. As my sister would say. (122-123)

After the vicarious lamentations, Nwapa's verdict on the institution of marriage in its present form, is that it is all vanity! Nwapa's call is for a marriage which has at its center, companionship and equality. In a closing authorial commentary on marriage, the view is:

Yes, companionship is important
We want someone
We can relate to
Like equals
Discuss issues with
Like equals. (140)

The play lingers on for a while after its seemingly inexhaustible dredging of the marriage institution and its vicious tributaries, providing the playwright further opportunity to extend the range of critical comments on areas and establishments such as Nigerian Women's Social Clubs (where tardiness, lack of punctuality and unclear sense of purpose and direction place them behind social clubs organized by foreign Nigerian wives), the penchant of Nigerians for imported foreign goods (to the detriment of local manufacturers), the hypocrisy in the activities of women's philanthropic organizations (where personal aggrandizement and exhibitionism overshadow simple acts of charity), the incongruity of Christian and Islamic religions on African soils (because they "are alien to our ways of life"), and irreverence and lack of humility in social manners. The social malaise in Nigeria seems endless and ubiquitous and Nwapa as a dramatist rotates her spray-and-cleansing funnel on all facets in all directions. Indeed drama as an art form seems to offer to Flora Nwapa more ease of communication and more exercise of the freedom of speech. It is likely that had Flora Nwapa lived after the publication of these plays, she would have moved more and more to drama as the most accessible medium to reach her target audience. Nwapa, the playwright, would have been inherently more successful than Nwapa the novelist.

As in *The First Lady,* the success of *Two Women in Conversation* is attributable as much to accessible themes and messages, as to the medium and strategies of conveying the messages. Nwapa employs as successfully as before the techniques of vivid stage directions, flashbacks, miming; cryptic and dramatic expressions which convey effectively the emotions of terseness and urgency. Her use of slangs at the right times and places adds authenticity to her setting and characters as in the use of "sabo" for saboteur when characters talk about the Nigeria-Biafra war. There is limited use of Igbo expressions which emphasizes the urban and cosmopolitan setting of the play. Instead, increasing numbers of Yoruba and Hausa expressions feature. This also accounts for the increased presence of Americanisms in the play. Despite the seriousness of the theme, there is humor no matter how fistfully clenched as in the scene where Juma's husband and his hired thugs run for their dear lives at the sight of Juma's hand-gun:

> And I thought I heard a fart
> Explode from one or two
> As fear drove them like a fiend
> You should see their haste
> To jump into their car
> Start up and flee. (95)

True to the format of the play, the language of the conversation rises at the highest peak of emotions and descends to placid and casual tones at informal moments. The playwright assumes the most rhetorical voice for the former, and instills Nigerian English expressions into the latter. There is this beautiful piece when Niki counters Juma's unthinking generalization that "Happiness is elusive":

> I know.
> It is a state of the mind.
> You cannot be happy
> When you are full of hate.
> Hatred inhibits happiness
> So you have to work
> At being happy
> At being reasonably happy.
> At creating a clean heart
> Within yourself.
> You are in a kind of bondage
> If you have an unclean heart.
> You are unproductive. (103)

There is also this rhythmic rendering of a scene after an air-raid:

> I ran as I saw
> People running up the hill
> I saw a damaged building
> Then I saw a body
> Without a head
> Then I saw a hand
> Without a body
> Another hand
> Then a leg

Oh my God.
Then I saw Onyeze's head. (92)

Nigerian English and local expressions are used to depict humor
and relaxed moods and the playwright uses them with confidence—"for where?" (101), "Agbalagba ako"—Yoruba for "Good-for-
nothing child" (78), "Kwashiorkor" (86), "Kunu" (Hausa drink-107),
"Ewoo air raid." (91) There is evidence of biblical influence in the
evocation of both imagery and character. When Juma's husband
returns to harass his wife after a disappearance of three years,
Nwapa casts him in the image of the biblical prodigal son, but with
a difference:

> Juma: Were you expecting a red-carpet
> treatment, or the fatted calf to be
> slaughtered for the prodigal son
> come home? Am I your father, to
> rejoice because my son who was lost
> has now been found? (77)

The contrasting imagery is apt, so is the irony and so is the satire.
Equally apt is the use of a conventional Christian hymn—

> God moves in a mysterious way
> His wonders to perform
> He plants his footsteps in the sea
> And rides upon the storm.... (113)

—to demonstrate how God intervened and granted Juma's prayer
to rid her of her husband who "living with him, was like living in
chains." Nwapa gives in the play a reasonable coverage of the
Nigeria-Biafra War and characteristically issues a verdict on the
entire cataclysmic experience:

> I want to forget that senseless war
> It was senseless
> Looking back now I can see
> It could have been avoided. (88)

There is a streak of ambivalence when Nwapa shifts from the rendering of dramatic actions to unveiled commentary on Women's
Liberation. Although the views which questioned the purpose of
Feminism or Women's Liberation as a movement can be said to be
the playwright's attempt to present a balance in the issues portrayed, they underscore Flora Nwapa's occasional stunning rebukes

of feminists and her repeated denials at various interviews that she was a feminist. At one moment she preferred the term "woman-ist" to "feminist" to describe her stance. In an obvious authorial comment, Niki, in an argument with Juma, expresses strong reser-vations on the Women's Liberation Movement:

Niki:	I was talking About this liberation thing.
Juma:	Yes?
Niki:	Isn't it getting too much?
Juma:	What do you mean?
Niki:	I mean, we should be women.
Juma:	Did I say I was a man?
Niki:	No, but sometimes we behave like men We do the work meant for men And leave them with nothing to do. Making them redundant And then we turn around And complain. (108)

Whatever Flora Nwapa's personal views about Women's Liberation Movements as an organized force to fight for the rights of women, it is clear from her plays that she is one with the movements in terms of ultimate goals and objectives. She has made clear through her fictional works her preferred strategies and modalities. It is very clear from *The First Lady* and *Two Women in Conversation* that no one can misunderstand or under-estimate Flora Nwapa's zeal for, and commitment to the female cause not only in Nigerian or African situation but as a global concern requiring the urgent action of all men and women. It calls for a re-examination of some fun-damental tenets of organized religions in their attitudes towards women as inclusive in the dictum that "before God all human beings are equal." Women as vessels of the infinite continuity of divine creation deserve the understanding and respect of the other half of humanity to ensure stability and harmony in every society. This seems to be Flora Nwapa's message to the human race. She has left a legacy!

III. The Sychophants

Using the sardonic title, *The Sychophants,* Nwapa takes umbrage at the social malaise in Nigeria which has deteriorated to the point of turning able-bodied youths into imbeciles and human caricatures. They are the victims of failures in sensible national planning, poor political leadership and a cancerous corruption which has eroded social values and human intellect. Nigerian citizens see no worthy purpose in defending Nigerian image and name because the country—their fatherland, seems to have nothing much to offer them to ensure their survival in life. As their name implies, the characters represent the so-called wasted generation in contemporary Nigerian society. They live from day to day on their reciprocal exploits of their country. They strive to deviously assist the leaders of the country in their vengeful waste and embezzlement of its abundant resources which invariably would elude them if they tried to secure them through honorable methods and ways.

The two characters Miki and Poki, both above forty in age, have resided in Lagos most of their lives. They are unemployed and cannot get employment because of a debilitating democracy. They have a grand design to make quick cash and become instant millionaires by defrauding the Federal Government whose key officials and functionaries have no rivals in their propensities and acumen for fraud, embezzlement, dishonesty and other forms of corruption. The government had issued a decree (which it cannot enforce) asking all Nigerians to *farm.* These include countless thousands of destitutes and the homeless who find refuge at the bus stops, under the bridges and market places. Included also is a large size of youthful humanity who are street traders in Lagos and who take advantage of the poor transportation system (where there is incurable traffic congestion) to turn the streets into market places and they themselves into irritating (to motorists) ubiquitous itinerant hawkers of all sorts of wares. The government had also issued a decree to rid the streets of Lagos of these menacing hawkers, without spelling out the modalities of implementing this law of eviction of the homeless from nowhere to nowhere. Miki and Poki, nincompoops with inflated self-delusions, have the grand design to persuade the government to award them a contract to build forty-two youth camps at five million naira each, to serve as homes for undesirable elements cleared from the streets of Lagos and other states of the Federation. This would indeed be a useful social service, except that in the classic style of the country's leaders and top government functionaries, no camps will be built when the contract is awarded. The money will find its way smoothly into private pockets after the government officials had got their kick-backs. Miki and Poki are day-dreamers who build their castles in the air but they never give up. They are totally bereft of all scruples and rational thoughts, thanks to a political system which now revels in the articulation and pursuit of lopsided values where appearance is mistaken as a routine, for reality. This play shows Flora Nwapa's immense capacities for classic and biting satire.

Miki and Poki are effigies and caricatures of the marginalized millions of Nigerians. They lack the capacity to understand their true predicaments of wallowing in an empty existence. But if, and when, we laugh at them, we are laughing at ourselves. Nwapa creates in this play a dry humor which chokes as it seeks to entertain. Every facet of Nigerian life is held up for derision and laughter: the government that makes decrees but has no plans for implementation; government officials who line their pockets with public funds with no desire for and no commitment to the responsibilities of their high offices; public servants who have no dedication to duty; military governors who indulge in all manners of gratifications with reckless abandon; unemployed sychophants who have multiple wives and countless mistresses. These are depressing situations for which as Nwapa puts it, "You've gotta laugh to cry." This picture of national decay and drudgery is not amusing to the playwright who urges immediate change to avoid a state of doom and cataclysm. But the change which the playwright seeks for her country is not external; it must come from within to make an impact. Nwapa's Nigeria is likened in the play to a wasteland. Thus, the greatest problem in our country today,

> ... is not so much
> Bribery and corruption.
> It is waste—
> Waste of human resources
> Waste of manpower
> Waste of food
> Waste of time and energy
> Doing the wrong things;
> Waste of words
> Waste of money
> In the name of grandeur;
> Waste of titles
> Given to the wrong people. (27)

It is an endless list of facilities and amenities which do not perform their essential functions to the detriment of the quality of life of the citizenry but no one cares. The playwright vehemently advocates change but adds,

> Change does not come
> Through the barrel
> Of a gun.
> Nor through a revolution. (31)

> The sought-after change is further qualified:
> Revolution is not the answer,
> Violent revolution,
> Shedding of blood
> Is not the answer.
> If we seek change
> For the good of our country
> We should start from within. (31)

Nwapa appeals to the sanity and conscience of her country men and women to heed her call for change for their own sakes. Nigeria can reclaim herself, and heal herself of all the maladies that have eroded her image and esteem at home and abroad:

> Change depends
> On the entire Nigerian people,
> Change, I think,
> Will come from within
> Not from without
> Nigeria will be great
> If you and I
> Realize that we have a part
> To play
> In bringing about this change. (30)

Nwapa's vision is apocalyptic and her voice prophetic. She seemed in a hurry to change the moral fibre of her fellow Nigerians, where all spiritual and sacred values seem to have fallen apart. The tragedy is that although the Miki's and Poki's in the country feel the anomie, the leaders and the powers that be have lost all sensitivity. Flora Nwapa speaks like the prophets of old and the reader senses her frustration that the repeated calls for Nigerians to redress their values and ways of life are falling on deaf ears. To the leadership, the country is divided into two camps—They and Us.

"They" refers in the play to the suffering masses and Nwapa draws attention to these marginalized people in the play by always capitalizing references to them—"THEY," "THEIR," "THEM," "THEMSELVES." In Lagos they are the inhabitants of:

Ajegunle
Amuwo Odofin
Maroko
Mile Two
Obanikoro
Oshodi
Orile
Bus stops
Under the bridges
Market places. (9)

But Nwapa points out that like the "US" in the society, THEY also have families, wives, children, feelings, desires, expectations and dreams.

The Sychophants presents a pessimistic and bleak portrait of Nigeria. The playwright holds up the puerile fantasies of Miki and Poki to shock Nigeria and Nigerians into a state of consciousness and realism. The Sychophants, Miki and Poki, have no home and do not want the country to change until they have defrauded her to their hearts' content. Will Nigerians understand and appreciate that Flora Nwapa is addressing the "sychophant" in each and every Nigerian? Like a constant refrain, the Sychophants who see no difference between civilian governments and military regimes in Nigeria, resign themselves to becoming empty vessels, the chattels of those in power forever. They will sing their praises in order to earn their daily bread, beyond that they have no ambitions and no loyalties:

I do not represent the masses,
I am not an ordinary person,
I am not a trade unionist.
I am not a worker,
I am not one of those
People who sell shirts
At Broad Street and
Tinubu Square
Those are the masses
I am a sychophant
No mistake about it. (65)

And further this elaboration is added:

> .... We are Sychophants
> We are always available
> To any Government in power
> To use as they think fit
> We sing their praises
> In order to win contracts
> And favours. (66)

Nwapa, anxious to touch on all aspects of national decay, draws attention to the break-down of the university system in Nigeria as a deep-seated concern because a country ceases to grow and develop when it blunts the edges of its educational system at any level but more so at the university level. Many Nigerian creative writers seem to be oblivious of the great danger posed by the consistent erosion of academic standards in Nigerian Universities by the insensitivity of those in power, to the need to recognize education as the most important investment that a nation can make in its youth. Nwapa describes the crises in Nigeria's higher education in sordid terms:

> Poki: So why say
> The students belong
> To the Ivory Tower
> When you know
> That that age is gone
> And gone forever?
> When you know
> That you no longer
> Distinguish them
> From the rest of the mob.
> That's why we talk of things
> Like zero-one-zero.
> What a country
> Where university students
> Can no longer have
> Three square meals a day? (54-55)

To cap it all, the sychophants voice the belief now popularly held in Nigeria because of the circumstances of the times that:

> Going to School
> Used to be a big deal
> Not anymore.

University graduates now
Roam the streets of Lagos
In search of jobs
They know are not there (66).

Flora Nwapa's three plays—*The First Lady, Two Women in Conversation* and *The Sychophants*—are parables on Nigeria for Nigerians. Nwapa must have intoned, "Those who have ears, let them hear." The tragedy is that she is not around to watch their impact, if any on Nigeria and Nigerians, as they make their entry into the 21st Century.

NOTES

1. According to Ejine Nzeribe (Flora Nwapa's first daughter), Flora Nwapa saw the published plays on her hospital bed at the University of Nigeria Teaching Hospital at Enugu, before her death.
2. See Chikwenye Okonjo Ogunyemi's scintillating essay, "The Invalid, Dea(r)th, and the Author: The Case of Flora Nwapa, *aka* Professor (Mrs.) Flora Nwanzuruahu Nwakuche."
3. Ifi Amadiume in "Beyond Cultural Performance: Women, Culture and the State in Contemporary Nigerian Politics" points to the power of women in Igbo society then and now.

WORKS CITED

Amadiume, Ifi. "Beyond Cultural Performance: Women, Culture and the State in Contemporary Nigerian Politics." *The Politics of Cultural Performance.* Ed. David Parkin, Lionel Caplan and Humphrey Fisher. Providence, Rhode Island: Berghahn Books, 1996. 41-59.

Nwapa, Flora. *Conversations* [Plays]. Enugu, Nigeria: Tana, 1993.

—-. *The First Lady* [A Play]. Enugu, Nigeria: Tana, 1993.

Ogunyemi, Chikwenye Okonjo. "Introduction: The Invalid, Dea(r)th, and the Author: The Case of Flora Nwapa, *aka* Professor (Mrs.) Flora Nwanzuruahu Nwakuche." *Research in African Literatures.* 26.2 (Summer, 1995): 1-16.

Evading Canonical Constraints: The 'Popular' As An Alternative Mode— The Case of Flora Nwapa

Jane Bryce

'I am not a single woman.
I am a very connected woman.'
— Efua Sutherland, 1991

The words in the epigraph cited above were spoken by a woman who, in terms of her work as a playwright and cultural activist over a period of three decades, occupies a position similar to that of Flora Nwapa as a ground-breaking African writer. Efua Sutherland was relating her reaction to a request that she join a panel investigating the status of the 'single woman' in Ghana. I have chosen to quote her response because, in a direct and personal way, it addresses much-theorized questions concerning the construction of marginality and centrality, and how these impact on perceptions of gender and identity. In rejecting the label, 'single woman,' along with its assumptions about how a woman is to be defined, it alerts us to an inherent danger in the critical enterprise of establishing paradigms and categories for the writing of African women. Efua Sutherland, mother and grandmother, an elder in her community and an internationally recognized writer, could not see herself in the category 'single woman' simply because of the absence of a husband. Her words, therefore, point to the inadvisability of assuming a transparent meaning for even so commonplace a word as "woman," a cultural signifier which is constantly shifting ground.

Like Efua Sutherland, and in concert with many others, Flora Nwapa seeks in her writing to examine, refute and redefine received notions of femininity, and thereby to interrogate wider issues of marginality and power. While she in is no way unique in her preoccupation with the prescriptions of gendered identity, Flora Nwapa as a writer offers a peculiar opportunity for an exploration of certain key contradictions, not only in her work but in its presentation, packaging, consumption and criticism. She is both central and marginal. As the first African woman to publish a novel and that novel (*Efuru*, 1966) and its successor (*Idu*, 1970) both forming part of the Heinemann African Writers Series, which was itself founded on the bedrock of her countryman Chinua Achebe's *Things Fall Apart* (1958), and is perceived as constituting the touchstone for "excellence" in African writing, she is central.[1] She is also marginal in that her subsequent novels, *One Is Enough* (1981) and *Women Are Different* (1986), self-published by Tana Press, Enugu, and appearing initially in the guise of cheap and gaudy paperbacks, have until recently been all but overlooked by the critics, for whom *Efuru*, in Susan Andrade's phrase, has "paradigmatic status." (Andrade 1990:97)

The centrality of *Efuru* as precursor of a woman's literary tradition in Africa seems by now to be well-established. The most cursory overview of critical readings of African women's texts is enough to demonstrate this centrality, whether endorsing its positive representation of a traditional woman or questioning its innate conservatism.[2] One of the effects of a novel's achieving "paradigmatic status" is inevitably that of the exclusion or marginalization of less obviously canonical texts. Hence the status of *Things Fall Apart*, summed up in Charles Larson's (1971) dictum that it may be considered the "Archetypal African Novel," has helped to shape the expectations of readers and critics as to what constitutes African writing. The tri-partite interrelationship between the popularity of this text, the self-professed mission of the publishing imprint to produce the "best" of African writing and questions of canonicity has implications for African writing as a whole. Women's writing, largely excluded from or some way down the league table of the canon, has been one obvious casualty of this artificial hierarchization process, along with "deviant" forms like popular fiction.[3] In the case of *Efuru*, which has achieved canonical status, critics have commented on its careful treading of the line between conformity and rebellion. Andrade (1990) observes that Nwapa herself appears to uphold the moral/immoral dichotomy in the representation of women by male writers, by celebrating "Efuru's modesty and adherence to appropriate models of Igbo femininity" (98). Nnaemeka (1994) questions why Efuru, as representative of tradition, is "placed at the center of the narrative, whereas the 'deviant,' less conformist characters remain marginal figures" (151). The inter-textuality of *Efuru* with *Things Fall Apart* (in the sense that it exists, as a narrative, in the gaps and interstices of Achebe's text) has been acknowledged by Nwapa herself. Indeed, she cited the appearance of Achebe's novel as marking the moment of realization for her that village life, tradition and an African subjectivity could provide appropriate material for "serious" fiction. *Efuru* may be read as falling within the "archetypal" purview as a quasi-realist novel of traditional life, even while it activates and makes explicit what is marginal or silent in Achebe's novel—the subjectivity of women.

Emmanuel Obiechina (1975) has described the West African novel as emerging from the movement towards "greater individuation and individual self-awareness, increased psychic and social mobility, and a new confidence in the individual self and also in the historical and cultural self," brought about by colonial capitalism's disruption of traditional culture (261). This highlights the important role played by the novel genre in the construction of a specifically African subjectivity, at the same time as it emphasizes its essentially innovative and syncretic nature. Achebe (1988) himself, in speaking about his early works as an attempt to show that the African past was not "one long night of savagery" from which it was rescued by colonialism, has articulated his consciousness of this interrelationship. He has pointed to the intertextuality of *TFA* with the representations of Africa afforded by such texts as *Mister Johnson* and *Heart of Darkness*.[4] Commentators on Achebe, furthermore, have demonstrated that it is not only in the content of his works, but in his radical reworking of the novel form itself, that he both counters negative myths about Africa and establishes the grounds for a revisioning and remythologizing of African identity

(Gikandi 1991, Obiechina 1975). It is therefore nothing unusual that Flora Nwapa, in her turn, should use the novel for an exploration of a specifically *feminine* African subjectivity. As Andrade (1990) points out in her article on African women's literary tradition, the ubiquitous trope of the Manichean allegory, "leaves no room for the necessary heterogeneity of the many facets of subject constitution" (93), and writing, therefore, becomes "one of the most powerful ways by means of which women inscribe themselves into history" (95). History, here, is taken to be a western-defined discourse, an essentially literary and literate construction which functions to exclude alternately the colonized and female subject. A slight shift in perspective, so as to acknowledge the existence of other, oral, historiographic forms, far from undermining Andrade's argument, would allow us to treat women's writing as part of a continuum with orature. In which case, women's writing, which Andrade sees as deconstructing Hegel's view of Africa as "only on the threshold of the world's history," assumes the power of representation not only through writing, but through its intertextuality with orature. Presence is thus doubly asserted, and Nwapa herself emerges as a writer who constantly demonstrates an awareness of her debt to that intertextuality.

What concerns me here, however, is why it is that we as critics do not hesitate to recognize this power in a novel like *Efuru*, while maintaining a discreet distance from such novels as *One Is Enough* and *Women Are Different*. Until very recently, the few critical glances at the latter two novels have seemed to concur that they are in no way comparable to Nwapa's earlier work. Susan Gardner (1994) focuses primarily on Nwapa's social context as a way of illustrating her achievement in writing at all, given the enormous disadvantages to women inferred from her discussions with Nigerian women professionals. Her conclusion that, "In the end, my admiration for Flora Nwapa is not so much for her artistry but for her entrepreneurship and commitment to women," suggests that it is what she represents ideologically that carries most weight (10). Yemi Mojola's (1989) description of Nwapa's work as "sociological in conception" appears to endorse this view, and raises again the conundrum of writing from a gendered perspective other than that of the "universal" or "neutral" (read "patriarchal"). In assuming a specifically feminine subject position, women writers today still lay themselves open to the charge of which Virginia Woolf (1979) complained: "How personal, they will say, rubbing their hands with glee, women's writing always is" (49). While both

Gardner and Mojola apparently take for granted the political nature of Nwapa's focus on women and the "personal," neither of them is seemingly very comfortable with how this manifests itself in the later novels. For Mojola, all the works, but especially *Efuru* and *Idu*, "constitute a solid structure documenting certain aspects of the traditional way of life of the Igbo people," and the two protagonists represent "Nwapa's vision of the ideal Igbo woman" (20, 21). However, Nwapa's attempt to expose the corruption and decadence of post-independence Nigeria is, for Mojola, undermined by "certain conceptual contradictions," specifically that:

> The depiction of the female characters on whose actions the author formulates her thesis about women's freedom is at variance with some of her criticisms of the society. The contemporary women who are economically independent through business such as Amaka and Chinwe, though hard-working, make their initial and even subsequent financial gains through debauchery and corruption. (Mojola 1989:25)

This critical censure is not far removed from that of Ernest Emenyonu towards the celebrated Lagos prostitute in Cyprian Ekwensi's *Jagua Nana*, and therefore partakes of the same moral/immoral dichotomy Andrade perceives in Ekwensi, Emenyonu and Nwapa herself (Andrade 1990:98). Mojola's objection that, "Chinwe's generation is making its wealth by dubious means and therefore is not worthy to be the yardstick for determining the emancipation of women. A generation of sober, self-respecting, industrious and determined women would have been more appropriate" (24) points to a prescriptive tendency in African literary criticism which overlooks the tensions and contradictions within the texts themselves.

Another line of criticism that needs to be addressed at this juncture is that of Chimalum Nwankwo, who in a thought-provoking essay discusses Nwapa's relative failure to act as carrier for "the Igbo word." In a sense, Nwankwo, by examining the relationship between Igbo language and culture, and Nwapa's personal style, endorses Andrade's claims for a process of writing-into-history. "The word," he says, "is the message, the kernel of every art," without which art becomes mere "frivolous spectacle" (Nwankwo 1995:42). The process whereby the "word" is transmuted into "art" he calls craft, and it is the success or failure of that craft that determines the long-term value of a work of fiction:

> The craft itself may come out and assert its image
> in reasonable comeliness and yet fail because of the
> *unfinished* wedlock between the word and the
> craft...The arts which endure somehow succeed in
> achieving the right chemistry between the word and
> craft.... (Nwankwo 1995:42)

Nwankwo maintains that Nwapa, partly as a result of a lack of self-consciousness and an absence of a determinate philosophical framework, fails to achieve "a gradual, consistent and coherent vision that taps successfully from history and culture . . . " in the manner of (and here he cites specific writers) Ngũgĩ, Soyinka and Achebe. It should not escape our notice that these writers constitute the very triumvirate which dominates Lindfors's (1990) league-table of canonical African writers (See note 3). One does not need to be a resistant reader to be suspicious. Is this a newer, more sophisticated form of the old masculinist criticism we have been fed *ad nauseam* and are still being asked to accept? Is Nwankwo's presentation of *Efuru's* lacunae [as, for example: "Nwapa's first two novels are in reality involved with the collapse of Igbo traditional structures and values during the early colonial period, but the historical markers for that trauma are unfortunately absent" (49)], a repackaging of the well-worn criticism of Nwapa as being too mundane, too domestic, too trivial, too insignificant? And, for this is really the focus of Nwankwo's critique, as a result of her "proliferating omissions," above all, the absence of "authentic veils of history and culture" (49), Nwapa's narrative fails to convince, to explain, to provide a sustaining and coherent vision, a "consistent development of thought" such as is found in the work of Achebe (50). Nwankwo illustrates this failure by articulating the previously noted hiatus between the two earlier novels and the later ones, asserting that: "After *Efuru* and *Idu*, the *word* becomes misshapen and loses the kind of beauty that historical and cultural environment impart on the truths in art" (46). After them, there is a "negative quantum leap in which the women float without their old anchors and bolsters and beauty and the men renege their old resolve" (50).

I am addressing Nwankwo's article in detail because, firstly, I find it a refreshingly unsentimental appraisal of a writer who, by virtue of her pre-eminence and attractive personal qualities, and the undoubted appeal of the mythological dimension of her work to African and diaspora readers alike, is in danger of becoming an icon. Nwankwo's brand of informed and combative analysis does more credit to the writer than the defensive or celebratory or romantic impulse of so much Nwapa "criticism." It is therefore not so much to refute as to question that I seek to engage with the underlying premises of Nwankwo's article (since to accept them is to agree with his conclusions, which I do not). In this spirit, then, my reading of Nwankwo's use of "craft" as a criterion for value-attribution is that "craft" constitutes a certain essential consistency of vision, a steady accretion of textual significations from one work to the next, culminating in a transcendent moment such as is found in *Anthills of the Savannah* (1987). Moreover, this vision must be self- consciously articulated and willfully deployed so as to produce "a clear ideological tendency" (Ngũgĩ), "a vibrant *ogun*ism" (Soyinka) or "a subtle cultural nationalism" (Achebe 1987:46).

Without denying these writers their exceptionally satisfying aesthetic qualities, does art as process *necessarily* work like this? Is it not all too often in the elisions and omissions, the gaps and dissonances and "infelicities" that the whole problematic of power and hierarchy, the relationship between "tradition" and "modernity" stand revealed? What I am asking, in other words, is whether an aesthetic evaluation based on coherence and closure is the most appropriate to Nwapa's texts, especially the later ones. When Nwankwo criticizes Nwapa for her confessed lack of ambition to be a writer, and "indecisive and tentative gambits when it seems she is on the verge of defining her literary practice" (43), what he reveals is a preconception about the role of a writer which may, indeed, not accord with Nwapa's articulations about her own writing. But is it necessarily a writer's job to *define* their practice? Let us not forget Achebe's own assessment of his literary aspirations as being "to help my society regain belief in itself and put away the complexes of the years of denigration and self-abasement" (1988:30). Furthermore, what one critic condemns as a lack of clarity ("the battle lines are not clearly drawn in a philosophical sense" Nwankwo, 50), another may commend for "ambiguity, complexity and paradox that point to the possibility of negotiation and compromise" (Nnaemeka, 108). Puzzling over the "omissions" of the later texts and their lack of clear connection to *Efuru*, Nwankwo once again makes the insidious comparison with Achebe when he argues that even after Okonkwo's death, "the core of Igbo political pragmatism remained intact," so that "we are able to rationalize clearly why certain things failed...while others survived" (50). This clarity of rationalization, he finds, is lacking in *One Is Enough* and

Women Are Different, in which the characters lack "the verve with which to combat metropolitan circumstances" (50).

My point of contention here is that to use Achebe's *oeuvre* as the touchstone for these novels is misguided and misleading, for two reasons. The first is that it unproblematically posits the value system which hierarchizes texts according to a monolithic and one-dimensional set of critical principles, without either interrogating the system itself or looking at what texts do. The other, related, is that, once we accept the multiplicity of textual production in Africa, the heterogeneity of audience, voice and function, we may be less anxious to invoke the critical pieties of "excellence," authorial vision and posterity. There are, after all, other categories than that of literary "excellence" in the sense invoked here, and Nwapa's later writing is more naturally equated with that of other less self-consciously "literary" writers, from Cyprian Ekwensi to Festus Iyayi and Funmilayo Fakunle. These are writers whose narratives may well give a clearer picture of the moral complexity and ambiguity of contemporary Nigeria than even *Anthills of the Savannah,* precisely because they partake of it. As Chikwenye Ogunyemi (1995) suggests, "The failed marriages and the resultant dysfunctional families are signifiers of the disruptive conditions of a nation that can be categorized as critically ill" (9).

I would therefore argue that, in moving away in her later work from the "traditional" (constituted as solid, secure, value-laden and conservative), and attempting to envisage a subjectivity which is contemporary, urban, self-determining and open-ended, Nwapa herself is unable to reconcile all the contradictions in the chaotic world of post-independence Nigeria. These show up variously as "conceptual" and stylistic flaws, as "inappropriateness" and even, at times, incoherence. To quote Gardner, "...the style, no longer creatively indebted to Igbo oral traditions as in the first two, is flat-footed, sterile and at times incoherent" (Gardner 1994:9). She points to one reason for this—poor editing—which returns us to the question of the packaging and marketing of Nwapa's work. It is no secret that publishers, (especially one, like Heinemann, in a position of ideological arbitration over texts produced in a specific and "other" cultural environment), do often engage in extensive amendments amounting, at times, to rewriting. The difference between Nwapa's earlier and later novels is surely at least partly attributable to this. To say so is not to detract from her artistry or her reputation, simply to recognize the way the imperatives of marketing impinge on, shape and re-order texts, so that the final product is in many cases a collaboration between writer and editor.[5] The Tana novels, then, being relatively unmediated by editorial intervention and not yet critically over-written, afford a kind of access to the raw material of fiction, in that the writer's struggle with representation, with how to write out of a discordant and fluid social context, has not yet been smoothed over or edited out. In this way, the nexus of forces which surround the fictional texts: the competitive capitalism and individualism underlying Nwapa's one-woman entrepreneurial publishing enterprise and her personal engagement, on a material level, with the very pressures and problems she explores in her writing—all these coalesce in the texts themselves in the form of stylistic "infelicities": shifts of narrative voice, inconsistency, repetition, contradictory impulses, moral uncertainty and at times unconvincing characterization.

The Tana publisher's blurb (presumably Flora Nwapa's own) describes One Is Enough as "the powerful and compelling story of a woman's struggle to find an independent and fulfilling life of her own," while Women Are Different is "the moving story of a group of Nigerian women, from their school days together through the trials and tribulations of their adult lives." The former appeared in 1981, the latter in 1986, but both deal with a clearly defined period. One Is Enough is set mainly in Lagos immediately after the Civil War, i.e., the early 1970s, ending with the coup which ousted General Gowon in 1975. Women Are Different covers the same period, but begins in 1945 and traces the lives of three women through the intervening years. The agenda is therefore clear: to provide a context for the "moving," "powerful," "compelling" stories of the female protagonists and thereby to examine the changes that came about in Nigerian social, political and economic life following Independence and the Civil War. In the process, the texts interrogate the role of tradition, the conventions governing marriage and sexual relationships, the strategies necessary for survival in a competitive urban context, the value system inculcated by colonialism and missionary education and its appropriateness in the new situation, the support system women provide for each other, the imperative of motherhood and the differences between the generations. It can be seen immediately that the concerns are both wider and more complex than those addressed in the earlier novels of village life, though there are correspondences, notably the preoccupation with motherhood and fertility, the role of tradition and women's relationships.

Both the critics I cited earlier agree that the Tana novels are stylistically "impoverished" by their distance from the "Igbo expressions" so characteristic of the earlier novels. It is certainly true that the simplicity and directness which strike the reader of *Efuru* and *Idu* as so judicious are replaced here by a simplistic use of language which borders at times on the banal. To dismiss the novels on this account, however, would be to beg the question of why Nwapa adopts this style. If we accept my earlier point about editorial intervention, then possibly what we have in the Tana novels is closer to Nwapa's own style, while the earlier novels benefit from tidying up and paring down to achieve a more "literary" effect. Alternatively, the difference is partly an effect of the narrative content. In this case, it could be argued that the purity of the Igbo-inflected expression, which so tellingly evokes the subjectivity of a non-literate traditional woman, simply fails to translate into a contemporary, urban context inhabited by educated, English-speaking characters. Nwapa, in other words, in looking for a language in which to render the new consciousness brought about by social change, opts for a colloquial, conversational, intimate tone which is both accessible and unobtrusive.

I want to refer here to Karin Barber's (1987) discussion of popular style, in which she confronts the criticism of popular arts as "unsophisticated," "crude," "simple," etc. by asserting what she calls "an aesthetic of immediate impact," which is designed "to appeal to a lowest common denominator of comprehension." It would be, she warns, "a condescension typical of elite criticism to suggest that the accessibility of these works is facile or simple-minded. Their kind of accessibility seems to me rather to be an art in itself" (43). Our various responses to Nwapa's style will therefore be determined partly by our categorization of the novels as "literary" or "popular" writing. This also raises the issue of readership: to whom are these texts addressed? Heinemann's African Writers Series, based in England, produces texts from writers all over the continent and markets them to both a European/American and a literate African readership, and is therefore international in scope. Heinemann texts (notwithstanding Marie Umeh's [1995] point about Heinemann's inconsistency of treatment of its authors), are more likely to be put on reading lists in schools and universities, discussed at conferences and critically assessed in journals, precisely because they are more easily accessible even within Africa itself (22). Indigenously published books are disadvantaged by problems of distribution, which make it all but impossible for readers outside the country to acquire them. On the other hand, a writer publishing at home may be sure of being read at home, in which case questions of the level of literacy of the readership and competence in English (which is a second language for many), will determine the degree of narrative and stylistic complexity. Furthermore, I have argued elsewhere that many women writers take refuge in popular fiction forms, like the romance, in order to evade precisely the kind of elite literary strictures pointed to by Karin Barber. In other words, by their choice of form they consciously place themselves outside the purview of "official" literary and canonical values, privileging instead a rewriting of both social and textual conventions. I wish to suggest that to approach Flora Nwapa's later work in this light is to emancipate both her and ourselves from the embarrassing necessity of celebrating her "good" novels while remaining silent about her "bad" ones. If, instead of literary "excellence" and its obverse, we employ the categories "elite" and "popular," I believe we come closer to understanding how these texts function differently from each other.

Emmanuel Obiechina's (1973) seminal work on Onitsha Market Literature has identified the multiple intertextuality of Nigerian (African) popular fiction. He cites the English canon, especially Shakespeare, the Bible, Indian films and chapbooks and self-improvement manuals, as well as, crucially, orature. More recently, he has made a case for the "elite" African novel as being structured and producing its meanings in accordance with aesthetic models drawn from orature, notably what he calls "narrative proverbs" (Obiechina, 1993). The African novel, he suggests, is "mediated by a communal consciousness and impulse in art and life" (125). Abiola Irele (1990) further insists on the paramountcy of orature and an attendant aesthetic of fluidity and participation in African written literature:

> Despite the undoubted impact of print culture on
> African experience and its role in the determination
> of new cultural modes, the tradition of orality
> remains predominant, serving as a central paradigm
> for various kinds of expression on the continent...(it)
> functions as the matrix of an African mode of dis-
> course...(and) thus represents the basic intertext of
> the African imagination. (Irele 1990:56)

This basic intertextuality, clear to see in *Efuru* and *Idu*, is less obvious in *One Is Enough* and *Women Are Different*, but it is nonetheless there. Both these novels employ proverbial sayings as a key to what Obiechina calls the "epistemological order within the novel" (Obiechina 1993:127). In *One is Enough* they are: "The day breaks for different people at different times" and "God's time is the best." In *WAD*, which urges foresight and looking to the future as the key to successful survival, the emblematic utterance is: "Any time you wake up is your morning." Taken together, these sayings elucidate the dominant preoccupation of the novels with change, a key feature of all popular culture.[6] In *OIE*, the utterance occurs at the end of a section of the narrative dealing with the changes in the life of the heroine, Amaka. When the novel opens, she is apologizing to her mother-in-law for her failure to conceive. In rapid succession, she learns of her husband's intention to take a second wife (by whom he already has two sons), quarrels with him, attacks him with a hammer, and leaves. Three years later, as she attempts to analyze the transformation she has undergone, she has just celebrated the birth of twin boys with a lavish party for her village people, paid for out of money earned as a successful contractor in Lagos. At the party, unidentified voices discuss the tradition of "throwing away" twins, now so thoroughly eradicated that Amaka's ex-mother-in-law even tries to claim them for her son. Her son, like everyone else, is impressed by Amaka's wealth and generosity, in which she conforms to traditional expectations and the belief that, "the whole village had a right to your wealth, because anybody among them could have had that wealth instead of you" (*OIE*, 116). However, Amaka's mother, who, though representative of an older, illiterate, traditional generation, is unconventional in many respects, takes pride in her daughter's countermanding of the traditional wisdom that, "You either had children or you had wealth," and that wealth for women "blocked the chances of having children" (*OIE*, 116). Nwapa here spells out the advantages to women of combining the traditional with the contemporary to create an identity of their own choosing. Amaka, far from being compensated for childlessness by wealth, achieves wealth and motherhood simultaneously. The "dictates of the time" (115), that Amaka should show off her wealth through ostentation, accord quite comfortably with the village's insistence on sharing. Furthermore, Amaka's reluctance to reveal the identity of her children's father is heartily endorsed by her mother, who states, "She would either have a husband or her business, she could not have both" (*OIE*, 118). Thus, the traditional emphasis on economic independence (which Amaka's aunt, for example, achieved only after giving her husband children and then finding him another wife so as to liberate herself from domesticity), is achieved by the younger woman without these constraints. Amaka's response to her ex-husband when he visits her is indicative of the radical, social and personal transformation facilitated by the new opportunities of post-Civil War Lagos. She finds herself confronting a series of questions, to none of which can she provide an answer:

> Was this the man with whom she had lived for six years?...Why did she feel nothing toward him anymore? Was it time? Was it Izu, the father of her twins? Was it her new position of wealth? The life she lived in Lagos? (*OIE*, 119)

Instead of answers, Amaka falls back on the philosophical framework of her village home: "She did not regret anything. Surely, God's time was the best. The day broke for different people at different times, according to the saying of her people. For her, day broke three years ago..." (*OIE,* 119).

I have used this passage as emblematic of the narrative impulse of the whole novel. It concerns change, personal and social transformation, and the specific way these processes impact on female subjectivity. There is no simplistic opposition in this text between tradition and modernity, older and younger, moral and immoral. Rather, each of these is a constantly shifting element in a larger configuration, characterized by a multiplicity of possibilities, identifications and complexities. The result, unlike the quasi-mystical closure of *Efuru*, is open-endedness, a sense that all solutions are at best temporary in a world of flux. However, an important point of continuity with *Efuru*, and another facet of the intertextuality with orature, is Jell-Bahlsen's (1995) account of Nwapa's protagonists as all being "guided and influenced by the water goddess," Uhamiri, the woman of the Oguta Lake. Jell-Bahlsen (1995), by enumerating the many conspicuous parallels between the characters of Efuru, Idu, Amaka, Kate, Dora, Rose and Agnes, makes a convincing case for Uhamiri as semiotic counterpoint of all of them, an explanation which also goes some way towards answering Nwankwo's point about the apparent rupture between the earlier and later texts (34). Although less convincingly demonstrated, a further intertextual example is provided by Brenda Berrian (1995), who cites "marketplace humor" or "*njakiri,*" as a "literary ploy" drawn by Nwapa from Igbo oral culture, defined as a "forward and backward exchange of dialogue," eliciting either "a celebratory response in support of the character's decisions and actions or a resistant response. . ."(57). These three examples together point towards one of the important ways in which these texts *function* as popular or unofficial writing. Through their interface with orature, although not as explicit as in *Efuru* and *Idu*, they use tradition as a way of commenting on social change, a change, moreover, in which tradition itself is implicated. (Amaka's mother's desire for her daughter's marriage to the father of her twins is an example of this, since the older woman, who is also closer to the traditions, is shown as adapting easily to the conditions prevalent in urban culture, her pragmatism only one step from Cash Madam materialism.) The questioning of tradition which began with *Efuru* provides a context within which we can begin to make sense of the disconcerting endorsement of corruption and immorality in *Women Are Different*:

> Chinwe had done the right thing. Her generation was doing better than her mother's own. Her generation was telling the men, that there are different ways of living one's life fully and fruitfully. They are saying that women have options...marriage is not THE only way. (*WAD*, 119)

It is important to notice that what is being juxtaposed here is not "traditional" and "contemporary" values, but the missionary school ideals which have so signally failed the generation which graduated from Archdeacon Crowther's Memorial Girls' School at the beginning of the 50s, and the strategies evolved by their daughters for survival in the 1970s. Moreover, these strategies, with their emphasis on industriousness, lack of sentiment, pragmatism and economic success, are seen as closer to indigenous "tradition" than the imposed colonial ideal of feminine self-sacrifice and submission. Echoing Ajanapu's contemptuous treatment of her sister, *Efuru*'s mother-in-law, when she taunts: "You wanted to be called a good wife, good wife when you were eating sand, good wife when you were eating nails...It was not virtue, it was plain stupidity,"(*Efuru*, 79) Amaka's aunt in *One Is Enough* uses her own story as a monitory image. After seven children, she got a new sixteen year old wife for her husband, and turned her attention to business. She measures her success by her daughters' happiness:

> My daughters have rich husbands. I planned all the marriages. They are happy with their husbands, but

I say to them, never depend on your husband. Never slave for him. Have your own business, no matter how small, because you can never tell. (*OIE*, 9)

When Amaka's mother endorses this view, and encourages her to take a lover ("Marriage or no marriage, have children," [11]), Amaka ponders the contradictions of her situation:

> That was not what she learnt in her few years at school. The good missionaries had emphasized chastity, marriage and the home. Her mother was teaching something different. Was it something traditional which she did not know because she went to school and was taught the tradition of the white missionaries? (*OIE*, 11)

This dilemma, which is central to both *One Is Enough* and *Women Are Different*, is symptomatic of the situation in which Nigerian women (the educated, middle-class, elite women of Nwapa's fiction, as well as those of Ifeoma Okoye's, Zaynab Alkali's, Buchi Emecheta's, Helen Ovbiagele's, et. al.) find themselves. It is to this dilemma—of how to piece together an identity and a modus vivendi out of the multifariousness of contemporary experience—that the texts speak. The dilemma is articulated by Amaka early in *One Is Enough* as follows:

> Was that really the end of the world? Was she useless to society if she were not a mother? Was she useless to the world if she were unmarried? Surely not. Why then was she suffering these indignities both from her husband and his mother?(*OIE*, 20)

Simply, almost naively expressed, the notion of an appropriate femininity which underlies the questions Amaka asks herself is one that nonetheless has great power to affect behavior, self-esteem and ultimately identity itself. Though the world that Achebe invokes in *Anthills of the Savannah* is as much in flux as Amaka's, the difference is that his characters (including the women), are possessed of certainties which direct them, even against the tide. Beatrice, in her identification with Idemili, has an archetypal fixity which precludes the open-ended questioning of Nwapa's women. I do not mean to imply a value judgement here, but only to point once again to the difference in textual function. Nwapa's texts raise questions about identity, and specifically constructions of femininity, which are, in a sense, unanswerable. *Efuru*, even with its mythic dimension, ends on a question: "Why then did the women worship her?" Twenty years later, the question which concludes *Women Are Different* has the same rhetorical suggestiveness. Rose, surveying her old school friends, sees herself as anomalous:

> Dora had come to terms with Chris and has her children; Agnes lost her lover but she has her husband and children; Olu always went back to his wife after each affair; Chinwe and Zizi had their youth to show. Even Tunde cherished a dear dead wife. But Rose, what had she? (*WAD*, 138)

As with *Efuru*, the question has already in one way been answered by the narrative itself: Rose has independence and good friends. She, like *Efuru*, is only a failure if, following Amaka's formulation, to be childless and unmarried is to be useless to society. The whole force of Nwapa's writing runs counter to this conventional construction of the value of femininity, but without being able to offer an utopian solution. The fact that all Nwapa's central protagonists end up alone suggests that happy endings are in some way seen as inauthentic or inappropriate. In her last two novels, Nwapa in fact sets up happy endings, but is apparently unable to carry them through. Amaka *could* have married Izu, Rose *could* have had a love affair with Tunde, but neither is permitted to do so. Is this symptomatic of Nwapa's assessment of the position of women in contemporary Nigeria—that there is no emotional satisfaction to be gained from men? Certainly a key factor in all the novels is the strength of women's relationships, whether teacher/pupil, boss/secretary, mother/daughter or as friends. Beyond this, however, there is no moral certainty (as in *Anthills*), to be ritually passed from one generation to another. Each generation remakes itself in a context of change, while taking what it can from the previous generation. This lack of certainty means that the moral world of Nwapa's later novels is ambiguous. She apparently endorses the new freedom of choice linked to educational and financial independence, while rejecting the new values accompanying that freedom. She endorses motherhood, but not marriage or emotional security through sexual partnerships. In other words, as I have already said, she deals pragmatically with the complexities of women's situation, without offering romantic solutions. What, if any, is the value of this?

It seems that as critics we feel impelled to make claims for the writers and texts we espouse. Often these claims are partisan, dictated by an emotional perspective which feels itself embattled or belittled by the hegemony of the critical establishment and its smug canonicity. Nwapa's novels, but especially the later ones, side-step the values of that establishment through ambiguity and lack of closure. It is in this sense that I find it useful to describe them as "popular" or "unofficial" writing. Like Ama Ata Aidoo's *Changes* (1991), Nwapa plays with the romance formula, invoking it through her characters' feelings, subverting it by avoiding the happy endings that seem close to hand. In so doing she, like Aidoo, acknowledges the powerful need for love and emotional satisfaction, but, again like Aidoo, suggests it cannot come from men under the current social system. Nwapa offsets the romance with realist elements of history and autobiography, most notably in the school section of *Women Are Different,* based on her own school experience. In seeking to represent "a world gone mad" (*Women Are Different,* 130), she freely crosses generic boundaries, disregarding conventions and creating new juxtapositions which highlight the fluidity of form and content. Whether this is part of a conscious authorial project is scarcely the point. The point is that the texts themselves, precisely because they do not prescribe a coherent philosophical and moral universe, are far truer indicators of the state of things in Nigeria than if they did. Meanwhile, what the texts celebrate is what most ordinary people struggle for—survival—an outcome which cannot be taken for granted, as the author's own death bears witness. In Nigeria, a writer may die needlessly for lack of drugs; another is in jail under threat of execution for defending his people against exploitation; another has his passport taken away and has to flee the country; another narrowly escapes death in a car accident on roads which are scarcely maintained. Nwapa's narratives function within this context of a crumbling infrastructure, a cynical, self-seeking government and daily violence perpetrated on ordinary people. There is no facile happy ending. As Comfort, the "township girl," says in *Women Are Different,* "To go Lagos no hard, na return" (5)—a statement that could stand metonymically for the life choices of all Nwapa's protagonists.

NOTES

1   See Adewale Maja-Pearce, "In Pursuit of Excellence: Thirty Years of the Heinemann African Writers' Series." *Research in African Literatures* 23. 4 (1992): 125-132.

2   See especially Obioma Nnaemeka' s "From Orality to Writing," *Research in African Literatures*. 25.4 (1994): 137-157, for a challenging discussion of the whole question of tradition and conformity, and *Research in African Literatures*. 26. 2 (1995), a special issue on Flora Nwapa co-edited by Chikwenye Okonjo Ogunyemi and Marie Umeh.

3   See Bernth Lindfors, "The Teaching of African Literature in Anglophone African Universities," *Wasafiri* 11 (Spring 1990): 13-16.

4   Chinua Achebe, "The Novelist as Teacher" 27-31, and "Racism in Conrad's *Heart of Darkness*" 1-13, in *Hopes and Impediments* (Oxford: Heinemann, 1988).

5   The case of another Nigerian writer, Buchi Emecheta, provides a useful point of comparison with Nwapa, in that her earlier and more celebrated novels, *The Slave Girl* (1977), *The Bride Price* (1976) and *The Joys of Motherhood* (1979) were extensively reworked by her London-based publisher, Allison and Busby. Her subsequent novels, produced by her own press, Ogwugwu Afor, exhibit the same stylistic deficiencies as Nwapa's Tana novels, but her reputation is such that all her work is in demand and highly marketable. Hence, the situation has arisen whereby her latest novel, *Kehinde,* (1994), was again extensively rewritten by Heinemann because she was seen as a financial asset to the company. The implications of this kind of intervention for the author, for women's writing and for African literature are serious, pointing as they do to the continued hegemony of a value system alien to the social configuration out of which the texts emerge.

6   Karin Barber (op. cit.) argues that popular art is "a new kind of art created by an emergent class, the fluid heterogenous urban mass. Located at the perceived source of social change, popular art was both produced by a new situation and addressed to it. The new urban mass faced both ways...the syncretism of their art...was therefore an expression and a negotiation of their real social position at the point of articulation of two worlds" (Barber, 14).

Works Cited

Achebe, Chinua. *Things Fall Apart.* London: Heinemann, 1958.

___.*Anthills of the Savannah.* Oxford: Heinemann, 1987.

___. *Hopes and Impediments: Selected Essays 1965-87.* Oxford: Heinemann, 1988.

Aidoo, Ama Ata. *Changes.* London: The Women's Press, 1991.

Andrade, Susan Z. "Rewriting History, Motherhood and Rebellion: Naming an African Women's Literary Tradition." *Research in African Literatures* 20. 1 (Spring 1990): 91-110.

Barber, Karin. "Popular Arts in Africa." *African Studies Review* 30, 3 (Sept. 1987): 1-78.

Berrian, Brenda. "The Reinvention of Woman through Conversations and Humor in Flora Nwapa's *One Is Enough." Research in African Literatures* 26. 2 (1995): 53-67.

Bryce, Jane. "Women and Modern African Popular Fiction." *Readings in*

Jane Bryce

African Popular Culture. Ed. Karin Barber. London: International African Institute and James Currey (Forthcoming 1996).

Cary, Joyce. *Mister Johnson.* Carfarx Ed. London: Michael Joseph, 1949.

Conrad, Joseph. *Heart of Darkness.* London: Norton, 1988.

Ekwensi, Cyprian. *Jagua Nana.* Oxford: Heinemann, 1987

Emecheta, Buchi. *The Bride Price.* London: Allison & Busby, 1976.

___.*The Slave Girl.* London: Allison & Busby, 1977.

___.*The Joys of Motherhood.* London: Allison & Busby, 1979.

___.*Kehinde.* Oxford: Heinemann, 1994.

Emenyonu, Ernest. Review of *Efuru. Ba Shiru* 1.1 (1970): 58-61.

Gardner, Susan. "The World of Flora Nwapa." *The Women's Review of Books* XI. 6 (March 1994): 9-10.

Gikandi, Simon. *Reading Chinua Achebe: Language and Ideology in Fiction.* London: James Currey, 1991.

Irele, Abiola. "The African Imagination." *Research in African Literatures* 21. 1 (1990): 49-67.

Jell-Bahlsen, Sabine. "The Concept of Mammywater in Flora Nwapa's Novels." *Research in African Literatures* 26.2 (1995): 30-41.

Larson, Charles R. *The Emergence of African Fiction.* Bloomington: Indiana Univ. Press, 1971.

Lindfors, Bernth. "The Teaching of African Literature in Anglophone African Universities." *Wasafiri* 11 (Spring 1990): 13-16.

Maja-Pearce, Adewale. "In Pursuit of Excellence: Thirty Years of the Heinemann African   Writers' Series." *Research in African Literatures* 23. 4 (1992): 125-132.

Mojola, Yemi. "The Works of Flora Nwapa." In *Nigerian Female Writers: A Critical Perspective.* Ed. Henrietta C. Otokunefor & Obiageli C. Nwodo. Lagos: Malthouse, 1989, 19-29.

Nnaemeka, Obioma. "From Orality to Writing: African Women Writers and the (Re)Inscription of Womanhood." *Research in African Literatures* 25. 4 (1994): 137-157.

Nwankwo, Chimalum. "The Igbo Word in Flora Nwapa's Craft." *Research in African Literatures* 26. 2 (1995): 42-52.

Nwapa, Flora. *Efuru.* London: Heinemann, 1966

___.*Idu.* London: Heinemann, 1970.

___.*One Is Enough* Enugu: Tana, 1981

___.*Women Are Different.* Enugu: Tana, 1986.

Obiechina, Emmanuel. *An African Popular Literature.* Cambridge: Cambridge Univ. Press, 1973.

___. *Culture, Tradition and Society in the West African Novel.* Cambridge: Cambridge Univ. Press, 1975.

___. "Narrative Proverbs in the African Novel." *Research in African Literatures.* 23. 4 (1993): 121-140.

Ogunyemi, Chikwenye Okonjo. "Introduction: The Invalid, Dea(r)th, and The Author: The Case of Flora Nwapa, *aka* Professor (Mrs.) Flora Nwanzuruahu Nwakuche." *Research in African Literatures* 26.2 (1995): 1-16.

Otokunefor, Henrietta C. and Obiageli C. Nwodo, eds. *Nigerian Female Writ-*

ers: A Critical Perspective. Lagos: Malthouse, 1989.

Umeh, Marie. "The Poetics of Economic Independence for Female Empowerment: An Interview With Flora Nwapa." *Research in African Literatures* 26. 2 (1995): 22-29.

Woolf, Virginia, "Women and Fiction." *On Women and Writing.* London: The Women's Press, 1979, 43-52.

"To the Root/of Dear Cassava": Rhetorical Style in Flora Nwapa's Poetry

Obododimma Oha

'Let us go back to our song
Where we have left it
To the root
of dear Cassava.'
—Flora Nwapa (*Songs*, 29)

INTRODUCTION: SINGING HER WORK

In Igboland, as in some other African cultures, it is common to see women singing while working.[1] Or to be more poetic: in Igboland, women *sing their work*. A woman who *sings her work* is assumed to be hardworking, even though the content of her work song may parallel the content of the spirituals sung by Africans who worked as slaves on white American plantations during enslavement. And within the context of patriarchal culture in which the woman's labor is exploited by the man, the work song itself seems to be a signifier of the gender oppression, suggesting the narrative of the woman as a "caged bird" that sings (to echo Maya Angelou).

Is the female writer in a male-dominated culture not also a representation of the woman-laborer who sings her work? Or precisely, is the female poet who is conscious of the deprivations and derogations the woman suffers in society, and who focuses on such in her poetry as a way of conscientizing society, not conceivable as a singer of her work?

Generally, the conceptualization of creative writing as "work," especially in the context of feminist commitment, is more than just a metaphorization. According to Helene Cixous in "The Newly Born Woman,"

> Writing is working; being worked; questioning (in) the between (letting oneself be questioned) of same and of other without which nothing lives; undoing death's work by willing the togetherness of one-another—not knowing one another and beginning again only from what is most distant, from self, from other, from the other within. (Cixous/Sellers 1994:43)

Singing her work (or *working her song!*) is a rhetorical endeavor; it is an activity in language, in which language is not just a means of persuasion but also a symbol of the female writer's perception of and attitude towards gender relations. As many feminist scholars like Julia Kristeva, Luce Irigaray, and Helene Cixous have pursued, the revolution in gender relations could also involve linguistic revolution (creative writing providing an opportunity for the feminist writer in this case). The revolution has been given a psychoanalytical dimension by the French feminists. Kristeva (1989), for instance, asserts that the "experience of the semiotic chora in language produces poetry" and insists on "searching for the inscriptions in language of the archaic contact with the maternal body which has been forgotten" (132-2). Rhetoric in this regard may not benefit much from logic—that is, logic as packaged and/or practiced in the masculine symbolic. In fact, Kristeva rejects "logical communication" as long as it suppresses or encourages the suppression "the maternal and the primordial link every subject has with the maternal" (131).

Poetry seems to provide a good opportunity for this celebration of link with the maternal body. Commenting on Kristeva's work in this regard, Sellers (1991) asserts:

> Poetic language in particular, since it incorporates the
> unconscious and body rhythms in a way other forms
> of language do not, offers a means of subverting the
> symbolic function by putting the subject into pro-
> cess—with himself as well as with the law. (51)

In "A Question of Subjectivity," Kristeva (1989) also argues that the
"experience of the semiotic chora in language" (which, according
to her, "produces poetry") could be seen as "the source of all stylistic effort, the modifying of banal, logical order by linguistic distortions such as metaphor, metonymy, musicality" (131).

Rhetorical style in Flora Nwapa's poetry seems to put these
feminist linguistic issues, particularly the "maternalization" of
style, to some test. Nwapa feminizes and maternalizes cassava (a
crop/foodstuff), thus stylistically calling attention to both the
m/other's role and suggesting relation of the feminine writer to
the maternal and the choric.

As a writer, Nwapa is better known as a novelist, and her novels have been given much attention. Her career as a poet, however, has not been much assessed critically, partly due to the
marginalization of women's poetry in Africa, and partly due to the
regard of poetry as being peripheral in her career. Existing commentaries on her poetry have also deplored her style. Mojola
(1989), for instance, has argued:

> While the ideas in these "Songs" are valuable, the
> compositions, especially "Rice Song," lack the essen-
> tial qualities of poetry such as economy of language
> and the use of symbolic expressions and imagery.
> "Cassava Song" possesses a few characteristics of
> poetry but "Rice Song" is absolutely prosaic. (27)

It is, first of all, significant that Mojola takes only one paragraph of
fourteen lines to comment on Nwapa's poetry and does not provide any analysis of the poems to back up the claims made.

It is my objective in the present essay to study rhetorical style in
Nwapa's *Cassava Song and Rice Song* (henceforth, *Songs*) more extensively and properly within the context of contemporary discourse on
feminine writing. The essay will be based on the conceptual frame
of the female poet as a singer of her work, in which case rhetorical
style will be discussed in relation to the rhetorical constraints of genre
and register, "authorial ideology," "aesthetic ideology" (Ngara 1990:11-
12), context of culture, and rhetorical purpose.

RHETORICAL CONSTRAINTS

In order to be able to appreciate Nwapa's style in *Songs*, it would be necessary to study the poems in relation to the constraints of genre, ideology, and social context in which they were produced. Such considerations, in fact, are already a given in non-formalistic stylistics and discourse analysis. As Ngara (1990) has pointed out in *Ideology and Form in African Poetry*:

> The understanding of poetry requires an understanding and appreciation of historical and social conditions, ideological factors, literary forms and devices and, of course, a sufficient mastery of the language in which the poetry is written. (16)

These considerations are necessary especially as the literary text is a part of the social process of communication and therefore presupposes the existence of context.

One of the elements that conditions Nwapa's style in *Songs* that we need to consider is the frame used. The frame chosen by the poet is "song." This frame, on the one hand, alludes to the early (performed) nature of poetry, and to (implicit) cultural posture. Poetry, as Denys Thompson (1978) has observed in *The Uses of Poetry,* has early origins as *Songs.* According to him, "Songs to accompany every activity of the working day are among the oldest of all poetry" (28). These work songs exist in every culture, but to those of us who are Africans, it would seem more than familiar as a very significant and socially-based art form. It would also be quite familiar to the African woman who has an authentic rural experience. In fact, one could say that work songs are a poetic genre that African rural women have used in demonstrating their skills in literary composition and performance.

Work songs also point to the social and psychological relevance of art. Being socially based and, of course, being constrained by the work context, the songs are often direct in communication. Use of language in the songs is not "sophisticated" and does not need to be. They are meant "to lighten labor and increase the efficacy of work" (Thompson 1978:34). Thus it is clear that it is in the nature of the genre of work song to "entertain" its singer and attend to her psychological needs, and to do so in simple style.

A (modern) written poem that takes the frame of work songs[2] thus requires a reconciliation of modal features. In some cases, some poets have settled for a predominance of the characteristics of written poetry. In some other cases too, some poets have tried to retain the simplicity of language which typifies the register of work song as being different. However, in both cases, we still have a "recontextualization" or "register-switching" (Wales 1990, 398) in which the oral song is made a written text that could be read privately. For some poets, as I mentioned above, this change of mode is so significant that the syntactic and semantic aspects of style would also have to change so that the text would not be that of the Barthesian *ecrivant*. The maintenance of closer relationship with the oral mode, however, does not necessarily diminish the integrity of the poetry. It is the inability of a critic to recognize the influence of the oral frame that rather threatens the integrity of the poem.

The point being made therefore is that Nwapa's *Songs* maintains this affinity with the work song (which frames it) and so we cannot judge its style outside of this relationship. Also, inside the text, our attention is called to this relationship in a subtle way and the poet's posturing as "singer" is also suggested, as could be seen in the epigraph to this essay.

Thus the "aesthetic ideology" or "the literary convention and stylistic stances adopted by the writer" (Ngara 1990:12) in *Songs* is that of orality. She chooses to return to the traditional and rural performance of poetry. Such a return is not surprising especially when, to many post-colonial writers in Africa, as Okpewho (1988) has stated, it seems to be an intellectual obligation:

> For no honest person would wish to forget the cultural traditions out of which he has grown and which constitute the background to his intelligent responses. (23)

Those who are familiar with Nwapa's works would easily identify such inclination and the fact that she often tries to locate the woman and woman's roles in the culture.

The last point made above leads me into a consideration of the role of "authorial ideology" or the "stance the writer takes" which "determines the poet's perception of reality" (Ngara 1990:11) in *Songs*. Nwapa's concern with the position of the woman in society is feminist, and in her novels she has sought to present her female protagonists not just as sufferers of masculine exploitation and abasement, but also as questing subjects. In *Songs* we also have the female voice and presence, which signify both the female writer as participant in social discourse (a de-silencing of woman) and revaluation of the image of woman in a male-dominated society.

Feminist literature, as Keitel (1989) has argued, is a mediating text in the sense that it seeks mainly to "consolidate group understanding" by presenting an authentic female experience (14). Such an inclination, says Keitel, often makes representation inevitable in the feminist text. However, being a mediating text, at least covertly, does not necessarily mean that *Songs* would be unchallenging stylistically to the reader. We should rather expect certain mechanisms of meaning that suggest the poet's attempt at deconstructing androcentric characteristics in the culture.

Indeed, the patriarchal nature of (the context of) culture significantly affects Nwapa's rhetorical style in *Songs*. Food culture in Igboland (at the time of the writing of *Songs*) shows some societal dislocation. Cassava is a cheap and rather local foodstuff in Igboland, particularly in Nigeria. However, due to some colonial mentality manifested in the preference for foreign things, (some) Nigerians tend to view cassava as inferior to (imported) rice. The preference, as I have argued elsewhere (Oha, 1997), also signifies class consciousness: even though these Nigerians are not that rich, they would not like to be seen as poor. To eat cassava is to mark oneself out as poor! Moreover, cassava is derogated as being odoriferous. And to be a consumer of what is odoriferous is to suggest oneself as uncivilized or backward. (However, some Nigerians who reject cassava openly, eat it in secret!)

What is even more useful to contemporary intellectual discourse is the dimension of gender presupposed in the cultivation and cooking of cassava in Igboland. For instance, the woman, as a farmhand, is identified with the roles of cultivating and cooking cassava in Igbo culture. The androcentric logic in this case seems to be a process of analogy. The supposed inferiority of the woman is linked with the inferiority of cassava; that is, the prejudice against the woman is extended, to enable the man to emphasize her otherness. In this case, it is not even a matter of seeing the woman in terms of the man as Simone de Beauvoir (1972) has stated in *The Second Sex:*

> She is defined and differentiated with reference to
> man and not he with reference to her; she is the
> incidental, the inessential as opposed to the essen-
> tial. He is the subject, he is the Absolute—She is
> the other. (16)

Seeing the woman in terms of cassava distances her further from the man, making her difference more significant.

Being a feminist writer, Nwapa naturally finds this signification of gender prejudice in Igbo food culture a useful material for
her poetry. Since the signification is androcentric, she tries to
undermine it by exploiting the notion of difference in a contemporary feminist sense, instead of rejecting it. The difference in this
case becomes an advantage for the woman—as caring, rational,
and capable of showing society the right path to take.

Thus, in *Songs,* she tries to elevate the position of cassava (as
mother) by eulogizing it and to condemn preference for and sub-
stitution of cassava with imported rice. In other words, rhetorical
style in *Songs* is modelled on binary opposition at the macro-level,
as shown in Figure 1.

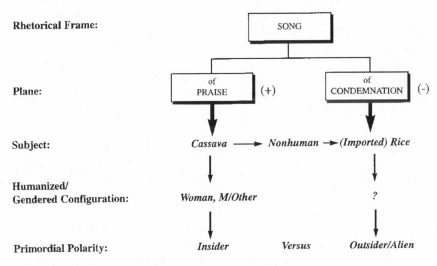

Figure 1. Binary opposition at the macro-level in
   Cassava Song and Rice Song

In the analysis that follows, I will identify and discuss the linguistic devices the poet has used in her rhetorics of praise and condemnation.

THE RHETORIC OF PRAISE

Praising, as an illocutionary act, is normally used in expressing
approval for a subject's (social) action, posture, etc., and to encourage such. Often what is praised is the positive, that is, in ideal rational situations. Praising, therefore, is an illocutionary act that

promotes the addressee's "positive face," or the desire to be seen as a desirable person (Brown & Levinson, 1978). In the words of Lim and Bowers:

> Positive face is people's want that their goals, possessions, and achievements be ratified, understood, approved of, liked, or admired. Positive face want, then, is supported by expressions of understanding, affection, of solidarity or by positive evaluation or of formal recognition of one's qualities. (420)

In *Songs* the persona (who is obviously female) accomplishes the task of praising cassava by *presenting* its (her) desirable qualities. In the main, cassava is humanized. In this case, linguistic elements that are naturally used in referring to human beings are used in referring to cassava, the most common being the word "mother":

> Great Mother Cassava
> Bear with us
> Henceforth
> We shall sing your praise.
> (*Songs*, 4)

To attribute human qualities to cassava (a non-human entity) is to elevate the latter, more so when those qualities are humane. Also, these qualities are feminine. This dimension of gender is indeed very significant, especially as it subtly evokes certain extra-textual conditions which would affect the reading of the text.

Certain questions become pertinent in this case. Why should cassava be configured as feminine or why should it be referred to as "Mother"? Why is it not perceived as "Father"? And would it also not make a difference if Nwapa had used no gender-indicating term?

The point then is that gender is crucial in the rhetoric (of praise) in *Songs*. On the one hand, the presupposed values of the (feminine) gender used are necessary for a revaluation of cassava which, as I pointed out earlier in the exploration of rhetorical constraint, is culturally derogated and associated with womanhood. On the other hand, the praise of cassava is (indirectly) the praise of the feminine. The metaphorical process, which indeed interfaces with personification, is such that it is bi-directionally "transformational." As Kirwan has rightly argued, in metaphorization,

> The subsidiary subject is also modified by its rela-
> tionship with the principal subjects....The subsidiary
> subject, then, both changes its significance under the
> influence of the principal subject and 'filters and trans-
> forms' the usual associations of that principal subject,
> selecting and emphasizing what might be ignored or
> even suppressed by a different metaphor. (51)

Thus, by mapping the qualities of "Mother" onto cassava, these qualities are equally being celebrated and motherhood also being authenticated.

It is also quite significant that the "goodness" of the M/Other or the feminine is presupposed in the persona's Mother-Cassava analogy. The persona (implied author) assumes that the reader appreciates (or shares with her) the knowledge of the values of motherhood, and on the basis of such knowledge would under-stand and appreciate the values of cassava. Of course, in the Igbo (African) milieu of the "Songs," motherhood, as Acholonu (1995) has argued, is a very important factor in the life of the individual and of the society:

> Motherhood and childbearing are central to the life
> of African people. It is not an overstatement that
> motherhood is the anchor, the matrix, the founda-
> tion on which all else rests in the African society,
> and especially the family. (31)

In *Songs*, therefore, there is an evocation of mother-bond and moth-er-love, which are essential reflections of the attempt by the femi-nine writer to "write the m/other's body," to return to the pre-symbolic *Chora,* to the "semiotic"—which, as Kristeva (1989) says, is repressed in the masculine symbolic (131). As Sellers (1991) has also said, Cixous, just like Kristeva, believes that the writing of the mother's body is "a primary factor in preventing the codes of the patri-archal symbolic from becoming rigidified and all-powerful" (140).

To further foreground the maternal presence, the word "Mother" is used quite frequently in "Cassava Songs." Reiteration of a lexical item in a text not only enhances cohesion, as Halliday and Hasan have shown, but it also calls a reader's attention to essential areas of meaning. To emphasize "Mother" is to empha-size the feminine, also to emphasize (the expression of) the bond with the maternal. Such emphasis obviously is a way of challeng-ing that which attempts to silence or suppress the maternal bond.

419

Indeed the reiteration device could be seen as performing a psychological function of producing memorability. In this case, the poet causes the audience to *remember* the m/other, a subject that patriarchalism struggles to suppress, to cause people to *forget*.

The rhetoric is thus designed to express affection for and solidarity with motherhood indirectly. It is an endeavor meant to increase *liking* for (mother) cassava—in other words, a rhetoric that must also necessarily use credibility tactics. Among the qualities of Mother Cassava are her simplicity and humility.

> We plant your stem
> Throughout the year
> We clear the farm
> Only once a year.
> In a short time
> You grow very big
> In a short time
> You are ready for harvest.
> You make no fuss
> You are most humble
> You are most kind
> To your children.
>   (*Songs*, 3)

The simplicity and humility of cassava are seen as qualities of greatness, not of subordination and inferiority. The persona, using the comparative degree, places cassava over and above other plants like yam and cocoyam that are often given a higher position in the Igbo agro-culture:

> The yam is great
> But you are greater
> The cocoyam is great
> But you are greater.
>   (*Songs*, 3)

The use of the comparative degree enables the poet to thematize difference, for generally we would say that all the plants are useful and desirable. But it is the feminine qualities of cassava that make a difference. Difference, in contemporary feminist thought, is no longer viewed as signifying feminine inferiority but as signifying feminine existential credibility.

To further substantiate the greatness of (Mother) Cassava—for indeed the poet attempts to provide rhetorical proofs—the poet identifies the role of cassava in a socio-historical experience. The rhetorical proof in this case is the role of cassava as a "savior" of many starving persons in the embattled Republic of Biafra:

> ...during the war
> The people looked for this
> Special species
> All over the place.
> You saved the people
> Mother Cassava
> From hunger and certain death
> Great Cassava.
> (*Songs,* 31)

It is indeed true that many people, particularly soldiers fighting in the bushes, lived on just raw cassava tubers for days during the Nigeria-Biafra war.[4] It is therefore not surprising that the poet configures the crop (Mother Cassava) as "Saviour" *(Songs,*9). A savior gives her life that the other might live. In the same way, (Mother) Cassava gave her life that people might live. The analogical reasoning may appear simplistic, especially if we do not take cognizance of the underlying "sexual politics" (Moi), but it does satisfy conditions for literariness.

Furthermore, the reference to (Mother) Cassava as "Saviour" (with capital "S") may also suggest a deconstruction of the rigid masculine interpretation of "the Messiah" as not-female. It liberates the term from androcentrism and, when viewed alongside the maternal role, enables us to understand better the basis for the feminization of cassava.

Lexical items like "Saviour" naturally give positive face to the referents mainly because of the intertextual relations involved. The image of the Savior, which in Christianity *reminds* us of Jesus Christ, is thus a trope that enhances persuasiveness, more so if the reader is a Christian who shares knowledge about somebody laying down his life for others (sinners).

Also, as schematized in Figure 2, the image of (Mother) Cassava is enhanced through the use of epithets (E). Cassava appears in the text alternatively as "Great Mother Cassava," "Great Cassava," "Great Woman," "Great Mother," "Mother Cassava," and "Dear Cassava." The most frequent epithet used in this case therefore is the word "great." The descriptive adjective is meant to positively modify our concepts of "Mother" and of "Cassava,"

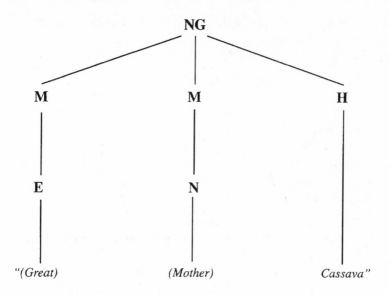

Figure 2. The Use of Premodification Tactic in Praising Cassava

Both "Great" and "Mother" pre-modify "Cassava," and both are optional elements of the Nominal Group (NG). This optionality explains why the poet uses "Great Cassava" and "Mother Cassava" alternatively (though not synonymously). The use of the "M" elements together is an intensification of emotional impact which the rhetoric of praise indeed necessitates.

At another level, the NG "Mother Cassava" (with capital "M") behaves like a proper noun/name; in other words, the status of "Mother" as a mere modifier is stylistically suppressed. "Mother" becomes a (natural) collocate of "Cassava." The gendering of cassava appears natural.

Generally, therefore, Nwapa's use of the rhetoric of praise in *Songs* at the surface promotes positive face for cassava as crop/foodstuff. But at the deep structure, it is indeed the positive face of the feminine (the woman) that is addressed. In other words, as an indirect speech act, *Songs* praises womanhood which could refer the reader to some background knowledge or "'members' resources" (Fairclough, 1989) about femininity and "Sexual politics" in society (to use Moi's [1994/1985] apt term).

I will now turn to another dimension of rhetoric in *Songs*—condemnation. *Why* does the poet condemn imported rice? What relevance has such rhetoric to the feminist position we find in "Cassava Song"?

THE RHETORIC OF CONDEMNATION

Condemning, the binary opposite of praising, naturally presupposes the performance of acts or maintenance of postures not approved by a subject. With the (background) knowledge we already possess about the feminine/feministic basis of the persona's praise of cassava in *Songs,* we could therefore approach her condemnation of imported rice with the hypothesis that: the position of imported rice threatens the feminine/maternal interests or positions figuratively ascribed to cassava.

The threat posed by imported rice to cassava is generally ideological. At the macro-level, nationalism (indeed decolonization) is signified by the call for a ban on imported rice in *Songs.* The reason proffered for the call is not only the foreignness of rice (as a symbol for coloniality) but the economic and psycho-social implications in the preference for the imported rice. Rice, to the postcolonial Nigerian society, has come to signify class superiority.

> ...the white man came
> Bringing their foods to us
> But rice was not their food
> Neither was it ours.
> Yet in no time, rice was elevated
> Rice became the food of the rich
> For in my home, there is a saying:
> "A poor man does not eat rice."
> So rice became a rich man's food
> Which he and his family
> Ate only on Sundays and on
> Festive occasions especially weddings.
>   *(Songs,*53)

The class-consciousness underlying the preference for imported rice is perceived as disruptive of social and political order, and particularly matrimonial order. The persona, in fact, presents rhetorical proofs which include prevalence of hoarding and suspicions, at the social level, and disaffection in the home. She cites examples with cases of wives quarrelling with their husbands over nonpurchase of imported rice, and a bride being unhappy at her wedding because she has not served rice to her guests, since she believes that "Marriage is no marriage without rice" (*Songs,* 55). Thus the importation of rice becomes the creation of economic and psychological forces that could ruin society.

Juxtaposed with the stability created by cassava as a maternal force, the rice situation seems to give us the impression about the threat to the maternal/feminine economy (Cixous/Sellers, 44). Indeed, the voice calling for the ban on imported rice is also maternal and it upholds maternal values, which in the traditional Igbo context include material sustenance of the family, prevention of wastefulness, etc. And the position of Mother in the family could be generalized to indicate the essence of the maternal voice in the guidance of the entire society.

The relationship between Mother Cassava and imported rice is thus that of binary opposition, with the latter possessing negative attributes, as represented in Figure 3:

MOTHER CASSAVA	VERSUS	IMPORTED RICE
+ domestic (insider)		+ alien (outsider)
+ familiar		– familiar (strange)
+ caring		– caring
+ accessible (affordable)		+ inaccessible

Figure 3: Binary distinctions in the representations of cassava and imported rice

The perceived strangeness of (imported) rice puts cassava (the familiar) at some advantage. Cassava is *known* (as maternal) while rice is the *unknown*. Indeed imported rice, unlike cassava, is not described in gendered terms. This *silence* over the sex of (imported) rice could be seen as strategic. It could be seen as signifying the unfamiliarity, the strangeness of rice. Thus, the silence for which Terry Eagleton (1976) suggests what a writer is "unable to articulate"—the *incompleteness* of a text (34-5)—correlates with the dominant ideology underlying *Songs* as both a post-colonial and a feminist text. The ideology in this case seems to rely heavily on the simplistic logic of the difference between the Self-Same and the Strange—Other, a logic which is not acceptable in feminist literary theory, since it is one of the devices used in the masculine Symbolic Order in constructing stereotypes of femininity. To perceive rice as stranger in the context of a gendered discourse is to reveal to a non-casual reader that the dominant logic being used is still the masculine. Or rather, we could say that the feminist voice in *Songs* is to some extent deconstructed by the presence of masculine logic of the Strange—Other.[5]

Unlike the rhetoric of praise to cassava, the persona uses satirical devices in condemning the importation and consumption of rice. In the following excerpt, for example, humor is employed in showing the strangeness of rice:

> In my village
> The old did not eat rice
> When the young ate rice
> They sneered:
> What are you eating?
> Bird's food?
> Are you a bird?
> To eat grains?
> How many grains
> Can you eat to fill your stomach?
> *(Songs,* 57)

The embedded narrative presents a rhetorical slant that demeans rice. Story-telling could be (and indeed is) an ideological and a strategic weapon, especially in contexts where binarity is taken for granted and/or established. Rice, in the satire, is referred to as "Bird's food," such that its eater also shares in the pejoration—the eater of rice suggests herself as a bird! The point is that the discursive functions of the questions supposedly asked by the old people[6] in the village are more than just the perlocutionary evocation of laughter. They are indeed important aspects of the general question raised by the poet (indirectly) about the integrity of imported rice. The questions could also be viewed pragmatically as an invitation from the poet to the implied audience (which includes Nigerians who have preference for foreign rice) to *laugh* at the foolishness of replacing the familiar with the unfamiliar.

The rhetoric of condemnation thus complements the rhetoric of praise in *Songs.* And since the illocutionary act of condemning is face-threatening, it could also be seen as a means of further creating the basis for the valuation of (Mother) Cassava, that is, using attack as a form of defense.

CONCLUSION

In exploring rhetorical style in *Songs,* I have tried to show Nwapa's concern with femininity. Indeed, the analysis has not covered every aspect of style, but I have tried to narrow the discussion to the major aspects that suggest the text as a mode of interaction between the poet and society (as the implied audience). Within the frame of song that I have based my essay, I have been mainly concerned with, as Mills would say, "the ways" that *Songs* "encode(s) gender and the ways that representations of gender...form part of the logic of the text" (202).

It is my contention that Nwapa's poetry indeed accomplishes much by taking the rhetoric of femininity further into aspects of post-colonial life where sexuality, though dominated by the masculine symbolic, has not been sufficiently interrogated imaginatively. The maternal image she has used coheres with her concern with womanhood in her prose works generally. But above all, she, in "Cassava Song" stimulates serious thinking on the relationship of feminine writing to the imaginary semiotic to which gender philosophers like Kristeva, Cixous, and Irigaray have constantly drawn our attention. Such an experiment by Nwapa is indeed instructive to feminine writers in Africa who have to seek alternative and authentic rhetoric for woman.

NOTES

1. Singing in this work context could be seen as an attempt at "lightening" the heart, of creating a more stable psychological state for self.
2. Contemporary written poetry in Nigeria bears evidence of the fact that some poets would want their readers to view the poems within the frame of "Song" (even though such indications may sometimes be outright deceptive). Examples of the use of the song frame include Ezenwa-Ohaeto's *Songs of a Traveller,* Pol Ndu's *Songs for Seers* and Niyi Osundare's *Songs of the Marketplace.*
3. This concept is sub-classified by Lim and Bowers (420) into "fellowship face" ("the want to be included") and "competence face" ("the want that...abilities be respected.")
4. I am using the term "Nigeria-Biafra War" instead of "Nigerian Civil War" in line with the argument that, historically speaking, Biafra once existed as a "nation."
5. This *interference* with the feminist voice in *Songs* is as a result of the poet's adoption of a decolonizing posture which, as we could see in the critical writings of Chinweizu, Ngugi, etc., often relies on the primordial distinction of them-versus-us.
6. Note the distinction made by the persona between the old and the young as perceivers of social experience. Apparently, the old are the voices/agents of wisdom while the young are the voices/agents of folly.

WORKS CITED

Acholonu, Catherine Obianuju. *Motherism: The Afrocentric Alternative to Feminism.* Owerri: Afa/LHHP, 1995.

Angelou, Maya. *I Know Why the Caged Bird Sings.* New York: Random House,

1969.

Barthes, Roland. "Ecrivains et Ecrivants." *Critical Essays.* Illinois: Illinois University Press, 1972.

Beauvoir, Simone de. *The Second Sex.* Trans. H. M. Parshley. Harmondsworth: Penguin, 1972.

Brown, P. and S. Levinson. "Universals in Language Usage: Politeness Phenomena." *Questions and Politeness: Strategies in Social Interaction.* Ed. E. Goody. Cambridge: Cambridge University Press, 1978.

Cixous, Helene. "The Newly Born Woman." *The Helene Cixous Reader.* Ed. Susan Sellers. London: Routledge, 1994:35-45.

Eagleton, Terry. *Marxism and Literary Criticism.* Berkeley: University of California Press, 1976.

Fairclough, N. *Language and Power.* Harlow: Longman, 1989.

Halliday, M. A. K. and Ruqaiya Hasan. *Cohesion in English.* London: Longman, 1976.

Irigaray, Luce. *Speculum of the Other Woman.* Trans. Gillian C. Gill. New York: Cornell University Press, 1985.

___.*This Sex Which is Not One.* Trans. Catherine Porter. New York: Cornell University Press, 1985.

Keitel, Evelyn. *Reading Psychosis: Readers, Texts and Psychoanalysis.* Trans. Anthea Bell. Oxford: Basil Blackwell, 1989.

Kirwan, James. *Literature, Rhetoric, Metaphysics: Literary Theory and Literary Aesthetics.* London: Routledge, 1990.

Kristeva, Julia. "A Question of Subjectivity—An Interview." In *Modern Literary Theory: A Reader.* Ed. Philip Rice and Patricia Waugh. London: Edward Arnold, 1989:128-134.

___. *Desire in Language: A Semiotic Approach to Literature and Art.* Trans. Thomas Gora, Alice Jardine and Leon S. D. Roudiez. Oxford: Blackwell, 1981.

Lim, T. and J. W. Bowers. "Facework: Solidarity, Approbation, and Tact." *Human Communication Research* 17.3 (1991): 415-450.

Mills, Sara. *Feminist Stylistics.* London: Routledge, 1995.

Moi, Toril. *Sexual/Textual Politics: Feminist Literary Theory.* London: Routledge, 1994/1985.

Mojola, Yemi I. "The Works of Flora Nwapa." In *Nigerian Female Writers: A Critical Perspective.* Ed. Henrietta Otokunefor and Obiageli Nwodo. Lagos: Malthouse Press, 1989:19-29.

Ngara, Emmanuel. *Ideology and Form in African Poetry: Implications for Communication.* London: James Currey, 1990.

Nwapa, Flora. *Cassava Song and Rice Song.* Enugu: Tana Press, 1986.

Oha, Obododimma. "Culture and Gender Semantics in Flora Nwapa's Poetry." In *Literature, Gender and Popular Culture in West Africa.* Ed. Stephen Newell. London: Zed Books, 1997.

Okpewho, Isidore. "African Poetry: The Modern Writer and Oral Tradition." *Oral Tradition and Written Poetry: African Literature Today 16* (1988):3-25.

Sellers, Susan. *Language and Sexual Difference: Feminist Writing in France.* London: Macmillan, 1991.

Thompson, Denys. *The Uses of Poetry.* Cambridge: Cambridge University Press, 1978.

Wales, Katie. *A Dictionary of Stylistics.* Harlow: Longman, 1990.

Never A Gain? A Critical Reading of Flora Nwapa's *Never Again*

Obododimma Oha

> Sometimes we live the wars between nations as personal events. Sometimes a private drama appears like a war or natural catastrophe. Sometime the two wars, the personal and the national, coincide. Sometimes there is peace on one side (in one's heart) and war on the other.
>
> Helene Cixous (1994, xv)

Many Nigerian artists who have based their writings on the Nigeria-Biafra conflict (1967 to 1970) have indeed tried to "live" the war between the two nations as "personal events." They often try to evoke certain emotions, either directly or indirectly. In some cases, however, such emotional involvement seems to be embarrassingly close to the rhetoric of the politician. Writing in *Fact and Fiction in the Literature of the Nigerian Civil War,* Akachi Ezeigbo (1991) observes that Flora Nwapa's technique in her war fiction *Never Again* (1975), for instance, is "similar to that of the politician or rhetorician who is bent on persuading his audience to see his point of view. The similarity surfaces in the rhetorical and declamatory devices, in the unmitigated emotional involvement and the tendency to moralize issues" (96).

Indeed, Nwapa's *Never Again* attempts to "seduce" the readers into seeing the situation on the Biafran side of the war through the "eyes" of her anti-war narrator, Kate. Since the narrator thus seeks to interfere with the reader's freedom of judgement and perception of reality, it invites a critical reading. We could, from a skeptical distance, call into question the narrative rhetoric, particularly the validity of regarding the narrative (which is itself propagandistic) as a realistic portrait of the propaganda situation in Biafra. In this essay, therefore, I examine critically the communicative implications of the selection and rejection of events in the narrative. I will first of all analyze the representation of pro-war propaganda in the context of *Never Again,* and then discuss the significance of gaps and silences in the narrative. I will also examine aspects of the narrative that the novelist could have gainfully developed.

Propaganda is widely understood to be a very potent weapon in war. Often its purpose, as Sam Keen (1986) has pointed out, "is to paralyze thought, to prevent discrimination, and to condition individuals to act as a mass" (25). Thus it could be seen as a psychological weapon: as an anti-enemy weapon, it may seek to demoralize; as a defensive weapon, it may seek to "immunize" and "inoculate" loyalists, and to engineer confidence (Oha 1995, 4). Hitler (1984), in fact, recognized these psychological dimensions of war propaganda in *Mein Kampf:*

> The art of propaganda lies in understanding the emotional ideas of the great masses and finding, through a psychologically correct form, the way to the attention and thence to the heart of the broad masses. (165)

The discourse contexts presented by Nwapa in *Never Again* show the effects of propaganda on social psychology and on the security of the individual. In fact, within the context of *Never Again,* Biafran propaganda could be said to be very effective inside Nwapa's "Biafra," but this effectiveness is portrayed as an undoing of personal security and interpersonal relationship.

At the general level, Biafrans are portrayed as being (re)modelled by Biafran propaganda to (a) perceive relationships in the war as binary oppositions, (b) possess an unwavering but misleading optimism of Biafra's victory, (c) prefer lies to *true* stories about the state of affairs in the war, and (d) exaggerate Biafra's efforts in the war.

The logic of binary opposition is seen by Sam Keen as typical of the psychology of the *Homo hostilis* (the hostile or enemy-making human) which is often demonstrated in war. According to Keen,

> The corporate identity of most peoples depends on dividing the world into a basic antagonism:
>
> | | | |
> |---|---|---|
> | Us | Versus | Them |
> | Insiders | Versus | Outsiders |
> | The Tribe | Versus | The Enemy |
>
> In other words, paranoia, far from being an occasional individual pathology is the normal human condition.
>
> (Keen 1986, 17)

The use of this binary logic in discourse in the war situation is therefore quite inevitable. Conflict itself presupposes the existence of opposites. In the context of the Nigeria-Biafra conflict, individuals on both sides of the war must define where they belong and automatically perceive other issues relating to the war in terms of the Us-versus-Them demarcation. As could be seen in the following excerpt from the speech made by Madam Agafa during a meeting in Ugwuta, the binary logic is a strong cognitive mechanism in the war.

Excerpt I

> "...Why are *we running?* We are running from the *Vandals.* Who are the Vandals? Who are *they?* They are nobody. *We* must continue fighting against them until we vanquish them. God is alive and *God knows* that *our cause is just.* They killed us in Lagos, they killed us in Zaria, in Kano, in Jos, in Kafanchan, in Ibadan, in Abeokuta. We said to them, it is enough. Leave us alone in peace. We are going to our home...." *(Never Again, 12)* (Emphases added)

The pronominal deixis—We/Us and They/Them—indicate the opposing identities in the war. The first-person-plural pronominal forms tactically suggest inclusion while the third-person-plural pronominal forms suggest exclusion. Thus "they" that are against "Us" are portrayed negatively as a way of alienating them and conjuring up hatred for *Them.* They are "Vandals"—haters and destroyers of civilization. As such, killing or opposing *them* is a noble and desirable thing! The pervasiveness of the decivilizing lexical item "Vandal" in the references of the Biafran characters to Nigerian troops in *Never Again* is a reflection of the influence of the rhetoric used by the Biafran leadership on public discourse in Biafra generally. Ojukwu (1969), for instance, uses the lexical item in his speeches, as could be seen in the following excerpt form *The Ahiara Declaration:*

> In the Ikot Ekpene, Azumini and Aba sectors of the war, *the Vandals,* whilst maintaining their positions in Ikot Ekpene and Aba with our troops surrounding them, have continued to suffer heavy casualties in their attempt to hold firmly on to Azumini. (3)

Loyalty, therefore, is demonstrated by Biafrans in their speaking like their leader, viewing the enemies as the leader views them. In other words, the use of the word "Vandals" by the characters is a way of signifying conformity or their acceptance of common thought and perception (produced by leadership rhetoric).

This also applies to other aspects of the binarity suggested in Excerpt [I], namely: (a) We are right; they are wrong, (b) God is on our side; the devil is on their side; (c) We shall win; they shall lose. The involvement of the supernatural is a moralization of the conflict situation, and naturally nobody would like to identify with *the bad* or the wrong. To identify with wrong (*as the enemies are doing*) is to be an enemy of God! Thus God is (re)created as warrior and politician. Sam Keen in this regard rightly argues:

> God and country may be quite separable in theory, but in day-to-day politics and religion they are fused. God sanctifies *our* social order, *our* way of life, *our* values, *our* territory. Thus, warfare is applied theology. Probe the rhetoric used to justify war, and you will find that every war is a "just" war, a crusade, a battle between the forces of good and evil. Warfare is a religio-political ritual in which the sacred blood of our heroes is sacrificed to hallow our ground and

> to destroy the enemies of God. (Keen 1986:27)
> (Emphasis added)

The reliance on supernatural support, which is further demonstrated in the sacrifice to the Woman of the Lake, Uhamiri, prevents people from making timely arrangements for the evacuation of women and children, and for the defense of Biafra's cities. Those who suggest evacuation or mention non-supply of arms are rather blackmailed and threatened with detention. They are labelled "sabo" (saboteurs). In the following excerpt, the lexical item is used in alienating the referent, Adigwe, for suggesting the evacuation of women and children:

Excerpt II

> "When we evacuate the women and the children, who will cook for the soldiers? Who will fetch water for the women to cook for the soldiers? My people, didn't I say that this boy is a *sabo?* We should hand him over to the Army. I am a man and I am going to defend Ugwuta. People like *this boy* should not be allowed to speak in a meeting like this...." (*Never Again*, 14) (Emphases added)

The word "sabo" in Excerpt [II] above profoundly threatens the "fellowship face" (Lim and Bowers 1991:420) of Adigwe, his desire to be seen as a member of community. Thus the narrator tells us of the collapse of (cordial) interpersonal relationships which is due to non-tolerance of views that do not conform to the univocality of leadership in the war context.

Moreover, this univocality endangers personal security. Adigwe in Excerpt [II] above is threatened with detention for his suggestion. Also, Kate and her husband, Chudi, cannot say anything that contradicts the popular voice of the Biafran propaganda. Independent thought in the war discourse is thus paralyzed—a situation parallel to that of the totalitarian Oceania in George Orwell's *Nineteen Eighty-Four* (1982).

However, this dominance of state propaganda is exploited by individuals further in public discourse through exaggerations and lying about the state of affairs at the war fronts. In this case, individual story-tellers in *Never Again* present fabrications which may seem to enhance Biafran propaganda, but which are indeed inimical to the sense of security of persons in the war zones. The narrative, as a matter of fact, calls attention to this ironical nature of permissible discourse in the war. Such fabrications are welcomed by the society because they represent what the people want to hear, not what they (have to) fear. What they (have to) fear—which are suggested as the real—are regarded by the society as the unreal, the demoralizing narratives that must be suppressed.

The narrator, in the following excerpt, for instance, finds such art of fabrication distressing and ridiculous:

Excerpt III

...as soon as Umuokane was mentioned, I paid attention at once. "Dead vandals, all over the place, Nigeria is depopulated. Without exaggeration, I can say that there were over seven thousand dead vandals there. The villagers were summoned to bury the dead. They were paid five shillings for every body buried. The Vandals will not attack for the next three months. Their losses were too much in that single encounter. They should not have struck when they did."

I thought, seven thousand soldiers! Villagers burying the dead. It made no sense to me. Where were the villagers? Where was the battlefield? How big was it to accommodate seven thousand corpses? There were some who thought that a battlefield was like a football field....(*Never Again*, 24)

To use the words of John Locke (1959) in *An Essay Concerning Human Understanding,* human beings "love to deceive and to be deceived" (146)—the more so in a war situation where telling the *true* state of affairs would be seen as a disservice to the state, and would also mean endangering personal safety. Thus, in war, the human predicament is further compounded by the ironical nature of discourse. Social values suddenly change: society now prefers lies to truth.[1]

In essence, *Never Again* is not just a text that exposes the nature of propaganda in Biafra, or one which presents the human predicament in wartime. It is principally an example of an *anti-war rhetorical text*. Unlike the pro-war rhetorical text which argues war (for instance, the stories and speeches of the fictional characters which at the micro-level call for the defense of Biafra), the *narrative* rhetoric attempts to discourage war. It therefore uses the events narrated to show the undesirable nature of war and the negative effects of pro-war propaganda.

The use of a female character in presenting the anti-war narrative rhetoric, in fact, seems to be of some significance. The narrator, as the "implied author" (Booth, 1975), does not focus directly on the role of gender in the warfare, since both men and women engage in the war propaganda. However, we could link the anti-war posture to the feminist deconstruction of the masculinist view of political life and culture. War is sometimes (if not often) seen as predominantly a masculine affair, even though histories of femininity show the existence of some amazonomachies in Greece and ancient Dahomey. Discourses on war often rank women and children together as the vulnerable groups,[2] and indeed, when war breaks out, the problems of femininity in a male-dominated context tends to increase. In *Never Again* for instance, the excuse given by the pro-war propagandists for objecting to the evacuation of women and children is that there would be nobody to cook for the men and for the soldiers. The security of the women (even pregnant ones) is of little or no concern.

Warcraft, which is seen as a masculine affair, is therefore seen as being in opposition to femininity. In *Never Again,* Ezeama expresses the sexist cultural view of femininity and war when he says to Chudi:

Excerpt IV

"You saboteur. Before the Vandals entered Port Harcourt, you fled. Yes, my people, two weeks before they entered Port Harcourt, this spineless man fled with his wife and children and all his property.... You are a *woman.* Are you going to die twice?..." (16) (Emphasis added)

The concept of war thus shows another avenue for masculine demeaning of womanhood, and an expression of the masculine stereotyping of femininity.

It is, however, this masculine culture of war and the glorification of such destructive culture that one of the voices of the text deconstructs. The anti-war posture is portrayed as being more sensible, preserving, not destructive. There is never a gain in such culture of warfare; rather we have strandedness, disease, hunger, death, and the collapse of interpersonal relationships.

The fact that Kate, a woman, is the anti-war protagonist in the context of the novel therefore seems to suggest the moral superiority of womanhood which men have tried to alienate from warcraft as a demonstration of sexual inferiorization.

However, in spite of this moral positioning of the female narrator, which is a credibility strategy, *Never Again* as a narrative mode of rhetoric does not transcend the emotional weakness of the pro-war propaganda that it seeks to undermine and demystify. In Excerpt [V] below, the narrator, in fact, uses emotional appeals that are comparable to those used in Excerpt [I] by Madam Agafa (whom she criticizes):

Excerpt V

"Why, we were all brothers, we were all colleagues, all friends, all contemporaries, then, without warning, they began to shoot, without warning, they began to plunder and to loot and to rape and to desecrate and more, to lie, to lie against one another. What was secret was proclaimed on the house tops. What was holy was desecrated and abused. NEVER AGAIN." (*Never Again*, 70)

Who are "they"—Biafrans or Nigerians, or both? The pronominal "they," as I have tried to show earlier, is tactically employed in alienating, and it is a linguistic expression of primordial sentiment of paranoia in political propaganda. It is obvious that the narrator is unable to make distinctions again in terms of the Agent and the Affected, and so uses the pronominal "they" exophorically, taking us "outside the text altogether" (Halliday and Hasan 1984:18).

Also, the narrator attempts to seduce the reader merely by appealing to emotions about security, sacredness, and brotherhood, instead of delving into the logical dimensions of these issues, at least through the use of "satellite events," i.e., those events that "amplify or fill in the outline of a sequence by maintaining, retarding, or prolonging the kernel events they accompany or surround" (Cohan and Shires 1993:54). It seems that what Nwapa cannot artistically fulfill in the narrative, she tries to make up through the (seemingly easier) means of appealing to emotions. Her narrator, just like the pro-war rhetor, Madam Agafa, merely plays on the emotions of her implied audience (the reader) in Excerpt [V] above. In other words, the narrative undermines itself as a critique on the demerits of sentimentalizing war.

On the other hand, the narrative also contains evidence that contradicts the claim by the narrator that "Words are impotent" (*Never Again*, 25). Words, in a war situation, are "weapons" as well as therapy. It is not only arms and ammunition that enable conquest. In fact, it is human *attitude*, more than anything else, that prepares the ground for victory or defeat; and *words condition attitude*. The negative aspects of propaganda, therefore, cannot completely cancel out the relevance of propaganda in warfare. Also, the excesses and abuses in the use of propaganda by some individuals in Biafra do not necessarily suggest the baseness of Biafran propaganda as the narrator would like us to believe. Furthermore, if "words are impotent," what is the relevance of the narrative rhetoric in *Never Again* itself?

On the other hand, the narrative contains testimonies that Biafra's war efforts have been grossly exaggerated by fictional characters. The deliverance of Ugwuta, for instance, stuns the narrator who had all along given us the impression that stories of Biafra's victories were (badly crafted) fiction. Commenting on the clearance of Ugwuta she says:

Excerpt VI

"Apart from some few soldiers on the streets, Ugwuta was desolate and empty. Only a few days before, the place was swarming with people. Now it was empty. It was a battle ground. Where were the Nigerian troops who entered it barely three days ago? Where was everybody? What folly? What arrogance, what stupidity led us to this desolation, to this madness, to this wickedness, to this war, to this death? When this cruel war is over, there will be no more war." (*Never Again*, 70)

Apart from the contradictions in the narrative and the emotional rhetorical postures that are designed to seduce the reader, there are also clear indications that Nwapa has deliberately left out some aspects of the war that could interrogate the authority of her ideological position and rhetoric. The setting of the narrative is restricted to Biafra. We are prevented from knowing what is happening on the Nigerian side of the war. Even when Nigerian troops enter the same setting where the narrative is based, nothing is revealed about their conduct, feelings, and interfactional behaviors. The restriction not only signifies the limitation on how much the narrator (as a human being) can know, but also how much the novelist, Nwapa, *would like* her audience to know. In other words, the restriction could be linked to the pragmatic position of the text as an expression of ideology. The absence, the "not-said," is therefore strategic and "eloquent" as well. As Terry Eagleton (1991) has stated:

> It is in the significant *silences* of a text, in its gaps and absences that the presence of ideology can be most positively felt. It is these silences which the critic must make 'speak.' (34)

The consumers of the text are provided with only those events that agree with the novelist's sentiments about the war. Just like the fictional characters—Kal, Madam Agafa, and Ezeama, whom the narrator criticizes—Nwapa avoids the narration of events on the Nigerian side which may free the reader from the tyranny of her ideological position. Indeed, the narrative offers the narrator the opportunity of making selections and rejections. As Cohan and Shires have pointed out:

> A story can be analyzed as an organization of selections and substitutions as well as one of additions and combinations. Paradigmatic grouping of events of one kind as opposed to another; or on actors, events involving one set of characters as opposed to another. (Cohan and Shires 1993:67)

What is selected therefore speaks loud and clear about what has been rejected. And, in order to understand and properly locate the ideology underlying what has been selected, we also have to listen to the not-said, to the rejected voice.

Telling stories, therefore, is also a political engagement, more so when the stories themselves are based on political experiences in society. Even when the stories are presented by their authors as being fictional, they still suggest ideological positions (either directly or indirectly). In the case of the literary texts that are based on the Nigeria-Biafra war, we cannot really expect objectivity from their authors, especially as such authors are members of the society in conflict and are bound to maintain certain positions. At best, the texts would seek to propagate certain convictions their authors hold about the war. In propagating such convictions, however, they do not have to resemble (or accommodate) what the politician in his (war) rhetoric would say. As I have argued in the essay, this closeness of Nwapa's style in her anti-war narrative rhetoric to emotionally-based political rhetoric is one of the weaknesses of *Never Again.*

I have also argued that an aspect of her narrative that could have made tremendous impact is the masculinization of warcraft. The anti-war position of her female narrator could have been developed to show how war, as a destructive force which men seem to be claiming as an expression of masculine superiority, is itself an abasement of the masculine. In other words, it is not necessarily Biafran propaganda that is misleading and destructive (to the individuals in Biafra) but the masculine (re)interpretation of warfare as an expression of masculine superiority and nobility. Such masculine propaganda is what drives individuals into self-destruction in their bid to convince other people that they are "men," not "women." It is such masculinization and glorification of war that is indeed never a gain.

NOTES

1. It could however be argued that (a) war, as in the Spartan society, is a normal human condition, and that it is peace that is temporary; (b) that life and every day activities, metaphorically speaking, involve "wars," and often do involve propaganda and preferences of lies to truths, as we have in actual war situations.
2. An inaugural conference of the African Society for the Psychological Study of Social Issues (ASPSSI) which was held from May 28 to May 31, 1995 at Ibadan, Nigeria, on the theme, "Addressing the Psychological and Health-Related Problems of War and Violence on Women and Children in Africa," identified women and children as very vulnerable groups in war situations and who require protection. In this case, men are indeed the (major) actors in warfare.

WORKS CITED

Booth, Wayne C. *The Rhetoric of Fiction.* Chicago: The University of Chicago Press, 1975.

Cixous, Helene. "Preface" In *The Helene Cixous Reader,* Ed. Susan Sellers. London: Routledge, 1994: xv-xxiii.

Cohan, Steven and Linda M. Shires. *Telling Stories: A Theoretical Analysis of Narrative Fiction.* New York: Routledge, 1993.

Eagleton, Terry. *Marxism and Literary Criticism.* Berkeley: University of California Press, 1976.

Ezeigbo, T. Akachi. *Fact and Fiction in the Literature of the Nigerian Civil War.* Lagos: Unity Publishing and Research, 1991.

Halliday, M.A.K. and Ruqaiya Hasan. *Cohesion in English.* London: Longman, 1984.

Hitler, Adolf. *Mein Kampf.* Trans. Ralph Manheim. London: Hutchinson,

1984.

Keen, Sam. *Faces of the Enemy: Reflections of the Hostile Imagination.* San Francisco: Harper & Row Press, 1986.

Lim, T. and J. W. Bowers. "Facework: Solidarity, Approbation, and Tact." *Human Communication Research* 17:3 (1991):415-450.

Locke, John. *An Essay Concerning Human Understanding. Book III* Rpt. New York: Dover, 1959.

Nwapa, Flora. *Never Again.* Enugu: Tana Press, 1985.

Oha, Obododimma. "The Therapeutic Uses of Rhetoric in War." Paper presented at the First Inaugural and Annual Convention of the African Society for the Psychological Study of Society Issues (ASPSSI), Ibadan, May 28-31, 1995.

Ojukwu, Emeka. *The Ahiara Declaration: The Principles of the Biafran Revolution.* Geneva: Mark Press, 1969.

Orwell, George. *Nineteen Eighty-Four.* London: Heinemann, 1982.

Flora Nwapa's *This Is Lagos:* Valorizing The Female Through Narrative Agency

Naana Banyiwa Horne

The issue is that there were indigenous feminisms. There were indigenous patterns within traditional African societies for addressing the oppressions and injustices to women.

—Molara Ogundipe-Leslie,
Re-Creating Ourselves (223)

[W]omen, as speaking subjects, have been transformed into written objects through the collusion of the imperialistic subject and the patriarchal subject . . . [Now] these beleaguered written objects are reinscribing their relevance as speaking/writing subjects.

—Obioma Nnaemeka, "From Orality to Writing:
African Women Writers and
the (Re)Inscription of Womanhood" (138)

[W]hat do Barthes and Foucault have to do with Flora Nwapa? In spite of their deaths, the two men continue to colonize literary theory from the other world, thereby disenfranchising many writers. By practicing their craft which is osmotically transmitted through ancestral worship as their texts are read as Revelations, part of the literary community has helped to put them magisterially in control. Nwapa's lucid, profound, and spiritually charged writing, like that of her many sisters with their different ancestors, has been one attempt to free ourselves from such intellectual and literary tyranny.

— Chikwenye Okonjo Ogunyemi, "Introduction: The Invalid, Dea(r)th, and the Author: The Case of Flora Nwapa, *aka* Professor (Mrs.) Flora Nwanzuruahu Nwakuche" (3)

Flora Nwapa, Nigeria's premiere female novelist, carved a niche for herself as the foremother of the Igbo female-authored text when she emerged on the literary scene in 1966 with *Efuru,* the story of an Igbo female whose life is strongly steeped in a female-centered universe. Her instrumentality in the institution of a Nigerian women's literary tradition is worthy of celebration.[1] Through valorizing the female voice in her works, Nwapa has had a tremendous impact on eroding the hegemony of male characters and male-authored texts in gendered discourses. A prolific writer whose legacy to the female literary tradition is attested to by her six novels and two collections of short stories, with their wealth of multifaceted female portraits, Nwapa conscientiously challenges a literary tradition dominated by male authors and their perpetuation of a phallocentric literary history[2] characterized by either marginalized female characters or stereotyped female portrayals inspired primarily by male sexual fantasies.[3] Her persistent valorization of the female voice acquires increased pertinence in the face of the blatant marginalization of the literary efforts of women authors themselves. In spite of her prolificacy, most of her works received little or no critical attention, an anomaly which, ironically, her death has helped correct by inspiring a burgeoning of literary responses in commemoration of her life and literary accomplishments.

Valorization is used here mainly in the context of the portraiture of females as agents of voice and subjects of action rather than

as mere receptacles/objects of other actors' designs, or as spectators who passively observe other/male subjects in action and initiate no actions of their own, except when they are incidentally
swept by the actions of male subjects. Indeed, the exercise of voice
itself constitutes action on a most fundamental level. The institu-
tionalized impositions on women's voices in many cultures, for
instance, and the political repression of writers' voices in highly
repressive political establishments testify to this reality.[4] Each of
Nwapa's novels, as well as collections of short stories, provides
ample evidence of a female-centered world that valorizes the
female voice; however, this essay focuses principally on Nwapa's
first collection of short stories *This Is Lagos and Other Stories*
(1971).[5] What renders this collection distinct is the valorization of
the female voices that ring through the stories. Regardless of their
station in life, Nwapa's women are imbued with voices that ring
out to be heard. They are vocal about their pains and joys; they
assert themselves through their determination to take charge of
their lives; they attempt to control their destinies as far as that is
possible, given the odds against them.

This fine collection, comprising nine stories, constitutes a
wealthy tapestry of the contemporary Nigerian world in general
and of intimate female portraits in particular, portraits that extend
from the ranks of the well educated to the barely educated and
embrace women's professional and domestic worlds. Nwapa presents female role constructs from a wide spectrum of women's
lives. Her characters are depicted in a multiplicity of roles, negotiating various spaces: maternal, conjugal, occupational, domestic,
social, political, and individual. The opening story "The Traveller"
aptly closes with Bisi, the female protagonist, asking her female
friend: "Musa, what kind of man is he? And what kind of woman
does he think I am?"(9) This rhetorical question frames the stories
within a gendered discourse of females operating from the consciousness that their actions emanate from a world in which the
perceptions of them are strongly informed by their sexual identity, more so than by other factors. So, Western-educated or not, they
interrogate their worlds and the powers or forces manipulating
their lives. Indeed, Nwapa's stories are filled with struggling
women; but they are no voiceless strugglers. Life is by no means
a crystal stair for these female characters who confront the
formidable task of negotiating barriers erected by the forces of
patriarchy that marginalize and objectify women's lives and bodies. And many of the stories arise out of the suffering encountered

by women in their efforts to negotiate these barriers.

An issue that is of interest to this paper is how Nwapa mediates the process of valorizing characters most of whose lives are rocked by tragedy and, above all, succeeds in capturing the pathos in their lives without rendering them mere victims who are mowed down by superior forces. This feat, in my view, is accomplished through the *agency of narration,* that is, the process through which the elements that comprise her narrative text are constructed. Narratology, the study of narrative theory, provides a key to a structural analysis of Nwapa's feminist ideology. Mieke Bal's *Narratology: Introduction to the Theory of Narration* (1986) illuminates this issue.[6] Narrative, Bal indicates, is a consciously constructed process—the careful ordering of the elements of a fabula ["a series of logically and chronologically related events that are caused or experienced by actors"(5)] into language signs that can be designated a text. "These elements are organized in a certain way into a story. Their arrangement in relation to one another is such that they can produce the effect desired, be this convincing, moving, disgusting or aesthetic"(7). The text, therefore, is the outcome of a careful process of structuring, or rather, of manipulation, that is, of handling, of combining, of presenting the elements in the fabula to produce a particular perspective. In Nwapa's *This Is Lagos,* she consciously manipulates the process of narration in furtherance of her feminist intentions: the valorization of women's lives and voices.

The distinction Bal makes between story and text, narrative agency and writer, and, above all, focalization, are worthy of notice. Two key terms, both of which are action verbs, are particularly insightful for this analysis. The first is *uttering,* the act of employing language signs to tell; and the second is causing to see, presenting to view or vision. These two key terms underlie the concepts of narration and focalization. Bal defines a narrative text as "a story that is told in language;" in other words, a story converted by a narrative agent into "language signs"(8); hence it is imperative not to confuse either the story with the text, or the author with the narrative agent. In the creation of a narrative text, the author "withdraws and calls upon a fictitious spokes[wo]man, an agent technically known as the *narrator*"(8) to perform the task of transforming a story into a text. This narrative agent, or narrator, then, is "the linguistic subject" performing the "function" of expressing itself through the language that composes the text(119). Emphasis is on the "function" rather than the "person" of the agent; hence Bal's choice of "it" in reference to that agent. It is on the process of producing the linguistic signs that are so integral to the presentation of the elements of the fabula; hence the significance of recognizing the narrator as "that agent which *utters* the linguistic signs which constitute the text"(120, emphasis added).

The narrative agent, in converting story into text, employs specific language signs to produce a particular perspective, thus manipulating the elements in the story, through narration, that is, through the way it presents them, to generate (a) specific perspective(s), or way(s) of *seeing* that influence(s) interpretation. This process of *seeing* "indicates an activity of focalization. . . .[It draws attention to] the relationship between the 'vision,' the agent that sees, and that which is seen"(104). The key word here is relationship, a relationship that is three-layered, involving actor (the agent causing or experiencing the events), narrator (the agent expressing itself through language signs), and focalizor (the eyes through which the elements in the narrative are viewed). These three layers have the potential to overlap or not depending on that which is performing which function at which point in time. "Because the definition of focalization refers to a relationship, the subject and the object of focalization, must be studied separately"(104). A key relationship is that between narration and focalization. While the "subject of focalization, the *focalizor*," constitutes "the point from which the elements [in the narrative] are viewed" (104), manipulating the way of *seeing* the text remains integral to the function of narration. The narrator can be outside of the events it narrates, an external narrator, or be inside "the fabula it itself narrates"(122), a character-bound narrator. Furthermore, the external narrator is often not the sole *teller of the tale;* direct speech is also usually employed. Anytime direct speech is employed in a text, narrative agency, that is, the function of employing specific language signs to influence perception, is temporarily taken over by an actor/character, one who is part of the fabula and is active in causing or experiencing events in the fabula. Hence the imperative of ascertaining who specifically is performing the function of narration at any point in time: "The narrator is the most central concept in the analysis of narrative texts. [Why?] The identity of the narrator, the degree to which and the manner in which that identity is indicated in the text, and the choices that are implied lend the text its specific character"(120).

Nonetheless, focalization provides "the most important, most penetrating, and most subtle means of manipulation"(116) in the narrative text. It "can lie with a character (i.e. an element of the fabula), or outside of it"(104). This implies that the narrator can double up as a character, that is, as an agent causing or experiencing events in the fabula(5); and that same character-narrator can function as the point of focalization: "If the focalizor coincides with the character, that character will have a technical advantage over the other characters"(104). The reader will be made to see with the eyes of the character telling about itself and possibly about others; and will, "in principle, be inclined to accept the vision presented by that character"(104). Given this potential for manipulation, "it is important to ascertain which character focalizes which object....because the image we [readers] receive of the object is determined by the focalizor. Conversely, the image a focalizor presents of an object says something about the focalizor itself" (106). If narration implies focalization, it is because "language shapes *vision* [emphasis added] and world-view.... [Consequently,] *seeing*, taken in the widest sense, constitutes the object of *narrating*"(120-21). So then why differentiate narration from focalization? "[O]nly the narrator narrates, i.e. utters language which may be termed narrative"(120). So should we perceive focalization to be the same as narration, we would fail to distinguish between "linguistic, i.e. textual, agents and the purpose, the object of their activity"(120). Purpose undoubtedly undergirds focalization.

How do these elements of narratology inform Nwapa's narrative technique and her ideation of feminist principles in *This Is Lagos?* The author consciously aligns gender with narrative agency and focalization to underscore her valorization of female lives and voices. That the writer is female informs her choice of narrative agents, and her focalization of the elements of the fabula to privilege female perspectives in the collection. Out of the nine stories, only three, "Jide's Story," "A Soldier's Return," and "My Soldier Brother" have male characters at the center. The remaining six have female characters dominating the action. Even then, the stories with males at their center still focus considerable attention on women. Notably, two of the three, "Jide's Story" and "My Soldier Brother," employ character-bound narrators who are male, thus falling within the context of first-person narratives. On the other hand, only one of the six stories with females at the center of the action, "The Loss of Eze," can be termed a first-person narrative because it employs a character-bound narrator who is female. It is my conviction that in Nwapa's collection, when males speak in their own voices, which is the case in the two first-person narratives, their voices issue out of a patriarchal need for hegemony. In both "Jide's Story" and "My Soldier Brother," in which focalization lies mainly with the male character-bound narrators, they operate out of the posture of the assumed superiority of men. This posture manifests itself in males presuming to know, better than females themselves, what is best for women.

The young male narrator of "My Soldier Brother" has rather disparaging notions regarding women; his depiction of his aunt, Monica, and his sisters brings out this propensity. Even when direct speech is employed in this story, as in the following case:

> 'When will I be of age?' I asked stubbornly.
> 'When you are in a University,' she said without thinking.... (124)

Male hegemony is what is accentuated. This is because the "I" asking the question (the actor of that event) doubles up as narrator and focalizor, thus causing all three to coincide in a single person. In other words, the young male character who is narrating most of the story also focalizes the events and their actors. So even when he steps aside to let his sister speak in her own voice ("'When you are in a University'"), he still manipulates our perception of the sister, and by extension, females, by providing a cue for interpreting the sister's action ("she said without thinking"), that undervalues her intelligence and, by extension, that of other females. Consequently, in this story, the impact which prevails is an overwhelming display of a lack of respect for females by males.

Of course, since the way the focalizor focalizes an object helps reveal the focalizor, male hegemony is also strongly mirrored by the young male character-bound narrator through whose eyes we see the women in the story. The following utterance which is uttered by him in reaction to something his sister says in jest, strongly speaks to this point. "I did not understand my sister. But who understood sisters, who took them seriously anyway? They were all women, weak creatures. Men were strong, especially men like Adiewere"(124). His depiction of his aunt, Monica, is even more telling. He makes her out to be a buffoon, an insensitive woman who always embarrasses the family. "Aunt Monica never *said* the correct thing"(126), he reports. At Adiewere's funeral, "[s]he sat *talking nonsense* all the time. I felt like asking her out of the house, but then I could not do it. I was a little boy"(133, emphasis added). Ironically, the male, character-bound narrator perceives his youthfulness as an impediment that prevents him from exercising the hegemonic privileges of his gender at this juncture.

Considering that he is just a little boy, it is evident that his denigrating remarks about the female members of his family, and his consciousness of patriarchal privilege, mirror the perceptions the male world around him holds of females and males and their relative positions in a male-dominated world.

In fact, the disparagement of the speech and actions of females by the young male narrator here recalls a common practice by male writers and critics alike. Florence Stratton affirms that "derogatory remarks about women's speech have been a staple of African (male) literature and criticism"(83). This hegemonic practice is particularly evident in Elechi Amadi, an Igbo male writer who has been the most compared with Nwapa, and his noted juxtaposition of the "dignified buzz" of his male characters with "the high-pitched, piping, market-chatter of the women" in his novel, *The Concubine* (43). Naana Banyiwa Horne discusses the manifestation of the male tendency to devalue women's actions, particularly their words, in her contrastive study of Amadi and Nwapa in *Ngambika* (1986, 119-129); and Stratton focuses on the phallocentric underpinnings of both the devaluative critiquing of Nwapa's *Efuru* vis-a-vis Amadi's *The Concubine* and the general manifestation of phallocentric criticism in the exclusion of female-authored texts by male critics from the African literature canon in her book (1994). Perceptively, in Amadi's novel, his elevated group of males is busily but silently crunching kolanuts that have been served by the women who are portrayed with no dignity (43), making it evident that while the male novelist has no problem commending the female world for service (the rendering of manual skills for the comfort of the male world), he is inhibited in acknowledging the intellect of the female world, particularly as it is reflected through speech. This shortcoming, however, seems completely oblivious to some male critics. Eldred Jones and Eustace Palmer, two forerunners of African literary criticism, both highly commend Amadi's skill as a writer who writes perceptively on human nature while dismissing Nwapa's literary engagements with the woman's world as trivia and the language of her female characters as "small talk."[7] Conversely, Naana Banyiwa Horne's discussion of the two authors' techniques accentuates Nwapa's skillful tailoring of her narrative to her feminist intentions.

> Nwapa depends almost exclusively on dialogue to
> reveal Efuru's world, and she proves very success-
> ful at it. Her novel is filled with the daily conversa-

tion mainly of women, a technique that captures
most effectively the oral-aural nature of the world
she unfolds. The constant banter of women reveals
character as much as it paints a comprehensive,
credible, social canvas against which Efuru's life can
be assessed. *The total world view is brought to life
through dialogue.* (121, emphasis added)

It is my conviction that Nwapa, the female author whose art has
not escaped the hegemonic ravages of the patriarchs of African literary criticism, writes/strikes back at patriarchy, in "My Soldier
Brother," a story with a very male title, in which male hegemony
is so blatantly displayed, thus signifying on her detractors through
the intervention of narrative agency. Stratton's feminist critiquing
of Jones's and Palmer's perceptions of Nwapa's artistry reveals
them to be phallocentric(5). And in this story, Nwapa, operating
with the smoothness of an adept puppeteer, mounts the controls
of her art, manipulating perception through narration and focalization to reveal that phallocentric criticism of her works is as questionable as the young male narrator's perceptions of his
womenfolk.

In the only other first-person narrative with a male central character, "Jide's Story," the story of a man who cheats on his wife, narrated and focalized mainly by himself, that the author is a female
whose interest is in making manifest both the suffering and
strength of women in confrontation with patriarchy becomes evident in the manner in which this first-person narrative (most of
which is spoken in Jide's voice) is intermingled with other/female
voices. While Jide narrates most of the text and focalizes most of
the events and their actors, the manner in which his role as narrator and focalizor is combined by the author bespeaks male chauvinism, callousness, and insecurity. At the same time, the female
voices that are interjected periodically, through direct speech,
intervene in bringing to the fore female strength in spite of the suffering caused by men's exploitation of the patriarchal conventions
of sexism. In functioning as narrator and focalizor, Jide's depiction
of women, while undermining them, also exposes him to be sexist. He womanizes shamelessly and focalizes events to rationalize
his objectification of women as instruments of pleasure and abuse.

When he first meets Rose, he perceives her to be "so cheerful
and *easy.* She was not *inhibited* in anyway"(21, emphasis added.)
Though not stated directly, that the next information communi-

cated in the narrative text is his getting married to Rose soon there-after focalizes her easiness and uninhibitedness as strong incen-tives in his marrying her. While on the surface this observation does not seem pertinent, it nonetheless serves to project the blame for his infidelity unto his wife. Jide being a good-time lover makes uninhibitedness an appreciable quality in his spouse; and he, sur-reptitiously, creates the impression that Rose has lost her fun-lov-ing nature and become "proper" following their marriage, an excuse men usually give for infidelity. But then there is strong indication that he is responsible for the change that occurs in Rose following their marriage. He admits disapproving of her wearing wigs or make-up, or making herself conspicuously appealing once he makes her his wife, thereby manifesting patriarchy's attempts to control the physical appearance of females under the guise of safe-guarding their sexual integrity, and its promotion of husbands' con-ception of their wives as property, as possessions that must be guarded from the gaze and forages of other men. This objectifica-tion of women as bodies/property becomes particularly pro-nounced in Jide's focalization of Maria:

> In the boat I met Maria. She was a glamour girl. You simply saw glamour written on her forehead. She carried a very large wig on her head. The make-up was revolting. But she was irresistible. She was every inch a woman. Very feminine. I fell for her. I fell for her make-up, I fell for her wig, I fell for her wiggle. All the qualities she possessed, including her make-up were what I told my wife I abhorred in women. (22)

The above perception is reductionist. To Jide, Maria is nothing more than a sexual colony to be invaded and conquered to spice up his boat trip from London to Nigeria, and to be discarded once her sexual usefulness has been exhausted. What he sees is not a human being but an asssemblage of flesh/meat that is effectively processed to subsume desire. And like a hungry dog, he falls greed-ily on the meat, consuming it with passion until he is forced to choke on the not easily digestible portions, the full package that makes Maria more than just a sexual object. His limited percep-tion of her as a seductress who is conspicuously decked out to sig-nal her willingness to be seduced is a male ploy that puts the burden of women's victimization on women themselves and frees men of responsibility. It mirrors a popular underpinning of masculinist discourse—that women exist to be possessed by men, and that women dress up and go to great lengths to make themselves conspicuously appealing in order to make it easy for men to be drawn to women's sexuality. It is this same rationale husbands provide for controlling the physical appearance of their wives in their conviction that they need protection from their sexuality and from the gaze of others.

Male chauvinism is also evident in Jide's persistence in reading his wife's mind, a quality he starts manifesting soon after marrying Rose. When, two months after their wedding, he leaves Rose in London to return to Nigeria, one can surmise he is disappointed at her composure despite his departure. There is a tinge of disappointment in his tone as he reports: "She took my departure calmly. For days, I knew it bothered her, but she was a calm girl by nature and she bore it very well"(22). The insecurity that starts to surface in him at this point mounts as his narration unfolds. When he entangles himself in a web of affairs following his marriage, and Rose shows no visible signs of torment on account of his infidelity, he starts casting doubt on her integrity, instead of confronting his own shortcomings. He becomes really disconcerted that, after he has stayed out all night one time, Rose would not ask the simple question, "'Where did you sleep?'"(31) He rationalizes: "If she had asked, perhaps it would have been better. I would have felt better and less guilty"(31). What follows attests to his mounting hysteria: "Was Rose up to something? Didn't she love me? Was she having an affair? Why did she not ask questions? *We had no child yet*. Was this worrying her?"(31, emphasis added) Jide, of course, picks on a time old source of patriarchal torment for women—the construction of childlessness as a diminishing index of womanhood—to insinuate Rose may be preoccupied with far more serious worries than his infidelity. His sense of superiority is affronted by his suspicion that he does not hold sway over his wife's emotions; for being *the man*, he considers it his prerogative to have the upper hand in the relationship. So when Rose gives him no satisfaction about being disturbed by his womanizing, he becomes truly distraught:

> For days I thought Rose would ask questions, but she did not. What was at the back of her mind? Was she being liberal? Did she love me at all? Why was she not jealous? Why did she behave in that superior manner? Was she making a fool of me? I was

told that when a married woman had an affair with
men, it was her husband who was the last to know.
(33)

While his reactions reveal his childishness and insecurity, they
nonetheless mirror men's disconcertion at any challenges to patriarchy's hegemony.

"Jide's Story" offers perhaps the most poignant example of the
high price women pay in seeking a full life, as well as bringing out
the resilience forged in them by the vicissitudes of life. Ironically,
it is not Rose, Jide's wife, through whom these attributes are made
manifest, but Maria, the woman who is sexually exploited by Jide.
In fact, Rose abets in her own victimization and that of Maria when
she chooses not only to close her eyes to her husband's infedelity
but to perceive Maria as a dereanged woman whose actions and
statements are generated by a deranged mind. A wife and a mother seeking further education in London, Maria is dogged by mental illness in her attempts to negotiate maternal, conjugal, and
professional roles. She is victimized as a wife, a mother, and a sex
object; yet she constantly negotiates professional and personal
spaces in ways that, even as they force attention to her victimization, underscore her strength of character. In London, her husband
beats her and locks her up leaving her with neither food nor money
for herself and her children for several days at a time. This mis-
treatment pushes her beyond the boundaries of sanity, forcing her
incarceration in a mental asylum and eventual forced separation
from her children. Notably, this dimension of the text is focalized,
not by Jide, the male character-bound narrator, but by Peju, a
female character in the fabula. When Peju temporarily assumes
the role of narrator, presumably to inform Jide how dangerous
Maria is, her focalization of events through narration reveals Maria
to be tragic and admirable at the same time, and men to be abusive and destructive to women. The following interchange between
Jide and Peju speaks to this revelation:

.. 'Tell me about Maria.'
'Maria, poor girl. It wasn't her fault. Men, they
doom we women. Very intelligent girl. And I hear
that she handles her job in the Ministry very well.
So they can't sack her, and of course, she has connections, and you wouldn't know she is not well.'
'Is she insane?' I asked gripping the chair.

'She is insane. She went off her head in London.
She was in the asylum for nearly a year. But being
a brilliant girl, she was able to finish her course
before she returned. Her husband was cruel to her
so it entered her head. Her two sons are still in
London.' (27)

Maria's tragedy evidently emanates from her efforts to negotiate
occupational, maternal, and conjugal spaces in a gendered context
that is weighted against women but privileging to men. Her maternal aspirations render her even more vulnerable. Because women
rather than men are orientated towards assuming primary care of
their children, separating a woman from her children serves as an
effective source of torture that unscrupulous men often resort to in
manipulating women. On the other hand, women's love for their
children and their desire to be good mothers often force them to
give in to the sexual demands of men in the belief that it can aid
them in negotiating their maternal roles. Women's sexuality provides the root cause for their victimization because men, not seeing beyond it, either attempt to take possession of it, like Maria's
husband who locks up her sexuality by locking her up, or to ravage
it, like the womanizing Jide who sees Maria as only a *playgirl*. The
truth, however, is that women are complex beings and also creatures of their environment. So, faced with a world that attempts to
reduce them to sexual objects for the gratification of men's sexual
desires, women oftentimes attempt to transform their own sexuality into a weapon to be employed in negotiating other spaces. Maria,
for instance, resorts to sexual ploys as a weapon with which to
retrieve the maternal space she has lost. Unfortunately, it only
makes her more vulnerable to sexual exploitation. Peju succinctly
sums up the sexual woes women face on account of the patriarchal
conventions of sexism in the statement: "Men, they doom we
women"(27). This sense of doom results from patriarchy's objectification of women as bodies/property, thus turning intelligence and
beauty, which should be a winning combination, into a deadly one
that spurs abusiveness in immature, jealous, and insecure men.

While Maria's case is sad and her suffering immense, she is not
without dignity and strength. Peju's focalization of events attests
to Maria's intelligence and strength of character, even as the former laments the latter's vulnerabilities. Though quiet throughout
most of the narrative, Maria retrieves her voice in the end when
during her showdown with Jide, her voice rings out loud and clear,
asserting her readiness to retrieve the maternal space from which

she has been dislodged. Jide's account of this confrontation is informative:

> 'I am ready,' she said.
> 'Ready for what?' I asked.
> 'Aren't we going to London to collect the two kids?' she said.
> 'Please, please go home.'
> 'I won't go home. You promised we were going to London tomorrow. I won't go home. I just won't go home,' she started shouting. 'You promised in writing. Here are the letters you wrote, you did. The letters are in this bag.'
> It was past midnight, but a crowd gathered. I had no words. (32, emphasis added)

Though it is evident Maria is confusing Jide with her husband, it is significant that as she regains her voice, inspired by her commitment to her children and her persistent attempts to reclaim her maternal space, Jide is rendered voiceless. Her commitment to the role of motherhood in spite of the obstacles erected in her path by patriarchy stands out starkly against Jide's lack of commitment to his conjugal role. It is an affirmation of Maria's strength that even in the grip of her illness, she manages to negotiate various spaces, occupational and gendered, and recognizing that she has been displaced from her maternal space, initiates action that, in her mind, will enable her to re-enter that space and regain control of it. In this female-authored text, it is no wonder that what is valorized is women's lives and experiences despite the fact that the story is narrated and focalized mainly by a male.

"The Loss of Eze," the only first person narrative in the collection that is narrated by a female, does not issue out of a hegemonic posture. In fact, in this story of female heartache, narrated by Amede Ezeka, the female character-bound narrator, hegemonic postures are critiqued and the roles of victim and victimizer are both rejected. Her narration is infused with the fluid essence of the blues, being, on the one hand, her lament of love-lost and the pain of desertion and loneliness she suffers on account of it and, on the other, her assertion of the will to pick up the pieces and embark on a search for new life and happiness in spite of the fear of betrayal. Amede's story is neither informed by the same ego-centricity that is evident in the two previously discussed stories narrated by male character-bound narrators, nor does she give much attention to reading other's/Eze's mind(s). Instead, she bares her soul to the reader, concentrating on the depth of her love for Eze.

Eze had been life itself to Amede, and when she loses him to another woman, she is devastated: "I thought my heart would break when I lost Eze. He was everything to me"(92), she confesses. She focalizes the events in the fabula in a manner that takes the reader inside her life, offering insights into female and male reactions to love:

> My interests were diverse, and I was never bored until I lost Eze. It was an affair that took everything out of me. The fire that burned inside me in those one and a half years of love was a consuming fire. I didn't know it would ever be quenched. I thought it would burn forever. Eze said it would burn forever and I believed him. It burned for only eighteen months, and when it was quenched, nothing could rekindle it. (92)

Love is presented as a powerful vehicle whose potentiality for fulfillment Amede embraces wholeheartedly only to be burned. Female love, as embodied in her, is portrayed as unconditional giving of the self in trust to a love partner. On the other hand, male love, as embodied in Eze, is depicted as self-love; it does not nurture the other. Fire aptly captures the essence of Amede's love. Her love blazes, once ignited by Eze; however, it is quenched, burning itself out, because Eze fails to fuel it. The double-edged quality of love as nurturer and destroyer is beautifully embraced by the fire imagery. It is invested with the potential to infuse an inspiring glow in those in love, or to burn, consume and destroy them, leaving misery in its wake. Amede first experiences the positive fire of inspiring love that is fulfilling; but then, Eze, the one stoking her fire/love, stops tending it, leaving it to burn her and eventually burn itself out, leaving her devastated.

Unlike Jide's narration, which is characterized by sexism and objectification among other problematics, Amede's narration centers her as the subject, making her feelings the primary subject of interest in the story. She explores the depth of her love, her heartache and fear, and, above all, her struggle to reclaim her life and find love again with an openness that is refreshing. Rather than engage in castigating Eze for deserting her, she delves deep into her own feelings, her vulnerabilities:

> For six whole months, I did not know what to do
> with myself. You are lost when you suddenly find
> yourself left out, when you suddenly discover that
> you have lost someone who was life itself to you.
> You grope in darkness, you ask yourself so many
> questions whose answers frighten you. You are
> empty inside you. (92)

Her heartache plunges her into the depths of fear: "I was gripped
with fear, fear of the world, fear of human beings, fear of all around
me"(93). The threat of despondency, however, forces her to critique the victim status that giving in to despair would foist on her.
It is this confrontation that leads her to reject the role of victim
and embark on reinscribing herself into the world from which she
has been shut out by Eze's desertion:

> The fear was getting too much, it was getting me. I
> was losing my self-confidence, a quality I thought I
> had in abundance. And that was what really fright-
> ened me and made me shake myself, and say, many
> times over, 'Amede come out of it, come out of it
> before you are destroyed.' I said this aloud to myself,
> I said it in my dreams, I said it everywhere. (93)

The process of reinscription, Amede recognizes, entails taking control of her life, venturing into the social world to reclaim the social
life and love she has lost through her loss of Eze. "So after six
months of losing Eze I said to hell with him. To hell with all I held
dear, I must live my own life the way I wanted it. Eze or no Eze, I
must live my life fully and usefully"(93).

It is this venture that ironically leads to a confrontation with
hegemonic postures. In her contemplation of how to achieve her
objectives, she finds herself moving from one extreme situation (a
victim status) to another (a hegemonic stance). She goes from being
"lonely . . . shut out of the world that was around" her, to a determination to go "out again to take" her "share of it"(93). Being female,
Nwapa implies in this text, entails a sensitivity towards hegemonic postures; for the moment Amede unwittingly slips into a hegemonic frame of mind, she immediately catches the folly in her
stance, making a commitment to avoid such postures:

> My share! I shuddered as the word 'share' came to
> my mind. Wasn't it because some people wanted
> their 'fair share' of wealth, without contributing to

this wealth that left the country in such a mess, that ushered in military rule, that had led to so much bloodshed and suffering. No, I mustn't use that word again. (93)

This scene is epiphanic, for it leads not only to a discourse on hegemonic attitudes, but culminates in an identification of the military, a clearly male-dominated organ of society, as the ultimate symbol of hegemony in Nigerian society; it is one that is associated with the utmost brutality and destruction of life and human dignity. Notably, going back out into the male-dominated world to find one's niche is conceived in masculinist terms: "The world, in many respects, was a bull. A mighty bull. You couldn't even seize it by its horns. The horns were too many and too dangerous to come near to"(93). Bull Fighting, as is well recognized, is a highly publicized masculinist activity, a macho symbol in contemporary Western aesthetics. It performed wonders for the literary career of Ernest Hemingway, the embodiment of male ethos in American letters, the writer who enshrined masculinism in his writing and in his life, choosing suicide over confronting life in the face of sexual impairment.

Patriarchy is symbolically conceived as a many-horned bull, ever ready to gore those/females who come into confrontation with it; and Amede is fully aware of her vulnerability in venturing into this male-dominated world. She is, nonetheless, equally aware that there is dignity in being female, and strength, for her, will come from tapping into her feminine self. That she finally dons her wrappa, the symbol of female dignity in Nigerian culture, and takes stock of her appearance, is an assertion of Nwapa's feminism, one that valorizes those distinct attributes of African womanhood and femininity—women taking pleasure in looking good and recognizing the enhancement to their sense of well-being that investing in their good looks brings them.[8] While patriarchy objectifies the female body, portraying it as something that is there for the gratification and control of males, Nwapa negates that perspective by depicting females like Amede who exude self-confidence, are intelligent and not only good looking but relish their good looks and other feminine attributes. In this story narrated and focalized by Amede, a female who is herself centrally engaged in the events in the fabula she herself tells about, her determination to pick up the pieces of her life and continue with her pursuit of fulfillment in spite of her emotional scars is valorized. Though burnt by love, she overcomes her pain and fear, finding the strength within herself to shake out of the stupor of despair, and venture back out into the world in search of happiness; and her courage is rewarded. Once she goes back out into the world and meets Tunde, she is willing to explore falling in love again despite her awareness that love can lead to heartache. Amede's story closes on a hopeful note, with her excitement at the impending date with Tunde and the prospect of a meaningful relationship. Her narrative replicates the prevailing tenor of most of the stories in Nwapa's *This Is Lagos*—women who are burnt by life's experiences, but, their scars notwithstanding, continue to meet life head-on with tenacity and resilience.

The three stories discussed thus far are the only ones in the collection with character-bound narrators, with the remaining six recounted by external narrators who are outside of the events of the fabula. What narratology underscores in *This Is Lagos* is that gender is an important index in narrative theory. Even in the stories with character-bound narrators who are male, the author being female intrudes in the manner in which the language signs constituting the various texts are uttered and the events of the texts are focalized by the male characters. Their intervention exposes their collusion with patriarchy in the victimization of the female characters, thus leading readers to a greater understanding of the plight of females in a male-dominated world. Moreover, in the six stories with external narrators, there is only one, "A Soldier's Return," in which the narrator can be perceived as gender neutral, and it happens to be the one story in the collection that is truly male-centered, as its very title suggests. Women are peripheral in this story in which the central character is male, with his (a soldier's) escape in the nick of time from the site of an ethnic-motivated coup forming the central focus for the events.

Notably, in the five remaining stories, women are cast as the primary actors, that is, as active subjects who are fundamental in causing and/or experiencing the events which are constructed into Nwapa's narrative texts, and the events in the fabula emanate from women's lives.[9] These women in *This Is Lagos* are best remembered as agents of action and change; they are not merely acted upon but actors in their own right. Like Ozoemena in "The Delinquent Adults" and Agnes in "The Child Thief," they are capable of initiating actions whose repercussions reverberate around them and even beyond. Furthermore, as much pain as characters like Agnes, Maria, Nwanyimma, and Ozoemena encounter, none of them can be categorized as a passive victim who has no hand in causing and/or experiencing the events going on around her. In fact, most of them are at the center of their worlds, functioning as movers and shakers who remain undeterred by the high price they often end up paying in human suffering for being agents of action. Life for them entails confronting the challenges of existence in a world in which gender constrains one to second-class citizenship. The female-centered themes and the casting of women as primary actors of the events in the texts strongly suggest a female identity for the external narrator who utters the language signs which constitute the texts and on the individuals through whose eyes we see the events that compose the texts. Nwapa focuses on all facets of women's lives, breathing life into a wealth of female characters peopling her world; yet it is in the depiction of women's maternal and conjugal roles that she is at her best. "The Child Thief" and "The Delinquent Adults" are, consequently, among the most poignantly rendered of her stories. These stories—about the plight of the childless woman and that of the young widowed wife who has to face the raising of her children alone—are in my estimation the most grippingly told, striking a chord of pathos to which all women, regardless of background, can relate.

As in "Jide's Story," the trauma that life can be for women as they attempt to negotiate conjugal and maternal spaces is situated at the center of "The Child Thief," a story in which Nwapa explores with pathos the pathological hold the need for children exerts on society in general and on women in particular. The protagonist, Agnes, starts off as a vibrant, spirited young female with very sophisticated ideas about marriage and "a lot of boyfriends whom she sacked at will"(48). She becomes engaged when still a student, in her last year; and she marries right after acquiring her School Certificate, filled with visions of conjugal and maternal bliss, and aspirations to live happily ever after as the ideal wife and mother. Her friend and classmate, Bisi, recalls:

> She [Agnes] had set ideas about marriage. She had said that she was not going to marry anybody who had no car. She dreamt of evenings out when her husband returned from work, and how he would open the car for her to enter, and all that. She was going to dress very well to please her husband. She was not going to be like other girls who refused to look good after the first baby had arrived. She was of course not going to breastfeed the baby, for it was

messy. All she was going to do, was take some
tablets and the breast milk would dry up. She was
going to have four children, two boys and two girls.
She was going to have a wonderful home. She was
not going to work. Looking after her husband and
children were enough to occupy her. (51)

Ten years later, Bisi, encountering her once cheerful, sophisticated friend is shocked to see instead, a fat, unkempt, highly distraught woman whose real life situation has proven antithetical to
the picture-perfect image she had taken straight out of Western
romance magazines with their emphasis on the bourgeois aspirations of Westernized females.

The struggle to negotiate conjugal and maternal spaces take a
heavy toll on Agnes, who fails to bear any children, after ten years
of marriage. Her problem is that conjugal and maternal spaces are
conflated in her society because the primary motivation for marriage is procreation. Igbo sociologist Ada Mere, explains this phenomenon in an interview with Gay Wilentz: "In pro-natalist society,
marriage is the expected end; when procreation doesn't happen,
that woman has failed an essential life goal"(11). Agnes's own envi-
sioning, as a young female, of her future, though lifted out of a
Western bourgeois context, is nonetheless strongly informed by
this intertwining of conjugal and maternal roles that weigh heavily on African marriages, making failure in the maternal sphere
spill into the conjugal realm also. Driven to desperation by her
estrangement from her husband on account of her childlessness,
Agnes, in the end, turns into a child thief, convinced that "bearing" a child for her husband is the only way to become reinstated
in his affections. Nwapa reveals so poignantly that an environment
that subordinates all facets of life for its women to their biological
function as breeders exacts a heavy toll on childless women. Since
fertility is considered the essence of womanhood, failing to bear
children renders a woman useless. It is sadly Agnes's rejection of
the persona-non-grata status forced on childless women that turns
her into a criminal.

In this story, Nwapa's juxtaposition of Agnes's envisioned
future—with its Western bourgeois aspirations of ideal womanhood—against the reality of her life as a childless African woman
provides a platform for interrogating the structures upon which
both Western and African patriarchies are erected. As appealing as
the picture of the bourgeois aspirations of ideal womanhood seems

to Agnes, it is, nonetheless, tainted by the commodification of females that underlies the overvaluing of their sexual roles and undervaluing of their talents outside of the sexual realm that patriarchy perpetuates to delimit women's power. What one confronts in Agnes's picture, therefore, is a conflation of the Western and African underpinnings of patriarchy: the commodification of women both as sexual creatures whose primary function is to cater to the sexual needs of their men while mirroring the success of their male mates by their comportment, and as breeders whose reproductivity is as essential to the survival of their worlds as it is in fostering the maleness/virility of their mates.

Education, which should stimulate women's interrogation of the factors controlling their lives, in Agnes's case, unfortunately, co-opts her into aspiring to a lifestyle that underpins Western patriarchy. By ensuring that through marriage she will have a car, a wonderful home, and a sophisticated, leisurely life, Agnes is condoning male hegemony through her total dependence on her husband, and confining her fulfillment to gender roles, a trap into which most Westernized African females fall. Her situation is indicative of the dependency syndrome that Ogundipe-Leslie cautions against.[10] Furthermore, in confronting her situation as an African woman, the same lack of interrogation of the factors colluding in her victimization result in her despondency. Rather than find self-worth through other means, she subscribes to the belief that her inability to bear children is diminishing and, therefore, does not entitle her to respect and affection from her husband. In fact, her sense of self is so eroded that she is willing to welcome another woman—one who has produced children for her husband—into her matrimonial home to make up for her failure to produce children of her own. Agnes's tragedy is that she does not look past gender roles in mapping out her future. Moreover, when she weaves her vision of a perfect future around conjugal and maternal bliss, that is, being the perfect wife and mother, she does not contend with the picture becoming flawed by a possible failure to bear children; neither does she consider combining gender with professional roles. So once she finds herself incapable of bearing children, desperation sets in. It is no wonder that when she fails to bear children, that is, fails as a mother, she also fails as a wife; the world she builds around gender definitions crumbles. Agnes seems to have missed the vital lesson that Efuru, for instance, learned in her own confrontation with childlessness: that one can still lead a meaningful existence and find self-worth in spite of not bearing a child. While in the end, Efuru learns to serve her larger community by becoming a worshipper of Uhamiri and to find self-worth outside of motherhood,[11] Agnes, in her despair, is driven into becoming a child thief.

The tragedy of childlessness has been written about by both male and female writers; however, what makes Nwapa's treatment of the subject in this story noteworthy is her transformation of Agnes from helpless victim to the initiator of an action that has far-reaching consequences. Refusing to be a passive sufferer, she embarks on action to reverse her childlessness and retrieve her lost conjugal and marital spaces. When the various medical interventions result in her developing a tumor rather than getting pregnant, she tells no one the truth; instead, she cultivates the pregnant look that the tumor gives her to perpetuate the illusion that she is finally going to have a baby and continues to make physical preparations for the coming of the baby. She then goes to a hospital to have the tumor removed and steals a child that she takes home as the outcome of her "pregnancy." Even though the actions she engages in lead to no real solution, their planning and execution show intelligence and daring, in addition to giving her control over her own life and destiny, thus bringing an unusual twist to the issue of barrenness and the way barren women are generally depicted. Desperation transforms Agnes from passive sufferer to a driven woman, one who thinks things through systematically and succeeds in hatching a plan that will enable her to experience motherhood, even if briefly, and to bring happiness to her husband by making him a father, even if only for a moment. Her composure at the height of the action is noteworthy. She remains calm and in control of the situation, even when the detectives show up, to the very last act of naming her crime and submitting it to her own process of interpretation:

> 'You see, I made provision for the baby. I am not just an ordinary child thief. I wanted to have him as mine, not to sell him.' 'You should have adopted one.' 'I didn't want to adopt a baby. I wanted to have a baby....' (64)

Agnes gives voice to the despair of so many childless women who are pushed to the limits of their sanity by societies that overvalue women's procreative capabilities. In the same way that too much valuation of material acquisitions instigates crime, too much pressure on women to produce children can push childless women over the edge. Nwapa adds depth to her treatment of the subject of child-

lessness by insinuating that men are not immune to being stigma-
tized in this issue. Evidently, women are not the only ones undermined by the inability to reproduce. Men also seem to feel
diminished because their wive's failure to bear children renders
them less manly, that is, less virile. Even though Agnes's husband
has children by another woman, which testifies to his fertility, that
his longstanding wife has failed to bear children evidently makes
him feel diminished. His eager anticipation of the birth of Agnes's
baby and the spark that anticipation injects into their conjugal life
confirm the gratification he himself will get from the whole process.

Motherhood definitely is a dominant influence in many of
Nwapa's women's lives, and it is in reaction to situations emanating from this role that many of the female voices in the collection
are raised. In "The Road to Benin," it is Nwanyimma's status as a
mother who is committed to bettering her family's circumstances
that is valorized. This story, unlike the ones we have discussed so
far, does not focus on conjugal strife. In fact, husband and wife are
portrayed as a team, working for a common purpose—the better-
ment of their children's future. The only central character who is
uneducated, Nwanyimma, is determined that her children will get
the best possible education to ensure they will not inherit the hard
lives that have been their uneducated parents' lot. With a husband
who is a daily paid laborer and, consequently, a highly precarious
source of support for the family, she becomes the family's main-
stay through her creative trading skills:

> Nwanyimma traded in fruits and all kinds of veg-
> etables. She had white women as customers. She did
> not go to school, but she was able to speak pidgin
> English which the white women were able to speak
> as well. She did not only sell to white women, she
> also sold to cooks of white men. Thus she was in a
> very privileged position. Her customers were those
> who were able to pay any price. (34)

She proves astute in her business dealings. When her produce does
not measure up to the standard of her high-paying customers, she
not only takes it upon herself to inform them but goes in search of
fresher produce for them, thus rendering herself invaluable to her
customers and securing their loyalty. Also, Nwanyimma exercises
foresight and critical thinking in planning her children's future.
Given the consistently high performance of her oldest child in
school, she and her husband start exploring financial options for

paying for his higher education before the time arrives. While Nwapa presents husband and wife jointly planning the financial obligations facing the family, it is, nonetheless, evident that it is upon the wife that the brunt of the financial support of the family falls. There is not much money to be earned as a day-laborer. On the other hand, the market women of West Africa have earned a reputation for being the economic backbone of their world; and whether they acquire reputations as "cash madams" or not, most of the food that ends up in the stomachs of Africa's children, men, and women is not only cooked by women, but acquired through women's mercantile efforts. Thus, through her depiction of Nwanyimma, Nwapa valorizes the commercial/business enterprising of African women as financial backbones of African families and even nations.

Nwanyimma's toil to better her family seems to be paying off once her oldest son wins a scholarship for higher education. But then her dream comes tumbling down because he ends up in jail for meddling with bad company and drugs. In this story of shattered hope, it is the mother, Nwanyimma, who voices the desperation of the family through her repeated questioning: "What does he want me to do with it? Tell me, what does he want me to do with it? (46) This question to which no answer can be given is prompted by the son handing the mother a rap of Indian hemp when the parents go to see him in jail. While both Nwanyimma and her husband are clearly shattered by the experience, it is her voice that rings out, poignantly reverberating with the pain of parents betrayed by a delinquent child on whom they have banked so much hope. On the symbolic level, this story bespeaks the betrayal of the African market woman/mother by the educated elite and politicians. While the market women put their toil into educating their children and supporting their politicians, they are often let down by both. Their children, often succumbing to bourgeois temptations, give in to selfish, individualist pursuits, that either lead to self-destruction like Ezeka, or render them too self-centered to mind anybody but their own pleasures; and the politicians, through corruption and ineptitude, make a mess of the nation, leaving a legacy of failed economic ventures and other messy problems, like the bag of hemp Ezeka squeezes into his mother's palm when she goes to visit him in jail.

In the story "The Delinquent Adults" too, conjugal and maternal roles constitute the basis of Ozoemena's suffering, with the dilemma posed by Western-oriented approaches to women's marital and kinship relations vis-a-vis the presumed security and sta-

bility inherent in an African-oriented organization of these relationships providing the dramatic tension which leads to the exploration of some of the fundamental underpinnings of patriarchy.
This tension runs through most of Nwapa's works.[12] What she
makes manifest in this story is that indigenous practices foster kinship ties; nonetheless, they are also often undermining of women's
rights and even well-being. This is poignantly brought home in the
life of Ozoemena, a woman who is widowed at a very young age,
and the turmoil she goes through in confronting customary practices which, in upholding kinship, also promote patriarchy. Her
experiences foster the conviction that marriage and its related practices oftentimes strip away women's selfhood, subjecting them to
the hegemony of patriarchy. Ozoemena drops out of school to
marry Chukwuma, the man she loves. Unfortunately, Chukwuma's
older brother, Uzonwane, who is not Western-educated himself,
objects to his brother marrying a girl with some Western education: "He was suspicious of girls who went to school. He feared that
they were too wise, wiser than their ages, and that Ozoemena was
going to dominate his brother, and make him not care for his own
family"(74). Underlying Uzonwane's suspicion, of course, is the
threat the Western-style nuclear family system poses for the survival and prosperity of the patrilineage of which he is the head.

It is ironic that in this story, wisdom should be connected with
Western education rather than with age, a factor that, in my view,
finds possible explanation in the fact that Nwapa, the Western-edu-
cated female author identifies with her protagonist, the young
Ozoemena, whose Western education is perceived as a threat by
Uzonwane, the symbol of customary and patriarchy in the story.
The irony is intensified by the recognition that Ozoemena's education is marginal. She does not even have her School Certificate.
She, in fact, passes that up to get married. Of course, Uzonwane's
initial distrust of educated girls naturally escalates once
Chukwuma dies suddenly, and Ozoemena resists the customary
practice governing the settling of her deceased husband's estate
which makes the deceased's property, widow included, the inheritance of the patrikin. Widowed at twenty-two years of age, the
threat of facing life without her loving mate and raising two young
children alone, with no independent means of support for herself
or the children, becomes Ozoemena's living nightmare. Her plight
is intensely complicated by the machinations of the delinquent
adults around her: her greedy and cantankerous brother-in-law,
Uzonwane, the one to whom Chukwuma's widow and children

should pass, along with his material possessions; and her equally cantankerous mother, Ntianu, who hatches counter schemes to thwart Uzonwane and ensure her daughter financial security.

In this tragedy, the odds are highly stacked against the young widow because circumstances connive to make her suspect in her husband's death. Being clairvoyant, a gift which earns her the nickname Joseph in school, she has a dream, the night preceding Chukwuma's death, that turns out to be a premonition of his untimely death. Confiding her dream to a neighbor friend, the neighbor twists the information into a vicious gossip that throws suspicion on Ozoemena. In a culture that regards no death as without possible cause, those closest to the deceased are the ones who come under the most suspicion. This factor is borne out by the vicious rumor spread by Uzonwane that Chukwuma had left behind bags of money which his widow had hidden. In Ozoemena's favor is her probing mind, which is reflected in her introspectiveness and constant interrogation of events and situations, coupled with a fighting spirit. Thus, even though shaken by the turn of events, she refuses to succumb to being a victim, interrogating instead the forces manipulating her life, and critiquing not only her attackers, but the factors conspiring to keep her under suspicion:

> So this is life she said to herself several times. She too was being accused indirectly of killing her husband. And the dream? It was only a dream. She had had so many dreams which were just dreams. But why did she dream that her husband was going on a journey, and it turned out to be that he was really going on a long journey of no return? Was it predicted then that her husband would die? Who predicted it? God? What God? Impossible. It was an accident, true and simple. It could not be anything else. Accident. The lorry driver was careless. He had no brakes, so he crushed her husband while he was returning from work, her husband who was returning to her and his children. But why must it be her husband. (77)

She explores the burden of her clairvoyancy, questioning fate, destiny, and even God's will among other things, including wondering if it is "destined after all that we must die at a particular time, and in a particular way"(79). This trend of thought challenges the traditional perspective of there being no deaths without cause, that is, without interference from others, thereby rendering bogus the suspicion that she had a hand in causing her husband's death.

Ozoemena's life is impacted considerably by the machinations of the delinquent adults around her; but it is, above all, her own role as actor, that is, as one who is primary in causing and in experiencing the major events of the fabula that shapes not only her life but the direction of the text. Her special gifts and personality traits provide the major indices that fine-tune the action. Compounding her clairvoyancy, intuitiveness, and introspection is an iron-clad will which often pits her against the forces surrounding her, and a strong ethical sense that compels her to stand by her beliefs and refuse to cave in under pressure. In a society that lays great premium on kinship rather than marital ties, Ozoemena's strong commitment to her deceased spouse and children pits her against her in-laws because her commitment is, in effect, an act of resistance against the hegemony of tradition and patriarchy, drawing attention to the intricate connection between the two. A close examination of the law governing the settling of the estate of the deceased makes this connection evident.

For the Igbos, as for most Africans, kinship rather than marital relationships forms the basis of human organization. Thus, kinship relationships and obligations supersede marital ones. By the same consideration, the contraction of marriages is perceived as an essential means for the family—in the case of the Igbo, the patrilineage—to enhance its holdings, both human and material. Consequently, rules governing marital relationships are set up to privilege the kinship unit, which is patriarchal in this case. Since marriages add to the patrilineage's total strengths, tradition favors practices that consolidate its assets. For instance, the property— human [children and spouse(s)] and material—acquired by a man in the course of his life, passes on to his kinship family. His matrimonial family (wife/wives and children), in fact, becomes part of the totality of the wealth he bequeaths to the patrilineage. Hence, upon the death of the husband, the woman's continued integration into her deceased husband's family is ensured by another male member of her husband's family marrying her.

While the hegemony of patriarchy is evident in this organization of affairs which clearly privileges the deceased man's kinship family, I suspect, nonetheless, that its engineers do not consider it devoid of sensitivity towards the widow. Since it is the law for the material possessions that supported the widowed woman's matrimonial life to pass from her control to her deceased husband's family anyway, positioning her through remarriage to continue reaping the harvest she contributed to sowing can be perceived as an arrangement she is likely to welcome. It can even be argued that the deceased's kinfolks' preference to integrate the widow into her matrimonial family reflects the patrikin's thoughtfulness in wanting to keep mother and children together, particularly considering that the children, by right, belong to their father's people, and their mother has no legal right to them. Therefore, patriarchy itself may even glean something redeeming in this practice which guarantees a woman's material well-being—financial and sexual—in spite of the fact that the husband in whom both were invested is no more. But then, can one remain totally oblivious to the fact that this practice has the potential to reduce a woman to nothing more than an object (like the material possessions that pass from the matrimonial family's use to become the common property of the patrikin)? This realization, in my mind, raises a fundamental question regarding this practice. Could it then be that the law making a packaged deal the passing-on of a widow and the material accoutrements that formed the material basis of her conjugal and maternal life prior to her bereavement was, in fact, devised with the intention of diverting the woman's attention from the reality that she is, in fact, being dispossessed of the wealth she has contributed to building as a spouse?

Unfortunately, it is not only the men whose actions validate patriarchy. Even women like Ntianu unwittingly promote the entrenchment of patriarchy by their actions. Ostensibly to help her widowed daughter realize her goal to continue her education, Ntianu, sends Ozoemena to a rich man she is made to believe will arrange for her to continue her education, but who, in actuality, is to establish a sexual liaison with the young widow to guarantee her financial security and protection from her vicious in-laws. Ozoemena is naturally outraged by this plot to sexually exploit her, and she challenges the perverted sense of ethics of her mother and the man with whom she connives. Her resentment against this blatant attempt at sexual victimization reflects strongly in her confrontation with the big shot:

> 'Why should you insult me in this way?' Ozoemena
> finally asked. 'What is my sin? Why do you want to
> exploit me because I lost my husband, and it would
> seem, have no means of livelihood. Haven't you a
> grown-up daughter? Would you like her to be thus

treated if she married and lost her husband in a motor accident?' (90)

She does not spare her mother either. When upon her return, Ntianu comments she had come back quickly, she retorts: "What exactly did you want me to go and do at Aba?...I would have thought that if I wanted to be a prostitute, I should go about it my own way, without the aid or the connivance of my own mother" (91). Her boldness absolutely disarms her mother, forcing her to acknowledge her error.

In fairness to Ntianu, it must be acknowledged that as ethically questionable as her action is, she is following established practice in resolving her daughter's dilemma, just as Uzonwane is following the established norm in insisting on marrying Ozoemena. For while the deceased's brother has the right to acquire the widow of his brother, she or her family also can organize her sexual and financial well-being to thwart the schemes of the deceased husband's family. Ntianu dislikes Uzonwane intensely because of his resentment of her daughter's marriage to his brother; so nothing will give her more satisfaction than to thwart him. Therefore, setting her daughter up with the rich old man, in Ntianu's mind, is a step in the right direction towards ensuring financial security for Ozoemena and her children as well as protection for Ozoemena from the harassment of a vicious in-law, topped by the personal satisfaction of preventing Uzonwane from enforcing his right. What she does not seem to grasp, however, is the recognition that this action entrenches the hegemony of patriarchy by giving the rich old man power over her daughter. Ethical considerations have no place in the machinations of all the elders involved in this drama either, the rich dirty old man to whom Ozoemena is sent by her mother included; hence the significance of the title "The Delinquent Adults." It is Ozoemena's consciousness of the lack of ethics informing the behavior of the adults, buffeted by her recognition that all their plans in the end enforce patriarchal norms that empowers her to reprimand her elders and reject their solutions to her problem. She never loses sight of the fact that it is, after all, her life that is at stake, and any true solutions will have to be ones that compromise neither her integrity nor her sense of self-worth. Despite her youthfulness and meager education, she proves herself an intelligent/wise woman indeed, conscious of the workings of patriarchy and its built-in snares for eroding women's self-worth.

In this story, Nwapa presents Western education as a positive force that arms women to interrogate their worlds because of its capacity to open the door to financial independence for them and their children, thereby liberating them from the manipulations of ruthless husbands and in-laws and from traditions which privilege patriarchy and underprivilege women. Indeed, Uzonwane's distrust of educated girls stems from his conviction that they pose a threat to patriarchy. This conviction, I submit, is the real reason underlying his fabrication of the tale that Chukwuma left behind bags of money that his young widow is hiding. Convinced that Ozoemena will not willingly submit to becoming his wife, harassing her about Chukwuma's property becomes Uzonwane's way of forcing her to submit to him. Since the young widow has no independent means of supporting herself and her children outside of her deceased husband's property, and the practice of marrying the deceased's brother or an available male relative is intended to guarantee the widow's economic well-being, Ozoemena's refusal to marry Uzonwane, which will constitute her contractual severance from her matrimonial family, will, in this case, be construed as a confirmation of her having hidden/stolen a valuable portion of Chukwuma's property to guarantee herself financial independence.

At the heart of her tragedy, she becomes keenly aware that her plight is worsened by her inadequate schooling. With a higher education, she could work and earn enough to support herself and her children. As she contemplates her plight, she recalls her principal's disapproval of her choice at the time: "Her principal had asked her what she would do if anything happened to her husband and she found out that she had no trade of her own"(84). She had, at the time, been indignant that the principal would even concede such a thing. But upon the death of her husband, she is reminded poignantly of the principal's words, and she even starts regretting not having waited to take her School Certificate Examination before marrying. However, as painful as her situation is, probing the factors surrounding her choice leads to the recognition that she had married for love and had a wonderful marriage despite the fact that it had been short lived:

> Chukwuma was a loving husband. He took great care of her, and when the children began to arrive, he took even greater care of them. From the onset, it was a happy marriage. She had no cause for regrets.

> She was more appreciative of her husband, because he was just an ordinary clerk....He was so young, and yet he did not behave in that irresponsible way common to young men of his age. And she was devoted to him. She loved and cherished him. She was determined to pay back by being a good mother and a good wife. (78)

Through introspection, she is able to separate her decision to marry young because she is in love from the issue of fate (the untimely death of her husband). She assures herself that "she did not make a wrong choice....It was fate, it was death that played tricks on her. She could not, no one could fight against death or fate. Her choice was right. It was death that was wrong"(84). Ozoemena's struggle to understand her situation brings her to the realization that as pressing as her present need for more formal education is, because it is essential in ensuring her economic well-being and that of her children, her having privileged love over education at the time she decided to marry Chukwuma constitutes a valuable part of her life that she cannot undervalue by regret, for choice empowers.[13] Therefore, rather than give in to despair, she tackles her problem by taking some practical decisions. She decides against remarriage and opts to go back to school in order to complete her education and become a professional teacher, "so that she would be able to look after her children"(84).

One of the strengths of Nwapa, the African woman writer, is her perspicuity in truly penetrating the African woman's situation in relation to education. Western education does have its place in Africa, as in the rest of the world: it enhances women's marketability especially in a global market. This is a realization to which Ozoemena and other Nwapa female characters come. It is this consciousness that makes her determine to go back to school regardless of the impediments in her way. Education presents itself to her as a liberating tool that will free her from the clutches of patriarchy and the confines of customary practices that erode women's sense of selfhood. It is significant to note that Western education as Nwapa presents it in *This Is Lagos* is a bettering influence in women's lives, opening up greater professional and social opportunities and fostering economic and personal independence, and thereby mitigating the constraints gender roles exert on women. This dimension of Western education must not be confused with the erroneous equation of it with innate intelligence.

Nwapa's depiction does not project Western education as a marker of intelligence. In fact, she shows that critical thinking is not a hegemony of the West (a misperception which is widespread in its perniciousness) for women in her text who have no Western education show evidence of this skill in their lives. What she projects to be the main significance of Western education is its ability to make women upwardly mobile economically and socially. To foster this distinction between true education and Western education, Nwapa makes her heroine a school drop-out who does not even have a School Certificate. Yet her limited education notwithstanding, that Ozoemena is a critical thinker reflects in the way she is focalized both by the external narrator and by Ozoemena herself through her utterances. On the other hand, her near destitution on account of her lack of marketable skills and the demise of her breadwinner raise consciousness in Ozoemena of the need for her to go back to school to acquire some professional skill.

It is evident that Nwapa's women with less education lead more confining lives while the better educated ones have more active social lives, with a broader horizon contextualizing their activities. Most have lived and studied abroad, and are apparently better situated to support themselves. Thus, characters like Bisi, Amede, Rose, Peju, Iyabo, and Maria, who are well-educated, some of them to the university level, enjoy considerably more independence, having apparently more options for fulfillment than their less educated counterparts, and enjoying a relatively high standard of living. Even Maria, in "Jide's Story," who is mentally unstable, is able to support herself financially through her professional skills upon returning home to Nigeria. Furthermore, travel seems to have expanded the more educated women's consciousness of patriarchy and its constraining impact on women's lives; hence, their prioritization of their professional and career lives over marriage. In fact, except for Rose and Maria (even then Maria's husband is not around), these higher educated women with professions in *This Is Lagos* are not married and do not seem too eager to jump into marriages. Most of them obviously consider marriage an impediment to their personal freedom, or reject the victim status that marriage seems to impose on most women. Peju and Iyabo in "Jide's Story" seem to have relinquished the encumbrances that come with marriage and settled for keeping a string of rich boyfriends who can buy them expensive presents and support their rich, fast lifestyles. That they have suffered from failed relationships can be surmised, for getting drunk has become a regular pastime for both. In fact,

Peju confesses she has been driven to drinking, and though she does not say what drove her to it, her commiseration with Maria hints she is a kindred spirit in suffering, albeit one who is too tough to go under, and survives through the adoption of a liberal lifestyle. Bisi in "The Child Thief" muses, "Marriage must be strange....It did not make any meaning to her"(55), upon encountering Agnes, her old school mate from ten years back, who used to be spirited, now transformed into a fat, unkempt, despondent woman who will literally do anything in the name of marriage. While they perceive marriage as confining, they seem to cherish meaningful relationships, which ironically are scarce in the text, and continue to aspire to entering into meaningful relationships eventually. Amede anticipates falling in love again once she meets Tunde and aspires to enter into a lasting relationship with him.

That education is perceived primarily as significant for upward mobility is reflected by some of Nwapa's characters with limited education who tend to perceive marriage as a means of bettering their socio-economic status. Agnes in "The Child Thief" falls into this category; so does Soha in the title story, "This Is Lagos." Both women have their school certificates, which prepares them for working lower-income jobs, making marriage an attractive choice for socio-economic advancement. In fact, Agnes marries right after taking her School Certificate and depends entirely on her husband for financial support. She becomes a seamstress and starts earning her own money only after her husband's people throw her out of her marital home on account of her childlessness. Also Soha, "who had all her education in a village"(10), goes to the city and acquires a teaching job in a primary school. She "did not like teaching, but there was no other job, and so, like so many teachers, the job was just a stepping stone"(10). She attempts to resolve the problem of making ends meet by finding a man with a brand new car, who can take her shopping, and consequently moves in with him once she becomes pregnant. That Soha's choice is influenced by her desire to move up the socio-economic ladder is strongly insinuated by the fact that she lived with her aunt in a crowded household and had to contribute to the household and child-care chores rather than have a good time the way city dwellers are usually perceived as doing. There is, on the other hand, Ozoemena who drops out to marry before she can obtain her School Certificate, but seriously considers returning to school after the death of her husband in order to ensure economic independence for herself and her children.

The world contextualizing the stories in *This Is Lagos* is urban and, consequently, more overwhelming in impact by contrast with the relatively close-knit, laid-back, rural context out of which *Efuru* emerges. The very title of the collection evokes the rather hectic, intimidating, hazardous context of city life, with the tangle of entrapments it harbors for women. This atmospheric condition is itself gendered, for the snares that threaten are primarily ones that target unsuspecting women's sexuality. In "The Traveller," Bisi is wily enough to recognize that Musa's solicitations are not merely platonic, and that he is seeking sexual gratification to spice up his life on the road. Indeed, in the title story "This Is Lagos," the story opens: "They say Lagos men do not just chase women, they snatch them"(10), a statement that situates the enhanced vulnerability of women within the context of slick, citified men's wiles of entrapment. Significantly, Soha's mother issues this cautionary statement to her daughter on the eve of her departure for Lagos, rendering the title a caution to Soha/women to be alert particularly in their dealings with slick, citified men. However, Nwapa's women in these stories, whether in the professional sphere or in the more private sectors of life, prove themselves capable of dealing with the complexities of city existence. They are movers and shakers who struggle to either combat patriarchy whenever possible and often pay a high price for attempting to retain control of their lives. And even when they go against insurmountable odds, they tend to go down with a bang rather than a whimper.

Notes

1. The special issue of the *Research in African Literatures* 26.2 (1995), commemorating Flora Nwapa's life and literary career testifies to the recognition she is finally receiving for her contribution to African literature, particularly African feminist discourse.

2. Florence Stratton in *Contemporary African Literature and Politics of Gender*(1994), engages in a detailed discussion of phallocentric criticism in African literature. Her introduction ("Exclusionary Practices") and Chapter Four ("Flora Nwapa and the Female Novel of Development") are particularly significant for this paper.

3. Of marginalized female characters in the Nigerian literary canon, refer to Chinua Achebe's early novels and Wole Soyinka's *The Interpreters* among others. Cyprian Ekwensi's *Jagua Nana* and Elechi Amadi's *The Concubine* provide classic examples of stereotyped female portraitures inspired by male sexual fantasies. Also for critical essays discussing this issue, refer to Naana Banyiwa-Horne's "African Womanhood: The Contrasting Perspectives of Flora Nwapa's Efuru and Elechi Amadi's *The Concubine*" and Carole Boyce Davies' essay, "Motherhood in the Works of Male and Female Igbo Writers: Achebe, Emecheta, Nwapa and

Nzekwu," both in *Ngambika: Studies of Women in African Literature*(1986).

4. The execution of Nigerian writer, playwright, political activist, Ken Saro Wiwa on November 10, 1995 is a most blatant manifestation of this repression.

5. Flora Nwapa, *This Is Lagos and Other Stories* (Trenton, NJ: Africa World Press, 1992). All citations are to this edition and will be referred to by page numbers only.

6. This paper draws considerably on Bal's ideas regarding narrative theory. I am grateful to 'Molara Ogundipe-Leslie for her readings of this paper, and particularly for introducing me to Mieke Bal's work, which has been extremely beneficial in providing me with insights into the connections between a writer's narrative techniques and the process of manipulating readers' perceptions. I am also thankful to Adlancy Horne and Marcia Kutrieh for being sounding boards for my ideas and for reading the drafts of this paper.

7. Cited in Stratton (81). Both Jones and Palmer are taken to task extensively for their phallocentric criticism of the two authors' novels by Stratton(81-86).

8. The position taken here on femininity, also referred to as feminine self, is empowering of females—females investing in their looks because doing so gives them pleasure and enhances their self-image. This position differs from the perspective presented in Ogundipe-Leslie's *Re-Creating Ourselves,* which, as her discussion of Simone de Beauvior and Patricia Spack indicates (62), reflects Western feminist perspectives on femininity which have issued forth in reaction to static patriarchal theories based in biology, theology and psychoanalysis. While these static theories have given rise to stereotyping females as passive, submissive and dependent on account of being racked by feelings of inferiority at having bodies that are imperfect, these theories also form the backbone of the patriarchal system, its victimization of women, and its construction of women's involvements with activities that enhance their self-image into engagements aimed at winning male approval and validation rather than as ones women engage in for self-gratification.

9. al (5). Note her identification of characters as actors, that is, as "agents that perform actions....*To act* is...to cause or to experience an event." It is noteworthy that in defining fabula, she specifies to cause or to *experience* the "series of logically and chronologically related events" that make up the fabula. Male-authored texts tend to cast females as objects rather than subjects. See Naana Banyiwa Horne(119-120).

10. This issue is discussed in *Re-Creating Ourselves.* See especially the essay "Beyond the Women's Decade: A Message to Middle-Class Women"(159-164).

11. Gay Wilentz discusses the alternative route to acceptance and fulfillment that Efuru finds through becoming a special worshipper of Uhamiri (17-19). See also Naana Banyiwa Horne (127).

12. Gay Wilentz (9) notes the presence of this tension in Nwapa's *Efuru.*

13. The theme of empowerment in relation to African women's ability to make choices that bring them happiness is a major focus of Edris

Makward's essay on Mariama Bâ's novels in *Ngambika*. This subject, to my knowledge, has had limited exposure in African literary criticism.

WORKS CITED

Amadi, Elechi. *The Concubine.* London: Heinemann, 1966.

Bal, Mieke. *Narratology: Introduction to the Theory of Narrative.* Translated by Christine van Boheemen. Toronto: University of Toronto, 1989.

Davies, Carole Boyce. "Motherhood in the Works of Male and Female Writers: Achebe, Emecheta, Nwapa and Nzekwu." *Ngambika: Studies of Women in African Literature.* Ed. Carole Boyce Davies and Anne Adams Graves. Trenton, NJ: Africa World,1986:241-56.

Ekwensi, Cyprian. *Jagua Nana.* London: Heinemann, 1975.

Horne, Naana Banyiwa. "African Womanhood: The Contrasting Perspectives of Flora Nwapa's *Efuru* and Elechi Amadi's *The Concubine.*" *Ngambika: Studies of Women in African Literature.* Ed. Carole Boyce Davies and Anne Adams Graves. Trenton, NJ: Africa World, 1986:119-29.

Makward, Edris. "Marriage, Tradition and Woman's Pursuit of Happiness in the Novels of Mariama Bâ." *Ngambika: Studies of Women in African Literature.* Ed. Carole Boyce Davies and Anne Adams Graves. Trenton, NJ: Africa World, 1986: 271-81.

Nnaemeka, Obioma. "From Orality to Writing: African Women Writers and the (Re)Inscription of Womanhood." *Research in African Literatures* 25.4(1994):137-57.

Nwapa, Flora. *Efuru.* London: Heinemann, 1966.

____*This Is Lagos.* Trenton, NJ: Africa World, 1992.

Ogundipe-Leslie, 'Molara. *Re-Creating Ourselves: African Women and Critical Transformations.* Trenton, NJ: Africa World, 1994.

Ogunyemi, Chikwenye Okonjo. "Introduction: The Invalid, Dea(r)th, and the Author: The Case of Flora Nwapa, *aka* Professor (Mrs.) Flora Nwanzuruahu Nwakuche." *Research in African Literatures* 26.2(1995):1-16.

Stratton, Florence. *Contemporary African Literature and the Politics of Gender.* New York: Routledge, 1994.

Wilentz, Gay. *Binding Cultures: Black Women Writers in Africa and the Diaspora.* Bloomington: Indiana Univ. Press, 1992.

Vision and Revision: Flora Nwapa and the Fiction of War

Theodora Akachi Ezeigbo

INTRODUCTION

In a paper entitled "Nigeria—The Woman As A Writer" which Flora Nwapa presented in Iowa City, Iowa, USA, in 1984, she began with a brief summary of the history of Nigeria from the colonial period to political independence in 1960 and also touched on the violence and misrule that characterized the post-independence period, leading to military intervention and a civil war that lasted for almost three years. At the end of this sobering introduction, she asked a pertinent question: "Now, how do women fit into this seemingly chaotic situation?" (3). A reader of Nwapa's fiction would discover that she spent much of her literary career exploring the experiences of women in "chaotic situation(s)" created by men, especially those who have power and influence in society.

Nwapa was one of the many writers in Nigeria who recreated the Nigerian Civil War in their own imaginative literature. Some of her contemporaries who found the war a fitting subject for literature are Chinua Achebe, Cyprian Ekwensi, Wole Soyinka, Chukwuemeka Ike, John Munonye and Elechi Amadi. The war also introduced new writers, some of them writing for the first time after the civil war. As I stated in an earlier study (Ezeigbo, 1991, 81) four levels of literary participation could easily be identified among the writers who based their work on the experience of the war. The first group is made up of writers who experienced the war first-hand and Flora Nwapa and all the writers whose names appear above (except Wole Soyinka) belong here. They also form the core of what could be regarded as the literary pioneers in Nigerian literature. Soyinka is the sole representative of the second group, having been incarcerated by the Federal Military Government of Yakubu Gowon and having experienced the injustice of the war, outside the war zone. A third group is constituted by writers who neither saw military action nor were anywhere near the disturbed areas; examples are Buchi Emecheta, Kole Omotoso, Isidore Okpewho, and J. P. Clark (who wrote poetry about the crisis and the war). In the fourth group are those—particularly the younger writers such as Eddie Iroh, Kalu Okpi, Andrew Ekwuru, Okechukwu Mezu—who participated in the fighting on the Biafran side or engaged in other risky official roles—like military intelligence—and who later became writers after the war. It is logical to state, therefore, that the Nigerian Civil War has been explored from many perspectives by different categories of writers.

In this paper, I intend to examine Nwapa's recreation of an aspect of the Nigerian civil war from a perspective that is unambiguously feminine.[1] Her war writing include the novella, *Never Again* (1975), a story entitled "My Soldier Brother" in the collection *This is Lagos and Other Stories*(1971), and the first three stories ("Wives at War," "Daddy, Don't Strike the Match" and "A Certain Death") in a second collection entitled *Wives at War and Other Stories*(1980). Nwapa in her unpublished paper, "The Role of Women in Nigeria" described *Never Again* as "a war novel dealing with the role Biafran women played during the civil war in Nigeria" (8). As in her other writing, women's experience and women's activities form the thematic core of Nwapa's war fiction.

THE USES OF AUTOBIOGRAPHY

Though Nigerian writers present a common front in their depiction of the civil war as senseless and destructive, yet it is apparent that they have not only adopted different approaches to recreating the war experience but have also highlighted and emphasized different aspects of those events. Nwapa, for instance, uses the autobiographical mode (a point of view she adopts for the first and last time in her novels) and imbues it with authenticity and clarity.

There is every indication that Nwapa assumes the identity of author, narrator and heroine in the novel. In essence, Kate, the heroine of *Never Again,* is Flora Nwapa herself. Jane Bryce (1991) described the novel as a "semi-fictionalized memoir"(1). This verdict seems apt, especially when one takes into account some parts of Ejine Nzeribe's unpublished article "Remembrances of the War Period," in which she records her own and Flora Nwapa's experiences of life during the war in Ugwuta and during the invasion of the town by federal forces.[2] Through the eyes of a seven-year-old child, she witnessed her family's flight from Ugwuta, after the invasion of the town. Commenting on their new location after the evacuation, she writes: "We went to Atta, a village where there was no water. The huts were bare and the mosquitoes unbelievable" (1).

This depressing condition recalls a similar dehumanizing scene in *Never Again* where the narrator, Kate, relates her impressions of Akatta (a fictitious name for Atta?):

> ... we saw the poverty of the people staring us in the face. There was barrenness everywhere ....

> On our way back I kept asking Chudi how long he thought we would be in Akatta. It was a God-forsaken place. There was no life. There was no water. (66)

Parts of Ejine's account corresponds closely to Kate's in *Never Again.* Compare, for example, her comment on her mother's (Nwapa) habit of listening to the radio:

> Mama...always has her radio on, listening to news of the war especially the BBC. Looking back now, she was always alert as if awaiting something.... I reckon she wasn't amused by the events of the war because she argued always with people and people, I believe, thought her to be a saboteur! (1)

Now see Nwapa's version:

> We fell to talking about the BBC and its news. The view of the others was that I was a fool to take it seriously. The BBC was simply lying. Nigeria was not preparing any attack on Ugwuta. It was just not possible. "But we heard Nigerians were making arrangements for flat bottomed boats for the attack," I insisted. "Lies, all lies," many said. (24-25)

It is interesting to observe the close resemblance between the two passages, underscoring a shared experience between Flora and Ejine—mother and daughter—and indicating that Nwapa drew heavily from her personal experience of the war to write *Never Again.*

The difference between the two points of view or accounts is that Ejine has "recorded" the events of the war while Nwapa has "recreated" and "transmuted" it into fiction. Both narratives adopt the autobiographical mode; the former assumes the identity of an historian while the latter wears the garb of a literary artist. However, we must not lose sight of the point being stressed here, which is that *Never Again* is Nwapa's experience of the civil war; and having this knowledge at the back of one's mind is important as it does, to a large extent, explain Nwapa's peculiar narrative technique in the novel, which is suffused with passionate utterances that all too often plummet down to moralizing statements and rhetorical questions:

> Where were the Nigerian troops who entered it (Ugwuta) barely three days ago? Where was everybody? What folly? What arrogance, what stupidity led us to this desolation, to this madness, to this death? When this cruel war was (sic) over, there will be no more war. It will not happen again, never again. NEVER AGAIN, never again. (70)

The autobiographical mode has often been described as a favorite of female writers (Betty Wilson 1989:vii). The autobiography is a genre of assertion which develops as a way of presenting the self-esteem of the protagonist (or author) who is portrayed as always better than the average person (Ezeigbo 1988:128). In *Never Again,* the protagonist's experience of the war and her responses to the events are manipulated to emphasize her triumph over the trauma and psychological dis-orientation that the tragic war inflicted on people. Kate's resilience, her courage and her survival strategies enable her to rise above the hypocrisy that characterizes the actions of the community leaders and the so-called elite as well as to consistently reject what Gloria Chukukere (1995) aptly called "the social and military propaganda which seeks to deceive rather than protect the people" (139).

Two non-fictional autobiographies—Leslie Ofoegbu's *Blow the Fire* (1986) and Rose Njoku's *Withstand the Storm: War Memoirs of a Housewife* (1986)—have also graphically explored the events of the Nigerian crisis and civil war.[3] Njoku's position as the wife of one of the most senior army officers in Nigeria and later in Biafra (before he was detained by the government) gave her access to not only the Biafran Head of State but also to other highly-placed individuals in the enclave. This unique position, hardly occupied by any other woman in secessionist Biafra, enabled Njoku to experience life from a wider perspective than most women in the area. *Withstand the Storm* gains from this, as can be seen from the richness of its account and the broader perspective covered than is the case in *Never Again* and *Blow the Fire.* Indeed, Jane Bryce denotes this distinction when she compares Njoku's book with Buchi Emecheta's *Destination Biafra* (1982) which, to my mind, is the most ambitious novel based on the events of the Nigerian crisis and civil war in terms of scope and the number of characters. Bryce states:

> *Withstand the Storm* makes an eloquent counterpoint to Emecheta's *Destination Biafra,* in the sense that Rose Njoku actually underwent many of the experiences Emecheta ascribes to the fictional Debbie Ogedemgbe. (39)

Unlike *Withstand the Storm* and *Destination Biafra,* the "universe" of Nwapa's *Never Again* is strictly limited to those events that have a direct effect on the lives of her protagonist and her family. There is every reason to believe that the limitation of Nwapa's account is consequent upon her limited experience of the war, for it does not seem that she ever left, Ugwuta where she lived throughout the period of the war after the evacuation from Port Harcourt (except, of course, the short period she fled to Atta with her family after the Federal Forces reached Ugwuta). Ejine Nzeribe's unpublished "Remembrances" confirm that Nwapa did not work all through the period she was in Ugwuta. She and her family lived a rather private life. Conversely, the writers of Nwapa's generation who lived inside Biafra during the war were deployed to one job or the other in the various ministries or departments. Some of them were sent on missions abroad to represent the "new nation." The extent of the writers' activities during the war is conveyed in a conversation Chinua Achebe had with Harvey Swados in 1969 in which he stated:

> My commitment isn't to a government but a cause. Fortunately, so far, although they've called on me for all kinds of service, the government has never asked me to do anything that I would balk at doing. (9)

Nwapa appears to be the only one of them who lived a private life, totally unconnected to government or public service. (Was this by choice, or could it be she was never approached by government, as the rest were?) Her uninvolvement in this regard might explain the absence of military action, political intrigues, or even diplomatic manoeuvers in her war writing. This limitation is also observable in Ofoegbu's *Blow the Fire*, (1986) which she describes as the account of her family's experiences, in the introduction to *Blow the Fire*, she confesses:

> We did not move in high circles so those looking for names of well known personalities will be disappointed. We were not close to the army so there are no military revelations. (Ofoegbu 1986:2)

In spite of this limitation, however, both accounts are gripping and represent two authentic views (factual and fictional) of the tragic war, narrated from women's perspective. For Nwapa the limitation has another implication. It empowers her vision in a different direction from that of the male authors but equally important. By focusing her searchlight on Ugwuta society—a microcosm of Biafra, a miniature canvas—Nwapa is able to give a profound insight into "the evils of war and the evils of propaganda in wars," to use her exact words in her interview with John Agetua (6). Ironically, by being truly local, her vision attains universal applicability. Anita Kern (1978), in her review of *Never Again*, commends Nwapa for giving the reader "a realistic glimpse into the tensions of the war on a basic level" (58) and providing "an enlightening account of plain day-to-day coping under fear, and a moving and strong plea that such a terrible clash should indeed happen never again" (59). Nwapa's main preoccupation in this novel—and in her short stories based on the war—is to explore women's ability to cope under the stress of a civil war and to depict the survival strategies women devised to counteract the mental and physical dislocations that prevailed. As in her other work, Nwapa's vision in the war writing is woman-centered and the main actions in the novel and short stories revolve around individual families as their destinies unfold in the events of the cruel war. Flora Alexander's 1989 comment on the fiction of contemporary women novelists writing in the tradition of the English novel is applicable to Nwapa's vision:

> The fact of their gender has had some effect on their experience and their perceptions of the world, and this in some measure reflected in the nature of the fiction they write. (x)

In her war fiction, Nwapa validates women's war experience as a fit subject for literature and an important addition to the canon of Nigerian civil war writing.

The Poetics of Survival—Women and War

The most prominent theme in *Never Again* and Nwapa's war short stories is survival. In a way, it is valid to say that this theme is the focus of everything women prose writers have written on the war. Njoku, Ofoegbu, Emecheta and Nwapa deal with women's struggle to survive the war with their families. According to Anaedobe and Ezenwaka (1982), women did not participate in the fighting in the wars fought in Igbo traditional society and in the era of colonization. Rather, they played supportive roles such as cooking for the fighting men, trading and providing food for their families, and taking care of the sick and the wounded.

From the numerous factual and historical accounts written about the civil war, especially by those who were in Biafra during the fighting, one could see that women remained consistent with their age-old role of providing support services, food, and sustenance to their families and the fighting men. It is rather disappointing that most male writers who recreated the events of the war in fiction neglected this aspect of women's contribution but chose to highlight and exaggerate women's moral laxity,[4] forgetting that this was insignificant in comparison with women's efforts towards winning the war, towards the survival of the family, and towards the rearing of children.

What makes Nwapa's war writing stand apart is her refusal to delineate women in the war situation as prostitutes whose personalities are crippled by malignant moral lapses. Nwapa has always seen women as independent, assertive, and economically active. Consequently, she does not perceive them as sex objects whose only means of survival is the selling of their bodies to the highest bidders. Neither does she indulge in sensational details of women's marital infidelities, as Ekwensi does in *Survive the Peace* (1976) and Aniebo in *The Anonymity of Sacrifice* (1974). Nwapa is fully conscious of the fact that women played their traditional roles as providers, nurturers, and caretakers and has chosen to recreate this aspect of the war experience. By portraying female characters who are strong, independent and assertive, Nwapa writes against the female stereotyping that has been a feature of most male-authored texts in Nigeria, before and after the civil war. In *Never*

Again, Kate's actions and responses are guided by her conviction that the war is needless and that the only reasonable thing to do is to ensure the survival of her family. Her decision is strongly conveyed in these words:

> I was determined not to see my children suffer. I would sell all I had to feed them if I had to. They were not going to be hungry. They would not suffer from kwashiorkor. (27)

Kate's determination to protect her family is shared by Ndidi Okeke in the story, "Daddy, Don't Strike the Match" in *Wives at War and Other Stories.* In spite of her husband's experiments, Ndidi knows Biafra will lose the war and "all she was interested in was the survival of her children, herself and her husband and her dear ones" (29).

Nwapa focuses on respectable married women who are determined to survive the war. Like Buchi Emecheta, she is engaged in a revisionary act to commemorate women's activities in the war—activities that are either trivialized or disregarded by male authors. Emecheta, in *Destination Biafra,* subverts male dominance by making Debbie Ogedemgbe not only an assertive and independent woman but also a soldier. Nwapa's form of empowerment for her female characters is more feminine than Emecheta's but no less effective. Her women, unlike Debbie, do not carry guns but rather look after husbands and nurse babies. Yet these roles do not limit them nor reduce them to appendages of men. They are not passive or colorless either. In addition to their domestic roles, they also organize themselves to care for the fighting men and prepare food for them. The remarkable thing about them is that they are totally aware, and know what they want and how to go about it. Kate's resolution to evacuate as soon as a town is threatened is a result of her experience in the places she had fled earlier:

> Death was too near for comfort in Biafra. And for us who had known no danger of this kind before it was hell on earth. I meant to live at all costs. I meant to see the end of the war. (5)

Consequently, she does little else for the duration of the story except take measures to ensure their survival.

Kate refuses to be confused by the propaganda of the Biafran authority and the hypocritical acts of the Ugwuta community leaders and intellectuals. Knowledgeable and intelligent, she sees through the deceit and incompetence of those who pretend to know the war situation and to lead the people. She reasons:

> Words were impotent. Biafra could not win a civil
> war by mere words. How I longed to say that to
> Ojukwu. But who was I? And besides, it was too late,
> too damn late. We had already lost the war. We lost
> the war when we lost Port Harcourt. It was sheer
> madness fighting after Port Harcourt. All right-
> thinking people knew this. What we should have
> done was surrender. (25)

Many of Nwapa's women share Kate's vision of the war as needless and futile. In *Destination Biafra*, Debbie and all the women who
are with her in the jungle are also aware that the war has achieved
nothing except the destruction of life and property. But like Kate
and the other women in Nwapa's short stories on the war, the
women are powerless to change the situation or influence the leaders. Their impotence is further compounded by their vulnerability, leading to their being subjected to injustices such as rape,
physical abuse, and exploitation—many of them even die, like the
pregnant woman in *Never Again* and Baby Biafra's mother in
Destination Biafra.

Faced with psychological and physical trauma as well as imminent death from the war machinery or starvation, the women in
Never Again devise various strategies of survival. Relying on the
Biafran forces, as Kate discovers, is precarious. The cowardly acts
of soldiers (officers and men) and of the civilian elite who are in
leadership positions alienate the women. The threat to life is worsened by insensitive and callous men who stall the evacuation of
threatened towns and cause the death of thousands of people,
especially women and children. Kate has no illusions about Biafra
and sees the enclave as a disaster area and a deathtrap. The result
is that she becomes cynical and is ever ready to move at the short-
est notice. The family's car is kept ready and clothes packed in case
they have to move quickly. Her accurate predictions about the war
situation, particularly the fall of some towns, win her husband
Chudi to her side, making him cooperate with her in practicaliz-
ing her survival strategies.

Other methods of survival adopted by women include farming,
buying and selling in the local markets, and participating in the
"attack trade" (which flourished at the borders between Biafran
women and traders from the Nigerian territories) and paying others to take the place of their relations in the army. The woman nar-

rator in "A Certain Death," a story in *Wives at War and Other Stories,* is determined to save her brother by paying someone to replace him when he is called up for active service. Her desperate desire to protect him from "a certain death" in the warfront without deliberating the moral implications of her action is born of her knowledge of the real situation of things, as illustrated by her reflections:

> Suicide it was. To be compelled to go to war and to be conscripted into the Biafran army was suicide, pure and simple. (37)

She prays fervently:

> God, please be merciful to us and save us from death, from air raids, from accidental discharges, from night-marauders, from deserters, from saboteurs, from white mercenaries, from overzealous leaders. (37)

In "My Soldier Brother," one of the stories in *This is Lagos and Other Stories,* it is the elderly aunt, Monica, who condemns openly Adiewere's enlistment. The young officer's subsequent death justifies her opposition and affirms her perception of soldiering as a dangerous and destructive profession and war as needless and senseless.

In *Never Again,* trading, buying, and selling are activities which take up much of women's time. War or no war, these preoccupations have to go on if the community is to survive. Unfortunately, the destructive potential of the war is felt in the marketplace when the enemy planes raid, strafe and rocket the markets, killing hundreds of people there. Nwapa captures vividly the imminent danger, in Kate's words:

> The market place was full of people. They were buying and selling. Tears filled my eyes as I picked my way among the crowd of market women. In a short time, there would be a stampede. Many people there would not be able to go back to their homes to take anything. Many would leave their wares behind and run empty handed. Many would die in the stampede. (45)

Women who take part in the "attack trade" are no less exposed to great danger. This form of trade made it possible for manufactured goods and certain food items to be available in blockaded Biafra.

Incidentally, Nwapa has not given enough space to discussing the activities of women involved in the trade in *Never Again*. It is to the next novel, *One Is Enough* (1981), that one has to turn in order to read about the brave and independent women who engaged in the trade. They are described as "the new generation of women contractors.... They were all involved in the 'attack trade' during the war" (52). The omniscient narrator in *One Is Enough* has this to say about one of the ten women being discussed:

> Yet there was another who was engaged in the 'attack trade' during hostilities. She went to one of the war fronts to buy cigarettes, batteries and 'guff' with some other women 'attack traders.' While they were sleeping in a hut, some stragglers attacked them at night. She was brave, she had a toy gun that belonged to her son. She always travelled with it. When the straggler woke her up demanding all she had, which was tied onto her waist, she pointed her toy gun at him. He panicked and she escaped. But the other women in the group were unprepared. They robbed them and shot one who resisted. (53)

Nwapa's exploration of women's survival strategies during the war reveals that women are exposed to as much danger as the fighting men in any war situation. In fact women and children are shown to be the main casualties of war. In times of crisis, it is the weak and the helpless who are the greatest victims.[5] The travail and death of the pregnant woman in *Never Again* who tries to escape from Ugwuta with her children evoke the biblical prophetic saying of Jesus Christ (St. Matthew's Gospel, Chapter 24, verse 19) that, in the last days, the weak will be sorely afflicted: "And woe unto them that are with child, and to them that give suck in those days!"

Women's struggle to survive and their disenchantment with the war bring them into conflict with some members of their immediate family or community. Kate's anti-war position, her cynical comments and strong voice of dissent attract reproach from family members and hostility from the rest of the town. At one point her family's movement is restricted, their home searched for infiltrators and their activities closely monitored to prevent them from fleeing the beleaguered town in their car. Kate, Chudi, and Bee are regarded as traitors; but, ironically, the real traitors are Madam Agafe, Kal and Aniche who, though fanatical in their support for the war, are the first to flee even before the federal forces enter

Ugwuta. Madam Agafe is a hypocrite and a liar and is the most obnoxious character in the novel, apart from Kal.

In the title story "Wives at War," Nwapa focuses her searchlight on women leaders and activists whose extra-domestic roles and survival strategies include mobilizing women to contribute to the war effort and provide food and care to the fighting forces; yet the strongest impressions one gets of them are those of rival groups competing with one another, pursuing self-aggrandizement and preoccupied with the projection of their self-importance. When a false report circulates that a group of women has been sent to London to brief the Queen on Biafra's problems, some women leaders storm the office of the Foreign Secretary demanding to know why they themselves were not the ones sent. They appear to be less concerned about the raging war than their petty rivalries and personal ambitions. As in the case of Madam Agafe, Nwapa satirizes these women leaders and contrasts them with Kate, Ifeoma, Ndidi, and Bee, who are more realistic, humane and practical in their wish for the end of the tragic war and in their care for their immediate families.

In the light of these observations, particularly those relating to Madam Agafe, it is obvious that Chidi Amuta is unfair to Nwapa in his criticism of *Never Again* when he states:

> As a result of the author's preoccupation with feminist propagandizing, the women in this story are the only courageous and sensible creatures. (95)

The least that can be said is that both male and female characters receive knocks from Nwapa, depending on their nature and activities.

AESTHETICS AND THE IDIOM OF VIOLENCE

Earlier in this study, attention was drawn to Nwapa's preference for the autobiographical mode in depicting war experience in *Never Again*. It was also noted that the novel is the first and only novel where Nwapa's 'self' and personal experience come closest in fiction.[6] Willfried Feuser's (1975) observation on Okechukwu Mezu's novel, *Behind the Rising Sun,* that the main characters "are drawn from real life and only thinly disguised" (41) is applicable to some of Nwapa's characters in *Never Again,* particularly the narrator and her husband. Her choice of the autobiographical mode might have been guided by the desire to make bold and direct statements to register her condemnation of the civil war which nearly destroyed her society. Furthermore, the search for a medium to communicate her vision of the Biafran holocaust terminated in the adoption of the realistic mode as the most suitable option to recreate her individual perception of the civil war.[7]

The task that confronts the writer who uses the realistic mode is maintaining a reasonable degree of detachment from his/her subject—that is, the extent to which the artist is able to stand aside, as James Joyce succinctly puts it, "paring his finger nails." Has Nwapa's use of both the autobiographical and the realistic modes provided aesthetic satisfaction to her readers? After a careful reading of *Never Again* and some of her short stories on the war, it is fair to admit that in some sections Nwapa seems to have allowed emotional involvement to overshadow aesthetic consideration. In such instances, her voice becomes strident; and her moralizing tone becomes all too apparent. Consider, for instance, this passage from *Never Again*, in which the author castigates the two sides (Nigeria and Biafra) for disagreeing and resorting to war:

> Why, we were all brothers, we were all colleagues, all friends, all contemporaries; then, without warning, they began to shoot, without warning, they began to plunder and to loot and to rape and to desecrate and more, to lie, to lie against one another. What was secret was proclaimed on the house tops. What was holy was desecrated and abused. (70)

This kind of emotional outburst is not seen in such other texts of hers as *Efuru* (1966) and *Idu* (1970), where the omniscient narrator's voice is restrained and judiciously modulated.

Having said that, there is, however, ample evidence in the novel that Nwapa has made creative use of the realistic mode in significant ways. Perhaps, Yemi Mojola's words best capture this virtue in Nwapa's narrative design:

> With deep insight, the psychological and physical stresses of the war on the individual and the collectivity are vividly portrayed. (20)

Her achievement lies in her descriptive power, her use of images and metaphors to give the reader insights into the basic themes that preoccupy her in the novel. Consequently, the reality of the civil war is imbued with new meanings and enriched with fresh perspectives that may not have been plumbed by other writers.

This is particularly the case with the theme of survival which Nwapa gives in depth and dynamic treatment.

In *Never Again,* madness or the threat of it in the mental state of characters becomes an important metaphor to denote the destructive impact of the psychological, emotional, and physical stresses that war inflicts on people. The prevalence of the metaphor of madness and insanity in the short stories based on the Nigerian civil war is the subject of this writer's unpublished paper entitled "The Taste of Madness: The Short Stories on the Nigerian Civil War" (Ezeigbo, 1992). A similar pattern is observable in Nwapa's war fiction. The metaphor of derangement is best explored in *Never Again,* where all the characters or, at least, most of them exhibit some form of mental disorder or instability. Kate's flight from different locations—Enugu, Onitsha, Port Harcourt, Elele, and Ugwuta—and her unrelieved and sometimes panic-stricken thoughts on how to survive the war with her family have adverse effect on her mind. Her fears are articulated in the words "madness" and "insanity" which she utters several times in the novel. Here are some examples:

> All day I was irritable. Perhaps I was getting mad. What was actually wrong with me? Was I really getting off my head? Everybody could not be mad at the same time. Perhaps it was I, I getting mad. (35)

> God, spare us this madness. This insanity, this hell on earth... (57)

> Why don't they tell them to march on and march on, so that, this madness will end. So the war will end. (58)

> Are you insane? Have you gone out of your mind? .... You are crazy, Agafe you are crazy. (59)

Other instances where the word "madness" or its derivatives recur are 41, 55, 60, 63, and 75. Ezekoro, whose madness has turned full circle, perishes when he chooses to return to the deserted Ugwuta. His anti-survival move is comparable to the decision of the boy in the Igbo folktale who stubbornly returned to the farm only to be killed by evil spirits.[8] Nwapa seems to be saying that war is like a corrosive which destroys people's minds, driving the victims to a state of extreme anxiety or self-destruction. The condition is manifested in a lesser or greater degree in many of the characters in *Never Again* and the short stories. Madness with its destabilizing influence is glimpsed in the melancholia suffered by Bisi in the story "Wives at War" and the unnamed narrator's brother, after the loss of his family in "A Certain Death."

Other motifs Nwapa explores in *Never Again* are barrenness and aridity, which literally describe the condition of Akatta but, on the other hand, are symbolic of the sterility of the whole of Biafra. It is a bleak situation which is detrimental to the survival not only of the new nation but also of the people. Disease imagery abounds in the novel as seen in the prevalence of leprosy, kwashiorkor and craw craw which are physical manifestations of the corruption in Biafra and the decadence in spiritual health and moral uprightness. The diseases are a threat to people's lives as they ravage the whole community, causing the environment to degenerate to what Chikwenye Okonjo Ogunyemi aptly calls "an amorphous mortuary" (6).

Marie Umeh's perceptive comment on Emecheta's evocative description of feelings of imprisonment in the female condition in *Destination Biafra* (201-2) reminds one of a similar treatment by Nwapa in *Never Again* of women's obsession with the feelings of entrapment and insecurity and their desperate effort to escape, even as Biafra seems to journey towards disintegration.[9] It is noteworthy that the female characters who are motivated by love for their families and strong desire for their survival are the most engrossed with seeking avenues of escape from the tragic situation of the war. The nature of war engenders constant movement and brings about, in *Never Again,* the journey motif similar to that which Marie Umeh (1987) and Grace Okereke (1994) have identified and discussed in their well-written essays on Emecheta's *Destination Biafra.* Nwapa's Kate and Emecheta's Debbie share some common psychological and mental experiences which Marie Umeh (1987) describes as a "passage from idealism to realism ... from innocence to experience, from ignorance to a profound knowledge of evil in the world " (199). After her flight, following the recapture of Ugwuta, the narrator in *Never Again* returns home to discover the enormity of the destruction the town suffered. Her sense of outrage and her despair are captured in bare and terse language, highlighting the violence involved:

> The place was desolate.... There was no life. No goats, no chickens, nothing. Everywhere one went, one saw clear evidence of battle.... The thatch-houses overlooking the Lake were all burnt down. There

was evidence of mass graves beside the burnt down
houses. The barns were empty. Bullet holes gaped
from the walls of the houses that were still standing. (79-80)

Her "knowledge of evil in the world" has become complete.

CONCLUSION

Nwapa's imaginative recreation of the Nigerian civil war leaves no
one in doubt about her perception of the war as a destructive and
dehumanizing historical event that should have been prevented
from occurring. She sees the war conflict as bizarre and tragic, without heroes, or even heroines. She does not view the ordinary soldier or the common people as heroes, as Festus Iyayi posits in his
novel, *Heroes* (1980), whose events are based on the civil war. From
Nwapa's point of view, the war brought out the worst forms of
sadism, selfishness, hypocrisy, and dishonesty in people. The odd-
ity of the situation is symbolically dramatized by the experience
of the previously deaf-and-dumb Ezekoro, who regains his power
of speech and becomes strangely empowered to articulate with
words the abominations in the land and, perhaps, sound a note of
warning to people.

The abomination in the land alienates the powerful and ubiquitous "Woman of the Lake"[10] who hardly lifts a finger to ward off
the invasion of Ugwuta and its destruction by the federal forces.
Could this be a reaction against her straying children who call on
her for succor and protection and at the same time cry to the Holy
Spirit for help (57)? This confusion in codes of religion which features on several pages of the novel—signifying a mingling of the
indigenous and foreign religions—naturally introduces insecurity
and uncertainty, which in turn jeopardize survival and unified
action on the part of the community. Interestingly, the Lake is "The
only thing that stood undisturbed, unmolested, dignified and solid
.... It was calm, pure, peaceful and ageless. It sparkled in the sun-
light, turning now blue, now green as the sun shone on it" (80).
What a contrast with the rest of the environment! It all seems to
symbolize that the evils human beings visit upon one another are
not allowed to taint or unbalance nature or the powers that control nature.

However, women appear to be a vital source of hope for the
regeneration of society. As agents of survival, they devote their
energy to countering the forces of destruction, hunger and starva-

tion, through their survival strategies. Nwapa's revisionary position implies that the war would not be won through violence or military campaigns, but through the termination of hostilities and the preservation of life. And women seem to be in the best position to bring about the realization of these ideals.

NOTES

1. Hers is a perspective that is steeped in women's culture and explores the war experience from women's point of view and with a female narrator's voice. This is the opposite of the dominant masculinist perspective which is represented in the war writing of male authors like Aniebo's *The Anonymity of Sacrifice* (1974), Ekwensi's *Survive the Peace* (1976) and *Divided We Stand* (1980), Ike's *Sunset at Dawn* (1976), Iroh's *Forty Eight Guns for the General* (1976) and *Toads of War* (1979), Iyayi's *Heroes* (1980), Munonye's *A Wreath for the Maidens* (1973), Okpi's *Biafra Testament* (1982), Omotoso's *The Combat* (1972), and Saro-Wiwa's *Sozaboy* (1985).

2. Ejine Nzeribe is Flora's first daughter.

3. Leslie Ofoegbu does not prevaricate in the least as to the impression her autobiography is meant to leave in the mind of the reader, the book's front cover itself announces: "The true story of a woman's fight for survival during the Nigerian Civil War."

4. This observation is true of texts such as Iroh's *Toads of War,* Ekwensi's *Survive the Peace,* Aniebo's *The Anonymity of Sacrifice* and, to some extent, Okpewho's *The Last Duty.*

5. It is no surprise that women's organizations all over the world work for peace and press for a world that is free from wars and other forms of crisis, knowing full well that women and children suffer most when such violent situations arise. Maintaining peace in the world at continental, regional and national levels was strongly advocated by women of all nations who attended the Beijing Conference on Women in China in September, 1995.

6. Nwapa published a total of six novels but she hardly dramatized herself or her life in her work as some writers have done. Compare her, for instance, with her compatriot, Buchi Emecheta, who has written herself into her work many times over, as is clearly evident in *In the Ditch* (1972), *Second- Class Citizen* (1975), *Double Yoke* (1983), and *Head Above Water* (1986).

7. A similar search is evident in the work of other writers: Soyinka settled for the symbolic and absurdist modes in *Madmen and Specialists;* Omotoso for the parabolic and allegorical modes in *The Combat* and Okpewho for what he himself described as "the collective evidence technique" in *The Last Duty.*

8. In the folktale, the boy forgot his *oja,* a musical instrument, in the farm, and insisted on returning for it against his parents' advice. He died in the attempt, like the mad Ezekoro. Metaphorically, Biafra's fate would, like Ezekoro's, lead to self-destruction.

9. The premature death of the woman and her stillborn baby in *Never Again,* just like the death of Baby Biafra and its mother in Emecheta's *Destination Biafra,* negates the philosophy of survival which women in both novels espouse. It also symbolizes the demise of the new nation, Biafra, which is destined not to survive.

10. Nwapa's strong and effective portrayal of Uhamiri, The Woman of the Lake, as the source of inspiration, protection and sustenance for the people of Ugwuta is not maintained in *Never Again* as in the preceding novels, *Efuru* (1966) and *Idu* (1970). What could account for this? Could it be a result of the decline in Uhamiri worship by the community—a direct consequence of Christian influence in Ugwuta in the period covered by the events in the war novel?

WORKS CITED

Achebe, Chinua. *Girls at War and Other Stories.* London: Heinemann, 1977.

Agetua, John. "In the Interest of Traditional Women—Interview with Commissioner Nwakuche." *Sunday Observer.* Benin City, Nigeria. (18 August 1974).

Alexander, Flora. *Contemporary Women Novelists.* New York: Routledge, Chapman and Hall, 1989.

Amadi, Elechi. *Sunset in Biafra.* London: Heinemann, 1973.

Amuta, Chidi. "The Nigerian Civil War and the Evolution of Nigerian Literature." *Canadian Journal of African Studies.* 17. 1 (1983).

Anaedobe, B. A. & Ezenwaka, T. N. *A Short History of Uga.* Onitsha: Ethel, 1982.

Aniebo, C. K. *The Anonymity of Sacrifice.* London: Heinemann Educational Books, 1974.

Bryce, Jane. "Conflict and Contradiction in Women's Writing on the Nigerian Civil War." Ezeigbo & Gunner, eds. The Literatures of War: Special Issue, *African Languages and Cultures* 4. 1 (1991):29-42.

Chukukere, Gloria. *Gender Voices and Choices: Redefining Women in Contemporary African Fiction.* Enugu: Fourth Dimension, 1995.

Clark, J. P. *Casualties.* London: Longman, 1970.

Ekwensi, Cyprian. *Survive the Peace.* London: Heinemann, 1976.

___. *Divided We Stand.* Enugu: Fourth Dimension, 1980.

Ekwuru, Andrew. *Songs of Steel.* London: Rex Collins, 1979.

Emecheta, Buchi. *In the Ditch.* Barrie and Jenkins, 1972

___.*Second-Class Citizen.* London: Allison & Busby, 1975.

___.*Double Yoke.* London: Ogwugwu Afor, 1981.

___.*Destination Biafra.* London: Allison & Busby, 1982.

___.*Head Above Water.* London: Fontana, 1986.

Ezeigbo, Theodora Akachi. "Artistic Responsibility: Male Characterization in Buchi Emecheta's Novels." *Lagos Review of English Studies* 10 (1988):126-138.

___.*Fact and Fiction in the Literature of the Nigerian Civil War.* Lagos: Unity, 1991.

___. *"The Taste of Madness: The Short Stories on The Nigerian Civil War."* Unpublished paper,1992.

Ezeigbo, Theodora Akachi & Gunner, Liz, eds. The Literatures of War. Special Issue: *African Languages and Cultures.* 4.1 (1991).

Feuser, Willfried. "A Farewell to the Rising Sun: Post-Civil War Writings from Eastern *Nigeria." Books Abroad.* 49 (1975): 40-49.

Ike, Chukwuemeka. *Sunset at Dawn.* London: Fontana, 1976.

Iroh, Eddie. *Forty-eight Guns for the General.* London: Heinemann, 1976.

___.*Toads of War.* London: Heinemann, 1979.

Kern, Anita. "Flora Nwapa. *Never Again." World Literature Written in English* 17.1 (April 1978).

Mezu, Okechukwu. *Behind the Rising Sun.* London: Heinemann, 1973.

Mojola, Yemi. "The Works of Flora Nwapa." *In Nigerian Female Writers: A Critical Perspective.* Ed.,. H. Otokunefor & O. Nwodo. Ikeja: Malthouse Press Ltd., 1989.

Munonye, John. *A Wreath for the Maidens.* London: Heinemann, 1973.

Njoku, Rose. *Withstand the Storm: War Memoirs of a Housewife.* Ibadan: Heinemann, 1986.

Nwapa, Flora. *Efuru.* London: Heinemann, 1966.

___.*Idu.* London: Heinemann, 1970.

___.*This is Lagos and Other Stories.* Enugu: Nwamife, 1971.

___.*Never Again.* Enugu: Nwamife, 1975.

___.*Wives at War and Other Stories.* Enugu: Tana, 1980.

___.*One is Enough.* Enugu: Tana, 1981.

___.*"Nigeria—The Woman As A Writer."* A paper presented in Iowa City, Iowa in 1984.

___.*"The Role of Women in Nigeria."* Unpublished paper.

Nzeribe, Ejine. "Remembrances of the War Period." Unpublished manuscript.

Ofoegbu, Leslie Jean. *Blow the Fire.* Enugu: Tana, 1986.

Ogunyemi, Chikwenye Okonjo. "Introduction: The Invalid, Dea(r)th and the Author: The Case of Flora Nwapa, *aka* Professor (Mrs.) Flora Nwanzuruahu Nwakuche." *Research in African Literatures.* 26.2 (1995):1-16.

Okereke, Grace Eche. "The Nigerian Civil War and the Female Imagination in Buchi Emecheta's *Destination Biafra." Feminism in African Literature: Essays in Criticism.* Ed., Helen Chukwuma Enugu: New Generation, 1994.

Okpewho, Isidore. *The Last Duty.* London: Longman, 1976.

Okpi, Kalu. *Biafra Testament.* London: Macmillan Press Ltd., 1982.

Omotoso, Kole. *The Combat.* London: Heinemann, 1992.

Saro-Wiwa, Ken. *Sozaboy.* Port Harcourt: Saros International, 1985.

Soyinka, Wole. *Madmen and Specialists.* London: Methuen, 1971.

Swados, Harvey. "Chinua Achebe and the Writers of Biafra." *New Letters.* 40.1 (1973).

Umeh, Marie. "The Poetics of Thwarted Sensitivity." *Critical Theory and African Literature.* Ernest Emenyonu Ed. Ibadan: Heinemann, 1987.

Wilson, Betty. "Introduction to Juletane" (A novel by Myriam Warner-Vieyra). Oxford: Heinemann, 1989.

Orality and the Metaphoric Dichotomy of Subject in the Poetry of Flora Nwapa's *Cassava Song and Rice Song*[1]

Ezenwa-Ohaeto

Flora Nwapa's poetry in *Cassava Song and Rice Song* has received minimal critical attention even from those critics who claim to have paid the most attention to either the writings of Nigerian female writers or the writing of Flora Nwapa. In as much as few published Nigerian female poets exist, the impression from such minimal attention is that not all the writing (or, to be specific, poetry) by female writers deserve critical attention. That impression is certainly disturbing, for it could be assumed that either the poets are incompetent or that the poetry is substandard. But that is far from the truth, for there are poems by Nigerian female writers that are competently executed and their subject matter relevant to the realities of Nigeria and Africa. Thus, it has been relevantly observed elsewhere that "the study of contemporary Nigerian poetry may never be complete without the assimilation of those feminine poetic impulses" (Ezenwa-Ohaeto 1988:668).

However, it should be recognized that the silence over Flora Nwapa's poetry is significant in several ways precisely *because* the corpus of criticism on modern African literature has been generated by both female *and* male critics in substantial numbers. And it is almost a decade since Carole Boyce Davies (1986) insisted that "Flora Nwapa, [is] clearly a victim of literary politics. Cavalierly dismissed by many critics as unimportant, after revision she is credited with recreating that oral culture that African society is noted for and making important contributions to her genre." (14) But that revision has not been extended by those "critics of revision" to her only poetry collection. Perhaps they have over-imbibed Katherine Frank's (1987) injunction with reference to African female writers that "it is to the fiction that we must continue to turn. Imagined worlds are more potent than real ones, refined as they are of hesitation, doubt and compromise" (33). Although it is possible to explain that the number of prose works by African female writers exceed the number of their poetic works, the relegation of Flora Nwapa's poetry to a nebulous literary zone that is neither fiction nor poetry has not helped matters. One critic refers to the *Cassava Song and Rice Song* as a "book of prose poems" (Umeh 1995, 23) while the diligent Brenda F. Berrian (1995), who has compiled several bibliographies of African women writers, classifies it as part of Nwapa's children's literature (126). Thus, it is not surprising that Aderemi Bamikunle's (1994) study of feminism in Nigerian poetry by women[2] does not refer to Flora Nwapa's poetry.

It may well be that the various aspects of the literary politics attending the works of Flora Nwapa have metamorphosed beyond the issue of the gender of the writer, for neither the insistent female critics nor the male critics have bothered to provide reasons for their literary silence over her poetry and children's literature. It is unconvincing to interpret this issue as does Florence Stratton, (1994) who argues that "Nwapa's reputation as a writer is only just beginning to recover from James's (1971) and Palmer's (1968) assault, despite the early efforts of Ernest N. Emenyonu (1975) who has from the outset defended her" (81). That recovery has not been portrayed in the criticism of Nwapa's children's literature and poetry, which may mean that there is more to it than meets the eye of the reader. Perhaps the poetic talents displayed by Nwapa are not significant, but this contention cannot just be implied; all literary views must be analyzed adequately before conclusions are reached. All the same, it has been demonstrated that, even if the poetic talents are inadequate, the *subject matter* of the poetry is significant and insightful because Nwapa "uses the poetic medium to comment on sundry issues. The two songs could be perceived as symbolic, for while the "cassava song" symbolizes tradition, the "rice song" symbolizes Western influence. In the process, Nwapa indicates the subtle manner through which corrupt and destructive influences could be introduced into the society" (Ezenwa-Ohaeto 1988, 664). It is in confirmation of this observation that Chimalum Nwankwo, in his very brief but relevant statement (1995) on the poem, comments that in *Cassava Song and Rice Song,* through "a subtle and seemingly innocuous manner, the old conflicts in the center of the crisis of Western contact re-emerge. This emergence is immediately in stride with the Igbo *word.* The old Igbo values of family and motherhood, along with the resilience of the Igbo woman, become adroitly woven into the role of the rather new cassava staple in Igbo life" (50). It is that poetic delineation of subject and its metaphoric dichotomy that is the essence of Nwapa's poetic achievement and one of the major issues for exegesis in this essay. Nevertheless, Nwankwo also observes that in the poetry of Nwapa in *Cassava Song and Rice Song,* "we know that we are dealing here with the healthy competition between the matrimonial values of tenderness and the hardly mentioned arrogance of patriarchy. We also know that as *Rice Song* delicately suggests, the traditional is fighting to survive the taunt of alien ways" (50). It is this duality in the utilization of subject that adds insightful dimensions to Flora Nwapa's poetry.

The major purpose of this essay is to show how that metaphoric dichotomy of subject is combined with the poetics of orality to widen the focus of Nwapa's poetry while indicating equally wider implications for multi-cultural societies as well as societies that have been subjected to violent socio-historical forces. The orality of Nwapa's poetry emerges through the title which indicates that the work is a song about "cassava" and "rice" which implies the awareness in Igbo society of the reality of an oral performance with songs as well as the Igbo perception of duality in life. This idea of duality is portrayed by a proverb which states that "when one thing stands, another stands by its side," and it is this consciousness of a dual imperative that makes the metaphoric use of "cassava" and "rice" significant. Equally significant is Nwapa's use of food items as basic poetic icons and this is important in the sense that the Igbo accords respect for hard work, and in the society that existed before the disruption of Western presence such instances of hard work are demonstrated through farming. The prominence of farming in that society is hinged on its importance as an essential occupation on which the survival of the individual depends. That primary task was made much more problematic during the civil war in Nigeria when it became extremely difficult for many people to acquire food. Thus, it is not strange that Nwapa in *Cassava Song and Rice Song*[3] locates the historical markers in the poetry to both the pre-colonial and civil war periods thereby making the issue of survival a relevant poetic factor.

All the same, that historical marker implicated through the reference to the war is subsidiary, for the subject in the poem is historically ubiquitous since cassava is a staple food which has been consumed in varied forms for a long time in Nwapa's society. Thus, it is the ever present value of that staple food which motivates this panoramic panegyric to cassava and it confirms the statement of Donatus Nwoga (1979) on the orality of modern African poetry, which is that "one major fact of this domestication is the localization of the influences that now impinge on the poets" (44). Thus, the *Cassava Song* reflects that localization of influences that impinge on the writer as well as illustrating an exploitation of the Igbo praise poetry tradition. The poet with a four stanza chant writes:

> We thank the almighty God,
> For giving us cassava
> We hail thee cassava
>
> The great cassava.
> You grow in poor soils
> You grow in rich soils
> You grow in gardens,
> You grow in farms.
>
> You are easy to grow
> Children can plant you
> Women can plant you
> Everybody can plant you.
>
> We must sing for you
> Great cassava, we must sing
> We must not forget
> Thee, the great one. (1)

The appellation(s) that the writer uses to identify cassava (like "the great one") are part of the oral tradition of praise poems in which the object of the praise is presented in superlative terms. Although this poem commences with a reference to the "almighty God" which is indicative of the influence of Christianity on the writer, the poem still proceeds to give cassava the type of qualities associated with that "almighty God." Particularly instructive and clearly appreciated by the poet is the fact that cassava is easy to grow and can be planted by everybody. Perhaps it is significant that Nwapa mentions the ease with which women and children plant it in order to indicate the close affinity between the plant and women, as well as the view that it is a non-discriminatory plant that can yield itself even to children.

That reference to women and children becomes a metaphor when the writer elaborates on the value of cassava later in the poem. The persona asks for forgiveness for the neglect of cassava—thereby personifying and imbuing it with divine qualities. However, in the process of making that reference, the symbolism attached to cassava becomes apparent as the poet writes:

> As children, you fed us
> You were like a mother
> You fed us fat
> But we easily forget.
>
> You must pardon us
> Great Mother Cassava
> Great Mother Cassava
> You must pardon us. (2)

First, the writer acknowledges that cassava is "like a mother" but she laments that the children "easily forget" that fact, and then it is addressed clearly as "Great Mother Cassava." The metaphoric depiction of cassava as mother reminds the reader of all those attributes associated with motherhood and the selfless service of mothers who provide food for children who "easily forget." Thus, Nwapa succeeds in weaving into her poetic vision the familial essentialities of motherhood that are usually taken for granted.

This poem is interestingly developed into an artistic juxtaposition of cassava and motherhood, with the common characteristics that they "make no fuss," "are most humble," and are "most kind" to their "children" (3). These are undoubtedly the qualities appreciated in Igbo women, and they portray the resilience of those women who can effectively create matrimonial comfort through hard work and selfless acts. In effect, the praise in this poem—literally a praise of cassava—is figuratively a praise of motherhood. The frequent repetition of the praise appellation ("Great Mother Cassava") which occurs in the poem is used to perform the twin function of emphasizing the orality of the poem and the symbolic essential view of motherhood. Thus, Nwapa writes that cassava is a "Great woman"(5). That attribution of the female gender to cassava also illustrates that it is not just the state of motherhood that Nwapa is examining and emphasizing but also the state of womanhood. In spite of any resistance in the reader to appreciate that fact, its metaphoric mode of presentation subverts an implied patriarchal view of reality. In recording the processes for the preparation of cassava into "fufu," the poet regards the odor as feminine, as well as part of the food's identity and also as something that emphasizes that it is a "Great Mother" who is a "lover of children" (7). The glorification of cassava is not made with exaggerated images. They are appropriately chosen images that reiterate the high value placed on cassava. The patriarchal dimension to the glorification of "mother cassava" emerges when the writer laments:

> We sing for the yam
> We have yam festivals
> Why, oh why are
> These denied you? (9)

It is a significant question because yam is regarded as the "king of crops" in Nwapa's Igbo society and, accordingly, many men and few women cultivate it. Thus, an interrogation of its importance is a metaphoric interrogation of its patriarchal image in the Igbo society. Later in the poem, the poet makes the patriarchal association between yam and men clear when she writes that, while everybody digs out cassava:

> Not so the great yam
> Who need
> The great might
> of men. (19)

Since men are thus associated with yam, patriarchal attitudes could easily be associated with yam cultivation, which the poet hints as part of the culture of yam festivals. It is equally this idea that informs the reference to yam as arrogant and incapable of yielding itself to preservation for extended use over a period of time.

These subtle forms of subversion of reality make the *Cassava Song* a subtly gendered but arresting poem. Furthermore, Nwapa does not leave her poetic objectives ambiguous for she utilizes multiple images to illustrate her metaphoric vision. She also focuses the issue of cassava and motherhood through an illustration of the fact that after the cassava has been cooked and the utensils need cleaning, no one would want to accept that task except a mother:

> Who will wash the pot?
> Who will wash the mortar?
> Not me
> Not me.
>
> "Never mind"
> Mother says quietly,
> "I'll wash the pot
> Mother and cassava are one."
>
> Yes, they are one
> One loves her children
> The other
> Also loves her children.
>
> Both you and Mother
> You are long suffering
> You love your children
> You are wonderful. (11)

The comment that mother and cassava love children and are long-suffering reinforces that poetic linkage that Flora Nwapa has already fashioned. It reflects the values that the Igbo attach to the family as well as the expected beneficial effects to the members of that family. More importantly, the appreciation of children normally associated with mothers establishes that linkage between cassava and motherhood and the values the Igbo attach to the family. Thus, when the protagonist praises cassava, it is at the same time a praise for motherhood as well as an implicit praise for "Mother Earth," since cassava has "that quality which Mother Earth gave" (15). The implication is that "Mother Earth" (referred to as *Ala*, the Earth goddess) has imparted some of that reverence to motherhood and mother-cassava.

Nevertheless, part of the orality of *Cassava Song* lies in the structural presentation of this poem through anecdotes and rele-

vant tales. Such anecdotes and stories are part of the Igbo folk tradition, for when the poet-protagonist recollects the experience of
students in boarding schools who drink garri "after lights out" with
coconuts and "sometimes with sugar"(23)—contrary to the instruction of the white Principal—that information and its presentation
highlight the orality of a poet-cantor/singer regaling listeners. Such
anecdotal tales as the tale concerning the politician who "will not
eat at Government parties"—but rather tells his wife to "keep [some]
food" in the house for him to eat after "watching the menu"(28)—
are not digressive. They cast insight on the reason for the major
metaphor which is the service of cassava which in turn reiterates
the indispensability of women. Thus, the exhortation of the singer-
character in this poem is significant, for she asks:

> Why should you not
> Give praise to
> All mighty cassava
> Who has saved you? (35)

The implication is that praise should be given to mothers who have
"saved" and served the society since the cassava metaphorically
refers to them. In that exhortation, the listeners are even asked to
worship "cassava" as they "worship the gods" and as they worship
the "Beauty underneath the lake" (36). That divine quality perceived in cassava is further portrayed by the poet through the narration of the incident concerning the women who defied the might
of the British army with only cassava stems:

> The women, they stood
> Defiantly they stood
> Armed only
> With Mother Cassava. (41)

In effect, the "protection" provided by cassava in a time of war
becomes a metaphor for the protection women provide even when
they appear inadequately prepared, like those women armed only
with pieces of the cassava stem. This incident also conveys a subtle indictment of the British for the crisis that emanated from their
contact with Africans. This instance of colonial oppression and
colonial high-handedness reiterates that women were not spared
in the oppressive acts of the colonial era.

That protection that cassava offered probably accounts for the
conclusion in this section of the poem that cassava should be
regarded as divine:

So the oracle
Must select
The best day
For mass purification.

Then we thy children
Especially the women
We shall wear black
Before your shrine. (46)

The wearing of black clothes here signifies expiation and repentance, which will make the "mass purification" culturally significant. However, the fact that Nwapa insists that "especially the women" shall "wear black" before the shrine of cassava re-emphasizes her association of "mother cassava" with women and the metaphoric importance of that association in the poem.

Similarly, in the *Rice Song* the subject is equally metaphoric, for the poem is full of delicate suggestions concerning the society and its inhabitants. In this segment of the poem, the traditional parallels the modern and the foreign parallels the indigenous and that parallel becomes the fulcrum on which the poem rests. Interestingly, both food items, "cassava" and "rice" are metaphors for either groups or individuals who either refuse to or decide to live according to the norms and mores of the society in which they exist. Cassava has become part of the society; rice, on the other hand, is alien and its importation, consumption and production are fraught with risk. Thus, in *Rice Song* Nwapa condemns the society's penchant for rice although she acknowledges that rice was not originally so supreme in people's diet but that dishonest businessmen surreptitiously introduced it until "our people gradually/Began to have taste for rice" (60). And like all acquired appetites, rice managed to subdue the desire for locally grown food. The reputation of rice changed, as the poet states:

So rice became a rich man's food
which he and his family
Ate only on Sundays and on
Festive occasions especially weddings. (53)

Rice was a delicacy
All they knew was
That it was rare. (58)

The elevation of rice into a status symbol and its perception as "a delicacy" are ironical because it is not the contents of rice that confers the status that it is a rich food, but rather the fact that its scarcity affects the financial and material resources of the individual seeking to acquire it. Furthermore, it is not rice that makes itself a delicacy, but rather the stew prepared separately with which it is eaten. Nwapa thus shows—through the misplaced priority in the elevation of rice as a "rich man's food"—the subtle manner through which corrupt and destructive influences could be introduced in the society.

Interestingly, the poetics of orality in the poem artistically represent anecdotes, and the anecdotes concerning the manner through which beer and rice are made to displace indigenous drinks and staple foods portray the decline of values in the wider society. The poet-protagonist recollects that her mother tells her of a bride "who got married" in the church after she had had five children since she and the husband were converted in old age but she refused to be congratulated because:

> You talk of marriage, you talk
> of marriage, did we cook rice?
> Marriage is no marriage without rice. (55)

In the same way, the protagonist perceives the insinuation of rice into the center of events as part of the devious manner beer is used to replace palm wine:

> Many drank palm wine
> And did not have
> The palate for star beer
> So salesmen went to drinking houses
> Brought star beer in cartons
> and gave them free
> To whoever was there
> and now, how many
> Breweries have we in Nigeria
> In 1984? (60)

It is symptomatic of a confused society with misplaced priorities that such displacements occur. Thus, these anecdotes woven into the poem reiterate the argument that rice must be banned and local growers of other food items encouraged. This use of anecdotes also portrays what has been explained as the effective exploitation of an oral literary tradition where "the modern writer makes only a

selective use of elements of oral tradition. The form of presentation bears a limited relationship to the oral tradition, and even
when familiar characters are used they are deployed" in a "somewhat altered order of relationships" (Okpewho 1992:314). In effect,
this anecdotal tale prepares the reader to respond to the other critical issues like the criticism of the military leadership, especially
when the poet-protagonist says:

> Attention soldiers
> Attention rulers of our great nation
> Attention great generals
> Who have fought in the civil war
> Careful, you poets in uniform
> We cannot eat words... (66)

The reminder that words are not edible is a way of cautioning the
military rulers and reminding them that they must act decisively
in the interests of the society. It also hints at the incompetence of
those soldiers. This direct address to the soldiers is part of the direct
address technique in oral poetry and the orality of its poetic use is
to emphasize for the reader the reality of the issue that is being
explored in the poem. Moreover, the satirical implication in reference to the soldiers becomes clearer towards the end of the poem
when the poet insists that "soldiers are precise/Soldiers are not
vague/They are no nonsense men" (72). But that is an ironical comment, for these soldiers depicted in the poem are vague, impre-
cise, and incapable of providing the necessary leadership vision.
The introduction of this element in Nwapa's poetry is part of her
metaphoric use of subject in the poem.

Certainly the orality of *Cassava Song and Rice Song* emerges
through that metaphoric dichotomy of subject, although Nwapa
does not appear to care a great deal for the vehicles of poetic technique. Nevertheless, the devices of orality that she adopts contribute
immensely towards the success of the poem. For instance, Nwapa's
use of appellations and attributive epithets in the praise of cassava
reflects the qualities of the *Oriki* (praise) poetry which are actually
equivalents to names—names which can indicate undesirable qualities as well as desirable ones and which "are seen as being in some
way the key to a subject's essential nature." It is further stated that
in addition to human subjects, "animals, plants, and all kinds of
inanimate objects possess *Oriki* too" (Barber 1984:503). Thus, the
Oriki of the cassava establishes its positive culinary value, while the
rice is perceived as socially, politically, culturally, and economi-

cally destabilizing. In addition, the implicated use in this poem of what has been described as "the oral tradition of insults and name-calling" (Nwachukwu-Agbada 1993:94) is part of that orality that Nwapa has exploited. For instance, rice is personified as disuniting the ties of marriage since it causes rancor between married partners, creating avenues for economic corruption and—at least potentially—generating civil disorders and wars.

All the same, one critic points out a minor aspect of the technical flaw in Nwapa's poetry when he argues that her work "as it is projects flashes of inchoate and unsustained wisdom. The word is there, but very often it is blighted by unbearable incongruities" (Nwankwo 1995:51). This critic gives an example of the protagonist's statement concerning the need to appease Mother Earth and the Earth goddess by making sacrifices, which is linked to the idea of an angry God in heaven who sends the plague to torment the people as in the time of Pharaoh. Thus, this critic feels that "an item for Igbo world view is rather unwisely yoked to another from a questionable biblical history to produce a certain kind of unsettling wedlock. Syncretism of any kind has something to do with engendering a new harmony from the otherwise culturally disparate and not the grafting of mutually unassimilables" (Nwankwo 1995:51). All the same, in spite of that unpolished syncretism and the sometimes unintegrated digressions that occur in Nwapa's poetry, *Cassava Song and Rice Song* demonstrates the orality and the use of a metaphoric subject for poetic focus.

Flora Nwapa may not have produced technically arresting poetry, but her deployment of the elements of orality, her ironical and satiric symbolizations and the major metaphoric focus of her poetical work transform it into a noteworthy body of literature that persistently interrogates reality, subtly revises prevalent notions, and humanistically reflects on life and society. Flora Nwapa, in spite of whatever poetic infelicities may be gleaned from her work, is one of those who created that tradition of modern Nigerian poetry which has effectively absorbed the features of oral traditions that are used to make both the language of their writing and its subject matter elastically amenable reflecting the poetics of orality.

NOTES

1. The author is grateful to the Alexander Von Humboldt Foundation, Bonn,

the University of Mainz, and University of Bayreuth, Germany.
2. Aderemi Bamikunle entitles his work "Feminism in Nigerian Female Poetry," which I take indicates that he meant poetry written by Nigerian women because it is problematic to categorize works themselves as either "female poetry" or "male poetry."
3. *Cassava Song and Rice Song* will henceforth be referred to as either *Cassava Song* or *Rice Song* in the text, depending on the section of the poem under discussion.

WORKS CITED

Bamikunle, Aderemi. "Feminism in Nigerian Female Poetry." *Feminism in African Literature* Ed. Helen Chukwuma. Enugu: New Generation, 1994:1-21.

Barber, Karen. "Yoruba Oriki and Deconstructive Criticism." *Research in African Literatures* 15.4 (1984):497-518.

Berrian, Brenda F. "Flora Nwapa (1931-1993): A Bibliography." *Research in African Literatures* 26.2 (1995):124-129.

Davies, Carole Boyce. "Introduction: Feminist Consciousness and African Literary Criticism." *Ngambika: Studies of Women in African Literature*. Ed. Carole Boyce Davies and Anne Adams Graves. Trenton, N.J.: Africa World Press, 1986:1-23.

Emenyonu, Ernest N. "Who Does Flora Nwapa Write For?" *African Literature Today* 7 (1975):28-33.

Ezenwa-Ohaeto. "The Other Voices: The Poetry of Three Nigerian Female Writers." *Canadian Journal of African Studies* 22.3(1988):662-668.

Frank, Katherine. "Women Without Men: The Feminist Novel in Africa." *Women in African Literature Today* 15 (1987):14-34.

James, Adeola. "*Idu*: Flora Nwapa (Review)." *African Literature Today* 5 (1971):150-53.

Jones, Eldred. "Locale and Universe." *The Journal of Commonwealth Literature* 3 (1967):127-31.

Nwachukwu-Agbada, J.O.J. "The Eighties and the Return to Oral Cadences in Nigerian Poetry." *Matatu* 10(1993):85-105.

Nwankwo, Chimalum. "The Igbo Word in Flora Nwapa's Craft." *Research in African Literatures* 26.2 (1995):42-52.

Nwapa, Flora. *Cassava Song and Rice Song.* Enugu: Tana Press, 1986.

Nwoga, D.I. "Modern African Poetry: The Domestication of a Tradition." *African Literature Today* 10 (1979):32-56.

Okpewho, Isidore. *African Oral Literature: Backgrounds, Character and Continuity.* Bloomington: Indiana University Press, 1992.

Palmer, Eustace. "Review of *Efuru.*" *African Literature Today* 1 (1968):56-58.

Stratton, Florence. *Contemporary African Literature and the Politics of Gender.* New York: Routledge, 1994.

Umeh, Marie. "The Poetics of Economic Independence for Female Empowerment: An Interview with Flora Nwapa." *Research in African Literatures* 26.2 (1995):22-29.

PART FOUR
THE BODY POLITIC:
LOCATING FEMALE SEXUALITY

In the 1996 annual conference of the African Literature Association hosted by the State University of New York at Stony Brook, I was invited to join in a panel discussion on "The Migrating World of Flora Nwapa's Fiction: *The Lake Goddess*, chaired by Dr. Marie Umeh. I was in august company. The others on my panel were Professors Ifi Amadiume, Chimalum Nwankwo, Linda Strong-Leek, and Gay Wilentz. The presentations and discussions took off.

What surprised me was the reaction to Professor Strong-Leek's paper on "The Quest for Spiritual and Sexual Fulfilment in Flora Nwapa's *Efuru* and *The Lake Goddess*, published in this section. At the surface, Dr. Strong-Leek's analysis of the relevance of female circumcision and its psychological and physiological impact on Nwapa's Igbo communities seem perfectly logical, indeed, sensitively sympathetic. Imagine my surprise, then, when several African women attending the discussion took the panelist to task, challenging her observation that female circumcision resulted in negative psychological and physiological impact. When Dr. Strong-

Leek referred them to Nwapa's novels, they challenged her to show where the author had said that circumcision had a negative impact on Efuru. What had hitherto seemed to be a routine scholarly conference exploded into a passionate discussion on clitoridectomy. African women did not agree with Professor Strong-Leek. African-American women did.

Now, as a male, the only non-African individual in the room, I thought it best to keep my mouth shut and listen! What emerged from that panel discussion, two papers from which are produced in this section, was an important perspective on "locating female sexuality."

After the controversial panel ended its discussion I met an elderly African woman scholar who had just finished participating actively. As our glances met, she broke into a friendly smile that seemed to ask me my opinion about the discussion. I ventured. It seemed to me, I hesitatingly observed, that what was being discussed was "What is sexuality?" Is it something physical, psychological? Is the pleasure derived from it physical? Mental? Finally, it seemed to me, that what was at the core of the controversy is whether culture played a vital role in "locating" sexuality.

To my relief, the lady agreed with me. Her eyes, hitherto excited, at the edge of a challenge, suddenly turned soft, radiating wisdom. Yes, she nodded her head kindly, sexuality is not sex, she said. It is a way of being. We talked some more, as participants do in conferences, over lunch. What I learned from my wise friend from Africa that afternoon forms the theoretical basis for this introduction.

What does locating female sexuality mean? In semiotic terms it is a question of sign. What represents female sexuality? In societal terms, it is a question of rights. Does the woman have the right to determine her own sexual needs? In art, it is a question of form. How does sexuality transcend into energy so that, as desire, it can transform not only characters, but the very medium of writing itself? If, as my wise African friend says, sexuality is a way of being, how does this being express itself and transform the quality of life? Is sexuality a quest? Fulfilment? Empowerment? The energy that transforms women's way of being and knowing the world?

Linda Strong-Leek burst the controversy open by using female circumcision as a metonymy for sexuality. Dr. Strong-Leek is right. Clitoridectomy is not desirable under any circumstances. If it were, where would such logic take us? In India *sati* was practiced in the nineteenth century, a custom according to which the wife of a deceased man had to be consumed to death in the funeral pyre of her husband. It was said that women went to heaven if they did that; that, in fact, many women *wanted* to burn themselves with their husband's dead bodies. Did that make the custom right? Thus, if the argument hinged strictly on the issue whether female circumcision was justified or not, most women, I think, would side with Dr. Strong-Leek.

But let us, for a moment, extend the argument. For that, we have to go to Professor Amadiume's paper, "Religion, Sexuality and Women's Empowerment in Nwapa's *The Lake Goddess*," also published in this section. Dr. Amadiume's exegesis on the erotic relationship between the Woman of the Lake and her priestesses brings into sharp focus two related but very different facets of being a woman in a Third World country: religion and sexuality. Religion helps them to get in touch with the spiritual power that is internal or external to them so that ultimately they attain freedom from material and worldly preoccupations and achieve salvation, the ultimate freedom. Sexuality, Amadiume argues, creates a tension in African women in Africa, that ultimately denies them authenticity. "What is this sex that men so much fuss about?" Amadiume asks in her paper in a bold attempt to determine whether authenticity lies in the imagination or action. She does not give us a conclusion. But reading her paper, we may at least try to formulate a paradigm, one that links imagination with religion; action with sex. And therein lies our tentative answer to the controversy on clitoridectomy and to the question on locating sexuality.

What I have learned from reading these two essays, and with the help of my wise African lady friend is, that, since sexuality is a way of being, it is steeped in culture and myth. Authenticity can be realized if we "locate" ourselves within a given culture and myth, *and* are honest with ourselves. This does not necessarily mean blind acceptance of a custom. It does mean being aware of the implications of a culture or myth, and positioning oneself with a delicate sensibility so that we do not offend those who practice the custom, nor blatantly join those who don't. Sexuality is being. It is being not only in this psycho-social world, it is also partaking in the spiritual dimension of ourselves. Just as religion helps to get in touch with a supernatural force external or internal, so does authentic sexuality moves the individual toward self-discovery. Dr. Strong-Leek observes that Nwapa's protagonists find redemption after the painful "bath" by retreating from traditional roles as wives and mothers to become worshipers of Uhamiri. Dr. Amadiume, on the other hand, sees religion as an extension of sexuality, supplementing a way of being in the world. The relationship between women and the Lake Goddess is spiritual *and* erotic.

It may be important that, unlike the previous two, my paper in this anthology is not the same as the one given at the conference. At ALA, I spoke about desire, violence, and the post-colonial woman. I discussed Amaka in *One is Enough* with the protagonists of Buchi Emecheta and those of an Indian-American writer, Bharati Mukherjee. Here, I am writing about the role desire and sexuality play in the transition of Efuru and Amaka from the private to the public domain. It is my contention that Nwapa locates herself in an important feminist position by not seeking a reversal of roles from the private to the public, as most western feminists do, but enjoining the public to their already acceptable role in the private domain. In other words, Efuru and Amaka continue to value the private domain of womanhood —love, marriage, children—but do not want all of them for themselves. By adding their public role of the priestess, helper of other women or by staying public in business, Efuru and Amaka supplement the role of the private with the public. I try to show that desire and sexuality play vital roles in this transition. This addition of roles in women may be seen as redemption as Dr. Strong-Leek does, or as spiritual as Dr. Amadiume does, nevertheless, Nwapa makes a strong statement for Nigerian, and ultimately, all women, that they do not have give up or reverse their roles as woman in order to be authentic. They have to extend themselves. Fulfilment comes with integrating oneself into society, not uprooting society or even assimilating with it.

Whither female sexuality? From the chapters that follow it seems to be in the mind-body and spiritual connection.

Shivaji Sengupta

Religion, Sexuality, and Women's Empowerment in Nwapa's *The Lake Goddess*

Ifi Amadiume

With Flora Nwapa's untimely death and the anticipated publication of her last novel, *The Lake Goddess* (unpublished manuscript), one faces the problem of how to interpret a novel published posthumously. Nwapa may not have intended it, but this very religious novel, focused on her icon, a water goddess, appears to me the writer's final statement on a persistent "feminist" concern with the question of female deviation from convention and the search for alternative means of women's "empowerment." I intend to show how the focus on this water goddess persona—with its fixed, stereotypical sexuality—has affected Nwapa's work by limiting her plot, while enabling her to expand the boundaries of feminist discourse.

Nwapa tried to pull together many strands in *The Lake Goddess*, tackling issues related to women and marriage, that is, men and women and sex, daughter-and-mother relationships, daughter-and-father relationships, women's friendships, tradition and modernity, "deviant" sexuality, happiness and contentment. The main force of the novel remains Nwapa's MammyWater spirit, a Water Goddess she calls Uhamiri or Ogbuide, and her relationship with women who experience possession by her.

MAMMYWATER SPIRIT POSSESSION SICKNESS

Some background understanding of the MammyWater phenomenon is central to our reading of Nwapa's work. Belief in this spirit is widespread along the West African coast. This is particularly so in communities near rivers, creeks, lagoons, oceans, and lakes. In these same communities, there also exist several other female water deities who are not MammyWater. This is not surprising since Africans generally deify rivers as Goddesses.

Although scholars disagree on the identity and origin of MammyWater, there is an area of convergence in both the physical descriptions and character attributes of the MammyWater of popular culture, and the identification of symptoms of possession. It is generally claimed that these spirits have luxurious homes deep inside the waters, are usually very beautiful, assuming various human shapes, but manifest themselves particularly as half-woman and half-fish, that is, mermaid-like. The idea of an enticing, naked, beautiful woman combing her hair by the water is fairly standard. It is also believed that this beautiful woman can cause tragic accidents and can give her followers riches but will deny them children. Although it is sometimes claimed that male MammyWaters exist, "She" seems the natural gender used by all the writers on MammyWater. MammyWater spirit and the associated cult in this particular register of possession sickness are very much woman-specific. As Madam Grace Joe of Oron says in Jell-Bahlsen's film (1991), "when I began to bear children, MammyWater hooked my child, that is, Mary Magdalena. I got it from my mother. Then, my daughter got it from me." All of these women end up priestesses and healers, guarding "secret knowledge" taught by this water goddess.

MammyWater possession afflicts a selective choice of pretty females or males for wives or consorts. Those afflicted show symptoms of abnormal behavior, such as talking to oneself and experiencing periodic seizure. To further situate the MammyWater phenomenon, I find Chinwe Achebe's (1984) study of the *Ogbanje* "spirit being" phenomenon most insightful. I have found her version of Igbo cosmology a fascinating contrast to the overwhelmingly patriarchal representation of a world view dominated by a male God, lesser gods and ancestors as the only divine sources of moral conduct, grace, and spiritual happiness. Although her version also retains the idea of a remote, sky-dwelling Supreme Deity "Chiukwu," it gives prominence to two Goddesses, Onabuluwa, a forest goddess, and Nne Mmiri, a water goddess, who control the movements of both humans and spirits into and out of the human world and the spirit world. Complex as Igbo world views are, the key to their understanding is the principle of relationality, which is usually referred to as holism, hence the constant cyclical movements between worlds, as for example between the human and spirit worlds, in a constant recycling. Only thus can the phenomenon of *Ogbanje* be understood as human beings whom the Igbo claim to be really half-spirit, or 'changelings' or 'repeater babies'; they are only half in this world and half in the spirit world. They do not stay; they are sickly and usually die young only to be born again to unfortunate parents as they go on causing heartbreak and havoc to those who love them dearly!

Chinwe Achebe's research points to an explanation of this phenomenon in relation to Onabuluwa, the forest goddess with whom *Ogbanje* elu possession is associated and Nne Mmiri, a water goddess said to be responsible for *Ogbanje* mmiri possession, the latter being by far the most common. In describing the forces involved in the *Ogbanje* phenomenon, Chinwe Achebe refers to Nne Mmiri as "Mami Wota" (MammyWater). According to Chinwe Achebe, this Nne Mmiri "is the mother of a group of water spirit beings who journey to and from the land of the living and the dead. Her abode and empire is in the water. Her life begins and ends there" (15).

CONTESTING IGBO WOMANHOOD

It is in her very first novel, *Efuru* (Nwapa 1966), that Nwapa first introduced this religious phenomenon of women's spirit possession. This is a device cleverly used by Nwapa to subject certain aspects of Igbo traditional expectations of womanhood to criticism and at the same time "problematize" women's relationship with the Goddess Uhamiri, the spirit that Nwapa calls The Goddess of the Lake or The Woman of the Lake. This religious phenomenon also features in *One is Enough* (Nwapa 1981), where Nwapa further develops her ideas on marriage relationships into a sharper feminist critique of the marriage institution itself, with Amaka at her most radical. Compared to these two novels, *The Lake Goddess* is the most religious, and in my view, a radical feminist reconstruction of woman's place in Igbo traditional religious beliefs, particularly the idea of spiritual submission.

In Chinwe Achebe's Igbo cosmology the two female deities mentioned are represented as temptress goddesses out to entice people to change the pre-natal choice and destiny (*Iyi Uwa,* made with each individual person's own *chi,* variously interpreted in studies of Igbo religion to mean spirit double, spirit counterpart, guardian spirit, etc.) and make a worshipping pact with these Goddesses for worldly wealth or success (which is how Nne Mmiri acquires her spirit children as she herself has no children). Misfortunes and sickness are explained as the breaking of this pre-natal contract in the case of an *Ogbanje,* for those who make a pact with Nne Mmiri before continuing the boat journey into the world show signs of possession by her. They are usually extremely beautiful, often show some signs of mental illness, and are unhappily married—if they marry at all. They usually have difficulty conceiving or keeping alive the children they have and, more generally, invariably succeed by whatever means they choose to pursue wealth. Think of Efuru, Amaka, and certainly Ona, as Nwapa constructs each character "like a woman possessed"! This, in my view, means that her plots are already restricted to a limited representation of women in a certain stereotype.

In *Efuru,* when the *dibia* Enesha Agorua divined Efuru's calling to be a follower of the goddess, he said, "Uhamiri is a great woman. She is our goddess and above all she is very kind to women" (153). This emphasis on female exclusiveness of Uhamiri and her worshippers lays the basis for Amaka's rejection of marriage in *One is Enough,* but more directly the character of Ona in *The Lake Goddess,* for Nwapa has implied that women can find fulfilling companionship with other women, thus moving towards a radical separatist agenda.

On the personal side (that is, in marriage), Nwapa's heroines meet with tragedy, but in business they are successful. In every case, it is suggested that Nwapa's Uhamiri is locked in the background; Nwapa's wealthy and beautiful Goddess lives in the bottom of her Ugwuta Lake and gives selected women beauty and wealth, but no children. This choice between wealth, marriage, woman's independence, and motherhood is the tension in the novel *One is Enough* and the theme of *Efuru.* It is much more complex and fully developed in *The Lake Goddess,* where the satire and attack on Christian missionary education on these issues—already begun with Amaka—are intensified. In *One is Enough,* Nwapa has Amaka wondering about her ignorance of traditional knowledge as a result of her Western education: "Her mother was teaching something different. Was it something traditional which she did not know because she went to school and was taught the tradition of the white missionaries?"(11). I saw this as a hint that Nwapa felt a need to revisit the tradition-versus-modernity theme and would most likely seek a better understanding of Igbo women's traditional knowledge systems.

Therefore, I am not surprised to see that Nwapa's *The Lake Goddess* (unpublished manuscript) is yet again another woman-centered novel and an extremely radical one at that. Here, Nwapa characteristically dives head-on into controversies, introducing in the first few pages a women's network of widows including the mother of the powerful *dibia* Mgbada, who is going to be the father of Ona, the protagonist of this novel. Contrary to normative claims of patriarchal Igbo social rules and etiquette, in Nwapa's narrative women own kolanut trees, women break kola, women initiate marriage, and they cheat on the circumcision rite. The mothers bribe Mgbeokworo, the midwife, to pretend that Ona has been circumcised as they perform all the necessary ceremonies; a clever proposal here by Nwapa on how to preserve and "respect" custom in a modern context.

I find *The Lake Goddess* almost anthropological in its narrative on traditional culture and religion, which suggests that Nwapa had carried out extensive research on the topic of MammyWater possession that had been her obsession in previous works. Research of the topic must have enabled her to finally recognize the phenomenon she had been dealing with as Ogbanje, which I have discussed in reference to Chinwe Achebe (1984). Spirit possession and Ogbanje(ness) center around Ona, the "strange" new daughter and Idenu, her equally "strange" older sister.

In *Efuru* and *One is Enough,* the two troubled women Efuru and Amaka had to resort to the services of a dibia for both divination and healing of their barrenness, and in the case of Efuru for analysis of her dreams and the consequent confirmation of her call by Nwapa's Uhamiri. In contrast, Nwapa does something absolutely brilliant in *The Lake Goddess* by making Ona's father a strong dibia. Like Efuru, Ona is very close to her father and it is in him that she confides her innermost secrets, and she also runs to him when in need. Here I think Nwapa finally reveals her "project for the Igbo man"! It is this construction of father-daughter relationship written in slow and detailed narrative with genuine affection and tenderness. It is Ona's father who reads all the signs of his daughter's mental illness and possession by Uhamiri. It is also this father who makes the necessary sacrifices (made incidentally to the Earth Goddess, to the Lake Goddess, to the Goddess of the Road) in order to counter some of the destructive forces as much as he could and, when he could do no more, prepares Ona to assume her role as priestess of Uhamiri. Even after she is priestess, he hovers around his daughter, protective and supportive.

Ona's father is a sharp contrast to her mother Akpe who ends up "mad" herself due to her Christian fanaticism and total rejection of traditional religious beliefs, something she inherited from her Christian fanatic mother, Mama Theresa. By so "splitting" mother and daughter, what started off as a women's tale with strong matriarchs ends with the author dividing women on the basis of religion, showing women involved in church and school in a negative light. Madam Margaret, who ran the convent in Ugwuta that Ona attended, is dubbed "a religious fanatic" and described as "mean, hard, and evil." It is in fact Mgbada who finds a middle ground between Christianity and traditional Igbo religion by embracing both comfortably. Not even his daughter, Ona, is able to find this same balance, as she ends up outside society and squarely in the domain of MammyWater when all attempts at normative behavior have failed, including schooling and marriage. We learn from Nwapa as a possible explanation of Ona's troubles that, unlike Efuru who was called late in life, Ona was called too early (132).

SEXUALITY AND THE RECONSTRUCTION OF WOMANHOOD

With the intensification of her mental illness, the goddess becomes increasingly Ona's only source of contentment as she says to her father, "I love to see her. I have a sense of well being when I dream my dreams." It had been assumed that marriage and disvirgining would cure her of "madness" and the ability to "see," that is divine, for Ona could point out curative plants to her father even at the age of eighteen months. Efuru may have had good reasons to end her marriages. In the case of Ona, she is surrounded by loving and devoted men in her father and her wonderful husband, Mr. Chukwukere Sylvester, described by Ona's father as "her good husband, the husband that was given to her on a platter of gold" (165). But as "a woman possessed" the call of the Goddess of the Lake compels Ona to leave her husband and their three children. If we look at the context in which the final rupture takes place, we can see further proof that Nwapa had moved towards radical separatism. Ona, having had three babies, seemed to be settling down and seriously trying to be a good wife and mother. Nwapa writes:

> Her husband did not have to struggle with her before he could make love to her. She accepted it as her duty to surrender herself to him whenever he wanted her to do so. Sex was something that was strange even alien to her. Why did men make so much fuss about it? What was it that men enjoyed? It was dirty as far as she was concerned. What with the slimy stuff between her legs when it was all over. If she had her way, she would have none of it. (170)

Some feminists, both African and Western, would be quite quick to link clitoridectomy and the absence of it to the sexualities of Nwapa's characters, Efuru, Amaka and Ona. Amaka's enjoyment of sex is well expressed, although we are not told whether or not she had been circumcised. We know that Efuru had been circumcised. We are not told of her receptivity to sex or her feelings about it. We do know however that she was intensely in love with the two men she married, Adizua and Gilbert. We know of the cheating to have Ona remain uncircumcised, and we know that Ona with her intact genitalia did not enjoy sex with her husband. Even when she tried to tolerate sex and please him, she had felt nothing!

Nwapa is therefore raising a controversy in *The Lake Goddess* and I believe that she intended to expand the discourse on women and sexuality by forcing a dialogue that we cannot shy away from. By eroticizing the image of the Woman of the Lake in order to freeze or disengage a man's wife from him in the intimacy of their marital bed, what was Nwapa implying? Is this an echo of Amaka's despair: "I am in prison, unable to advance in body and soul. Something gets hold of me as a wife and destroys me" (Nwapa 1981: 127). Is it an echo of Efuru's dreams about "an elegant woman, very beautiful, combing her long black hair with a golden comb"? The woman beckons and Efuru follows "like a woman possessed" (Nwapa 1966:146).

Of all these personae, it is from a non-fictitious one—Eka Ete in Jill Salmon's article on MammyWater—that we get an insight into a problematic sexuality that is sometimes linked to MammyWater spirit possession. Describing Eka Ete's confused experience of MammyWater, Jill Salmons writes, "She always sees the spirit in the form of a woman, but she thinks that it could really be a man, for sometimes she dreams that they sleep together, and in the morning she actually feels as if she has had sexual intercourse" (1977:11). This certainly suggests a discontinuity or contradiction between perception and vocalization, resulting in a kind of mental imprisonment and silence, perhaps due to new experiences in societies that are changing, but whose traditional cultures taboo and forbid same-sex sexuality and probably have no terms for it. In Jell-Bahlsen's film, problematic sexuality is also echoed in confessions by an Ugwuta priestess, Ezemiri, about her failed marriage experiences and the loss of eleven children due to something like a lover's jealousy by allegedly the male water spirit Urashi, even though she is a MammyWater diviner and healer (Jell-Bahlsen 1991). As priestess of Uhamiri, Ezemiri would have experienced possession by that Goddess and not by the Goddess' s supposed consort Urashi—in which case, one can understand jealousy by Urashi. It should also be admitted that Igbo genderless pronoun using *ya* for he or she and him or her made it unclear if Ezemiri meant "her" and not "him" in referring to the spirit troubling her.

Linked with the controversy about deviant sexuality is the issue of motherhood, about Ona's feelings for her children. Nwapa writes ironically:

> Did she really care whether she had a child or not? Thinking seriously about it, she did not care one way or the other for her children. Did she not possess the instincts of a mother? Why was she not excited like everyone else about her children, such beautiful children? (171)

But then Ona's husband was a good and kind husband who pampered her and showered her with gifts and tenderness. Again, about Ona's response to this husband's kindness and love, Nwapa writes, "She had to make an effort to be good to this man who was her husband. But when she tried, she failed woefully. There was something beyond her, some force that prevented this closeness. Try as she could, it was impossible" (171-2). Ona, in her determination to reciprocate her husband's kindness, tries initiating love-making once in a while—which pleases her husband—but at the end of it all, she is left sleepless while her husband snores with contentment. It was during one such occasion that Ona "saw a naked woman standing in front of her. She had very long hair dripping with water" (172).

Strangely enough, it is not in her mother or even her grand-mother that Ona confides. It is her father who confirms her call to be a priestess of the Lake Goddess. The naked and heavily beaded majestic Lake Goddess herself confirms this call directly in Ona's vivid and electrifying account of her vision, finally witnessed by her husband. As Ona recounts, she journeys to the abode of *The Lake Goddess* who tells her:

> I have waited for a long time for you to be my priest-ess. I have chosen you. I want you but I don't want to force you or hurry you. Don't wait too long. Give this message to the man who lives with you. Tell him that you belong to me. Tell him to . . . . (184)

It is ironical that a man who has just witnessed a profound spiri-tual experience by his wife, a man who even wondered whether he had married a spirit, should soon after this have a conversation with his wife's father in which both men masculinize God, for Ona's husband says, "Everything is in the hands of God. The day I make good sales, I thank God. The day I don't make good sales, I thank God. We are mere mortals in the sight of the Almighty. We must continue to thank Him for being born at all and for breathing the air He gave us in abundance." To which Ona's father replies, "Good talk, my daughter's husband. Whoever does not acknowledge the power of Almighty God is a fool. Who are we to question His action?" (186). This conversation between the two men seems out of character, for it suggests their willingness to decontextualize Ona's "gift" to "see." They seemed more interested in the "gift" than the source of the "gift." Even after Ona says to her husband, "You are going to sell everything in the market today," a message he knew came from the Goddess of the Lake, we hear him attributing market sales to Him who is beyond understanding and beyond comprehension!

Perhaps Nwapa intended this men's conversation, in which God is masculinized, to contrast with that between Mgbeke and Ekecha, the two women fish sellers, in order to highlight two contrasting gender worlds or religions. In any case, the two women seem cru-cial to the completion of the story as I think that through them Nwapa recaptures the traditional woman-centeredness of her nar-rative, raising feminist issues and giving voice to women as Ona says, "Ogbuide wants all women to have voices. Women should not be voiceless. Ogbuide hates voiceless women." Mgbeke and Ekecha certainly show the Goddess of the Lake as the center of women's cosmos in this conversation:

> Mgbeke and Ekecha washed their faces and
> evoked the Goddess of the Lake to be with them.
> "How fresh the water is."
> "May She be praised."
> "Our Mother is supreme."
> "Who can question that?"
> "Year in, year out She is the same."
> "The hairy Woman."
> "The elegant Spirit."
> "Bring fishermen to us."
> "Fishermen who have caught fish."
> "We are your children, and we are women like
> you."
> "We sell fish caught in your water."
> "Isn't She great?" said Mgbeke at last. (193)

It is not surprising that it is to these women that the beautiful and ageless Lake Goddess, who is partial to women, gives her message to women through her gifted priestess Ona; the essence of the message is that women should submit themselves to Ogbuide and return to goddess worship!

Ona embodies this submission and return. With full acknowledgement of her status as priestess, freed from domestic duties and fully relocated to a shrine built by her father, Ona's sickness is ritually contained and managed as she is free to move from "normality" on certain days to "abnormality" on other days, keeping the ritual rules and taboos of The Woman of the Lake. In this ritual context of controlled liminality, Nwapa is able to have her heroine Ona do what Efuru was determined not to do—show outward signs of madness: "She sometimes knocked her head on the wall or on the ground, rolling over and over. No one interfered for she was doing what the goddess had commanded her to do" (221).

THE LIMITS OF MAMMYWATER

In looking at Nwapa's main characters, my intention is not to simply point out the importance of the theme of possession by MammyWater and how that has affected the development of Nwapa's narrative. I think it is also important to understand the limits of this Water Goddess phenomenon and her place in relation to other Water Goddesses, and to consequently try and tease out how the focus on this single goddess limited Nwapa's plots while, on the other hand, it enabled her to be creative and to expand the boundaries of discourse on certain controversial issues.

In the last seven lines of Ona's last song to the Lake Goddess (and I hope this will not be edited out of the text, since this particular section of the song is still hand-written in the manuscript that I read), we see the final coming together of experience, thought, and metaphor in the powerful symbolism of giving birth when she says, "Mother and water are the same. Without water who can live. Without mother who can live. Our beautiful mother come closer." I find this in contradiction to the anti-motherhood representation of the Lake Goddess as MammyWater in Nwapa's work. It is therefore valid to ask the question: Is there a difference between MammyWater as a phantom or temptress spirit and the more normative fertility mother goddesses such as earth, birth and river Goddesses—central to Igbo religions and social structure? If so, is Nwapa not appropriating attributes of more central Goddesses for a marginal phenomenon? Is Nwapa not also appropriating attributes of the mother for the daughter?

Certainly in the Igbo community of Nnobi that I studied, the Goddess Idemili does not invoke MammyWater kinds of fears, nor does she appear in the market place to charm children into following her, or entice bachelors in the cities into marrying MammyWater incognitos who turn into killer ghosts and demons in their beds, or cause trains to fall into the River Niger or boats to capsize in order to get followers. Idemili is not the temptress spirit living under the water. Every Nnobi village has a visible Idemili shrine with a *Chi* Idemili shrine right inside Afor marketplace.

MammyWater has more generally to do with deviance in gender construction and social constructions of sexuality. The phenomenon seems more linked with "anomalous" social categories and "deviant" behavior. In comparison, Idemili is normative for her human representatives, the Ekwe titled women led the village Women's Councils. They were the most respected and most honored women in their lineages and in their towns. You didn't gossip about them or make insinuations about their possession and their wealth, since these were positively viewed and women themselves recognized the leadership qualities of such women and enthroned them as leaders. The "normativeness" of the Goddess Idemili does not however mean that she is not a subversive phenomenon in terms of gender power relations at the center of socio-cultural constructions of power. There is certainly a lot of gender flexibility in roles and statuses in social processes, as I have demonstrated (Ifi Amadiume 1987). The gender culture of the Goddess Idemili is therefore not a conservative kind of normativeness, but an embodiment of female solidarity based on an ideology of a collectivist human empowerment.

Although these Goddesses embody contrasting sexualities, there is some meeting point in these constructions in relation to wealth. In the Nnobi construct, while possession by the Goddess Idemili is a positive sign for wealth and leadership, in contrast, possession by *her daughters,* the goddess Ogwugwu, is viewed as negative since it is characterized by female infertility, disorder, and failure. Therefore Ogwugwu is said to possess women as daughters. I expect the Ugwuta Igbo, like all the other Igbo communities, have their earth goddess and other fertility mother goddesses such as Nne Omumu. The Goddess Idemili of Nnobi has her continuity in her daughter Edo Nnewi, the founder goddess of a daughter town Nnewi, who, resourced by her mother with gifts of knowledge and prosperity, repeated her mother's fame as recounted in myth.

In Nwapa's narrative, Uhamiri's beauty, wealth, and barrenness set her in opposition to Ugwuta norms of womanhood. Reading a puzzle in Nwapa's treatment of Uhamiri, Jell-Bahlsen (1995) concludes that Nwapa is going against local belief by separating children and wealth in the idea of the goddess not giving children and even taking them away, hence neutralizing the notion of the goddess as the giver of fertility. She sees this as Nwapa deconstructing the myth of the goddess and re-constructing it to "voice her own concerns and ideals of womanhood."

It is significant that Nwapa herself did not know the answer to the question of whether, in true life, women had to give up children to follow the goddess (Jell-Bahlsen 1995:36). As I have argued, the belief by some that the goddess denies children or takes away children should be explained contextually in terms of local conceptual systems, as shown by Chinwe Achebe. As we can see, Ona had three children and possession by Uhamiri was used as an explanation for her maternal failings. Efuru's possession was established very early in the novel before the conception of her daughter, and came in handy when necessary as explanation for her misfortunes. It was towards the end of her second marriage that Efuru began to worship Uhamiri. It is therefore not the bereavement resulting from the death of her only child that drives Efuru to the worship of the goddess.

The question of children, I think, is irrelevant since the answer

derives not from the goddess but from an analysis of the social processes in which the goddess herself becomes a category of explanation, a paradoxical embodiment of both norm (wealth) and
deviance (barrenness). Efuru's riches came before she bore a child.
Igbo culture permits that motherhood need not be only biological.
Wealthy women had other ways of fulfilling this social expectation
through the manipulation of the gender flexibility in their system—
as, for example, the institution of woman-to-woman marriage. They
also need not get married in order to have a child, for they also had
the practice of daughters remaining in their natal homes as "male-
daughters" and having children by visiting or "invisible" lovers.

Nwapa, in her mining of Igbo traditions, could easily have used
these "liberal" Igbo institutions to situate her heroines without
resorting to a consistent negative representation of Igbo men as husbands or killing them off entirely—as when husbands either go off,
are impossible to live with, or are killed. Nwapa's treatment of man-
woman relationships is indeed problematic. Chimalum Nwankwo
(1995) picks up this same issue of Nwapa's construction of male
consorts, which he castigates as a "blighted image of Igbo manhood."
If restoring the power of the Water Goddess and women's full recognition of her and submission to her is Nwapa's final answer to the
question of women's empowerment, there seems no restitution for
the Igbo man as husband, but only as father.

In this context, one might ask if the "wayward" MammyWater
indeed has a consort in the deity Urashi, in the sense of the "divine
pair" described by Sabine Jell-Bahlsen (1991, 1995). Such claims
that the term "MammyWater" transcends gender pose a problem,
since we are not clear about later histories of, say for example, the
kind of syncretism which came with Christianity and
Europeanization. We are not really clear on women's worship of
indigenous male water deities or even exactly how the notion of
the supposed MammyWater divine pairness is elaborated in cultural or religious practice. If Urashi-and-Uhamiri's divine pairness
represents social ideals of "complementarity," Nwapa's "poetic
license" seems unable to reproduce this in developing conjugal
relationships in her fiction. The closest she gets to it seems to be
in a father-daughter relationship, that is, in an asexual pair.

CONCLUSION: SEXUALITY AND NWAPA'S RADICAL SEPARATISM

I pointed out earlier what seems a prescription for gender separatism in *Efuru*. By the same token, I see the treatment of sexual-

ity in *The Lake Goddess* as suggesting an expanded discourse in which Nwapa is opening up the option of choice in erotic lesbian sexuality and happiness, and consequently proposing a psycho-sexual solution to the question of women's autonomy as opposed to an economic solution. Her "pampered" heroine of *The Lake Goddess* achieves autonomy and happiness, but does so outside society in the domain of the "wayward" MammyWater! Ona was not really interested in economics. In contrast, for Nnobi women, the Goddess Idemili is about economic empowerment, women's solidarity, and organized power. These are struggles that go on inside society.

This new development in Nwapa's work as exemplified in *The Lake Goddess* leads to the question of audience determinism. *The Lake Goddess* seems very much Nwapa's own say on the side of Igbo tradition, but the West still looms large in determining the agenda of discourse. While the limited representation of the Water Goddess can be exciting for the creative imagination, it could only be achieved through the isolation of the Lake Goddess from an elaborate system of explanation, thus making her appear a paradox to women when this need not be the case since she really functions only within a limited register in the traditional setting as a category of explanation for illness and misfortune. Women resort to other goddesses for other needs. Central or normative goddesses, like the earth goddess for all Igbo societies or the Goddess Idemili for some Igbo societies, are usually the mother goddesses of fertility for farmers; indeed everyone needs to consult them for good health and prosperity.

Nwapa did not exploit the possibilities of Igbo gender flexibility in the institutions of woman-to-woman marriage as an answer to the problems of barrenness because her icon MammyWater limited that choice. In contrast to the marginality and atomization of women in Nnobi, Idemili represented women's involvement in society through the Ekwe leadership title which she gave women, so that Ekwe titled women became the earthly embodiments of the mother goddess.

Possession by Uhamiri signalled anxiety. Efforts were made to cure such a person and free her from the water spell, which was the role of the *dibia* in these novels. Nwapa's characters that I have mentioned, namely, Efuru, Amaka, and Ona, all express qualities associated with *Ogbanje*(ness). To be an *Ogbanje* or one possessed by water spirits and to belong to a cult as a result of such a possession are looked on negatively in all Igbo societies, indigenous and contemporary, precisely because an *Ogbanje* is an ephemeral being, just like MammyWater's slipperiness. Unlike Efuru, Ona is hardly a creative balance between considerations of contemporary individual choice and women's solidarity in the context of traditional Igbo women's knowledge and power network systems, solidarity and resistance strategies that have enabled mass mobilization by African women in historical social movements (Ifi Amadiume 1995).

Works Cited

Achebe, Chinwe. *The World of the Ogbanje.* Enugu: Fourth Dimension, 1984.

Amadiume, Ifi. *Male Daughters, Female Husbands: Gender and Sex in an African Society.* London: Zed, 1987.

___."Gender, Political Systems and Social Movements: A West African Experience." In Mahmood Mamdani and E. Wamba-dia-Wamba. Ed. *African Studies in Social Movements and Democracy.* Dakar: Codesria, 1995.

Jell-Bahlsen, Sabine. "The Concept of Mammywater in Flora Nwapa's Novels." *Research in African Literatures* 26.2 (Summer 1995): 30-41.

___. *MammyWater: In Search of the Water Spirits in Nigeria (Film-video, 59 min.)* Berkeley: University of California Extension Center for Media and Independent Learning, 1991.

Nwankwo, Chimalum. "The Igbo Word in Flora Nwapa's Craft." *Research in African Literatures* 26.2 (Summer 1995):42-52.

Nwapa, Flora. *Efuru.* London: Heinemann, 1966.

___. *One is Enough.* Enugu: Tana, 1981.

___. *The Lake Goddess.* (Unpublished manuscript)

Salmons, Jill. "Mammywater Wata." *African Arts* 10.3 (April 1977):8-15.

The Quest for Spiritual/Sexual Fulfillment in Flora Nwapa's *Efuru* and *The Lake Goddess*

Linda Strong-Leek

In 1966, Flora Nwapa published her first novel. Gay Wilentz notes in her book, *Binding Cultures,* that the work, *Efuru,* was to be the first novel published in English by a Black African woman:

> With the publication of *Efuru*, Flora Nwapa brought a fresh perspective to traditional West African culture and modern Nigeria in literary works by exploring a woman's point of view and exposing a society close to its precolonial roots.... Nwapa tells us that she 'writes these stories about women because they are familiar to her....' (3)

Until the publication of *Efuru*, West African literature was dominated by male writers such as Chinua Achebe and Wole Soyinka. Now a woman's voice would add another dimension to the conversation.

Efuru is the story of a young Igbo woman who, unable to flourish in the traditional Igbo world of marriage and motherhood, turns to the river goddess Uhamiri in order to find spiritual, sensual, and psychological solace. The title character is a woman living within a society that attempts to dictate almost every aspect of her life: from the man she will marry and the children she will bear, to the control she will or will not have over her own body, to her position within society and the community. This is a community of customs: of ancient, unquestioned traditions. A woman must marry in a certain manner; a woman must mourn in a certain manner; a woman must submit herself, willingly and happily, to the laws and the rituals of the community. Male and female roles are strictly defined: any deviation from those specified roles (for men or women) may mean total ostracism from the community.

One of the most important traditions in the community is that of the "bath," or clitoridectomy. It is the belief that a woman must undergo this operation before she gives birth because, as the novel suggests, any child who is born to a woman with an intact clitoris will become "unclean," and the child's chances for survival will be limited. Nwapa seemingly presents the simple story of the hardships and triumphs of the title protagonist; however, upon a closer reading, one finds a much more complex text. There is the underlying tone that "all is not right" with Efuru's world, especially after she undergoes her "bath." To begin to unravel the complexity of the novel, several questions must be considered.

First of all, what is the significance of female circumcision (the "bath") in this novel? Why is it difficult for Efuru to conceive after her "bath"? What is the purpose of this "bath" and the implication of the use of the word "bath" to denote circumcision? Does the operation change Efuru's mental or psychological self? Why does she "willingly" choose to worship the lake goddess, Uhamiri, if she knows that the women who worship her usually find monetary success, but not martial happiness, neither do they experience the "joy of motherhood"? Do Efuru and other women like her find their "joy" by escaping traditional societal roles dictated not only by patriarchy, but also by the fiercely strong community of older women who because of their earned position of authority within their respective communities have become the ultimate keepers of the status quo?

Nwapa continues to question the roles defined for women in her last novel, *The Lake Goddess* (unpublished manuscript). In the thirty years since her first novel, the status and responsibilities of

Igbo womanhood have not changed significantly. The novel begins in much the same manner as Nwapa's *Efuru*, with a marriage ceremony and a "feigned" circumcision. Although there are those who would wish to deny the continued practice of female circumcision in Nigeria, it is obvious from Nwapa's *The Lake Goddess* that the issue remains an important aspect of the womanhood ritual within some Nigerian communities. Note, for example, a recent response to the issue of female circumcision by a noted Nigerian writer, Buchi Emecheta:

> Well, I did not treat it [female circumcision] in any of my books because in our area it is not all that important. I was in Nigeria for three months and I saw only boys being circumcised. Like I said when people have left Africa for a long time, they get excited over irrelevant things. Reading some of her books [Alice Walker], you would think that as soon as you get to Africa, every girl is snatched and circumcised. That is the type of picture she is painting and we are very offended about it here in London because it is not that important.... In *The Color Purple,* she talked about how bush we are and we swallowed it and now she is following it up with this. We are not proud of her. Female circumcision in Africa is dying.... In Africa, female circumcision is no longer relevant culturally. (Ogundele 1996:454)

Although one may agree with Emecheta's assertion that Walker's presentation of female circumcision represents a monolithic voice, somewhat akin to the typical European missionary response to any and all things "traditionally" African, one must also question the importance of the ritual in modern Nigeria when not only African American, but also African women writers analyze the subject from a present-day perspective, within a contemporary context. The circumcision remains such an important element in Nwapa's *The Lake Goddess* that the older generation of females fabricate the actual ritual rather than meeting with the scorn of their community if the rite is not performed. Other African women, including Nawal el Sadaawi, Asma El Dareer, and Soraye Mire, speak of their own painful experiences with clitoridectomy, and the March 1996 edition of *Essence* magazine notes one woman's recent account of her encounter with the ritual:

> I feel the women grab me, gag me and lay me down upon a *matta*. "Be brave," they tell me. "Crying is a disgrace." Suddenly I feel an excruciating pain. My clitoris is sliced off! I try to pull away, but the women hold me. I scream, but no sound comes. Before my silent scream ends, a sharp blade has removed my labia majora and labia minora. As the women close my wounds with thorns and try to staunch the bleeding with scalding water, I faint from pain. (Barrie1996:54)

Hence, there can be no doubt that the ritual is a large part of life for women who live in various countries on the continent, as well as transplanted Africans living throughout the world who believe that they must maintain the traditions of the ancestors. Thus, the connections between Nwapa's first novel and her final one, far from being coincidental, attest to situations that have remained the same in the lives of women over several decades. What is different, however, is the protagonists' response to society's requisites for acceptance. While *Efuru* attempts to mold herself within the norm, Ona is destined from the womb to live in spiritual communion with the Woman of the Lake. The relationship with the womanspirit, thus, becomes the preferable alternative for both major characters. Nwapa ends *Efuru* with this arresting query:

> Efuru slept soundly that night. She dreamt of the woman of the lake, her beauty, her long hair, and her riches. She was happy, she was wealthy. She was beautiful. She gave women beauty and wealth but she had no child. She had never experienced the joy of motherhood. Why then did the women worship her? (221)

This question is of the utmost importance to the understanding of both of these novels: If women can find happiness only in the world of tradition that is set before them, why do they willingly gravitate towards the Woman of the Lake?—the childless, rich, beautiful, and happy woman? These works offer a compelling critique of the defined roles of women, while presenting women with another avenue to fulfillment.

On the way to this satisfaction, however, the women will confront many obstacles, including female circumcision, marriage, and motherhood. Although the main female characters in *The Lake*

Goddess are never physically circumcised, the mother and mother-in- law of Akpe/Maria (Ona's mother) both feel that they must simulate the operation in order to avoid any appearance of the abrogation of tradition.

Significantly, in a 1993 article in the *Harvard Law Review* entitled *"What's Culture Got To Do With It: Excising The Harmful Tradition of Female Circumcision,"* the author notes that "in Nigeria, the practice of female circumcision has persisted for numerous reasons, 'not the least of which is the insistence by elderly females in the various communities that the tradition must be continued'" (1949). Nwapa's mothers, then, are representative of the "elderly females" who believe that although they experienced the pain of circumcision themselves, the tradition must be maintained in order for the communities to survive in the face of the changes brought about by European colonization. Efuru and Ona must make many concessions to family and society before they find their true selves in spiritual communion with Uhamiri/Ogbuide.

Efuru attempts for some time to live within the realm of tradition, but it is only in her ethereal/erotic relationship with Uhamiri that she finds true peace; the same is true for Ona in *The Lake Goddess*. Physical sexuality is literally excised from Efuru after her bath, and the psychological self does not evolve into a complete being without the assistance of the womanspirit. Similarly, although Ona does not experience the physical circumcision that Efuru undergoes, and though her husband is a man who seems totally devoted to her and their children, it is not until she leaves her husband and children and acknowledges her role as priestess of Uhamiri that she is fulfilled. Society attempts to censure/circumcise her spirit by forcing her into an early marriage and, consequently, premature motherhood. The excisions then—one spiritual, one physical—bind the women to all other women like themselves, while connecting them on a spiritual/sexual level with Uhamiri. These correlations remove the "normal" constraints of life for women and allows them to live not merely within a physical/carnal world, but more importantly, a metaphysical realm with themselves and other women. With their sexual beings "excised," their supernatural essences take control and assist them in making their own choices which bring about complete fulfillment, outside the comprehension of men. Sexuality, then, takes on a philosophical meaning in the lives of these women. Sexuality is no longer the physical act of contact/intercourse, but it is associated with one's being as spirit, especially as womanspirit.

Society attempts to extinguish the existence of the powerful womanspirit by the act of circumcision. The protagonists takes center stage in these novels as both metaphor and participant. The act of "bathing" a young woman—a euphemism for removing her clitoris—is often an attempt to quell her desire for sex from any man other than her husband. As noted in the aforementioned edition of the *Harvard Law Review*:

> In addition to the religious arguments in favor of chastity, advocates point to the prevention of promiscuity as a separate and distinct reason to continue the practice of female circumcision. Because the clitoris is believed to provoke women to make uncontrollable sexual demands on their husbands—demands that will drive a woman to seek extra-marital affairs if her husband does not meet them—removal of the clitoris is presumed beneficial for women and for society. (1952)

Circumcision is, then, required to keep women's sexual desires under control. Men must remain in power, sexually, politically, economically, and otherwise. And if women begin to make "uncontrollable sexual demands" upon their husbands, women will then have the power, because it will be *their* (i.e., women's) needs and appetites that men must meet in order to maintain their own positions within their marriages and as the "heads of their households." The entire system of power will shift because men, not women, will have to compete to satisfy their partners. It is interesting that the issue of a woman's *ability* to quench her husband's desires is not an issue in Nwapa's texts. The issue is, rather, *stifling* the sexual *appetites* of women. This suggests that the male ego must be appeased at all costs in order for men to presume sexual superiority, even if it means silencing women both physically and psychologically.

However, in Efuru's circumstance, the circumcision has ramifications that, it seems, the community never really considered. How would the operation alter the way Efuru would see herself and her world? Would she ever be able to even enjoy a sexual relationship with a man? Would she be able to bear children? Efuru experiences great difficult in her attempts to conceive. Olayinka Koso-Thomas notes in her work, *The Circumcision of Women: A Strategy For Eradication,* that women who live in societies which practice female circumcision have at least a 25% decrease in birth rates, (as compared with women in societies that do not practice this ritual) as well as high rates of infant mortality. The concerns are somewhat dissimilar in *The Lake Goddess*.

Ona is never clitorized; therefore, it seems that she much more readily resists the norms of societal constraints. She often finds her mind, spirit, and body wandering away from her husband, home and children, into that spiritual/sensual world with Ogbuide, the Lake Goddess. It is in that world that she experiences wholeness. Her preoccupation with the womanspirit relieves her of those traditional "joys" women experience, and allows her to live within the metaphysical realm with the lake goddess. This is an analysis of the relationship between Efuru's "deadened" sexuality/spirituality after the removal of her clitoris and the rebirth of herself through her relationship with the Lake Goddess. I will also analyze the impact of the intact clitoris on the more able spirit/body of Ona in *The Lake Goddess* and her eventual emancipation from convention.

Efuru is viewed, from the beginning of the novel, as a strong-willed, intelligent, strikingly beautiful woman. But after she is circumcised, there is a noticeable change in her demeanor.

> So on Nkwo day, Efuru dressed gorgeously. She plaited her lovely hair well, tied velvet to her waist and used aka stones for her neck. Her body was bare, showing her beautiful breasts. No dress was worn when a young woman went to the market after the period of feasting. Her body was exposed so that the people saw how well her mother or her mother-in-law had cared for her. A woman who was not beautiful on that day, would never be beautiful in her life.... But underneath, something weighed Efuru down. (18-19)

Although Efuru remains in control of herself in various ways, and she is the supposed "apple" of her husband's eye after her bath, there is the notion that a "weight" is now a part of her existence. This weight continually pulls her back to the traditional world and forces her to remain in her first marriage, even after abandonment and suffering through the death of her only child, a female, alone. Efuru's clitoris, then, is not just a part of her sexual being, but it is, more importantly, the physical representation of her wholeness. As Paula Bennett asserts in "Critical Clitoridectomy: Female Sexual Imagery and Feminist Psychoanalytic Theory":

> ...the clitoris's presence or absence in the theoreti-
> cal treatment of female sexuality—as in the body
> itself—is of no small consequence to women. As
> Spivak (1981) has argued, 'The pre-comprehended
> suppression or effacement of the clitoris relates to
> every move to define woman as sex object, or as a
> means of reproduction—with no recourse to a sub-
> ject-function except in terms of those definitions or
> as 'imitators' of men.' (181)

> Investigating the clitoris's effacement is therefore a
> passage into understanding the historical and theo-
> retical suppression of women.... To the extent that
> male domination is based on women's sexual sub-
> ordination to and within the family, the "excess" this
> organ represents—the excess of absolute sexual
> autonomy—is a threat to individual males and to
> male rule generally. (118-19)

Thus, Efuru's excision is based on much more than the idea that a
woman must be "clean." The main ideology behind the use of circumcision, according to Bennett, is the desire within men to
remain in control of women, politically, economically, and (most
of all) sexually. The clitoris, left intact, allows women not only to
be desired by men, but also to experience passion themselves. This,
then, is the "danger" to men and society. Female sexuality—not
male sexuality—must be contained. The desires of women seem
to corrupt society, but the yearning of men must always be satisfied. Bennett continues:

> Psychologically speaking, as Thomas Laquer argues,
> (1990), Freud's theory of the vaginal orgasm does the
> same thing, albeit more humanely. It seeks to recon-
> tain the female sexuality by "Castrating" women,
> denying them the sexual agency and active power
> that would make them sexual subjects in their own
> right—and consequently terrifying indeed to this
> man who based both his career and his gender on
> men's genital superiority to women. (129-130)

Bennett's assertions here reaffirm the aforementioned assertion—
that women's sexuality must be governed, not merely in order to
maintain men's positions of political, economic, and social power,

but it also must be restricted—or eliminated—in order to keep women sexually subordinate to men. Although some women in societies which practice female circumcision maintain that they lead totally "fulfilled" sexual lives (Fanny Lightfoot-Klein.), one may easily argue that since many of these women never experienced sex with an intact clitoris, they have no other point of reference for their sexual experiences. Removing the clitoris, then, is somewhat akin to male "castration"—which is defined in various terms, but which basically implies the removal of the testicles, no longer allowing a man to give or receive sexual pleasure. If clitoridectomy has the same effect on women, it is of no little consequence that so many women turn to the barren, wealthy and beautiful Woman of the Lake for respite from their patriarchal worlds. Bennett notes the importance of the clitoris, or the removal of the clitoris as a means of control in traditional patriarchal societies, both in 19th century Europe and in the psychoanalytic theories of Sigmund Freud:

> Whether in the form of cauterization or actual clitoridectomy, medical assaults upon the clitoris were routinely performed by many nineteenth century gynecologists in order to bring under control socially undesirable forms of female sexuality such as masturbation, lesbianism, and "nymphomania".... It is only by rendering the clitoris accidental (as it were) to female sexuality (and thus eliminating it) that complementarity is made possible. In the binary terms of Freud's phallocentric definition, women are what men are not; as Cixous puts it, they are dark where men are light, passive where men are active, empty where men are full.... Thus deprived by Freud of that which makes her different but "equal" to men, the Freudian woman becomes man's negative obverse.... Her sexual and reproductive system thus neatly matched to his, she is, as Spivak observes, absorbed into the "uterine social organization" of family and state.... Under such circumstances, she can never also be what she is in herself—independent and autonomous in respect to her own sexual power. (129-31)

It is in this place, in this "uterine social order," the world of physical and emotional castration, that Efuru and Ona find themselves, and it is in this situation that the true essences of both women struggle to survive a calculated attempt at annihilation. Spivak and Bennett emphasize the importance of the clitoris, both physically and emotionally, to the fulfilled development of women. The removal of this organ, whether through the physical operation of clitoridectomy or through psychological means, subjugates women and assists in the maintenance of compliance. It is not then, merely the physical act of circumcision which places women in that sphere of uterine social order; in Ona's case at least, it is the attempt to quell her mind, body, and spirit.

Efuru is also circumcised in mind, body, and spirit. She remains in authority economically, but she is restricted to a great extent by her aspiration to prosper in the traditional manner—within the realm of motherhood and marriage. Her "sexuality" now properly maintained, she is able to focus on the fulfillment of the patriarchal standard. Efuru, so conditioned by society, even begins to question her own "femaleness" after she is unable to conceive two years into her first marriage. When she finally becomes pregnant and the baby is born, Efuru still cannot comprehend her own "success":

> Efuru lay there thinking of it all. 'Is this happening to me or someone I know. Is that baby mine or somebody else's? Is it really true that I have had a baby, that *I am a woman after all*. Perhaps I am dreaming. I shall soon wake up and discover that it is not real. (31, Emphasis added)

Efuru has been so psychologically damaged by the rumors and implications which surround her that she herself questions whether or not she is a *real* woman until after the birth of her child. The inability to reproduce takes on a much deeper meaning in this instance. It is not merely a matter of not having a child, but it is a matter of not even *being* a woman. Womanhood and motherhood are synonymous—a woman without a child, then, is not really a woman, and early in the novel other women echo this belief:

> A year passed, and no child came. Efuru did not despair. 'I am still young, surely God cannot deny me the joy of motherhood,' she often said to herself. But her mother-in-law was becoming anxious.... Neighbors talked as they were bound to talk. They did not see the reason why Adizua should not marry another woman since, according to them, two men

> do not live together. To them Efuru was a man
> because she could not reproduce. (24)

The villagers begin to refer to Efuru not as a barren *woman*, but as a *man*. She is a "man" simply because she does not have the ability to reproduce readily and rapidly. Woman as commodity and childbearer is presumed in this instance. Efuru's value and femininity lie in her ability to procreate.

Similarly, in *The Lake Goddess*, although much of the community has ascribed to the "new" religion brought to the village by the "strange people," Christianity in itself is not able to satisfy the needs of the people nor provide support for any of the traditions that remained necessities for the community. Although there is never a mention of the circumcision ritual in reference to Ona, her mother, Akpe, does take part in a so-called "circumcision." Both of the mothers of Ona's parents feel that it is necessary to—at least, on the surface—preserve the customs of the ancestors. Mgbada's mother, however, is not merely interested in the ceremony for the sake of ritual. She believes, like *Efuru*'s mother-in-law, that if a woman is not circumcised before the birth of her first child, that child will die:

> "We must put our heads together, Theresa,"
> Nwafor said. "You mean the circumcision?"
> "Yes, that's what I mean."
> "You know it is dangerous."
> "I know."
> "If you know, why must we think about it?"
> "It is the custom of our people. If it is not per-
> formed, Akpe will not be able to have children."
> "Theresa, you believe that?"
> "I do." (24)

This is an example of the archetypal confrontation between the pre-colonial system of values and the post-colonial merging of ideals. Many wish to maintain the customs and traditions of the ancestors; others fight for a place in the new power structure, to find that only a complete conversion to Christianity and total denial of one's ancestral past will be acceptable. Christianity could not, however, fill the void left by the abrogation of years of tradition. Thus, the mothers devise an alternative plan. There is a fear, usually justified, that the old ways and traditions will become a part of antiquity, and centuries of custom will be replaced. Compelled by the desire to remain faithful to their histories, the mothers formulate an alternative:

"We must devise a means, a way out..."
"I am listening."
"Of making everybody believe that circumcision
had actually taken place."
They fell silent.
"I see. Go ahead."
"We shall inform Mgbekworo."
"Yes, she will come on the appointed day..."
"She will bring everything used in doing it, and
everybody will know about it..."
"I understand. Then we bribe her. We tell her not
to do it."
"Exactly, and she would pretend that she had
done it...."
And so Akpe was "circumcised." The people were
happy and satisfied that their tradition was not
trampled upon by the new religion. (25-26)

This outward show of obedience preserves the status of the newly married couple. It is not coincidental, either, that the "circumcisions" take place early in the novels. Both Efuru and Akpe are young, marriageable women. Both are "beautiful" by their societies' standards. The traditions demand that a woman, before the birth of her first child, be circumcised. It is also important that neither Akpe nor Ona are ever "really" circumcised. And although Ona is easily able to bear children, she only makes a brief attempt to submit to society. Ona, clitoris intact, remains true to herself almost throughout her entire life. It is as though she is always in spiritual and sexual communion with the Lake Goddess. She never completely conforms simply to be accepted, nor does she allow society or her family to totally dictate her life.

Ona remains a spiritual/sexual being throughout her marriage. But her desire and fulfillment are never obtained within the confines of marriage and motherhood. Clitoris intact, power maintained, she is able to reach outside convention and find peace—beyond the expectations of any men in—her spiritual and temporal communion with Uhamiri. This is the only relationship that can meet all of her desires and satisfy all of her needs. Whereas Efuru's "damaged" body is only able to bring into the world a frail baby girl who will die in infancy, Ona's uncircumcised mother, Akpe, is able to bring into the world the child who would live in complete communion with Uhamiri.

Even before her birth, Ona's parents are aware of her unusual vocation. The "diviner" speaks to Mgbada as follows when he questions him about Akpe's difficult pregnancy:

> I know you are a diviner like myself. I know your father and your grandfather. Your child will be a strange child. That has been revealed to me. You cannot see it. A *dibia* does not cure himself. What is revealed to me about your unborn child cannot be revealed to you, because you are bound to your child by blood. (59)

Akpe experiences an extremely long and laborious pregnancy before a difficult delivery, and when Ona is finally born her mother is left exhausted. Before the age of three, young Ona begins to reveal her destiny to her mother. She is an avid swimmer who finds solace only in the lake. Ona tells her mother:

> "And last Nkwo when I went to the lake with grandmother, I saw Ogbuide. She was returning from the market. She was dressed in velvet and coral beads... Mother, I am serious. The day I saw her, she took me to her dwelling place.... Mother, it was not a dream. She told me to come with her. So I went with her." (73)

Akpe and her family attempt to force Ona into conformity. But she is rebellious in all matters. She does not succeed in school, and is sent away at one point to the resident "catechist" for an "exorcism" of sorts, but all to no avail. She is then married at an early age to a man who seems to genuinely care for her, but she can find no real happiness within the relationship. She continues to dream of the Woman of the Lake, who beckons to her to come and live with her and serve as her priestess. Her parents and society believe that if she is able to submit herself to the will of her husband, the spirit will go away and, like most women, her vigor will perish:

> Ona's parents believed that once she was married the evil spirit would leave her and then she would not be able to 'see' anymore. She was a virgin. Once she knew a man, they believed, she would lose the power to 'see.' (106)

Ona's power, then, is bound to her ability to function in the world without men. Once a man had "entered" her, her wholeness would

disintegrate—her connection with the spiritual world would cease —
and she would become the property of her husband—the commodity she is supposed to be in a world that values not spirituality in
women, but women's ability to please—and reproduce—other men.

But even after her marriage, Ona remains the chosen one of
Ogbuide. With Ona, there is not even a call to feign the circumcision ceremony. This may lend some validity to Emecheta's earlier quoted assertion that the Igbo in Nigeria are moving beyond the
need to continue ancient ceremonies, yet it does not leave one
with the impression that the ritual is near extinction. It is clear that
there is always an attempt to "cure" Ona—to quiet that realm which
yearns for something other than convention. This is a part of the
"revised" circumcision ritual—the cutting now is spiritual, but the
wound is no less difficult to heal.

However, Ona lingers within the metaphysical, and her sensuality reaches beyond the bounds of human perception. Even the
women in the marketplace acknowledge the special position of Ona:

> "I think it is Ogbuide herself who is troubling her.
> And unless something is done, she will never be a
> normal person....Even with a husband, yes. Her hus-
> band cannot compete with Ogbuide....Ogbuide will
> triumph in the end...." (131)

Ona's supernatural and erotic entirety is not found within the confines of convention. Her totality is located and experienced only
in her relationship with the Woman of the Lake. The women in the
marketplace are also reminded of an earlier call to worship by
Nwapa's first major female character, Efuru:

> "Efuru performed very well....You missed some-
> thing. She gave dignity to the whole cult. She was
> the leader and every member of the cult from far
> and near respected her. She was always calm...."
> "She never married again?"
> "Of course not. She was 'married' to Ogbuide."
> (132-33)

Nwapa is straightforward—Efuru gains respect and dignity only
when she begins to seek acceptance and refuge outside convention,
and she finds her soul with Ogbuide. She cannot, will not marry
again because she has joined her spirit with the woman who will
bring her peace. Karla Holloway in *Moorings and Metaphor* asserts,

> It is Efuru who is called upon to help the sick and to give advice. She becomes the center of the women's community in the way that the goddess centers the community's indigenous spiritual lives. At this point, Efuru's decision to dedicate herself to the service of the goddess, even in the midst of her marriage to Gilbert (Eneberi), articulates her intense dedication to the centering womanspirit that the childless goddess represents....Efuru's understanding that the goddess could not assure her children marks her dependence on the womanspirit she has nurtured and developed throughout her life. (177)

Free from the ties that bind women in traditional communities, Efuru and Ona are able to dedicate themselves fully to the worship of the womanspirit. And it is the womanspirit who is able to accomplish something that neither Efuru's nor Ona's husband(s) could—making these women complete.

Is it any wonder, then, that Efuru must "return" to the Goddess in order to find her whole being, while it is always apparent that Ona is complete, as a very young child, with the Woman of the Lake? Efuru had to endure a pain and shame that was omitted from the life of Ona—the circumcision ritual, the "disgrace" of not being able to produce viable children—while Ona remained intact physically and spiritually; but because of her desire to maintain the status quo, she too had to "return" to herself and her spiritual and sensual communion with the Lake Goddess.

It is quite obvious to Ona, to her husband, and to the entire community that she cannot easily position herself as a "normal" wife and mother, and that she does not belong in the world of the ordinary:

> Ona had two more babies...She appeared normal to everybody. She began to take her responsibility more seriously.... Her husband did not have to struggle with her when he made love to her. She accepted it as her duty to surrender herself to him whenever he wanted her to do so. Sex was something that was strange, even alien to her. Why did men make such fuss about it? What was it that men enjoyed? It was dirty as far as she was concerned. What with the slimy stuff between her legs when it was all over. If she had her way, she would have none of it. (170)

Even though the community is able to view outward manifestations of conformity, deep inside Ona is dying as she struggles to
maintain her position as a wife and mother. Sex is a "duty" to her,
a "slimy," strange function. The temporal world exists in her life
only as an attempt for others to find a location for her in a world
in which she cannot happily function. Sex, motherhood, and
domesticity are unnatural entities. Her only solace is in the spiritual world with Uhamiri. When Ona does attempt to initiate sex
with her husband, Uhamiri seems jealous. She appears before Ona
in the form of a naked woman, dripping wet with water.

> After what passed for a long period, she tried the
> intimacy again. It worked, but when it was over and
> her husband was snoring with contentment, Ona
> was tossing in bed. She could not sleep. It was at the
> cock crow that she eventually slept, but not for long
> for she saw a naked woman standing in front of her.
> She had very long hair dripping with water. She had
> on coral and agate beads on her neck and wrist. Her
> eyes were sharp and penetrating. They were pierc-
> ing her body. (172)

Uhamiri appears before Ona with "sharp and penetrating" eyes.
Ona has betrayed herself and the Goddess by initiating intercourse
with her husband. Ona realizes at this point that she can no longer
remain in the temporal world. She must heed the call of the
Goddess. Ona cannot rest within convention, and she begins her
new life with the Woman of Lake:

> There was no one at the lake front so she undressed
> and waded into the water. She immersed herself in the
> water.... She rubbed coconut oil all over her skin. Then
> she rubbed a mixture of the coconut oil and camwood
> all over again. She combed her long hair and plaited
> it with her fingers instead of with black thread. (225)

Ona prepares herself for the worship of Uhamiri by rubbing her
body with coconut oil, an erotic experience which enlivens the
physical as well as the psychological senses, and she confronts her
clad in nothing more than her own smooth, black skin. She has
come to the place her parents and society deemed unfit for her as
a female—the same place that Efuru is able to locate her soul—to
Uhamiri. As they give themselves in complete union and communion, Efuru regains that part which is taken away by force of

tradition, and Ona is now able to be true to the self which had beckoned her all of her life. Uhamiri restores Efuru and Ona, and moves them from the temporal world, represented by men, motherhood, and domesticity, into the spiritual realm of the worship of the womanspirit. Here, women are valued in their capacity to maintain—not merely create—life. They are in command of their own destinies. Uhamiri, as a woman, will not impose upon them anything detrimental to their existence. She knows their sorrows and does not seem to allow for any deviation from her world, which for Efuru and Ona almost resulted in the loss of their complete selves. *Efuru* ends with the rhetorical question, in *The Lake Goddess* Ona's life is the answer to the question. This is why the women willingly turn away from convention and freely give their lives to the Goddess of the Lake—with her they find complete solace and ultimately, freedom from the ties that bind other women to men and motherhood. Ona is the voice of these voiceless women. She can communicate on behalf of those less fortunate than herself—she speaks to the all-powerful Woman of the Lake:

> "Ogbuide wants all women to have voices. Women should not be voiceless. Ogbuide hates voiceless women," said Ona. "This is why the goddess whom I serve calls certain women to be her priestesses. She communicates with her fellow woman." (241)

And it is with their fellow "Woman" that Ona and Efuru find complete solace. Here, safe from convention, they exist in peace and harmony with the Woman of the Lake. She beckons to them and they answer. Women choose to worship her because only with her may they remain true to themselves as whole bodies and spirits. Within the spiritual realm, they do not need to subordinate their needs or desires to those of men. To the contrary, in this world, women reign supreme.

WORKS CITED

Barrie, Mariama L. "Wounds That Never Heal." *Essence*. (March 1996):54.

Bennett, Paula. "Critical Clitoridectomy: Female Sexual Imagery and Psychoanalytic Theory." *Revising the Word and the World: Essays In Feminist Literary Criticism*, ed., VèVè A. Clark, Ruth B. Joeres, and Madelon Sprengnether. Chicago: Chicago University Press, 1993, 115-139.

El Dareer, Asma. *Woman Why Do You Weep? Circumcision and Its Consequences.* London: Zed, 1982.

El Saadawi, Nawal. *The Hidden Faces of Eve.* London: Zed, 1993.

Holloway, Karla. *Moorings and Metaphors: Figures of Culture and Gender in Black Women's Literature.* Rutgers University Press: New Brunswick, 1992.

Koso-Thomas, Olayinka. *The Circumcision of Women: A Strategy For Eradication.* London: Zed, 1987.

Lightfoot-Klein, Fanny. *Prisoners of Ritual: An Odyssey into Female Circumcision in Africa.* New York: Haworth Press, 1989.

Mire, Soraye. *Fire Eyes.: Female Circumcision.* (A 60 minutes Film-Video), New York: Filmmakers Library, Inc., 1994.

Ogundele, Oladipo Joseph. "A Conversation With Dr. Buchi Emecheta." *Emerging Perspectives on Buchi Emecheta.* ed. Marie Umeh. Trenton: African World Press, 1996, 445-456.

Nwapa, Flora. *Efuru.* London: Heinemann, 1966.

___. *The Lake Goddess.* (Unpublished manuscript).

"What's Culture Got To Do With It? Excising the Harmful Tradition of Female Circumcision." *The Harvard Law Review.* 106.8 (1993):1944-1961.

Wilentz, Gay. *Binding Cultures: Black Women Writers in Africa and the Diaspora.* Bloomington: Indiana University Press, 1992.

Desire, the Private, and the Public in Flora Nwapa's *Efuru* and *One is Enough*

Shivaji Sengupta

> Efuru slept soundly that night. She dreamt of the woman of the lake, her beauty, her long hair and her riches. She had lived for ages at the bottom of the lake. She was as old as the lake itself. She was happy, she was wealthy. She was beautiful. She gave women beauty and wealth but she had no child. She had never experienced the joys of motherhood. Why then did women worship her? (*Efuru*, 221)

Dreaming of the Lake Goddess, the patron saint of women, Efuru in her sleep passes silently to the next phase of her life. Hereafter she will serve the Lake Goddess, symbol of fertility even though she never had children herself. Being childless has caused Efuru a whole life of grief, suffering at the hands of fallible people who could not appreciate her worth. Efuru decided not to belong to any man any more, nor to any family. Having been abandoned by one husband, accused of adultery by another, she has decided to forego the traditional joys of motherhood and family. She will be the priestess of the Lake Goddess. She would serve women. Nwapa ends one of her best novels with a question. Why do women who want children pray to Uhamiri, the Lake Goddess when she is herself childless? Fifteen years later when she wrote One is Enough, there were no more questions.

> "...I would like to thank [Father Mclaid] for this noble decision, ... I was the first to know that there would be no marriage... I shall forever remain grateful to him for proving to the world that I am a mother as well as a woman." (*One is Enough*, 154)

In *One is Enough* Father Mclaid, an honorable man who loves Amaka, renounces the Church to acknowledge being the father of her twins and marry her. But Amaka refuses, preferring to be independent. Her lover is outraged. But Amaka attains her freedom through the intervention of fortune. An auto accident in which Mclaid escapes certain death makes him realize the futility of giving up a lifelong priesthood for the love of a woman. He recants and is taken back to the Church. Amaka thanks the Father for making the noble decision to return to the Holy Order and for not insisting upon marriage. Father Mclaid has given her love. He has helped her business thrive. Most of all, he has proved to the world that Amaka was capable of having children. But having come out of a bad marriage, finding economic freedom, after being assured of the love of a man, she was not going to take on another husband. One was enough.

The two novels are about transition. Two women, Efuru and Amaka, go through difficult lives, struggle and humiliation, to eventually arrive at self-recognition. Efuru says to her book-learned doctor friend that, since her elopment with Adizua, after years of loving, having family, all she has done is come back home. Amaka ends *One is Enough*, satisfied that she has become a mother. Both heroines emerge from being confined to the house as wife, to become a part of the outside. Both, however, do not banish their roles of belonging to the interior entirely. Efuru becomes the priestess of Uhamiri, the lake goddess, but continues to serve women in their most feminine needs of being mother. Amaka is a successful business woman, but a proud mother.

These two novels of Flora Nwapa bring to the foreground changing concepts of the private and the public,[1] and the role desire plays in the transition,[2] and the implication this process has on female sexuality. Efuru and Amaka, both essentially private women at first, change with circumstances and enter the public domain. For each, then, the exit from the private and emergence into the public has been caused by the desire to live life with dignity, without accepting the traditional role of a wife.

This paper will examine the movement of Efuru and Amaka from the private to the public by deconstructing the two limits of the two locations, showing that what Efuru and Amaka attain—and hence, by implication, what Flora Nwapa stands for,—is not the reversal of the public and private when it comes to women, but an inclusion of the private into the public. This in turn, affects their sexuality. Although sex is never in the foreground of Nwapa's novels—there are no references to or descriptions of love-making in them—sexuality itself is an important theme in these novels.

But sexuality is not sex. It is a way of being in the world in relationship to others. Efuru's sexuality is different from Amaka's. In the former, it is subdued and diffused in her personality. Efuru is handsome, commanding, intelligent, sensitive, affecting men subtly almost inspite of themselves. Unlike Amaka she does not use it overtly to achieve her ends. She simply communicates with it, delicately within the parameters of her culture. In Nwapa's two novels sexuality links the private with the public. Amaka uses it to step into the public as it were. Efuru gives it up.

In the west gender specific literary and cultural criticism associates the public and the private with male and female roles, and views these two domains as not only separate but actually in contradiction with each other.3 Historically, the Western male views the world through oppositions, or dualism. Contradictions, dialectics or polarism are, therefore, the *modus operandi* of Western thinking. Whether the subject is love (male versus female), or labor (producers versus owners), human phenomena is conceptualized in dualistic terms, not surprising for a society accustomed to thinking about life in terms of Hobbesian conflict of interest.

Western civilization has, therefore, demarcated the public from the private. According to masculinist agreement, jobs, politics, business belong to the public domain. Love, family, the rearing of children, belong to the private. Religion in its organizational and political aspects belongs to the public; in terms of personal expression of devotion and faith, to the private. The domains of the public and private have been allotted according to sex roles. Males take the professional and public arena; women are given personal and private roles.

It is in this intellectual milieu that Nwapa's integration of the public with the private is significant. She eschews the traditional Western reversal of the private with the public, and attempts to blur the distinctions between them. Amaka uses love and sexuality, traditionally considered private, in the public domain to get men to give her business opportunities. Efuru's decision to serve Uhamiri brings her out into the public, but the protagonist's concerns are private, viz., women whose interests are marriage and children. Emerging into the public domain Efuru and Amaka use sexuality differently. Efuru is gorgeous as the server of the Lake Goddess who is herself beautiful. But she has given up sex in order for other women to have children and preserve their marriages. Amaka, after she has become a mother, tempers her sexuality and rejects the role of a wife. Hence, for these women there is a correspondence between sexuality, the private and the public.

This raises two critical questions: Do women view the world differently from men? As women, do they have a different way of relating to society, their outlook predisposed by their bodies? Does sexuality determine episteme, which as Foucault defines it, is a set of relations that interconnect at a given period, giving rise to particular kinds of knowledge and formalized systems?[4] Does difference in sexuality lead to subscribing to a different epistemology? Furthermore, do cultural, social and political differences, combining with gender, create a different world view? Do African women, specifically Nigerian women, have a different vision of the world from Western women? The answer to these questions is yes.

One of the more persuasive gender-specific epistemological theories that I have come across argues that women's epistemological experience has developed historically from "silence" to "voice," through the progression of the mind. Situating mind over body, the theory holds that the transition from the body to mind results in the movement from silence to voice. It happens to women through a process of gradual awareness of their own intellectual and/or affective potential and the ability to explain life both as concrete experiences and as part of a larger culture. The very act of deciding to voice one's position involves the mind and is an act of the will.[5]

This is also what Nwapa demonstrates in these two novels. Flora Nwapa questions the opposition between the private and the public by giving us protagonists, controled by the male establishment, break through these barriers, challenging traditional roles. Efuru becomes a priestess; Amaka rejects a second marriage. Both find their 'voice' after a series of traumatic experiences involving their very roles as women and concerning not only oppressive men but also onerous women who work as men's agents and perpetuate the public/private hierarchies and sexual stereotypes. As Efuru and Amaka move from 'silence' to 'voice' they readjust their roles as women and renegotiate for themselves the limits of sexuality, the private and public. The movement from the private to the public is critically significant in Nwapa's works as novels of protest.

These days Nigerian women have been compelled to join the workforce but with little change in the family protocols, particularly regarding motherhood. The fact that women bear children is a biological truth. That they are expected to bear children regardless of circumstances is control. Add to that the fact that male children are priority and a woman's worth is determined by how many male children she has, the control over women is complete because now they are held responsible for something that is biologically beyond their power. Thus, social control comes from expectations and is so much a part of the culture, ingrained so deeply, that it works in an individual from within. Efuru is a good example of this.

In this progression, desire plays a major role. Desire is energy, a vector with direction and purpose. In fiction desire unfolds in the theme and the author's style. Thematically, it is usually discussed in terms of energy of the characters, what creates their desire, where it takes them. It is a willful act that changes the course of events: Efuru's decision to transcend family and a world of men to live a public life of spirituality; Amaka's to "go it alone" in a public world of men without a man, as a woman.

In esthetics, desire is in the writer's language, its linguistic purpose. It is the text that an author creates within a given textual system (i.e., local English, standard English, the vernacular) and a given literary and intellectual milieu. We will examine these two aspects of desire in Nwapa's two novels, discussing the works not only from the thematic sociocultural point of view but also in terms of Nwapa's literary effort. We shall see that in structure as in style, Nwapa follows a rigidly closed pattern.

The structure is displacement. Nwapa uproots her protagonists from their own locations and makes them operate in another. For both, the displaced location is from the private domain: Efuru as a daughter in her own house; Amaka as the daughter-in-law in Obiora's house. Traumatic experiences like being cast out of a family in the case of Amaka and elopement and eventual abandonment by husbands in Efuru's case, bring about displacement. Displacement is not just being forced to go to a new place to live. It is also the consequences of such a move, an inverse relationship between the self and the new place. This displacement then causes the protagonist to struggle to adopt the new place as her own. Usually, this is achieved by marrying or finding a sexual partner in the new location. Here too, Nwapa's treatment of displacement in *Efuru* is more subtle and complex than in *One is Enough*. Finally, displacement usually ends with the protagonist's finding a "home," the latter arriving at a resolution. Nwapa shows the role of desire and sexuality in this transition and the adjustment the protagonists make in order to find their voice by discovering themselves. As mentioned earlier, in these works the finding of the voice is also the sign of transition from the private to the public and, ultimately, integrating the two.

Nwapa's language in these novels is a closed system, so that by leaving little to the readers' imagination, she removes any possible speculation about her social message. Unlike Nadine Gordimer or Toni Morrison, or, to a lesser extent, Buchi Emecheta, Nwapa is simply not interested in the "open spaces" in language. Hers is tightly knit, a means to get her message across. Her desire in language is not for "a little more truth," as Julia Kristeva sees the novel's endeavor to be, but, inasmuch as it is possible, the absolute truth. We shall see in the course of this essay how Nwapa uses language and the narrative structure to suit her end.

EFURU

The thematic structure of *Efuru* incorporates desire, displacement and integration. The three elements are themselves subsumed into two locations, the private and the public.

Following the story, we see a basic pattern emerging throughout the novel: something persuades Efuru to act. This may be an attraction for some one or idea, the will to change the present condition. Acting on it causes displacement, accompanied usually by hardship, suffering and trauma; finally, after a period of struggle and readjustment, there is integration, indicated by some kind of learning or resolution on the part of Efuru. Throughout, the displacement and integration would make Efuru's transition from the private to the private-public.

Efuru is the fictional life story of Efuru, the daughter of Nwashike Ogene, "the mighty man of valor" (11), the son of a great fisherman. In the beginning of the novel she falls in love with

Adizua, a poor fatherless man. Knowing that their respective families would object to their marriage on account of the great difference in station, they decide to elope. The novel begins with love, romance and defiance of reality:

> It was after the festival in which young men and
> young women looked for wives and husbands that
> Efuru first met Adizua. Adizua asked her to marry
> him and she agreed.
> One moonlit night, they went out. They talked of
> a number of things, their life and their happiness.
> Efuru told him that she would drown herself in
> the lake if he did not marry her. Adizua told her
> he loved her very much and that even the dust
> she trod on meant something to
> him....Efuru...went home very happy and light.
> "It's late, Efuru, where are you coming from?" one
> of her cousins asked her as she was opening the
> gate of the compound.
> "Don't you see the moon?"
> "Is that the reason why you should come home so
> late? I shall tell your father. A young woman like
> you should not be out so late."
> "I don't care whom you tell."
> (*Efuru*, 7)

In the beginning there is the end. Efuru's fateful utterance that she would drown herself in the lake if Adizua does not marry him turns out, ironically, to be metaphorically true. Efuru becomes the priestess of Uhamiri, the Lake Goddess, who lives in the bottom of the lake. Desire causes displacement. Marrying Adizua, Efuru has to leave home at her father's displeasure.

But for now it is defiance. "I don't care whom you tell." A young woman in love would even oppose her own father. Nwapa has already introduced her in terms of character: "She was a remarkable woman. It was not only that she came from a distinguished family. She was distinguished herself"(7). There was no denying such a woman.

Her father and brothers accept the marriage. Efuru begins the life of a married woman, as Adizua's wife, in earnest. At this point, two remarkable incidents take place, both indicators of Efuru's heroic struggle to adjust and cope. She undergoes circumcision and becomes a mother.

The first one is described in great ritualistic detail by Nwapa but the part worth noting is the conversation between Efuru, her mother-in-law, and the woman who performed the operation, Omeifeaku.

> "Do you know Nwakaego's daughter?....She did not
> have her bath before she had that baby who died
> after that dreadful flood...their folly cost them a son,
> a good son." *(Efuru, 14)*

Much, of course, has been written on the subject of clitoridecto-my. Western women see this practice as brutal. However, African women are not united against it. Notice that Nwapa's criticism of this practice is subtle and ironical. Despite the circumcision, Efuru's baby daughter, Ogonim, dies.

Nevertheless, Nwapa's treatment of female circumcision is cau-tious, justifying the importance of the ritual, but showing little faith in the cultural belief itself. Since the custom of clitoridectomy itself is too deep rooted to be eradicated, Nwapa shows in this episode the healing power of the family and the community rallying around the circumcised, and the power of rituals as a coping technique.

After the pain and the agony of the surgery subside, Efuru coop-erates completely. Her character shines through. She rubs cam-wood on her husband's singlet. "When his friends saw him, they knew at once that his wife has been circumcised and they teased him...Efuru grew more and more beautiful every day" (17).

Lest we pity Efuru, Nwapa makes it quite clear that the pro-tagonist herself wanted her circumcision to be known to others. Efuru grows more attractive after the circumcision. After the cus-tomary feasting is over, she is restless and wants to go out with her mother-in-law to trade in the market. She is tired of being confined, she says, besides money is short. "...we have not paid the dowry yet" (18), she says referring to the bride price, that Adizua was required to pay to Efuru's father. So Efuru prepares to go to the market. Flora Nwapa describes the "bathed" Efuru's entrance into the public with one of the best descriptions in the novel:

> So on Nkwo day, Efuru dressed gorgeously. She
> plaited her lovely hair very well, tied velvet to her
> waist and used aka stones for her neck. Her body
> was bare showing her beautiful breasts. No dress
> was worn when a young woman went to the mar-
> ket after the period of feasting. Her body was

exposed so that people saw how well her mother or her mother-in-law had cared for her. A woman who was not beautiful on that day would never be beautiful in her life. (*Efuru*, 18)

Nwapa dexterously combines tradition with the characteristic boldness of Efuru. To go public after only one month of feasting was Efuru's decision. To go as she did was traditional. To Western readers her partial nudity may also seem sexually provocative, but the purpose of Efuru's coming out in this manner is made clear by Nwapa: for the public to appreciate the care the mother-in-law showed in nursing her back to normalcy. The final sentence in the paragraph is a cultural signature informing readers of the meaning of beauty in Nigerian womanhood. From the Western perspective being beautiful for a woman after circumcision seems ironical because the circumcision renders the woman partially incapable of enjoying sex. But from Nwapa's perspective for women, beauty is not entirely tied up with sex. Efuru's coming out in the public enhances her beauty, her sexuality, as she is the symbol of every thing that is desirable in a woman. It may not, however, necessarily connote the ironical futility that circumcision makes one incapable of enjoying sex. To this situation, Nwapa adds one more crucial detail: the steady decline of the importance of Adizua who literally fades away from Efuru's life by disappearing.

Nwapa thus sets up the following situation: a family of women without a male to look after them is therefore forced to fend for itself. Efuru, with her daughter, Ogonim, and her mother-in-law, and the little maid-servant, Ogea, cannot any longer depend upon Adizua. Although Efuru in this respect was quite self-sufficient in the sense she was mostly successful at selling, and, in fact, has a good business head, Nwapa, nevertheless, draws our attention to the relative ineffectiveness of men in terms of managing their families. Between her mother-in-law and herself, Efuru was able to successfully maintain her family. Nevertheless, Adizua, failure as a breadwinner, is emotionally important to Efuru. Earlier during their first flush of courtship, Adizua had told Efuru that the very ground she trod on "meant something to him" (7). Now he deserts that very ground. In this, Nwapa's novels are "non-male" like Buchi Emecheta's novels. Men have a limited and functionally symbolic role (Efuru's father and Ona's in *The Joys of Motherhood*). Together these novels form a sharp contrast with Achebe's *Things Fall Apart* which is dominated by male characters.

Hitherto, Efuru's situation has been essentially private. Her elopement, marriage, even the reconciliation with her father are all considered internal family affairs, up to the time when Efuru comes out in public after her "bath." The clitoridectomy itself, according to Nigerian tradition, is meant to raise Efuru above the level of ordinary physical pleasure. The fact that the circumcision of the clitoris actually deprives women from receiving physical pleasure, leaving them with no choice over the matter, is, of course, the problem. Nevertheless, at least in this novel Nwapa presents both a criticism of this practice and a sympathetic and sensitive portrayal of the ritual connected with the "bath." The structural juxtaposition of the "bath" and Adizua's departure forever may also be a comment on the significance of sexual relationship and the role of the male in the private lives of women. Significantly, the circumcision defines the limit regarding male and female relationship, forcing Efuru to re-adapt her sexuality. Her coming out ritualistically in public, attractive to every one, yet without a part of herself implies that although her sexual appeal is not reduced her body to some extent belongs to the community and not to herself. Since the clitoris is a symbol of autonomy, at least in terms of sexual pleasure, it is seen as an impediment to belonging to the community. The "bath" is a cleansing of individuality, the ultimate sign that a woman can give up a particular kind of sexual pleasure for the sake of her family, progeny and society at large. By transcending sex, the woman makes the transition from the private to the public.

In the market place, therefore, Efuru is praised by a lot of people, some of them giving her money. They call her beautiful and congratulate the mother-in-law for looking after Efuru so well. Later, when Adizua goes to work Efuru rubs camwood dust on his body so that his friends would know that his wife had had "the bath." It is obvious that Efuru does not resist or even resent the clitoridectomy, or else she would not have voluntarily rubbed camwood on Adizua. She accepts the tradition with all its implications. After the ceremony she goes to the market place to trade and pay back the bride price on Adizua's behalf. But despite all this Efuru is not happy. "Underneath something weighed Efuru down." She is childless.

The birth of Ogonim fulfils Efuru's desire for motherhood. Adizua does not come back to appreciate fatherhood and the three women, Efuru's mother-in-law, Efuru, the servant-girl, Ogea, live together with the baby girl, Ogonim. Their strongest supporter is the feisty, hard-headed Ajanapu, sister of Efuru's mother-in-law, aggressive to the end, but one who understands quality when she sees it. Her bonding with Efuru despite her harshness is something that the latter cherishes throughout the life of the novel. It is Ajanapu who keeps Efuru rooted in reality.

Efuru's tryst with motherhood is short lived. The baby dies and she never has another baby. After a year of waiting she leaves Adizua's house only to fall in love with Gilbert, marries him but, again, does not conceive. Like Adizua, Gilbert also grows distant from her, staying away more and more whole nights. For the want of a baby, Efuru supports her husband's desire to take another bride, and graciously accepts the role of the senior wife. All this, however, cannot hold Gilbert. He goes away to Onicha, fathers a child and goes to jail. When Efuru's father dies, Gilbert is unable to attend the funeral because of his incarcerated condition. When he does return, Efuru falls seriously ill. The *dibias* unable to cure her declare her adulterous. Efuru is asked to confess and only then she would be cured. When Gilbert believes the *dibias's* conclusions Efuru is devasted. Ajanapu brakes a mortar on Gilbert's head and takes Efuru to a medically qualified doctor who cures her. Efuru "proves" her innocence by submitting her physical body to the goddess, Utosu, defying her to strike her dead if she were an adulteress. Her innocence proven, she leaves Gilbert. Her educated doctor friend asks her to go back to Gilbert because it is hard to live a full life without a husband. Efuru refuses. She has decided to give herself up to the service of the Lake Goddess, Uhamiri.The novel follows the displacement of Efuru from her father's house, back to her father's house.

As mentioned before, displacement is the inverse relationship between the self and the place. Mostly, writers force their characters to leave the place they belong to, as Amaka leaves her village and goes to Lagos. But sometimes the displacement happens through disowning, as Efuru first gives up her parent's home and then is herself disowned by Adizua and Gilbert. The displacement creates trauma, shaking up existing values, and, usually through action and adventure the protagonist ultimately comes to his or her own resolution.

Efuru's freedom from male institutions, no longer having a family, may be seen by some feminist critics as a reversal of the masculinist system. However, I think what Nwapa is giving us is an integration of the private with the public through displacement, not only of location but also in female sexuality.

Reversal suggests a revolution through which the location of power is reversed. Displacement, on the other hand, is not a revolution but a self-corrective measure that the individual and the society goes through when new social orders replace the old.

In the third world countries, the hierarchies regarding the private and the public have been introduced by Western colonization. Prior to colonization, women belonged to the private domain but in that arena they had authority over men. With colonization, hierarchies were introduced and the public domain became more important. In the post-colonial era, women have had to reconcile to the stresses and strains of traditionality and modernity. They have had to work and continue to be housewives, making their public lives as workers adjuncts to their private lives as housewives. Also, being in the open, third world married women have had to adjust their sexuality, often at the displeasure of their husbands. Their jobs are regarded by their families strictly as a source of income, and not only do not raise their status but, in fact, are seen as impediments to being a "good wife."

ONE IS ENOUGH

One is Enough begins with Amaka the protagonist, being summoned by her mother-in-law who proceeds to criticize her sadistically blaming her for being childless and arrogant. She will command her son, Obiora, to take a second wife. Amaka, of course, will be allowed to stay in the family out of the kindness of the matriarch, but she will lose her rights as her husband's primary sexual companion. Technically, a senior wife, Amaka will be secondary.

The restrictions put on Amaka threaten her very essence. Hitherto, she has been the main breadwinner of her family, but her role in the public domain is seen as an impediment to her success in the private sphere as a housewife and mother. Being childless was, of course, the problem. In Nigeria motherhood is everything, the all redemptive virtue of wives. As a mother, a wife commands respect. Without being one she runs the risk of being a social outcast.

So Amaka is cast out, a nasty fight precipitating her departure from Obiora's family. She runs away to the big city, Lagos. The marriage ends. Dispalcement begins.

Thus, though the material cause of Amaka's displacement is the fight she has with Obiora in self-defence, the real reason why her marriage broke down is because Obiora could not impregnate her. In Nigeria motherhood is not something women covet. They need it to survive.

Once out of the village, though, Amaka's fortune turns. Her attractiveness and good looks make influential men fall in love with her and help her. Encouraged by her sister who is a mistress of a Nigerian business man, she prospers. Then comes the Father Maclaid episode.

Although Amaka's initial displacement was forced, there was in her always the force of desire to better her and her family's condition. While married, she was the principal breadwinner and even saved money to buy her husband an expensive car. Obiora never acknowledged the gift in public, preferring to give the impression that he had bought the car. Sensitive to her husband's inferiority complex, Amaka went along with the deception. She had a good business head and was continuously trying to improve the family financially. In Lagos, she was the same, this time concentrating on improving her own lot, and her twins' when she became a mother.

Consequently, readers of *One is Enough* hardly ever feel the conventional agonies of the struggle from displacement that one feels reading Efuru. While Efuru's efforts at improving her lot was frustrated every time, Amaka found success in the material sense. And this is an important difference. Bearing in mind that *One is Enough* was written thirteen years after *Efuru* one might be tempted to conclude that Nwapa moved away from philosophical sadness about life to successful purposefulness. But there is a need to further analyze the trajectory of the heroine in the two novels by using the displacement paradigm.

If we examine the structure of Efuru's and Amaka's struggle a particular kind of meaning emerges that helps us not only to understand these struggles but the struggle of the post-colonial Nigerian women faced with the tension between the private and the public and one in which desire and sexuality play significant parts.

The overarching factor in both novels is motherhood. Being childless exerts a kind of pressure on women that, in fact, can only be understood as control over women. The two novels evince the following narrative structure that helps us to understand the control of the oppressor over the oppressed. The structure is laid out in the following stages: blame, guilt, emotional or physical force, displacement, trauma and struggle, integration, independence and the final resolution. The two novels pass through these phases and ultimately bring the protagonists to their resolutions. I have mentioned before that while all of these stages are easily discernable in *One is Enough*, the reader has to actively "discover" these stages in *Efuru*.

In Amaka's case the displacement is essentially positive. Her struggle to put her life together in Lagos is built on compromises but Amaka does not seem to be overly bothered by the moral issues pertinent here. Consequently, she does not show any tension even in the face of her girl friend's shock at discovering that she has had twins by Father Maclaid. Clearly perceiving her goal, Amaka goes about it resolutely, methodically, soberly and with compassion.

By contrast, Efuru is frustrated at every turn. Almost as a result of an original sin—marrying Adizua without even giving her father a chance to give his consent—Efuru, despite sterling qualities of character does not find success in family life. Beset by poverty, the married life of this woman from a very well-to-do family worsens progressively. There is, however, no guilt on the part of Efuru as Amaka felt not being able to conceive. Amaka's response to being charged with childlessness was nervousness and fear in front of her mother-in-law. Efuru's was sadness, not worry.

The physical force on Amaka is clear in the form of the fight with her husband. In Efuru's case this force is mostly emotional, coming in the form of abandonment from her two husbands. While the fight serves as a wake-up call to Amaka, the husbands' behaviors take a long time to sink in for Efuru who, being essentially a kind and empathic person, tolerated her husbands without being compensated. The clitoridectomy, absent from *One is Enough*, plays a major role in Efuru's displacement process, forcing her to read-just her sexuality and, in fact, her very role as woman and wife.

Similarly, Nwapa gives us two very different responses on the subject of integration with the "new" society. Firstly, Efuru's society is not new, just as her displacement did not require her to go to a distant location as it was in the case of Amaka. While Amaka's integration comes through her sexual relationship with Father Maclaid, producing for the offsprings thereby justifying her controversial actions, Nwapa denies that kind of integration—and satisfaction—to Efuru.Consequently, the resolution for each person is also distinctly different.

In *Efuru*, the integration comes not with a geographical location but with her gender, the female sex. Her decision to serve Uhamiri and women of her village so that they do not have to suffer childlessness and bad marriages makes Efuru an integral and important part of women in her village. As she lies sleeping, quietly dreaming of the Lake Goddess, Efuru is perfectly at peace. The similarity with the Goddess cannot escape the reader. Both are beautiful, wealthy and without children. Yet both are public figures, looking after the welfare of women in the private domain of marriage and motherhood. Uhamiri, the Goddess; Efuru, her priestess. Amaka's integration, by contrast, is materialistic, less spiritual. Her sense of satisfaction makes us happy but does not please us because her success has been dependent on men. Efuru's please us but leaves us with a quiet sense of sadness. In *One is Enough* the joy is polemical. In *Efuru* it is philosophical.

This brings us to the language of the two novels, specifically to the desire of Nwapa's language, her esthetics. What is desire in language? Kristeva explains it as the writer's urge for exploring truth.[6] To that I will add that desire in language is also the writer's affective will that actively seeks out something by committing pen to paper. It is more than a philosophical longing. It is creative hunger.

I have mentioned before that Nwapa's desire in language is in fact her desire for women's liberation from the male dominated Nigerian establishment. In both novels but particularly in *One is Enough*, the style is compact, giving little room for the reader to imagine. This is clearly evident in *One is Enough*. It seems to me that for certain writers symbolic representation are more fixed than others. To an author like Nwapa symbols assume universal proportions, unchanged and predetermined like mythology. For Nwapa, brought up within strict societal boundaries, mythical thought operate with fixed symbols. They are restricted tales like folklore and epics having only one meaning for those who receive them as a fixture in the culture. They do not receive mythology allegorically but literally. Nwapa handles the language in *One is Enough* more like closed symbols making only one meaning rather than opening up multiplicity of meanings. Consequently, the stages of the novels are single units, each having only one function in forwarding the story. To them, Nwapa combines universal symbols of struggle, courage, heroism, sexuality so that *One is Enough* becomes a straight forward action oriented tale. In this novel she is not interested in comparative truth or as Kristeva calls it a "little more truth."[7] Being entirely sociologically and politically motivated to raise the consciousness of Nigerian women, her language in *One is Enough* does not seem to be interested in relative truth. When it comes to women's suffering Nwapa's motivation is absolute truth. Her discourse is unidirectional, predetermined and programmed. Spurred on by her experience of the suffering of Nigerian women Nwapa's purpose in this novel does not seem to be esthetic or philosophical. It is instead a straightforward success story. By comparison, *Efuru* is a little more open-ended, esthetic and philosophical.

I submit that the progression of Nwapa as a writer from *Efuru* to *One is Enough* has been from the philosophical to polemic, from esthetic to political. In her final work, *The Lake Goddess*, this trend has been reversed. But for now, Nwapa seemed to have something urgent to say, a clear statement to make. In *Efuru* she created the protagonist as a subtle, sophisticated character, given to empathy and dreams.A powerful element in *Efuru* are Efuru's dreams. The novel is generally much more descriptive than *One is Enough*.

NOTES

1. Professor Gayatri Chakravorty Spivak deconstructs the opposition between the private and the public, concluding that what the deconstruction teaches us to mark the limits of the binary oppositions between the private and the public. See *The Spivak Reader*, ed. Donna Landry and Gerald Maclean, (NY:Routledge, 1996). This is a collection of critical essays by Spivak. "Explananation and Culture: Marginalia," (pp. 31-51) was originally given at a conference on "Explanation and Culture" given at the University of Southern California Center for the Humanities in 1979.
2. Elsewhere, I have explained my use of this term as "a longing which gives life its force. It asserts individuality, changes needs to wants. It is pure energy. In the complex interplay between control, hunger, and anger, desire is the invisible mover." See my essay, "Desire and the Politics of Control in *The Joys of Motherhood* and *The Family*," in *Emerging Perspectives on Buchi Emecheta*, ed. Marie Umeh. (Trenton, N.J.:Africa World Press, 1996): 227-246.
3 . See Spivak, p. 33.
4 . Michel Foucault, *Archeology of Knowledge* (NY: Pantheon Books: 1972), p. 191. See also his *The Order of Things: Archeology of the Human Sciences,* (NY: Pantheon Books, 1970).
5. See Mary Field Blenkly, Blythe McVickar Clinchy and Nancy Rule Goldberger, *Women's Ways of Knowing: the Development of Self, Voice, and Mind* (NY: Basic Books Publishers, 1986).
6 . Julia Kristeva, *Desire in Language* (N.Y.:Columbia University Press, 1980).
7. Kristeva, xiv.

WORKD CITED

Blenky, Mary Field, Blythe McVickar Clinchy and Nancy Rule Goldberger. *Women's Ways of Knowing: The Development of Self, Voice, and Mind.* New York: Basic Books, 1986.

Foucault, Michel. *Archeology of Knowledge.* New York: Pantheon Books, 1972.
____.*The Order of Things: Archeology of the Human Sciences.* New York: Pantheon Books, 1970.

Kristeva, Julia. *Desires in Language.* New York: Columbia University Press,

1980.

Landry, Donna and Gerald McLean, ed. *The Spivak Reader.* New York: Routledge, 1996.

Umeh, Marie, ed. *Emerging Perspectives on Buchi Emecheta.* Trenton, New Jersey: Africa World Press, 1996.

PART FIVE
(RE)CONFIGURATIONS OF
CHILD SYMBOLISM

Children's literature in Nigeria has over the years suffered abysmal neglect by writers, publishers, literary critics and even by educators and policy makers in education. Writing for children was not considered a matter of national priority. What seemed more urgent and worthwhile was to tell the story of the white man's tactless and high-handed incursions into Africa, the African struggles for liberation and the sour taste of the fruits of independence manifested in post-colonial socio-political anarchy and psychic disorder. Publishers gave children's literature the least attention partly because there was ready children's literature to be imported from abroad. Educators and educational planners had more crucial items to worry about in the curriculum than children's literature. Although literary critics identified the huge debts which contemporary writers owed to their oral heritage, they failed to maximize this insight in promptly calling for the development of children's literature as a matter of urgent national priority. Yet a careful study of the African mind in the traditional society would reveal that oral literature was carefully

designed and manipulated for acculturation and socialization purposes. Myths existed to explain to children the origins of the universe. Folktales were designed to teach and inculcate morals and
values as well as entertain and instruct young minds about societal
norms. Riddles existed to train the mind to unravel mental puzzles
and develop a sense of logic and rationality. In general, oral literature was useful to developing mental capacities, exciting imagination, inculcating cherished values and in the process, situate and
stabilize children in the unique circumstances of their environment.
One can conclude from all these that one of the primary aims of literature in the traditional society was to foster children's social and
mental welfare. Viewed from this perspective, children's literature
becomes a viable pragmatic instrument for social stability, as the
inculcated cherished values are transmitted from generation to generation. If this had been fully appreciated by early Nigerian writers
and critics, the critical statement that art for art's sake was irrelevant in the African situation would not have opened itself up to
undue irrational challenges in the academy.

In a country like Nigeria without a reading culture, with a very
high illiteracy rate, with prodigious problems of school children
being unable to read with good understanding of what they read,
the intrinsic values of an aggressive commitment to writing for
children would have been seen as the next logical battle after the
defeat of colonialism as an institution. Approached from the perspective of a mission for national survival, the development of children's literature would have served as a ready tool for inculcating
the reading habit, developing a reading culture and resuscitating
lost values in the contemporary society. The late realization of
these attributes has led to the attention now being paid albeit belat-
ed, by writers of children's literature in Nigeria in particular and
Africa in general. Some of the writers addressed the issue by writing first and foremost for their growing children who until then,
had nothing but literature imported from abroad whose themes
and content projected non-African realities and experiences. It is
in this sense that writing for children in Nigeria is seen as a rescue mission and something which requires far greater commitment and dedication than writing for adults.

The essays in this section of the book have explored and
assessed Flora Nwapa's contributions to children's literature from
the point of view of the above scenario. When Nwapa published
her first adult novel *Efuru* (1966), it was heralded as a rebellion
against male writers' stereotypes of womanhood in Africa. Her fic-

tion for children has equally turned out to be another milestone in her role of a pioneer literary revolutionary. She not only wrote to entertain young minds, but through her characterization of her child-heroes, she has also set the stage for positive children-empowerment in fiction and real life.

In her essay, "Conquerors of the Universe: Nwapa's Kiddies on the Move," Julie Agbasiere illustrates Flora Nwapa's fascination for creating literature that endows children with the qualities of independent inquiry, creativity, and evaluation. She uses for illustration a number of Nwapa's writings for children where young characters are given ample opportunities to assume responsible positions in their early childhood, because adults are not necessarily the sole repositories of knowledge. Agbasiere portrays Nwapa as quite successful not only in creating a link between oral and written literatures for children, but also in blending fantasy and realism, ultimately achieving a balance between the past and present.

In the essay, "Flora Nwapa's Legacy to Children's Literature," Ifeoma Okoye discusses Flora Nwapa's literary techniques in one specific work, *Mammywater,* and concludes that one of the most enduring contributions of Nwapa to children's literature in Nigeria is her innovation of the use of question and answer as a technique for writing stories for children. Okoye shows through the analysis of the narrative that Nwapa used the technique to advance plot and delineate characters.

Ernest Emenyonu's essay, "Flora Nwapa's Writings for Children: Visions of Innocence and Regeneration," discusses Flora Nwapa's commitment towards the restoration of good moral values in children. For Nwapa, children's fiction is a veritable organ for the projection of role models for growing children. This involves holding up for emulation the virtues of truth, innocence, honesty, courage, hardwork, humility, loyalty, and obedience. Flora Nwapa was intrigued by the state of innocence in childhood and sought through her fiction for children to perpetuate such positive virtues for growing children.

Flora Nwapa's mission to teach Nigerian children the cultural values of family, community, education, work, self-realization, and ingenuity is the topic on Marvie Brooks' essay, "'Either Pound in the Mortar or Pound on the Ground': Didacticism in Flora Nwapa's Children's Books." The methods used by the illustrators to effectively illuminate the themes, characterizations, and underlying lessons in Nwapa'a fiction are also explored by Marvie Brooks.

Ernest N. Emenyonu

Conquerors of the Universe: Flora Nwapa's Kiddies on the Move

Julie Agbasiere

There is a close relationship between African oral traditional literature and modern African literature. This is especially apparent in children's written literature. Just as the oral tradition aims not merely to entertain but also—at the same time—to install the child in his/her proper milieu by instilling the traditional society's norms, mores and world view, so does our written literature situate the child in the present day world with all its problems, conflicts, and hardships, thereby-hopefully-goading our youth to socialize, integrate, and become a mature and useful member of our society. In the modern written tradition, elements of folktales—notably the classic components of initiation structure, fantasy, and didacticism—are discernible. In terms of content, some writers such as Chinua Achebe and Buchi Emecheta establish a link with traditional literature by setting some of their works in the past and spinning their own "contemporary" folktales.[1] They also circumscribe the present day-to-day life activities.[2] Cyprian Ekwensi depicts the child in the cosmopolitan environment and runs him/her through a whole gamut of crimes and evils that are the bane of city life.³ Some literary works of Ifeoma Okoye and Chukwuemeka Ike track the activities of the child, sometimes at school and sometimes in the home.⁴

The socializing function of children's literature, especially the oral tradition, puts pressure on the child to conform. Adults' views on behavior patterns prevail and, consequently, the well-behaved child is he/she who, having internalized societal norms and values, conforms their behavior to societal expectations. In a study of Buchi Emecheta's children's books, Agbasiere observes that "for the child to grow into a responsible adult and a leader, he/she has to accept the elders' leadership role and humbly learn from the elders" (8). This observation merits some revision. Children should be given wider scope for exploration and self-actualization. Children's literature should, as Ahukanna points out, nurture "in the child the habit of independent inquiry, creativity and evaluation" (45). This is also the conception that appeals to Flora Nwapa. To her, children should be given ample opportunity to assume responsible positions as early as possible. In her children's stories, most of the child-characters are bold, intelligent, and possess a high sense of responsibility. They are actors whose actions make a positive impact on their immediate environment and on their own personality. They are hero-voyagers and the world is seen from the perspective of the child.

Flora Nwapa sets her stories in diverse locations—on land, under the waters, and in outer space. This paper assesses the development of the child as he/she carves a niche for himself/herself. It sees the child as an important factor in harnessing the forces of nature and in giving meaning to the universe. The children's books studied in this essay are *Emeka—Driver's Guard* (1972), *Mammywater*, (1979), *Journey to Space and Other Stories* (1980), "The Magic Cocoyam" (1980), and *The Miracle Kittens* (1980).⁵

I. THE CHILD AS A CREATIVE FORCE IN THE UNIVERSE

Children do not feature much in etiological tales of the oral narrative tradition. The main characters are usually gods, spirits, animals, elements, men and women. Tales of origin give an account of the genesis of the world and supply an explanation of various beliefs and practices. These tales serve to help the people realize "how everything they are told to do has its precedent and pattern in the past" (Nwaozuzu, 4), and enable "the traditional man to come to grips with a hostile and often bewildering environment" (Egeonu,

101). Nwapa in her etiological tales sets out to "correct" the popular tradition by giving children major roles to play in her cosmogonic set-up. "The Magic Cocoyam" is an etiological tale spun around the people's totemic beliefs, for it explains why the people of Womo village do not eat cocoyam. The story centers on two orphans, Kele and Lala, who are brother and sister. As the two children are swimming in the river, Mother Tortoise carries Lala away to her farm. With the help of some animals, Kele gets to the Tortoise's farm, works for Mother Tortoise and in the process discovers her secret and frees his sister. Both of them return to the village.

At the outset, Nwapa presents a village suffering from poverty and ignorance. They can fish but they cannot farm because they do not know how to farm and they have nobody to teach them. Kele's mission becomes twofold: to rescue his sister and to rescue the Womo people from starvation. It is the Bird who makes it clear to Kele on the eve of their journey to Tortoise's farm: "If you work hard, you will learn how the Tortoise farms. You will not only win back your sister, Lala, but also teach your people how to farm and reap good harvest" (46). Kele has the qualities to make him succeed. He is strong, hard-working, determined, and brave. Once in Tortoise's farm, he settles down and works hard. He quickly learns how to plant different types of crops and how to tend them. He is observant and notices that there is one crop—the cocoyam—which the Tortoise does not teach him how to plant. He becomes suspicious and plants it in his host's absence. When it germinates and grows rapidly, it talks and turns out to be Lala. He learns, therefore, that the Tortoise turns her victims into cocoyams and can be destroyed only if she is beaten with a cocoyam leaf. Kele beats her with a cocoyam leaf and Lala is freed.

Kele successfully accomplishes his mission. Firstly, he learns how to farm. Secondly, he destroys his adversary, the Tortoise. Thirdly, when he returns home triumphantly with his sister, the two children are given a tumultuous welcome. Thereafter, they teach the villagers how to farm. This story underscores the fact that children can solve adults' problems and teach them what they do not know. Adults are not the sole custodians of knowledge. The Womo villagers learn from the two children and are grateful to them. The adventure of these children gives rise to the people's totemic observance. The villagers have now become good farmers: "But Kele and Lala told them not to plant or eat cocoyam. Thus, from that day henceforth, cocoyam became the totem of the entire village" (54).

In this story, Nwapa takes the liberty of distorting the content of and characterization in traditional tales. Already the central characters are children. The attributes of Mother Tortoise are a clear departure from the "authentic" Tortoise. Here, the Tortoise is female, not male; she is killed and her shell breaks into a hundred pieces. All these distortions point to the revolutionary stance of Nwapa as an interpreter of traditional tales. This becomes quite apparent and striking when the Tortoise's delineation in oral literature is enunciated. Emenanjo expatiates:

> Mbekwu, the ubiquitous Tortoise, without whom no folktelling session is complete represents amamihe (wisdom) in the widest sense of that word. Down toearth and as selfish as Mbe might be, and occasionally trapped to stew in his own juice, he always exploits his wisdom and the folly of his co-animals to wriggle out of the tightest corners. Mbe is rarely killed. The greatest punishment he gets is that he is thrown out of the company of the other animals from above—and that is why his exoskeleton is as rugged as it is now. (xvi)

This Tortoise is certainly not the same one that Nwapa introduces in her tale. To Nwapa, therefore, stock characters should change with the times.

In *Mammywater*, Nwapa again explores the repertoire of folk lore, precisely into the legends of Oguta people. With the lake as setting, she unfolds the myth of the lake goddess, a deity that pervades her adult novels notably *Efuru* (1966) and *Idu* (1970). The protagonist of this story is Deke, a boy of ten, who goes to the home of *Mammywater*, under the lake to rescue his sister, Soko, carried away by the goddess.

The choice of a child as protagonist to confront the goddess appears to be apt because, shielded by innocence, the child can afford to be brave, forthright, and tempestuous without incurring the wrath and a possible reprisal of the goddess. Deke is, however, respectful, determined, and abides by the advice given him by the aged worshipper of the deity as to the code of conduct to adopt before the goddess.

Deke's encounter with the goddess reveals the nature and attributes of *Mammywater.* She is a spirit, ageless, superlatively beautiful and fabulously wealthy. She assumes the form of a beautiful young woman but can change into a fish. She reigns supreme in her home, where fish serves all purposes, from food to fuel to work force.

Critics of Nwapa's adult novels highlight the contradiction in the goddess's behavior towards the community under her influence. Uhamiri, Woman of the Lake as she is called, or *Mammywater,* approves of marriage but denies children to her devotees, in a society where marriage without children is considered an abysmal failure (Ikonné, 97; Nnolim, 48).

In *Mammywater,* the goddess goes a step further—she takes from Deke's mother not only her children but also her husband. The enigma that *Mammywater* embodies deepens as it is observed that in her home she keeps children. These children will always be with her, having been changed into her aquatic essence. It is possible that these characteristics of the lake deity are derived from popular beliefs. As Chukwuma posits:

> Nwapa in this children's book delineates the features and attributes of Uhamiri, the goddess of the lake. These features are drawn from popular beliefs, worship and legend of the Oguta people. (127)

In this story, there is a double quest, one encapsulated in the other—that of Deke to rescue his sister, and that of *Mammywater,* to make peace between the people of the lake and Spirit Urashi who threatens and reproaches the people for being disrespectful to him. The goddess makes Deke's quest fit into her design thereby turning the boy into the instrument of realizing her project. She reveals her strategy to the river-god, Urashi, as she presents Deke to him:

> I had thought that the boy's father would be useful. But when he arrived he was overwhelmed by the riches in my home. So he failed the test . . . I had to wait for ten years. I told this boy's mother that her husband had let me down and I must take her daughter, Urashi, the Great One. She wept and wept. Women, how they love children. But I had to do it to save my people from a certain death. (29)

She knows that once she takes Soko, Deke will come to seek her release.

Deke is not just a pawn in the hands of the water deity. Ignorant of the goddess's design, he distinguishes himself by being courageous, fearless, and daring. Above all, he rejects all invitations to touch the goddess, thus heeding the warning given him by the old woman. Before the two deities that decide the fate of the people, Deke comports himself with courage, meekness, and humility that impress *Mammywater,* and Urashi. He pleads for clemency and forgiveness on behalf of his people, as well as the release of himself and Soko. He promises to carry their message faithfully to his people. The deities are disarmed and disposed to grant his request. Deke becomes the redeemer of his people. His success is that of *Mammywater,* too. He brings about peace between the spirits of the waters and the lake community, as well as a reunification of his family. The goddess, however, spells out the terms to Deke before he is allowed to return to his mother:

> [Y]our mother, my good friend, must come to my
> home to be entertained. We shall perform the dance
> of the maidens for her and she will meet her hus-
> band. There, she will meet you and Soko also. Then
> you will all return to your world in peace. (31)

Deke's adventure gives some insight into the enigmatic attitude of the goddess towards the marriage institution. By denying children to her worshippers, she is trying to save the community from destruction. These children act as hostages in the aquatic spirit world. It is, however, through the instrumentality of the child that the community will have a new lease on life. Through Deke, the goddess makes peace with the spirit, Urashi. The reunification of Deke's family is a pointer and a guarantee that henceforth her worshippers will have children and live a normal life.

II. THE CHILD AT GRIPS WITH SPACE

From adapting traditional tales and folklore, Nwapa delves into science fiction. In *Journey to Space,* she situates the child in an urban environment which is receptive to modern scientific and technological advancement. Ngozi and Eze, the protagonists, are city dwellers, living with their parents in a flat. The house, probably a block of flats, has a lift and it is this lift that launches the children into outer space, under a spell cast on the children by a stranger.

As a result of this spell, the children are no longer actors; they do not initiate actions and do not take decisions to solve problems. They do not have a mission. As the lift shoots up into outer space, they do not even know their destination. Ngozi thinks that they are going to heaven and that they will see God. Eze thinks they are going to the moon. They do not go to any of these two places; they simply float in outer space. Their situation is contrasted with that of their companions, Doggy and the two fairies. Doggy has the mission of going to the moon to collect a precious stone that will confer immortality on the animals. The fairies have the responsibility of helping outer space travellers who are in need.

The occupants of the lift are drawn from the world of men and women, the world of animals and the inhabitants of outer space. They all maintain a friendly and cordial relationship and freely exchange ideas. By this, Nwapa pleads for harmony and better understanding in the world.

In this story, Nwapa appears to be saying that there is a dearth of substantial information on the moon. Doggy appears to the children to be knowledgeable, but his information is neither corroborated nor confirmed to be true. Doggy maintains that the moon has flowing diamonds, which makes the earth people who go to the moon ill and unable to settle there. In addition, fairies come from the moon and go in search of water. Finally, it is the earth people who drive the animals away from the moon with their noisy machines. The fairies refute the information that concerns them but cannot confirm the opinion about the moon since they claim to know little or nothing about that star. What is obvious and sure from their conversation is that no living thing can inhabit the moon, that the lift is not a proper vehicle to take people to the moon, and that the journey to the moon is hazardous.

In *Journey to Space,* Nwapa creates a "scientific lore" for children. She tries to explain why the moon is not inhabited by people from other planets, as well as the origin of death in the animal kingdom. According to Doggy, the animals inhabited the moon in the earliest of time and they never died. They followed man's journeys to the moon in his space ships. Man drove away the animals to another location and thereafter, the animals began to die. In fact, Doggy is sent to the moon to fetch a precious stone that will restore immortality to the animals. As for man, having driven away the animals, he found that he could not live in the moon and had to go back to earth. The story ends on a moralizing note. The fairies are going to take the children back to their flat and they admonish them, "Next time, don't be naughty. Don't play in the lift" (20).

III. THE CHILD'S DOMESTICATION OF NATURE

Flora Nwapa's children are ubiquitous. From the land of the animals, to beneath the lake, and then to outer space, they bestride the land of men and women. In her children's book, *The Miracle Kittens,* the children are involved in solving domestic problems in the home. Their house is infested with rats and this has reached an alarming proportion. Their mother, Madam Bum Bum, is helpless:

> The rats behaved as if they owned Madam Bum
> Bum's household. They ate up all the food in the
> kitchen. They danced and played as if the place
> belonged to them and them alone. And what's more,
> they frightened her children. (2)

Madam Bum Bum has no idea of how to get rid of the rats and finds her maid's suggestion of using rat poison dangerous.

The story shows the maturity of the children as they give suggestions that eventually solve the problem. Their names are not given, suggesting collective responsibility. The first idea is on the use of traps. This works marvelously the first time but gradually becomes ineffective. The next suggestion is on the use of cats. For this to work, the children work on their mother's phobia of cats in order to make her better disposed to allow a cat in the house. As attempts to buy a cat continue to be unsuccessful, the children hope for divine intervention.

Miraculously, the kittens arrive and by then Madam Bum Bum has undergone some psychological transformation. She happily and gratefully accepts the kittens and looks after them. The kittens perform the magic and rid Madam Bum Bum's house of rats.

Nwapa, in this story, shows the competence of children in solving problems in the home and in bringing about a change in the attitude and behavior of adults. Again, she insists that adults are not the sole repository of knowledge, for children can come up with original and even brighter and better ideas towards the solution to problems. Adults should, therefore, involve children whenever there is a problem in the home that needs a solution.

Emeka—Driver's Guard is different from the other stories in two regards. Firstly, the stories already analyzed involve children in early adolescence. Secondly, emphasis here is more on the psychological development of the child. The story centers on Emeka and Ngozi, twins and orphans, who live with their aunt in town. Ngozi adjusts quickly and does well at school, while Emeka is led astray by bad company, drops out of school and becomes a driver's guard. The risk involved and the vicissitudes of such a life force him to re-think what he wants to do in life.

The story focuses on the behavior of children within their peer group. Adults play the role of watch dog to save the children from destroying themselves. From a happy home in a village environment, Emeka comes to town where he meets unfriendly boys who bully him and tell lies about him. His encounter with Sokari and others yields mixed results. On the positive side, this association triggers off the desire to go to school among the children who are cared for by his aunt. On the negative side, Emeka gets the nasty experience of being lied about, and of not being trusted. For example, when Sokari later becomes Emeka's best friend, he exerts a bad influence on Emeka. The latter learns to tell lies, becomes a truant, and drops out of school, to the chagrin of his sister and aunt. He opts to become a driver's guard.

Emeka's new life, exciting at the outset, does not bring him lasting satisfaction and happiness. Other boys make life rough and difficult for him. He explains to his friend, John:

> ...The driver is kind....but the other guards are always bigger and stronger than I am. They bully me. When I fight them, I always lose. (38)

This is the problem that Emeka has to solve. He has tasted life in the home, at school with all its demands and restrictions, and currently on the job as a carefree guard and wage earner. Being strong willed, he does not easily accept other people's advice that runs counter to his own ideas and wishes. Now he is poised for a conquest of self.

Having cut himself away from home, Emeka always thinks of his sister and aunt each time he experiences a traumatic incident. Three major events affect him deeply and make him review his lifestyle. The first is his first fight, in which his teeth are nearly shattered. The second is an accident scene in which the driver's guard is among the dead. The third is his last fight, after which he decides to quit the motor job and become a house boy. The two women—Ngozi and the aunt—maintain a stabilizing force in Emeka's life. To him, Ngozi is a symbol of affection, serenity, and success. She is a type of magnet that draws him away from the wrong path. He sees his aunt as a constant reminder of the need to go back to school. At each stage, he resists the temptation to go back home until he cannot hold out any longer. Once he resolves to go home and back to school, he makes contingency plans in case his aunt refuses to pay his school fees.

Emeka's resolution is a conquest of self. He appraises and learns from his past behavior. He conquers his weaknesses and shortcomings. He confesses to his sister, "Ngozi, I am going back to school....I have learned a lot, Ngozi. Without an education you are nowhere" (43). He has discovered his mistakes and limitations and he is resolved to turnover a new leaf. As an Igbo adage says, "When an ant stings the buttocks, the latter become wiser."[6]

IV. FLORA NWAPA'S NARRATIVE TECHNIQUE: THE QUEST MOTIF

With the exception of *The Miracle Kittens*, Nwapa's stories recount the adventures of children. There are usually two protagonists, a boy and a girl, who are brother and sister. Two patterns emerge. In the first one, the girl acts as a catalyst that triggers off the quest. She is abducted and the boy goes on a mission to rescue her. In the second, she acts as a stabilizing force which checks the excesses of the boy. The structure of the stories is like that of initiation rites. There are three phases: "separation, transition and reintegration" (Kreamer, 58). This gives the story a circular movement—the child leaves home, faces hazardous circumstances, overcomes obstacles and problems, and returns home from where he took off. He undergoes some kind of transformation in the course of his/her adventure, becomes more mature and brings home happiness to the people around him.[7]

Nwapa spices her stories by weaving into them the fantastic and the supernatural, thereby blending the realistic and the folkloric. Butts observes:

> Such stories with their mixture of realism and the marvelous, their narration of the hero's journey as a quest, and the frequently happy endings have elements in common with the myths and legends. . . .(73)

By creating a link between the oral and the written, by a blending of fantasy and realism and by striking a balance between the past and the present, Nwapa can be likened to "minstrel singers" who according to Egeonu, "draw their inspirations from both the depot of folklore and contemporary lived-experiences, and their tales and other literary pieces which have strong moral and pedagogical undertones" (106).

In conclusion, it can be seen that Flora Nwapa diversifies the life situations within which her child-characters operate. These characters are delightful and interesting and each protagonist is "a very normal and identifiable kind of person" (Butts 1992:70). They all come from good homes and have a happy childhood. Family ties are strong, especially between brothers and sisters. Together, they face and conquer the world. Nwapa's children are explorers in the universe.

NOTES

1. See for instance, *The Drum* (1977) and *The Flute* (1979) by Chinua Achebe and *The Moonlight Bride* (1980) by Buchi Emecheta.
2. For this category, examples are *Chike and the River* (1985) by Chinua Achebe and *Nowhere to Play* (1980) by Buchi Emecheta.
3. Such works include *The Drummer Boy* (1960) and *Samankwe and the Highway Robbers* (1980).
4. These include *Eme Goes to School* (1979) and *The Village Boy* (1980) by Ifeoma Okoye, *The Potter's Wheel* (1973) and *The Bottled Leopard* (1985) by Chukwuemeka Ike. *The Bottled Leopard* touches on the dual personality in the mystical beliefs of traditional African society.
5. Flora Nwapa co-authored this book with Obiora Moneke. Her stories in this book are *Journey to Space* and *The Miracle Kittens*.
6. The translation of this adage from Igbo to English is mine.
7. *Journey to Space* does not embody some of the features mentioned. This is probably because the children are under a spell and can only take instructions from and chat with their companions.

WORKS CITED

Agbasiere, Julie. "Toward A Cultural Symbiosis: A Study of Buchi Emecheta's Children's Books." *Emerging Perspectives on Buchi Emecheta.* Trenton, New Jersey: Africa World Press, 1996.

Ahukanna, J. G. W. "Children's Literature in Africa: Myth or Reality." *WAACLAS.* 1.1 (1989):42-51.

Butts, Dennis. "The Adventure Story." *Stories and Society: Literature in Its Social Context.* London: Macmillan, 1992:65-83.

Chukwuma, Helen. "Flora Nwapa Is Different." *Feminism in African Literature: Essays on Criticism.* Enugu: New Generation Books, 1994:155-130.

Egeonu, I. T. K. "Traditional African Literature and the Convergence of Art and Life." *Readings in African Humanities: African Perspectives in World Culture.* Owerri, Nigeria: Vivians and Vivians, 1988:95-106.

Emenanjo, E. N. "Introduction." *Omalinze: A Book of Igbo Folktales.* Ibadan: Oxford University Press, 1977:vii-xx.

Ikonné, Chidi. "The Folk Roots of Flora Nwapa's Early Novels." *African Literature Today* 18 (1992):96-104.

Kreamer, Christine Mullen. "Transformation and Power in Moba Initiation Rites." *Africa* 65.1 (1995):59-78.

Nnolim, Charles. "Mythology and the Unhappy Woman in Nigerian Fiction." *Critical Theory and African Literature.* General Ed., Ernest N. Emenyonu. Ibadan: Heinemann, 1978:45-53.

Nwaozuzu, B. S. C. "Igbo Folklore and Igbo Worldview." *Nsukka Studies in African Literature* 3(1980):1-12.

Nwapa, Flora. *Emeka—Driver's Guard.* London: University of London Press, 1972.

___. *Journey to Space.* Enugu: Flora Nwapa and Company, 1980.

___. *Mammywater, .* Enugu: Flora Nwapa and Company, 1979.

___. "The Magic Cocoyam." *Journey to Space and Other Stories,* ed. Obiora Moneke and Flora Nwapa. Enugu: Flora Nwapa and Company, 1980.

___. *The Miracle Kittens.* Enugu: Flora Nwapa and Company, 1980.

Flora Nwapa's Legacy to Children's Literature

Ifeoma Okoye

Flora Nwapa will be remembered for her many achievements. First, she will be remembered for being Nigeria's first female novelist with the publication of her novel, *Efuru*, in 1966. Speaking about the publication of the novel, she said:

> Well, I had the courage to give it to Chinua Achebe who read it and said it was good. In fact, he sent it to Heinemann and about a month later, they replied and said that they had accepted it for publishing. I was very lucky indeed because I was spared the ordeal of reject slips.[1]

Second, she will be remembered as the first female publisher in Nigeria and the first female publisher of children's books in Nigeria. According to her, she set up Tana Press Limited in 1977 and Flora Nwapa Books later. Of these she said:

> My companies print and publish children's books as well as novels and short stories. We have published some women writers as well as a few male writers.

We are not going to discriminate against the men.
But we are partial to the women.[2]

Third, Nwapa will be remembered as a writer of children's books. She was one of the few prominent Nigerian and indeed African writers who thought it worthy to write for children. She does not belong to the crop of African writers who, according to Asare, seem "to think that writing for children was below their dignity," and who feel that they are clever if they publish "brilliant adult novels with a complex plot and difficult philosophical content."[3] Finally, Nwapa will be remembered for her legacy to children's literature in Nigeria which is the use of questions and answers as a technique in story writing, exemplified in *Mammywater.*[4] The exploration of this technique is the topic of this short tribute to a good friend and colleague.

Mammywater was first published in 1979. According to the author, it is a folktale, and typical of stories in that part of the world (Oguta), it is woven around water and its mysteries. The story is about Deke, who on learning that his beloved sister has been taken away by Mammywater goes in search of her and rescues her. His ability to ask the right questions and his courage to insist many times that his questions be answered stand him in good stead and help him succeed in his rescue mission.

Mammywater is a children's story based to some extent on question and answer 'movements.' It is an embodiment of a well-known Igbo proverb: one who asks questions never gets lost. This proverb says a few things about questions. The first is that people are encouraged to ask questions. The second is that asking questions is valued because it brings direction and knowledge and prevents people from walking into danger or from making mistakes. This second point is buttressed with another Igbo proverb which states that eating something without first asking questions about it brings instant death.

In this brief tribute to Flora Nwapa, I will first discuss the extent of and type of questions in *Mammywater* and then show how she uses this unique technique of question and answer, of which she is the originator, to advance her plot, and delineate her characters.

Altogether, there are 92 questions in *Mammywater,* a story written in about 500 words. The questions and their answers form nearly half of the story. This proportion is unique in a story where you are expected to have narration and descriptions of characters and setting. Herein lies Nwapa's innovation which is her legacy to children's literature.

Out of the 92 questions in *Mammywater*, Deke, the protagonist, asks 44 and his sister, Soko, who is the next important child character in the story, asks 20 questions, although she appears only very briefly at the beginning of the story. Mammywater, who gives the book its title, asks 24, while Urashi, the king of the river, and Mammywater's rival asks three. Deke and Soko's mother who is not given a name asks only one question.

The questions in *Mammywater* are varied. There are closed questions, that is questions which are expected to be answered by *Yes* or *No*, by a single word or very brief responses which can be interpreted as *Yes* or *No*. An example of such questions is this one by Deke to *Mammywater* as they enter a room and behold some coral beads:

> "Can I touch them?" asks Deke.
> "No you can't touch them." (25)

A second type of question in *Mammywater* is the rhetorical question. Rhetorical questions are those that expect no answers but are used to appeal directly to the readers, willing them to share the writer's or speaker's feelings. They are used to express strong feelings, opinions, and impressions. *Mammywater* asks such a question while talking to Soko:

> "Unbecoming of me? I am laughing at what you said.
> Isn't it funny?" (3)

Nwapa, in *Mammywater*, uses questions and answers to advance her plot. First, we learn about *Mammywater* and her mission through a question and answer session between her and Soko:

> ". . . who are you?"
> *Mammywater* replies, "I am Mammywater. I have come to take you."
> Soko is puzzled. "Why do you want to take me?" she asks.
> "I want you as my daughter." (3)

Similarly, from Soko's questions and her mother's answers, we learn that *Mammywater* has taken Soko, Deke's sister, to her home. And that Soko will be back only when *Mammywater* lets her go and that Soko is not a prisoner in Mammywater's house. From such sessions we also learn how Deke arrives at Mammywater's domain:

> "Now, now, child, tell me how did you come
> here?"
> "Quite simple. I dived." (16)

And where Soko is:

> "Where is she?" (She refers to Soko.)
> "She is at Urashi River, but she will come back."
> (17)

More of such revealing questions abound in the story.

Nwapa also uses questions and answers to delineate her characters in *Mammywater* instead of relying more on the more common authorial report. She reveals Deke to her readers through the questions Deke asks and the answers he gives to his mother, to Mammywater, and to Urashi, the king of the river. Deke is a courageous boy. This comes out from his answers to Mammywater's questions as he arrives at the latter's domain:

> ". . . Tell me, how did you get here?" (That was
> Mammywater.)
> "I dived if you must know."
> "You dived and surfaced here? Aren't you afraid?"
> "Of what?"
> "Of the lake, child."
> "Why should I be afraid?" I know the lake well."
> (15)

Later, he asks Mammywater:

> "Did you think I was going to cry? I never cry, you
> know, I am a tough boy." (17)

More questions and answers reveal some more information about Deke's character. He is ten years old. He does not know his father and he is an intelligent boy:

> So he asks *Mammywater* quietly but firmly: "Tell
> me, why did you take my sister away from my
> mother and me?"
> *Mammywater* replies quietly too: "Son, you are too
> fast for me. Eat your food, we'll talk later." (21)

Deke is also indomitable as the following question and answer session reveal:

> "Am I a distinguished visitor?" (He asks of

Mammywater.)

"Now, now, you are asking too many questions again."

"Answer my question, Mammywater. Am I a distinguished visitor?"

"Of course you are. Small boys don't come to me the way you came. You are distinguished."

"Am I the first distinguished boy to come to you?"

"No, many have come, but failed the test I set for them. . . ." (35)

Nwapa delineates Mammywater's character through questions and answers too. *Mammywater* is independent. She says so in answer to Soko's question. Soko wants to know who sent *Mammywater* to come and take her. *Mammywater* replies: "Nobody sends me, I send myself." (3) Again a question and answer session reveal much about Mammywater. She is not a woman but a spirit:

"Are spirits good, Mother?" (Soko asks her mother.)

"Yes, some are good and some are evil."

"Is this one good, Mother?"

"*Mammywater* is good." (7)

And spirits do not die:

"You will not die?" (Deke asks Mammywater.)

"Die, no. I am a spirit. Spirits don't die. We live forever." (21)

Also revealed by questions and answers is the quarrel between the two giants, Mammywater, the Spirit of the Lake, and Urashi, the king of the River:

"Why, why should you cross the line drawn by the pen of our God?" replies Mammywater.

"And why, oh why, must you send your fishes to molest me

and my household?"

"And why, oh why must you send your reptiles to kill my dear ones, the people of the Lake?"

"And that's why I am here, that's why this boy is here.

Shall we talk business now?" (40-41)

Nwapa's question-and-answer techniques in story writing is not quite sustained in *Mammywater* and needs some fine-tuning. Perhaps if she had lived longer, she would have been able to improve on this innovation that can come only from her.

Flora Nwapa as a pioneer has blazed many trails. It is left for us to tread the paths she has beaten, making them wider and deeper and extend them to other horizons. No tribute can be better than this.

NOTES

1. See "Interview with Chike Akabogu" in *Satalite,* Enugu, Nigeria (October 14, 1981):8-9.
2. See *The Guardian,* Lagos, Nigeria (August 17, 1988):16.
3. See Meshack Asare's "Writing for Children," in *A Handbook for African Writers.* Compiled and Edited by James Gibbs.
4. See Flora Nwapa's first children's book, *Mammywater.* (Subsequent references to this text will be referred to by page numbers.)

WORKS CITED

Akabogu, Chike. "Interview With Flora Nwapa." *Satalite.* Enugu, Nigeria (October 14, 1981): 8-9.

Asare, Meshack. "Writing for Children." *A Handbook for African Writers.* Ed. James Gibbs. Oxford: Hans Zell Publisher, 1986, 82.

The Guardian. Lagos, Nigeria. August 17, 1988, 16.

Nwapa, Flora. *Mammywater.* Enugu: Flora Nwapa Books, 1979. (Subsequent references to this text will be referred to by page numbers.)

Flora Nwapa's Writings for Children: Visions of Innocence and Regeneration

Ernest N. Emenyonu

The first of Flora Nwapa's books for children, *Emeka—Driver's Guard*, was published in 1972 after three main adult works—*Efuru* (1966), *Idu* (1970), and *This is Lagos and Other Stories* (1971). It was to take practically a decade before the publication of her second children's book, *Mammywater* in 1979, virtually three decades after Nigerian fiction had come to full local and international attention with Amos Tutuola's *The Palmwine Drinkard* (1952), Cyprian Ekwensi's *People of the City* (1954), and Chinua Achebe's *Things Fall Apart* (1958).

Between 1952 and 1979, very many works of fiction had been produced and published for adults by Nigerian authors. Very little attention was paid to writing for children. It was apparently not seen as something deserving national priority by the more established Nigerian authors, with the possible exception of Cyprian Ekwensi whose *Drummer Boy* (1960), *An African Night's Entertainment* (1962), *The Great Elephant Bird* (1965), and *The Boa*

Suitor (1966) forcefully established the beginning of a reluctant trend. Nigerian writers in general seemed to have been so zealously concerned with telling the ugly stories of colonialism with its heinous extensions of cultural alienation and neo-colonial outrages that they ignored creating literature for the young minds.

In many advanced countries of the world, children's literature serves as a ready tool for inculcating the reading habit and developing a reading culture. In the aftermath of colonial dislocation of indigenous values, children's literature in Nigeria would, in addition to paving the way for instilling the reading habit, also serve the purpose of acculturation and value re-orientation. It would create models, fashion heroes, and project virtues held sacred in the ancestral heritage; such virtues as truth, innocence, honesty, courage, hard work, humility, loyalty, obedience, courtesy, care for the young and the aged, respect for the laws, customs and traditions of the land, and patriotism as a crowning communal and national emblem.

Children's literature comes in a variety of forms—fiction, fantasy, biography, poetry, drama, folktale, and picture books. Carefully crafted and programmed children's literature would teach morals, satisfy inherent curiosity and excite imagination in children culminating in properly adjusted, well-socialized adulthood. To write appropriately and effectively for children in the Nigerian situation, therefore, tantamounts almost to a commitment to a socio-cultural crusade. The works must have substance which is relevant to the environment of the children and contain messages which would be apt for the mental and psychological well-being of the children in their unique circumstances.

This seems to be the motivation and impetus behind Flora Nwapa's fiction for children. Some of the stories are simplified versions of myths and legends well known in the local environments. Some are recreations of folktales, embedded in the culture, while others are simple stories told for the purpose of educating and entertaining the mind as well as challenging the children in special directions and tasks.

This essay focuses attention on *Emeka—Driver's Guard* (1972), *Mammywater* (1979), and *Journey to Space and Other Stories* (1980). It is important to mention that soon after the publication of *Mammywater* (1979), Nwapa published in 1980, *The Adventures of Deke* which, on the surface, seemed as a logical sequel. In actual fact, it is the same story with minor retouches of plot and story line. Deke as a child-hero was quite successful in *Mammywater,* such that Nwapa had to distill him from the larger narrative and recast him in a more circumscribed tale where the focus is entirely on his individual talents and cleverness which serve as role models for children of his age.

Emeka—Driver's Guard (1972) is partly set in a village location and partly in an urban environment. Each setting is characterized by its typical inhabitants whose outlooks on life, belief patterns and social behavior are as a result of and influenced by the corresponding environment. The themes also rotate from the supernatural to the human. The supernatural is dominant in the rural location where the elders, the sage and the diviners are the harbingers, while man's physical actions and reactions dominate the urban location with the products of the emerging western civilization (schools, vehicles, etc.) in evidence.

The story takes off from the myth of the birth of twins as a bad omen for the entire community unless prompt action is taken to eliminate them.

> When Emeka and Ngozi were born, everybody in
> the village talked about it.
> "Have you heard?" they said.
> "It is a bad day when twins are born in the vil-
> lage."
> Most people believed that evil would come to the
> village if the twins stayed there.
> "This is a bad thing," they agreed, "but what can
> we do? In the days of our grandparents we would
> have let the twins die."
> "But those days have gone," other people said.
> "We can do nothing." (4)

Nwapa adds an interesting but intentional ambiguity whereby the community is willing to tolerate the taboo if one of the twins is eliminated through a process or deliberate starvation. The obvious expendable choice is, of course, the baby girl.

> A woman of the village came to visit the twin's
> mother.
> "I want to advise you on what to do. Everybody in
> the village agrees that this is a bad thing."
> One of the twins, the boy, began to cry and his
> mother picked him up to feed him.
> "Go on. I am listening to you," she said.

"Feed the boy but do not feed the girl," the woman
said. "Don't feed her at all. Give her nothing.
Absolutely nothing."
"Why?" the farmer's wife spoke gently, although
she was horrified. "I must feed both my babies
when they are hungry."
The woman did not answer her question. "You
know why," she said. "In the days of our grandparents we did not allow twins to live. Today everything is different. So remember, feed the boy but
not the girl."
"When my husband comes," the farmer's wife
replied, "I shall tell him what you said."
Then the woman was afraid. "No, please don't tell
him. This is a matter for women, not for men."
"I must tell him. They are his children."
"Please!" the woman begged, for she was afraid of
the farmer's anger.
The farmer's wife picked up the baby girl and
began to feed her. "Very well," she agreed. "I shall
not tell him. I shall forget what you said." (4-6)

The parents of Emeka and Ngozi refuse to yield to communal pressure but instead determine to raise their twin babies as normal
children.

"We'll call the boy Emeka and the girl Ngozi," they
told everybody. "And we'll let nobody take them
away from us." (4)

Apprehension rules the land and when an epidemic of smallpox
ravages the village nine years later, no one is surprised or in doubt
as to its source. To preserve the twins, their parents send them to
live with an aunt in town. The epidemic is quite devastating and
claims many lives, including the parents of the twins. The story of
the twins Emeka and Ngozi begins at this point at the home of their
aunt/foster parent, in town—away from the misty environment of
the village. The past thus becomes the bridge to the present.

The relocation of the twins serves two purposes. first, the
author extricates them from the village setting where individuality and free will are fettered. In the village setting where their misty
origin is known to everybody, superstition will forever haunt their
existence. Secondly, in the urban setting the author is able to pro-

ject them for what they are as individuals who are answerable for their actions, and who have the freedom to make their choices in life and live with the consequences of those choices. Furthermore, the focus of the story is on childhood and the author had to put Emeka and Ngozi in a children's world. They interact with other children as children before Emeka is isolated in order to tell his story as a "driver's guard." Emeka made the choice and the reader, especially the young reader, follows his life through many trials and errors and the price he had to pay for them before he finally reforms and reaps the rewards which the author holds up for all children in similar circumstances.

To emphasize a point, Emeka and his twin sister are put in a school environment where even in the urban location the society expects the male child to excel over the female. The society also expects, indeed desires the female pupil to drop out and join the mother in farm work, learn household chores and finally get married. Nwapa deftly undermines these chauvinistic norms. While Ngozi manifests a high sense of discipline and is well-behaved at all times, Emeka is a bully. While Ngozi is prompt in her school work and always regular at school, Emeka is a loafer and truant, cutting classes at will and inattentive in class. While Emeka is poor at his school work, Ngozi does well and even lends Emeka a helping hand.

> He is very worried about the examination. He has not worked at all and he was not very good at his lessons. He had stayed away from school too often. Ngozi always tried to help him with his homework, but he knew she could not answer the examination questions for him.

> "Ngozi is a good girl!" his aunt went on. "Everyday, when she comes home from school, she reads her books. Why don't you work hard like Ngozi? Eh? Why can't you sit down and read? You're always running around and playing...." (16)

In addition to all these, Emeka steals and tells lies and it is Ngozi, the girl, who tries to correct him. To crown it all, Emeka tries to cheat his way through the final examination; unlike his sister who works very hard for the examination and not only comes first in her class, but wins a scholarship to go to college.

At last the great day arrived. Emeka wore the ring on the third finger of his right hand. The first paper was arithmetic. He did the first three sums very quickly, but could not do the other seven. He tried to do them, but he knew the answers were wrong.

Then the bell went and he gave in his answer paper. The rest of the examination was like that as well. He did not worry, because he knew he would get high marks. The ring would charm the teacher....

Then the day came when the results were announced. The headmaster started with the youngest class. He called out the names of those who had passed and those who had failed. It took a long time. When he came to Emeka's class, he called out Ngozi's name first. She beat the second person by nearly a hundred marks. There were thirty in the class and seven had failed. Emeka did not wait to hear his name. When the first twenty names were called and his name was not among them, he knew that the ring had not helped him.... (21-23)

Against all persuasion to the contrary, Emeka chooses to drop out of school in the carefree world of a driver's guard. Focusing on Emeka, Nwapa weaves a story of a prodigal who roams fruitlessly and relishes in easy life and reckless frivolities. He makes bad friends, gets into troubles but finally reforms and comes home a hero of sorts.

Although Nwapa makes a case for female industriousness and excellence even in competition with males, her ultimate goal is not to discredit her male characters in the book. She underscores the point that girls could be as good, and even better than boys, when they apply themselves assiduously. But what comes through as the enduring moral of the story is that excellence, high sense of responsibility and success are not gender bound. Nwapa's artistic intention, therefore, was to make good out of evil. Emeka, at the end, looks back and is sad about the life he has lived. He realizes that indiscipline and truancy do not pay. He had wandered off like the prodigal son but when he learned his lesson, he was determined to return to the path of hard work and honor as the only answer to his predicament. He does not accept failure and he no longer wants short cuts and false and easy solutions.

> Ngozi and Emeka had a lot to tell each other. They talked for a long time and at last Emeka said, "Ngozi, I am going back to school."
>
> Ngozi was very happy to hear him say this. "But will our aunt agree?" she asked.
>
> "If auntie refuses to send me to school I shall ask one of the teachers to let me be his houseboy, and he will send me to school. I have learnt a lot, Ngozi. Without education you are nowhere." (43)

This is the point of awareness and wisdom derived from within; knowledge acquired from experience. This is the climax which the author aimed at in the story. Children can go astray. Spanking or beating them may not make them change. But hard lessons learned from experience become for them the best teacher, not only for the characters in the story, but for children everywhere.

Nwapa projects the funny, the good and the beautiful in her child heroes. She does not suppress their short comings, but rather shows how children develop through trial and error, mistakes and misadventures. They learn from them and move from the point of negative behavior to positive attributes. Thus, the ugly sides are used as a canvas for the illustration of the process of growth and development. Nwapa's heroes are, therefore, carefully crafted to serve as role models and the stories of their lives become parables for their peers and all youth everywhere.

In style, plot, and artistic purpose, *Emeka—Driver's Guard* is a story of all seasons for children. Nwapa is able to come down to the level of children to capture appropriately their speech patterns and mannerisms, their instinctive rationalization of things and events and their innocent perceptions of events in the adult world as they relate to children. When they perceive differently, they react differently. When they set off knowingly to do wrong, it is because of external or peer influences.

The story is replete with dialogues all appropriate and convincing at the level of children. Nwapa is conscious of the language ability of the young readers for whom the book is written. Therefore, she provides explanations of strange and difficult words at the bottom of the page wherever they occur. At the end of the story, she adds three pages of further vocabulary exercise in which she draws attention to some words which had been used in the story which the reader should study for practice in communicative competence. Thus, while the reader is entertained by the beauty of the exciting story, his or her imagination is equally excited and at the same time their communication skills are improved.

Mammywater (1979), Nwapa's second children's book, is based on the myth and mystery of the goddess of Oguta Lake which dominates actions and events in Nwapa's adult fiction. In Oguta, the "Woman of the Lake," (The Lake Goddess), known in local parlance as Uhammiri/Ogbuide/ Mammywater, is a living reality, beyond fantasy and beyond myth. Adults of all ages accept her physical and spiritual existence and do not see such belief as a violation of any aspect of their religious faith or practices.

In her adult fiction Nwapa has shown the effect of the presence of Mammywater on the lives and actions of people. She can permeate the lives of people in the community and alter them for better or for worse. Conceptualization of the Lake Goddess (who inhabits the bottom of Oguta Lake) remains perpetually at the level of mystery, not to be probed and not to be challenged. Whoever she "calls" is expected to "follow" and serve her without question and without protest. When people mysteriously disappear from their homes, as is often believed, they are assumed to have been "taken" by Uhammiri and they henceforth inhabit the bottom of the lake with her. The lake, therefore, is both an object of relaxation as well as fear—fear of the unknown.

In its aura of mystery, the Oguta Lake is viewed by parents as an environment injurious to the health and welfare of unguarded children. They must stay away from it especially at certain hours of the day, but definitely at all hours of the night. It was in the bid to create a portrait of the Lake Goddess at a level harmless to children's imagination and mental grasp that Nwapa wrote *Mammywater*. It turned out to be quite a popular storybook for kids.

In *Mammywater* the mystery of Uhammiri is preserved intact. Her aura of inscrutability is also left intact but devoid of fury and malignancy. She is demystified sufficiently for her to be pleasantly accessible to children without devaluing her supernatural essence and dignity. Nwapa casts Mammywater in the mould of a mother, a caring and sympathetic mother who would feed a hungry child, tuck him/her into bed when sleepy and urge the child to rest fully when tired after an arduous task. But because Uhammiri is a supernatural figure and dwells in the world of spirits, human beings—no matter the age—who seek any form of communication or interaction with her, must go through certain rituals and stick to and obey some laid down laws and codes of conduct.

Any violation leads to imminent death or non-return to the human world, if ever. Such recalcitrant individuals may get trapped at the bottom of the Lake until a ritual of appeasement is duly performed.

The story *Mammywater* can be said to be in three parts. The first section introduces the reader to the background of the story. The setting is the Lake where a young girl, Soko, is enjoying herself swimming. Mammywater suddenly appears to her and, after some verbal exchanges, carries her away with her mother's consent. The second stage of the story takes place first at the home of the "kidnapped" child where there is a heated exchange between her elder brother, Deke, and their mother. Deke is so incensed by the disappearance of his younger sister that, after waiting for one whole week, he decides to go to the bottom of the Lake to rescue her from the Mammywater.

The action then moves to the bottom of the Lake where there is prolonged interaction between Mammywater and Deke. It is through this interaction that the reader gains insight into the personality of Deke, on the one hand, and the phenomenal nature and ubiquitous personality of Mammywater.

The third and final stage of the story takes place at the shrine of Urashi, the towering god of Urashi River. Here the reader learns for the first time why Soko and other people before her were carried away by the Goddess of the Lake. The reader also learns about the divine quarrel between Urashi and Uhammiri which has brought sufferings to the human beings in the community from time immemorial. Urashi and Uhammiri state their individual grievances and, through the intervention of Deke, peace is restored between the two supernatural powers and mankind is promised eventual harmony and equilibrium.

The ending of the story provides the necessary clue to the understanding of the beginning of the story. Soko is carried away in a manner that raises questions in the mind of the reader.

> Soko is swimming in the lake. Suddenly
> Mammywater appears. Soko looks intently at her
> and asks, "Who are you?"
> Mammywater replies, "I am Mammywater and I
> have come to take you."
> Soko is puzzled. "Why do you want to take me?"
> she asks.
> "I want you as my daughter," she replies.
> "Interesting, but who sent you?" Soko asks.

> "Nobody ever sends me, I send myself."
> "Well, in that case, you cannot take me," Soko
> says. But Mammywater insists, "Why can't I?"
> "Because," Soko goes on, "you are not my mother.
> Only my mother can take me." (3)

The dialogue is inconclusive until Soko's mother appears out of nowhere and, in what seems like an unmotherly act, shatters Soko's hopes and confidence.

> "You are not serious. My own mother wants me to
> go with you?" asks the puzzled Soko.
> "That is so, why not ask her..."
> "Do you want me to go with her?"
> "Yes, child, go with her."
> "But, Mother, I don't understand. Don't you want
> me any more?"
> "Of course I want you, my daughter, but go with
> her..."
> "No, Mother, I don't want to go..."
> Mammywater is annoyed. She turns her back and
> says, "Then I must go without her. Goodbye..."
> "No, don't go. Take her, take Soko."
> "Mother, you mean...? Are you not my mother?"
> Soko asks.
> "Go with her, child, and my God bless you." (5-9)

The author portrays Soko with sensitivity even if Soko's tone in her verbal exchange with Mammywater seems not very realistic, at best over-sketched at her age. It is interesting to observe Mammywater's final reaction as she carries away Soko.

> ...Mammywater grabs Soko with both hands, wav-
> ing her in the air, and saying to nobody in particu-
> lar, "You have heard, you are my witness. She says
> I must take her. I am taking her. I am taking her."
> (9)

It would seem as if Mammywater is taking Soko against her (Mammywater's) wish but the reader knows better at the end of the story when Mammywater absolves Soko's mother. It was thus the author's way of depicting the Goddess as motherly, conciliatory, and sympathetic.

Although Deke is mentioned in absentia by his mother during

the Soko-Mammywater encounter, it wasn't until the second stage of the story that Deke makes full physical appearance and indeed takes the center stage. Deke, in this part of the story, leaves home in search of Soko, his only sister. This departure from home at a point of crisis in order to return later as a hero, is a familiar pattern in many African folktales. But at the material time the future is uncertain and the hero wades into the unknown propelled (as is the case with Deke) only by his/her courage and commitment to a chosen cause. There is a noticeable technical flaw in Nwapa's introduction of Deke into the story.

> Deke comes home and does not see the sister. He is suspicious. "Where is Soko? If you don't tell me where Soko is, I'll cut off your head," he says. "My child, you can't cut off my head," says his mother.
> "And why can't I? Here is my knife."
> "You can't because I am your mother. And if you cut off my head, you will be accused of, of...." (10)

There is no suggestion before then in the story that Deke is a violent child nor is the reader shown in what way Deke's mother acted suspiciously. This threat to cut off a mother's head rings unnatural in a ten year old child, especially as it comes without any tangible provocation at this stage. Deke does not give his mother even the chance of answering his question as to Soko's whereabouts before issuing his violent threat. Perhaps Nwapa wanted to deck Deke with a "bravado image" to prepare the reader for his macho role later in the story. Perhaps, too, Nwapa wanted to draw forceful attention to Deke's impatience which becomes much more pronounced during his later encounter with Mammywater. Or, furthermore, it could be the author's way of emphasizing Deke's great love for his only sister. Either way, Deke's utterance to his mother is unnatural and unconvincing in a cultural setting where respect for elders, especially in a parent-child situation, is almost held as sacrosanct.

The encounter between mother and son gives the reader an insight into the type of adult veneration of the Lake Goddess in the culture. No evil word can be said knowingly by anyone against Uhammiri—not even through innuendo. So Deke could not expect to get a cut and dried, straight forward answer to his question of why or for what Mammywater took Soko. His mother's answer—

"Only Mammywater can tell," is apt and expected in the circumstance.

In spite of the false beginning, the author's subsequent characterization of Deke, the child-hero, is quite successful and symbolic. A face-to-face encounter with Mammywater requires
extraordinary courage in an adult and much more so in a child of
Deke's tender age. Nwapa does not artificially or mechanically
induce this in Deke. Instead, we see a psychologically designed
process whereby Deke braces himself up for the purpose. He constantly reminds himself and the reader that he is strong and courageous. He is bold when he talks to Mammywater, but not
indecorously so as was the case when he talked with his mother.

> "Tell me, how did you get here?"
> "I dived, if you must know.".…
> "Aren't you afraid?"
> "Of what?"
> "Of the lake, child."
> "Why should I be afraid? I know the lake well.".…
> "A child like you should be afraid of the lake. She
> is full of mysteries."
> "No, I am not afraid. I am the son of...." (15)

And later Deke tells Mammywater:

> "Now I see that I cannot leave, Mammywater. Did
> you think I was going to cry? I never cry you know,
> I am a tough boy." (17)

…and later still,

> "Now, now hold it. First things first."
> "But I am not a little boy." (20)

Another apt quality which the author endows Deke with in pur-
suance of his mission, is an inquisitive and curious spirit. With it,
Deke comes through in the story as a precocious, vivacious and
alert child. The author achieves this by making Deke active and
moderately loquacious in every situation and by giving him the
free spirit to ask many questions.

The Igbo have a saying that "the child who asks questions never
goes astray." Deke asks an abundance of questions which pay off
in the object of his mission because they help to create the right
impressions about his personality and behavior. Before eating the
food offered him in Mammywater's house, he is cautious and wants

to know who cooked it (19). Before he could make himself at home in Mammywater's house, he first wanted to know where his sister was (17) because "that is why I am here. It is purely business" (21). When he is served a delicious fish menu, he wants to know who caught the fish (19). When Mammywater asked him how old he was, he also wanted to know Mammywater's age and whether she would ever die (20).

He also asked some questions which bordered on morality. For instance, he wanted to know why Mammywater took his sister away from him and his mother, more so when Mammywater described his mother as her friend (21 & 27). He wanted to know why he was fed only with fish and no yams or cassava (23), And then he wanted to know why he still remained hungry even after he had eaten Mammywater's food (23). He had endless questions for Mammywater, some of which even seemed to catch her off guard. Even Mammywater had to admit that Deke's pace was too fast for her: "Son, you are too fast for me. Eat your food, we'll talk later" (21). Through his over-active personality and relentless questioning, Deke endeared himself to the Lake Goddess who proclaimed him "a distinguished visitor."

> "Distinguished visitors! Am I a distinguished visitor?"
> "Now, now, you are asking too many questions again."
> "Answer my question, Mammywater. Am I a distinguished visitor?" he insists.
> "Of course you are. Small boys don't come to me the way you came. You are distinguished."
> "Am I the first distinguished boy to come to you?"
> "No, many have come, but failed the test I set for them. Now, no more questions." (35)

Nwapa has a good ear for good dialogues. The conversation between Mammywater and Deke is so real and enchanting that the reader easily visualizes the situation. Deke comes through as a clever and likeable child who livens up a social atmosphere with innocent fun and humor which can only come from children. The reader wants to hear more questions and conversations with Deke and indeed one of the strongest qualities of *Mammywater* as a work of fiction is the beauty of its dialogues.

The theme of search (quest) which Nwapa explores here at the level of the child-hero is featured at the adult level in such classic works of fiction as Amos Tutuola's *The Palmwine Drinkard* (1952). But there are interesting differences. In Tutuola's spirit world the dead exist in physical form but walk backwards. In Nwapa's, the spirits are transformed into fish but retain their human qualities in every other sense. They can receive visitors and interact with them...talk freely with them, move about with them and even feed them. The only restriction is that human beings should not touch them or else they will be transformed into fish. Tutuola's hero, goes in search of his dead palmwine tapster and, when he finds him, cannot come back with him (the tapster) because the dead cannot return in their physical forms as people to the human world. Unlike Tutuola's hero, Nwapa's child-hero Deke is, because of his humility and good manners, promised that his sister—trapped in the spirit world of Ogbuide—could return to the physical world with him.

> The Great Spirits are touched by Deke's humility. They confer in whispers. Then Mammywater speaks: "We are completely disarmed. We are happy that you have shown humility. But before Soko is returned to your mother, she, your mother, my good friend, must come to my home to be entertained. We shall perform the dance of the maidens for her and she will meet her husband. There she will meet you and Soko also. Then you will all return to your world in peace." (45)

In Tutuola's spirit world, death is a finality. In Nwapa's, there is a leeway for interaction on specified terms.

In the final stage of the story, Deke and Mammywater are at the shrine of Urashi to settle an immemorial conflict between the two gods. This section of the story is based on another popular myth of a conflict between Uhammiri/Ogbuide which inhabits Oguta Lake, and Urashi, the male god of Urashi River. Nwapa arrays the two gods in anthropomorphic frames and opens up an encounter between them which best typifies a battle of the sexes.

> "You have come at last, Great Spirit of the Lake. It does not pay to be stubborn, Great Spirit of the Lake."
> "We have a guest here, Great Spirit of the River. Today we must not quarrel, Great Spirit of the River," replies Mammywater.
> "You always want to quarrel, you proud Spirit of

the Lake," Urashi continues.

"I don't quarrel, Great Urashi. What I have, I
hold."

"And what I have, I hold, my Great Spirit. But you
want to take that which does not belong to you,
you arrogant Spirit."

"The Lake belongs to me, the River belongs to
you. There has always been a dividing line. Why,
why should you cross the line drawn by the pen
of our God?" replies Mammywater. "And why, oh
why must you send your reptiles to kill my dear
ones, the people of the Lake?"

"Your people are sometimes stubborn, they don't
listen to me," says Urashi.

"And that's why I am here, that's why this boy is
here. Shall we talk business now?"

"Let's have kola first, my good spirit. There, you
over there, bring me kola!" Urashi calls. (39-41)

This is one of the best prose passages found anywhere in Flora
Nwapa's fiction. The two gods state their individual grievances with
candor and sensitivity. Oguta Lake is the domain of the goddess
Uhammiri while Urashi, a male god, inhabits Urashi River. Urashi
River and Oguta Lake converge at a confluence, but—quite inex-
plicably—do not intermingle or flow into each other. Each goes its
way. This rare geographic phenomenon is a center of attraction
for tourists and perhaps could be explained from a geographical
point of view, but to the local community the mystery is explained
in terms of the age-long conflict between the two gods who inhabit the Lake and the River, respectively. Nwapa gives the two gods
the human voices to explain the age-long rift between them in a
form comprehensible to human ears (adult and child).

"My water is muddy, yours is sparkling clear. Your
people laugh at my water, and I get furious," complains Urashi.

"So you push in, into my own water so as to pollute
it. But you are not a child. You know it is impossible for my water to be polluted by yours. Now let
us begin real business. This is the boy." (41-42)

This is, therefore, the explanation of a mysterious geographic phenomenon. The reader notices authorial manipulation in the reso-

lution of the conflict. What begins as name calling with each god trying to assert its supremacy (in an obvious battle of the sexes), ends in a mutual compromise where each accords the other his/her deserving respect. The conflict is portrayed by the author as a personality clash more than a gender war. At the center is ego and personal aggrandizement more than anything else. This is akin to Flora Nwapa's insistence that feminism is not necessarily an issue of the supremacy of one sex over the other and the struggle to liberate the under-dog. Instead, it should be a process to achieve mutual understanding, acceptance and respectability, human qualities which have no gender.

The "real business" referred to by Mammywater (42) is the resolution of the misery which the divine conflict had inflicted on the local communities. This is done through the instrumentality of the child-hero, Deke. Deke is portrayed as a model of a disciplined child, compassionate, sensitive, and adventurous. His journey to the bottom of the lake is ostensibly to rescue his sister from Mammywater, but in the process, he accomplishes a larger symbolic task of "rescuing" his entire community from the clutches of the gods. Deke is obedient, humble, respectful, and disarmingly polite. He is also firm, responsible, and courteous. The author uses him to project a model for other children in their inter-relationships with other human beings.

It is noteworthy that Nwapa's child-heroes are always given the task of saving and rescuing people in trouble and distress—children and adults alike. This conferment of responsibility is Nwapa's way of saying that the children of today are not only adults of tomorrow but also leaders of tomorrow, and leadership entails taking responsibility for the welfare of the community.

Deke performs the feat of obtaining atonement for his community. He not only wins for his people the forgiveness of the gods, but he also brings about a rapprochement between the two "warring" gods. Deke achieves this by his civility (to the gods) which seems to have cost him nothing but has bought everything for him and his people.

Mammywater was written with a definite sense of purpose for children. The narrative is lucid, the language is appropriate, the story teaches, tickles, and inspires children, especially those who are excited by mystery tales. The illustrations are vivid and captivating and the print is uniquely bold and attractive.

The last story in this study, "The Magic Cocoyam," was published in *Journey to Space and Other Stories* (1980) which Flora

Nwapa co-authored with Obiora Moneke. It is a (re)creation of an animal tale involving tortoise, the cunning trickster of the animal kingdom. It shares with *Mammywater* the common theme of concern and sacrifice for a loved one. In both a brother sets out to rescue a sister. In both cases the sister is younger. In both tales the loved one (victim) has been abducted by a strong supernatural force. The Lake Goddess abducts Soko and Deke goes to rescue her in *Mammywater*; in "The Magic Cocoyam," Lala is abducted by a huge tortoise and Kele goes to the land of the spirits to rescue her. The abductions take place in a river while the girls are swimming. Unlike Deke, Kele was physically present when the tortoise carried away his only sister.

Kele and Lala are orphans and have no other family. Kele's cry for help was not answered by anyone in the community. The community has been introduced earlier by the name of "Womo Village," which is poverty-stricken. No one there knows how to fish and no one farms because nobody had taught the people how and, in addition, their soil is not fertile. The story is simple and uncomplicated and is meant for children just beginning to read, or those who are read to by their parents.

A fish appears to Kele at the river-side and leads him to Mother Frog who leads him to a friendly Monkey who leads him to a kind-hearted Bird who leads him to the home of Tortoise where, following the advice and directives of his benefactors, Kele rescues his sister and the Tortoise is vanquished. They return to their village through the instrumentality of the animals in reverse order. The community receives brother and sister very well and, in turn, the two child-heroes teach their community how to farm, a skill they learned on Tortoise's farm in the spirit land. There is prosperity in the land and hunger and poverty are eliminated. A "magic cocoyam" was the instrument of Lala's final liberation from the imprisonment at Tortoise's house. Therefore, Kele and Lala forbade their community from cultivating cocoyam which then became the totem of the village.

"The Magic Cocoyam" focuses on the same virtues which Nwapa had highlighted and espoused in the previous stories: love for others, courage, service to community, obedience and mutual trust. Children should not be afraid to take on positions of responsibility and care, especially for the less privileged in society.

Nothing belies "art for art's sake" more than Flora Nwapa's fiction for children. Nwapa wrote as a concerned mother (writing first and foremost for her growing children), teacher, educator, and social worker. Her commitment was towards the restoration of good moral values and the provision of role models for contemporary youth. "Catch them young" was her social philosophy towards children. She was intrigued by the state of innocence in childhood and sought, through her children's books, to perpetuate positive virtues through adolescence. Nwapa had faith in children as future leaders of their society. Properly nurtured, children of today can rescue the adult world of tomorrow. The formula is laid out in the various plots in Nwapa's children's books. Each character, good or bad, is a mirror for other children elsewhere.

It is significant that Flora Nwapa's first children's book (*Emeka—Driver's Guard*) was published shortly after the Nigerian Civil War (1967-70) when Nigerian children had been unwillingly exposed to various types of violence, corruptive influences, indiscipline, and all forms of moral disorders. It is even more significant that this book, which reprimands anarchy in youth and espouses a return to moral rectitude, was written and published when Flora Nwapa was the Commissioner of Health and Social Welfare/Lands, Survey, and Urban Development, in East Central State of Nigeria.

WORKS CITED

Nwapa, Flora. *Emeka—Driver's Guard*. London: University of London Press, 1972, 4. (Subsequent references to the text will be referred to by page numbers.)

____."The Magic Cocoyam." In *Journey to Space and Other Stories*. Enugu, Nigeria: Flora Nwapa Books & Co., 1980.

____.*Mammywater*. Enugu, Nigeria: Flora Nwapa Books & Co., 1979, 3. (Subsequent references to the text will be referred to by page numbers.)

"Either Pound in The Mortar Or Pound on The Ground": Didacticism in Flora Nwapa's Children's Books[1]

Marvie Brooks

Children can either accept wisdom or good advice from the parents and elders in the community or they can reject it and learn the realities of life the hard way. Flora Nwapa, through her children's books, provides moral lessons for youth which they can heed or reject. However, she does not inculcate her gems of wisdom in dry didactic prose. Instead she uses the oral tradition, folktales, proverbs, traditional beliefs and customs to both enrich the story and to teach as well as entertain the child. Nwapa understands the need for Nigerian children to learn about family life, the community, self realization, school and friendship, choices and responsibility, and personal initiative and creativity. She very wisely accomplishes this in her works through themes, plots, settings, language, and characterization that reflect Nigerian culture, history, tradition, daily life, and social issues. Ezenwa-Ohaeto points out that Flora Nwapa "situates the child figures in her works in social and cultural situations that reflect the interactions and intricacies of life in an African society" (77).

The following themes, depicted in her stories for children, will be analyzed: respect for elders, gender balance, the use of folklore to teach values, wit, pluck, humility, and perseverance to overcome obstacles, "fostering" by kin, and the importance of education. The stories which will be examined in this article are the following: *Mammywater* (1979), *The Adventures of Deke* (1980), *The Miracle Kittens* (1980), "The Magic Cocoyam" (1980), *Emeka—Driver's Guard* (1972), and *Journey to Space* (1980)[2]

Many times, folktales and fairy tales about mermaids are taken from European and Near Eastern sources (*Encyclopedia Americana* 18:727).[3] However, T. Uzodinma Nwala, a professor of philosophy, informs us that in the Igbo traditional belief system "Rivers [are] associated with certain gods like Imo, Idemili, [and] streams like miriHaba"(41). Nwala reports on the religious calling and the duties of a priest who serves the water goddess(mermaid)—Nkpetine (1983:123). Chimalum Nwankwo also explains that "some of the most powerful deities of Igbo country are female and associated with water or underwater dwellings such as the goddess Idemili of the Anambra area" (44). Flora Nwapa's story, *Mammywater* (1979), is about a beautiful mermaid. The story is based on the Igbo lore of the lake goddess—called Uhamiri—in her town of Ugwuta, who is responsible for good fishing and safe travel on the lake.[4] The story involves transformation, a magic spell, a perilous task, a taboo, and a test of character. The folklore or custom transmitted to children is the tradition of the spirit or goddess of the lake, the worship of her by young women whom she has chosen, and also the dispute which she has with Urashi, the river spirit. In the novel *Efuru*, Nwapa informs us that Uhamiri, the lake goddess, and Okita [Urashi in *Mammywater*] owner of the river "were supposed to be husband and wife, but they governed different domains and nearly always quarreled" (Nwapa 1978:201). In the children's story, the mermaid/Mammywater takes as her captive a young girl who is swimming in the lake with her mother. The mother advises the frightened child to go with the mermaid, and she tells Mammywater to take her (Nwapa 1979:1-4). Deke, the girl's brother, loves his sister and takes upon himself the challenge of finding her. An old woman who worships Mammywater tells Deke how to get to the mermaid's home. She also informs him of a taboo—never to touch the mermaid, even when she invites him to do so. The old woman warns Deke to cultivate the character trait of humility (8). Thus, Deke dives into the lake and begins his underwater journey to find Soko who, unbeknown to him, has been transformed into a girl fish (9, 20). At the end of the story, Deke succeeds in his quest and lives up to his name, "a man of strength," because his personal courage and resoluteness have been tempered by obedience, humility and patience. The lake goddess/Mammywater/mermaid helps Deke, the hero, on two levels: with his immediate task of finding his sister and with his deep underlying family/village problems. For example, Deke learns the reason why the spirit of the river, Urashi, is quarreling with Mammywater, the lake goddess. The people of his village, who worship the lake goddess, have made fun of the river spirit because the river water is muddy and not clear and sparkling blue like that of the lake. In turn, the offended Urashi, the river spirit, sends reptiles to kill the followers of the lake goddess (28). Deke's humility and fervent cries on his knees for forgiveness touch Urashi and Mammywater (30-31). Mammywater assures the river spirit that Deke and Soko, when they return to earth, will inform the villagers of their error (30). Hence, Deke's willingness to practice humility and to obey the taboo not to touch Mammywater enables him not only to reclaim Soko, his sister, but also his father, whom he believed dead, from captivity in the underwater kingdom as well as the lives and well being of the people in his village (30-31). Thus, the example of Deke instills in youthful readers the desire to practice humility, to respect the wisdom of elders, and to be obedient. Children are also informed about an important belief in their culture, the worship of the lake goddess.

The Adventures of Deke (1980[a]) is an adapted version of Mammywater (1979) for younger children (Ezenwa-Ohaeto, 1995:78). As in the original story, this adaptation for younger children also includes the values of humbleness and obedience. In both versions of the story, the author shows that wisdom is not equated with gender through the characterization of Mammywater, the lake goddess, and the old woman who worships her. The Adventures of Deke (15 pages in length as compared with Mammywater's 32) is a shorter version of the original. In the adapted version, the story's ending is very neatly resolved so that younger children will have no trouble understanding the outcome (Ezenwa-Ohaeto 1995:69). For example, in The Adventures of Deke, the young hero's humble, obedient, and respectful manner touch both Mammywater and

Urashi. The two spirits agree to give Deke's sister back to him (1980[a]:14). Soko and Deke are immediately reunited. On the very last page the two children are shown hugging each other (Nwapa 1980[a]:15) and the last line of text states that "The golden canoe is waiting to take them back home" (ibid.: 15).[5] The golden canoe is paddled by the boy fish. In the original story, however, Deke sees his sister before this scene, while Mammywater is showing him around her home, which occurs before they visit Urashi. At this point, Deke recognizes Soko who has been transformed into a girl fish, and tries to get his sister to speak to him (Nwapa 1979:19-20).

The Miracle Kittens (1980[b]) is a delightful, rollicking picture book about the efforts of Madame Bum Bum and her family to get rid of the rats which have taken over the house. The story emphasizes family cooperation in solving a problem and the particular insight, persistence, and help that children can bring to a household problem. The smart rats ignore the poison set out for them and soon figure out how to get the food out of the traps without getting caught (Nwapa 1980[b]:12). The children suggest a cat to take care of the problem, but none could be found at the market (13). However, the young people insist that a cat will arrive either from a friend or from God (14). The miracle occurs when upon returning home from a trip, Madame Bum Bum finds a cat and kittens in her bedroom. They came in through the open bedroom window (15). Afraid of cats and rats, Madame Bum Bum's screams bring the entire household into her room, and the frightened mother cat jumps out of the window and leaves the kittens behind. However, Madame is not afraid of babies and so the baby kittens become the pride of the house. And of course, where there are cats the rats soon leave (19).

Children sometimes forget that pets need love and care just like humans. Hence, the lesson of being responsible for pet animals is depicted very effectively in this story. Madame Bum Bum immediately involves the entire family in setting up the type of care which will allow the kittens to thrive. "Now children," she said, "get some milk and water and fish, let us feed the kittens. Now Suzy, get the blanket over there to cover the little ones" (Nwapa 1980[b]:18). Thus, Nwapa shows that it is never too early to show children that love involves nurturing. The unexpected bonus is that someday one's pet may render a needed service. In this story, the healthy, well-cared-for kittens get rid of the rats. Flora Nwapa also demonstrates that out of the mouths of babes oftentimes comes good suggestions.

"The Magic Cocoyam" (Nwapa & Moneke 1980) is a lively folk-tale with a moral, a test of character, a task to perform, animals assuming the qualities of humans, a totem, and a transformation. Kele and his sister Lala are orphans. While they are swimming in the river Mother Tortoise kidnaps Lala (43). Fish advises Kele to see Mother Frog and follow her instructions. Mother Frog directs Kele to see Monkey who takes him up a tree where he finds Bird (45-46). Bird sets up the following specific tasks which Kele must perform in order to find his sister: be persistent in finding Mother Tortoise's farm and above all work hard and find out how she farms. Bird also warns Kele that he must be brave (46). Bird flies Kele directly to the home of Mother Tortoise. There Kele learns how to plant vegetables, fruits, yams, potatoes, and cassava, but the wily lad soon notices that Tortoise did not show him how to plant cocoyams (48). When Mother Tortoise goes to the river, Kele searches and finds the cocoyam. He plants it in a secret spot and is surprise to see how fast and big it grows before the Tortoise returns (48). While Kele is resting, the cocoyam plant tells him to "Dig me! Kill the Tortoise! Set me free!" (50). He recognizes the voice of his sister Lala, who advises him to use the cocoyam plant to kill tortoise and set her free. Kele takes a cocoyam leaf and confronts Tortoise with the kidnapping of Lala (50). Despite, the plea of the Tortoise, Kele strikes her with the leaf and her shell breaks into a hundred pieces (52). Lala is transformed back into herself upon hearing the news of Tortoise's death (52). Now that the main task is over, the author uses another technique to show that the story is going into the second half. In reverse sequence, each of the animals—Bird, Monkey, Mother Frog, and Fish—help the two children to cross the river and get back to their village. Kele and Lala warn the people not to plant or eat cocoyam (54). The cocoyam becomes the totem or living emblem of the clan/people of the village (Maquet 1974:317) because by means of this plant Kele saves Lala's life and punishes Mother Tortoise.

Emmanuel Obiechina explains that among the Igbo people, folktales are instructive as well as entertaining (qtd. in Nwala 1985: 207-208).[6] Thus, in "The Magic Cocoyam" the author draws the child's attention to the value of obedience. By listening and heeding the wisdom of Bird, Kele is rewarded not only with the return of his sister but also with the honor of introducing agriculture to his community. Another moral/proverb in this story which is very important is the lesson that "The cheater was cheated at last" (Nwapa & Moneke 1980:52). It is interesting to note that in this story, Tortoise the cunning was tricked; this animal is usually the hero in Igbo folktales (Nwala 1985:73, 157-158). Thus, Nwapa teaches youth that human beings may suffer punishment, just like Mother Tortoise, when there is an "abuse of intelligence" (159).

Emeka—Driver's Guard (1972) is a realistic fiction story about fraternal twins, a girl and a boy. The story is set in the real contemporary world where life is not perfect. The story encompasses family relations and problems, friendship, school, extended family and kin, the struggle to meet economic, social, and educational needs, familial love, growing up and facing responsibility, gender, and multiple births.

Nwapa immediately draws the reader's attention to cultural issues of gender and multiple births in the beginning paragraphs of the story. A woman in the village reminds the parents that according to the traditional view, the birth of twins will bring misfortune to the community and indirectly counsels the mother to let the girl die by advising her not to feed the female infant (Nwapa 1972:4-5). In former times, the Igbo believed that twins were contrary to the laws of nature and an abomination (Nwala 1985:58). According to Otonti Nduka, the lives of all in the community were of more value than one individual(qtd. in ibid.:161). Thus, "it was necessary to propitiate the gods, either by killing the twins or by sacrificing other human beings to the gods, or by performing some rites" (qtd. in ibid.:161).[7] At this point, Nwapa takes a resolute, brave stand on the issues of gender and twins. In her characterization of the mother, Nwapa allows her to firmly refuse to deny food to either the boy or the girl. Thus, the brave farmer's wife tells the village busybody that she will feed both children and threatens to report what she said to her husband (Nwapa 1972:5-6). The frightened woman begs the farmer's wife not to tell her husband. The mother agrees to forget everything as she feeds the baby girl right in front of the woman (6). When a smallpox epidemic strikes the village, the people remember the twins. The apprehensive parents send the twins to the mother's sister in the city (8). The children's names—Chukwuemeka (God has done well) for the boy and Ngozi (blessing) for the girl—reflect the mental viewpoint of their parents and serve to underscore their struggle and determination to keep both the boy and the girl.

The aunt never varies in her duty and love toward the twins even when their parents die in the small pox epidemic (Nwapa 1972:9, 12). Thus, through aunt's unfailing love and tender care, Nwapa is very careful to ensure that the lesson she inculcates—

about the strong helping less unfortunate kin to pull themselves out of poverty or a disaster—is not spoiled by a "fostering" situation, which is sometimes cruel and psychologically devastating to the child. This is achieved by making the aunt a childless widow and financially independent through her cloth business in the market. There is no competition from other children or a spouse.

The willingness of the kin, the maternal aunt, to assume responsibility for the children's food, clothing, shelter, educational fees, and moral instruction is a part of the Igbo moral value system which emphasizes generosity and the blood connectiveness of kin (Nwala 1985:152-153; Odejide 1992:150). Among the Igbo people, the upbringing of children is a social responsibility of the entire village (Agbasiere 1992:127,129). The children of your brother(s), and sister(s) are considered to be your children, so the immediate family has the responsibility to help bring them up. As Nwala (1985) points out "the Igbo have a legendary reputation for being their brother's keeper. The extended family and kinship system is an important part of the social system with its duties and obligations" (152).

The male twin, Emeka, suffers through the maturation process as he begins to grow up and makes foolish mistakes. He fails his school examinations because he listens to Sokari, an immature friend, and skips school to roam around the town (Nwapa, 1972:14-15). He turns down his aunt's offer to pay the fees for him to repeat the class. Emeka foolishly decides to leave school and become a driver's guard (23). Nwapa vividly describes the real world which Emeka faces when he leaves home and school. For example, Emeka gets into a fight with a guard, who is an older boy, and receives a hard blow in the mouth which leaves him unable to chew for a while (27-28). Another time, frightened passengers stop the truck because Emeka is fighting with another guard who tries to bully him. The angry owner of the truck beats both boys with a stick (30-31). While on the road, Emeka sees a bad accident and notices that among the dead bodies is a guard about the same age as he (28-30).

Emeka leaves the job and through a friend he gets a position as house boy to a family with three children (Nwapa 1972:37-39). Experience is the best teacher and Emeka slowly realizes this as he compares his former life at home with his aunt and sister to his present situation of unproductive wandering. He also finds out that Ngozi will be attending college (42). Formerly, Emeka did not like to obey rules. The author writes that at school and at home, there had been so many things he had not been allowed to do. "This life is better than school," he thought. "But he still could not sleep" (29). However, after being out in the world on his own, Emeka begins to understand that discipline is necessary in order to be successful and have a happy life.

After being away for two years, he returns home with nothing to show for the rough experiences that he has had while out in the world on his own (Nwapa 1972:43). The chasten Emeka humbly asks his aunt to take him back and allow him to return to school. He is willing to serve as a house boy to a teacher and get his school fees paid in this manner. The lesson is driven home to youthful readers when Emeka admits that "Without education you are nowhere" (43).

On the other hand, Ngozi is not weak in character and makes a good adjustment to town life, her new home, and school. She even achieves first place in the same examination which Emeka fails and wins a scholarship to college (Nwapa 1972:22,42). Kalu and Kalu (1993) explain that "the Igbo have been very open to female education" (236). The story allows children to see that success and the attainment of goals is possible for male and female as long as they are sufficiently motivated to work hard. The story also illustrates that elder or senior status can also be derived from maturity of mind and not just from actual physical age (i.e., the twin which was born first).

Journey to Space (1980[c]) is a fantasy about two naughty children who go on a journey through the sky and the clouds because of a spell placed upon them by a mysterious stranger. Eze and Ngozi ignore the stranger's warning not to play in the lift/elevator: "You children, you have been warned, don't play in the lift. You could be trapped and nobody would rescue you" (Nwapa 1980[c]:1). The children ignore the warning and saucily decide to press a button on the control panel. Suddenly, the lift goes right through the roof of the apartment building and into the sky (2). On the journey through the clouds, Eze presses another button and a friendly dog suddenly appears in the lift and becomes their helpful companion (6). "Doggy" is traveling to the moon to find and bring back a precious stone which would enable the animals to live forever (10). Nwapa presents folklore about the moon through the device of a story within a story. "Doggy" tells Eze and Ngozi the tale of how animals once inhabited the moon where they did not die but lived forever. The animals left the moon because the humans who were exploring space were so annoying. The animals now reside in the "animal world," where they die (9-10).

Marvie Brooks

The tale of why the animals no longer live on the moon allows the author, through the voice of "Doggy," to pose the following philosophical query for young readers which encourages them to think and to ask other related questions: "The Earth people came to visit the Moon instead. They were foolish. Why come all the way from the Earth to the Moon? What were they looking for?" (11).

The children are rescued by a pair of fairies who intercept the space travelers (Nwapa 1980[c]:13-14). The fairies provide further directions for "Doggy" so that he can continue on his mission to the moon, and return the children safely back to earth and into their own apartment/flat (16). The fairies caution the children: "Now we are going to take you back to your own flat. Next time don't be naughty. Don't play in the lift" (20).

THE VARIED WAYS IN WHICH NWAPA INSTRUCTS

Nwapa uses ceremonies, proverbs, questions-and-answers, actions or behavior, direct speech, and personal reflection or thoughts of the characters in her stories to directly instruct or to imply what is the best mode of conduct for a young person to follow and what are the important values to internalize. A prominent mode of instruction for children in *Mammywater* (1979) is the question-and-answer technique which is used to give guidance, disclose essential information, reveal the circumstances behind the characters' actions and behavior, and to describe the personalities of the characters in the story.[8] Mammywater/lake goddess/mermaid for example, transforms herself into a huge fish and cunningly inquires if the astonished Deke would like to touch her. In the ensuing query-and-answer, Deke keeps his poise, remembers the warning of the old woman, and saves himself from being turned into a boy fish:

"You said I should be like a fish, I am a fish, and
you are afraid, will you like to touch me?

Deke replies in the negative twice and escapes transformation.

"No."
"I won't hurt you."
"No."
"My clever boy, if you had touched me, you would
have turned into a fish." (Nwapa 1984:19)

615

By asking questions and remembering the good advice of the old woman, Deke escapes the trap set by the wily Mammywater and is able to complete his mission.

Through Deke's behavior and thoughts, Nwapa teaches youth to be obedient and remember their home teaching even when subtly presented with an opportunity to do otherwise. For example, the old female votary of Mammywater did not explain to Deke the reasons behind the stern admonition: "But above all she warns, 'Don't ever touch her even if she says you can'" (8). Deke remembers and takes heed because he respects her age and wisdom. From Deke's actions, children learn that it pays to respect their elders, and that wisdom comes from "listening to and learning from elders who are custodians of both morality and wisdom" (Nwala, 1985:158).

Nwapa uses a traditional ceremony to teach good manners. Children learn the Igbo rites of hospitality and treatment of guests in the kola-nut ritual depicted in *Mammywater*. Deke observes that Urashi, the river spirit, serves kola-nut to his visitor, Mammywater, before the two spirits discuss the reason for their quarrel: "Let's have kola first, my good spirit. There, you over there, bring me kola," Urashi calls (Nwapa 1984:28). Obiora Udechukwu, the artist, devotes an entire page to the illustration of a fish bringing the kola-nut in a dish to the two distinguished spirits (28). Nwala points out that the kola-nut "is part of every reception of a visitor" (151). In "The Magic Cocoyam" (1980), through the words of Kele, Bird, and Mother Tortoise, Nwapa clearly shows children that hard work and discipline are necessary when you want to obtain an objective. Kele followed the advice of Bird and accepted the fact that he would have to toil very hard on Mother Tortoise's farm in order to find his kidnaped sister and learn how to farm so that he could teach his village. The author writes: "Yes, I carried your sister away," confirmed the Tortoise. "But you must work for me before you can take her away." Readily Kele said he was prepared to work (46-48).

Through the actions of Kele, Nwapa instructs youthful readers to keep their eyes open for the opportunity to be courageous. By remaining alert, Kele finds out the secret weapon which ultimately defeats Mother Tortoise and enables him to rescue his sister. "The Tortoise taught Kele how to plant vegetables and fruits. Yams, potatoes and cassava too. Kele learnt very fast" (48). However, by keeping his eyes open he realizes that Mother Tortoise did not show him how to plant a particular vegetable. "But the Tortoise did not show Kele the cocoyam. She did not want the boy to plant it. So

Kele was suspicious" (48). When Mother Tortoise goes away, the quick-witted and courageous Kele makes use of the opportunity to find the cocoyam and he secretly plants it. "He murmured to himself:'Perhaps the secret of the Tortoise's power is in the cocoyam. Who knows?'" (48).

The animals in this story assume human qualities and Bird serves as the personification of human wisdom. Nwapa demonstrates to youth, through the example of Kele, that obedience is important when an elder advises a young person how to take care of a problem. Kele obeys the following instructions from Bird to the letter: "If you work hard, you will learn how the Tortoise farms. You will not only win back your sister, Lala, but also teach your people how to farm and reap good harvest" (Nwapa 46).

Through the behavior of Emeka and the interaction between him and his friend, Sokari, in *Emeka—Driver's Guard*, Nwapa makes it explicitly clear to young readers that it is unwise to thumb one's nose at school and discipline. School is very important in Igbo society, for education is one of the means by which economic and social progress is made. Nzimiro states that "those parents who cannot read and write strive desperately to save their children from 'the curse' of illiteracy and those who have already benefited from modern education do not want their children to slide down the social ladder from the heights they themselves have attained" (Nzimiro 1971:175). Thus, Flora Nwapa is careful to allow youth to see, in Emeka's own words, just how he foolishly makes a bad decision:

> His aunt sat by him and put her arms around him.
> "Please, Emeka," she said. "Don't worry about the
> fees. I'll pay the fees if you go back to school and
> repeat the class."
> "Please, Emeka!" said Ngozi.
> "No."
> "What are you going to do then?" asked his aunt.
> "I'm going to be a driver's guard," he said without
> thinking. He had often talked about it with Sokari,
> but not seriously. (23)

In the end, Nwapa also makes sure that children are aware of the thoughts and behavior of a remorseful Emeka who reverses his former careless thinking and actions by bravely deciding to return home and go back to school no matter what it takes:

> But if Emeka went back to school, he would have
> to start in Standard Three again. People would
> laugh at him.
> "Why are you in Standard Three, when your twin
> sister is in Standard Five?," they would ask.
> Well, it didn't matter what people said, Emeka
> decided.
> He would go to see his aunt and ask if he could
> live at home again. (40)

The story also instructs young people to be wary in their choice of friends (Ezenwa-Ohaeto 1995:75) for Sokari proves to be Emeka's downfall from the first time that they meet. Sokari characterizes Emeka as a "farm boy" (Nwapa 1972:12) while he plays the part of the "smart town boy" who involves the good but weak Emeka in foolish mishaps. Unwisely, Emeka selects Sokari as his best friend and thus unwittingly takes the first step toward leaving school. Through the verbal interchanges and the behavior of the two boys, Flora Nwapa shows children how thoughtless friends can spoil another individual's good character. At their first encounter, Sokari unjustly earns Emeka a whipping from his aunt because he falsely accuses Emeka of picking a quarrel with him and hitting him. The gentle Emeka's true explanation of the incident is overpowered by the bold, yelling Sokari who convinces his mother and Emeka's aunt that it is Emeka who lies.

> "Emeka, are you telling the truth?" his aunt asked.
> "Yes, aunt."
> "No, you're not. You're lying!" Sokari shouted
> again.
> "Mother, he's lying!" (11)

Children also see that on account of the pattern of behavior chosen by his friend, Emeka begins the downward spiral that culminates in his decision to leave school. Sokari entices Emeka to skip classes and hang around the town. Nwapa explains, "One day, on the way to school, Sokari said, 'I'm not going to school today. I'm going to the river instead. Are you coming with me?'" (14) At first, his conscience bothers Emeka and he reminds Sokari that they should be mindful of the displeasure of the school authorities. However, Sokari replies that he is not afraid and tells Emeka that "We're not babies any more. Come on, Emeka! Let's go to the river!" (14) The next time that Sokari proposes skipping school, children begin to see the influence of his friend upon Emeka through the answer that Emeka gives. This time, Emeka does not hesitate to take Sokari up on his offer. "Emeka laughed. 'Yes, let's! I don't like school'" (15).

Journey to Space (1980) allows Nwapa to make use of the question-and-answer technique, that she skillfully employs in *Mammywater* (1979) to tease out hidden information, other relevant questions, actions, and circumstances so that the individual seeking enlightenment will have a clear picture of the problem before they act.[8] The query posed by "Doggy": "Why come all the way from the Earth to the Moon? What were they [earth people] looking for?" (Nwapa 1980:11) presents a wonderful opportunity for children and parents or teachers to explore together the issues posed by space exploration which has had a tremendous impact on youth since the 1960s when Edwin Aldrin, Jr. and Neil Armstrong landed on the moon on July 20, 1969 (Puttkamer 1992, 17:76, 97).

Betsy Hearne (1990:159) points out that "sweetness and light is an order long past in children's literature." Through the use of questions, Nwapa stimulates her young readers to begin to think about, to read and to study, and to pose other queries about global issues that have an impact upon the lives of people everywhere. Elizabeth O. Ashimole explains that "only Flora Nwapa and Naiwu Osahon have made attempts at recognizing the impact of technology in their books for children" (Ashimole 1992:73). "Doggy's tale" opens up other issues that children could explore, e.g., the purpose of space travel, the benefits and the cost of space colonization, the contribution of space exploration to science both pure and applied, the possibility of life on other planet(s), the use of space for military and security purposes, new jobs and markets created by space exploration now and in the future, the question of balancing priorities when planning the national budgets for domestic projects, such as medicine, education, and social welfare, and for space exploration (Puttkamer 1992, 17:71-106; White 1996;1:294-296).

PICTURES ENHANCE THE TEXT

Illustrations are important in all books for youth. The book selection principles outlined by Marshall include the benefits of good artistic quality in the pictures accompanying the text. For example, a good illustrator should make certain that the pictures "enhance the text,... increase visual perception,...provide visual information, and... tell the story in books with and without words" (Marshall 1982:98). Obiora Udechukwu, the well-known Nigerian illustrator, is empathetic to the mood and content of *Mammywater* and this can be seen in the choice of colors, style, pictorial treatment of the story, and placement of the pictures in relationship to the written text.

Nodelman, in *Words About Pictures: The Narrative Art of Children's Picture Books* (1988), discusses the emotional significance of particular colors, the level of intensity which the hue of a particular color may arouse, and the symbolic value of the hue of a color in terms of light and darkness (60-61, 64). The majority of the black pen-and-ink line drawings in *Mammywater* are overlaid with washes of vivid sky blue or lemon yellow. Nodelman points out that the color blue is associated with sophistication, calmness, and serenity as well as melancholy. The color yellow suggests cheerfulness. The author also points out that the cover of a book may also suggest the conflict and characters involved (Nodelman 1988:49).

The cover of *Mammywater* does foreshadow the approaching conflict between Soko, the young girl who is the sister of Deke, the hero, and the mermaid or lake spirit called Mammywater.[9] The mermaid occupies the top left of the picture and she is looking down upon Soko who occupies the bottom right of the picture. Soko is looking up at Mammywater and pointing towards her. A diagonal line could run right through the figures of the lake goddess and that of Soko. A vivid sky blue circle emanating from the figure of the mermaid encloses the two figures. Thus, it seems that a relationship between the two and the power of Mammywater over Soko is implied through the diagonal position of the two figures, the blue circle enclosing Mammywater and Soko, and the fact that the blue circle emanates from the lake spirit down to Soko (Nodelman 1988:125-157).

The illustrations by Obiora Udechukwu are black pen-and-ink line drawings which depict pattern, action, setting, and the inner feelings of the main characters. Most of the black pen-and-ink line drawings have washes of sky blue or lemon yellow. One illustration employs the color red. The color blue is used in the melancholy scene which depicts the unhappy Soko standing between the lake goddess, Mammywater, who wants to take her and her mother who reluctantly tells her daughter to go with the mermaid (Nwapa 1979:2). In contrast, the color yellow is used in the scene where the elderly woman, who worships Mammywater, tells Deke how to find Soko, the taboo to avoid, and gives him a blessing (Nwapa 1979:8). The same cheerful lemon yellow is used in the illustration which depicts the old votary appearing to Deke in a dream and telling him that he is doing well and that Mammywater is a good spirit (23). Dark red, which is intense and makes one feel hostile, warm, or excited (Nodelman 1988:63) is used in just one picture in *Mammywater* and that is the illustration of the angry, scaly-skinned Urashi, the river spirit, seated on his throne in his kingdom where there are big crocodiles and long snakes (Nwapa 1979:26).

The text's verbal description of the underwater palaces of Urashi, the river spirit, and Mammywater, the lake spirit, are brought to life in the illustrations. For example, there is a gorgeous illustration of Mammywater and Deke looking through a door made out of live fish, into a room filled with sumptuous cloth, "...silks, jorges, velvet, damask, lace, akwete, [and] abada" (18). The illustration depicts luxury by showing pattern on top of pattern using black and white plus a loose lemon wash or spot color to make decorative patterns of all types that blend and differentiate at the same time. As Nodelman (1988) points out, the details in the pictures tell much about the character and in this case it is the riches of the mermaid, Mammywater, which Deke must resist (105-108).

At the end, both the river and lake spirits appear in an illustration that is similar to the pictures on a cameo brooch. Just the face and neckline is shown. The reader is able to see their faces in full and their expressions (Nwapa 1979:32). Nodelman (1988) characterizes this type of picture as "the unguarded face" (285-286). Deke's humility and tearful plea for the return of his sister touch the two spirits, and the reader sees them conferring in whispers over the release of Soko and his father (Nwapa 32). It seems that the color yellow, symbolizing cheerfulness, serves as a portent of the mood of the two spirits toward Deke.

The illustrations in *The Adventures of Deke* (1980) are drawn by the same artist who worked on *Mammywater* (1979). However, there are some changes. The book and its illustrations are smaller in size than the picture-book format of *Mammywater*. Thus, in *The Adventures of Deke*, the pictures are on the same pages as the text which appears below. On the other hand, the illustrations in *Mammywater* are on a separate page (to the left) and the text is on a separate page (to the right). The colors yellow and blue are used for both books, but they do not always appear in the same pictures. One major difference is the color wash used for the picture of

Urashi, the angry river god. In *Mammywater*, this is the only picture with a color wash of red, but in *The Adventures of Deke* this picture has a color wash of yellow. The pictures on the covers of the two books are also different. The cover of *Mammywater* which shows the water goddess and Soko, the sister of Deke, foreshadows the conflict between the water spirit and Deke's family, as well as the feminine aspect of the worship of Mammywater/Uhamiri. In contrast, the cover of *The Adventures of Deke* depicts Deke striding down to the lake to dive in and find the underwater home of Mammywater and rescue Soko. Hence, the cover foreshadows the quest of Deke. The golden yellow of the cover and the blazing flaming red sun, which is reflected in the water of the lake, emphasizes the danger and the excitement inherent in Deke's task.

The Miracle Kittens (1980) is lively and full of rhythm which the illustrations reflect. The artist, Emeka Onwudinjo, uses black pen-and-ink line drawings that are bold, strong, animated, and humorous. The illustrations have washes of scarlet red and sky blue, which helps to keep the excitement and energy in the story moving. The dust jacket is also in this wonderful blue. The warmth of the family and the droll humor underlying their mishaps that occur with each attempt to get rid of the rodent problem comes through in the text and picture.

In "The Magic Cocoyam" (1980), the pen-and-ink line drawings by P. S. C. Igboanugo are vigorous, energetic, and detailed. There are excellent scenes depicting Kele and Lala in their village and the daily activities of village life: Kele's discovery of the cocoyam plant, and Kele's surprise upon hearing a mysterious voice telling him to use the cocoyam plant to kill the Tortoise. Furious movement, noise, and chaos are depicted in the illustration which shows Kele using the cocoyam plant to strike the frantic, scared Tortoise(Nwapa & Moneke, 1980:53). The sudden bursting apart of the Tortoise's huge shell sends flying pieces of the shell exploding like projectiles into the edges of the picture. This is in contrast to the prior confrontation scene (51) where the pompous Tortoise angrily demands her supper from Kele, and then immediately dissolves into fear and dismay as she realizes that Kele knows the dangerous secret of the magic cocoyam plant:

> "You good for nothing child, where is my supper?"
> "You wicked Tortoise, where is my sister?" replied
> Kele who stood in front of the Tortoise holding up
> the cocoyam leaf. He threatened to smack the

Tortoise with it. At that the Tortoise began to trem-
ble. (51-52)

Roslyn Issacs' striking black pen-and-ink line drawings with touches of red in *Emeka—Driver's Guard* (1972) portray in a sensitive
manner the inner turmoil and experiences of the main character
as he learns about decisions and consequences. The book's cover
foreshadows the problems which Emeka will face as a driver's
guard because it depicts the scene in the text where Emeka is bul-
lied by an older boy who also works as a guard (30). The illustrations by Chinwe Oraeki in *Journey to Space* (1980) are lively black
pen-and-ink drawings with a color wash in earth tones of
orange/brown. The book's cover is in this same color. The drawings give the sensation of floating through space, and the faces of
the children reflect the fear and wonder of their adventure.
"Doggie," the talking canine, is depicted as a charming, friendly,
knowledgeable companion who helps Eze and Ngozi to relax and
to question man's exploration into space. Most children would welcome him as an ideal pet. The artist portrays the three companions looking out of the window of the magic lift at the sky and
clouds (Nwapa 1980[c]:9). Children will long remember the
enchantment scene which shows the mysterious stranger/wizard
with puffed-out cheeks expelling a mighty breath of air that sends
the lift through the roof into the sky (2). The fairies who rescue
the children have big wings which have a wash of blue/orange.
Their wings and their rescue in space of the children suggest more
angels than fairies. The fairies are depicted as cute, robust, lively,
"no nonsense" types, who are active workers, and not ascetic,
ephemeral, otherworldly creatures.

CONCLUSION

The family, societal, and achievement values, such as obedience,
modesty, endurance, discernment, courage, respect for elders,
helping the less fortunate, education, responsibility for pets, love
and care of family members and respect and equal opportunity for
male and female are depicted in Flora Nwapa's books for juveniles.
The stories are not "preachy." Youthful readers will enjoy good stories, in a Nigerian setting, that are filled with adventure, excitement, humor, folklore, family life, pets, and characters that are
realistic.

The stories will help young people to extend their experience
beyond their immediate environment, to redefine their sense of

the variety of choices available in life, and to develop an appreciation of the relationship between the method chosen to solve a problem or reach a goal and the ensuing desirable or undesirable outcome. The stories also provide Nigerian children with an understanding and appreciation of the culture of Igbo people. The books depict characters who employ Igbo principles, values, and rituals. Hopefully, Nigerian youth will learn that many of these principles are still valuable in today's society. Through her children's literature, Flora Nwapa enables Nigerian youth to avoid cultural anomie. Nwapa aids them in the socialization process, and at the same time gives them the joy of reading a good story.

ACKNOWLEDGMENTS

I am grateful to Professors Chikwenye Okonjo Ogunyemi, Ted Pontiflet, and Kema Chikwe for their suggestions and feedback.

NOTES

1   Many thanks to Professor Kema Chikwe for help with the wording of this title.
2   The following children's books by Flora Nwapa are not included in this essay because they are concept books, without words, for very young readers and do not contain the themes treated in this article: *My Animal Number Book* (1981), *My Tana Alphabet Book* (1981), *My Animal Colouring Book* (1979), and *My Tana Colouring Book* (1979). The titles are printed by Flora Nwapa Books and Company.
3   The information about mermaids is rich and interesting. H. L. Stowe explains, in *The World Book Encyclopedia*, that a mermaid puts a magic cap on the head of the man that she wants and this enables the human to live in the sea with her (Stowe, 13:343). Jobes states, in the *Dictionary of Mythology, Folklore, and Symbols*, that "In popular tradition [the mermaid] symbolizes beauty, regeneration, sea, sea mother, seduction, [and] wisdom" (Jobes, 2:1093).
4   The worship of Mammywater/Uhamiri, and her physical appearance plus the benefits which she bestows is discussed in *Efuru* by Nwapa (1978:146-154,165,195,199,201-202).
5   *The Adventures of Deke* (1980) depicts the two reunited, hugging children in the middle of the lake; whereas in *Mammywater* the artist does not show Soko either wet, brushing off water from her person, or in the middle of the lake when Mammywater first confronts her while she is "swimming in the lake" (Nwapa, 1979:1-3).
6   See Emmanuel Obiechina's *Literature for the Masses: An Analytical Study of Popular Pamphleteering in Nigeria* (1971), p. 70, qtd. in Nwala, 1985:207-208.
7.   See Otonti Nduka's *Western Education and the Nigerian Cultural*

Background (1964), p. 91, qtd. in Nwala, 1985:161.

8. Ifeoma Okoye analyzes Nwapa's use of the question-and-answer technique in her children's books, e.g., *Mammywater* (1979) and *The Adventures of Deke* (1980). Okoye states that "it is an embodiment of a well-known Igbo proverb: one who asks questions never gets lost" (Okoye 2). See "Flora Nwapa's Legacy to Children's Literature" by Ifeoma Okoye in this volume.

9. For a thorough and sensitive analysis of the character, the physical appearance, and the benefits which flow from Uhamiri/lake goddess/mermaid, and Flora Nwapa's treatment and usage of Uhamiri in her adult works, see "Flora Nwapa: Genesis and Matrix" (131-182) and "The Mammywata Myth as Gendered Insurance" (29-35) in Chikwenye Okonjo Ogunyemi's *Africa Wo/Man Palava: The Nigerian Novel by Women* (1996). In Flora Nwapa's home town of Ugwuta, most of the people call the lake goddess by the name of Ogbuide. The less used name of Uhamiri is the one fictionalized by Nwapa in her novels (Ogunyemi, 1996: n. 13, 140). Mammywater combines two concepts: the Igbo goddess tradition and the physical appearance and role of the biracial female children born to Igbo women and white colonial rulers. Many times, these female children help to bring good fortune to their families through the relationship that they had with the whites. Thus, the physical appearance of Mammywater, as depicted by Obiora Udechukwu in the children's books entitled *Mammywater* (1979) and *The Adventures of Deke* (1980) is that of a biracial female. However, the powers which she holds and the favors which she can bestow derive from the Igbo traditional belief system regarding the female water goddess (Ogunyemi, 1996:30-31). Now, Mammywater is no longer associated with a biracial person for any beautiful young lady may be dubbed a "Mammywater" (Ogunyemi, 1996:30). For example, in *The Joys of Motherhood* (1979), Nnu Ego, who is brought by her brother-in-law from Ibuza to Lagos where Nnaife her new husband lives, is called "Mammywater" by the neighbors. "The cooks and stewards and many of the Ibuza people living in that area came to congratulate Nnaife, and to remark that they had sent him a "Mammy Waater," as very beautiful women were called" (Emecheta, 1979:43).

Works Cited

Agbasiere, Julie. "Social Integration of the Child in Buchi Emecheta's Novels." *Children and Literature in Africa*, Ed. Chidi Ikonne, Emelia Oko, and Peter Onwudinjo. Ibadan, Nigeria: Heinemann, 1992:127-137.

Ashimole, Elizabeth O. "Nigerian Children's Literature and the Challenges of Social Change." *Children and Literature in Africa*, Ed. Chidi Ikonne, Emelia Oko, and Peter Onwudinjo. Ibadan, Nigeria: Heinemann, 1992:70-81.

Bleeker, Sonia. *The Ibo of Biafra*. New York: William Morrow, 1969.

Cullinan, Bernice E. *Literature and the Child*. 2nd ed. New York: Harcourt, 1989.

Emecheta, Buchi. *The Joys of Motherhood*. New York: George Braziller, 1979.

Encyclopedia Americana International ed. 30 vols. Danbury, Connecticut:

Grolier, 1993. 18, 727.

Ezenwa-Ohaeto, "The Child Figures and Childhood Symbolism in Nwapa's Children Fiction." *Research in African Literatures.* 26 (1995):68-79.

Hearne, Betsy. *Choosing Books for Children: A Commonsense Guide.* New York: Delacorte, 1990.

Idewu, Olawale and Omotayo Adu. *Nigerian Folktales.* Ed. Barbara K. Walker and Warren, S. Walker. New Brunswick, N.J.: Rutgers UP, 1961.

Ikonne, Chidi, Emelia Oko, and Peter Onwudinjo, eds. *Children and Literature in Africa.* Ibadan, Nigeria: Heinemann, 1992.

Jobes, Gertrude. *Dictionary of Mythology, Folklore and Symbols.* 3 vols. New York: Scarecrow, 1961-62. 2:1093

Kalu, Wilhelmina and Ogbu U. Kalu. "Nigeria." *International Handbook on Gender Roles,* Ed. Lenore Loeb Adler. Westport, Connecticut: Greenwood, 1993. 228-243.

Kurian, George Thomas and Graham T. Molitor, eds. *Encyclopedia of the Future.* 2 vols. New York: Simon & Schuster, 1996.

Maquet, J. "Totemism." *Dictionary of Black African Civilization,* Ed. Georges Balandier and Jacques Jerome Pierre Maquet. New York: Leon Amiel, 1974:317.

Marshall, Margaret R. *An Introduction to the World of Children's Books.* Aldershot, Hants, UK: Gower, 1982.

Nduka, Otonti. *Western Education and the Nigerian Cultural* Background. Ibadan: Oxford UP, 1964.

Nodelman, Perry. *Words About Pictures: The Narrative Art of Children's Picture Books.* Athens, Georgia: University of Georgia P, 1988.

Nwala, T. Uzodinma. *Igbo Philosophy.* Ikeja-Lagos, Nigeria: Lantern, 1985.

Nwankwo, Chimalum. "The Igbo Word in Flora Nwapa's Craft." *Research in African Literatures.* 26 (1995):42-52.

Nwapa, Flora. *The Adventures of Deke.* Enugu, Nigeria: Flora Nwapa Books, 1980[a].

____. *Efuru.* Reset ed. Portsmouth, New Hampshire: Heinemann, 1978.

____. *Emeka—Driver's Guard* London: University of London P, 1972.

____. *Journey to Space.* Enugu, Nigeria: Flora Nwapa Books. 1980[c].

____. "The Magic Cocoyam." *Journey to Space and Other Stories,* Ed. Flora Nwapa and Obiora Moneke. Enugu, Nigeria: Flora Nwapa Books, 1980.

____. *Mammywater.* Enugu, Nigeria: Flora Nwapa Books, 1979.

____. *The Miracle Kittens.* Enugu, Nigeria: Flora Nwapa Books, 1980[b].

Nzimiro, Ikenna. "The Igbo in the Modern Setting." *The Conch* 3 (1971):165-179.

Obiechina, Emmanuel. *Literature for the Masses: An Analytical Study of Popular Pamphleteering in Nigeria.* Enugu, Nigeria: Nwamife, 1971.

Odejide, Abiola. "The Abused Child Motif in Nigerian Children's Realistic Fiction." *Children and Literature in Africa,* Ed. Chidi Ikonne, Emelia Oko, and Peter Onwudinjo. Ibadan, Nigeria: Heinemann, 1992. 150-158.

Ogunyemi, Chikwenye Okonjo. *Africa Wo/Man Palava: The Nigerian Novel by Women.* Chicago, Illinois: University of Chicago P, 1996.

Puttkamer, Jesco von. "Space Flight." *McGraw-Hill Encyclopedia of Science & Technology.* 7th ed. 20 vols. New York: McGraw-Hill. 17(1992):71-106.

Stowe, H. L. "Mermaid." *The World Book Encyclopedia.* 7th ed. 22 vols. Chicago, Illinois: Field Enterprises Educational Corporation. 13 (1972):343.

White, Frank. "Extraterrestrial Life—Forms." *Encyclopedia of the Future,* Ed. George Thomas Kurian and Graham T. Molitor. 2 vols. New York: Simon & Schuster, 1996. 1:294-296.

PART SIX
CONVERSATION(S)
WITH FLORA NWAPA

The character of an author, perceptions of self, complexities, and attitude toward her work, are evoked through spontaneous conversation. Interviews give shape to an author's vision not readily available in her complete *oeuvres* or even an (auto)biographical work. Interviews are glimpses of an author's inner-feelings through a discussion of the creative process—Why she writes? When did she start writing? Who influenced her writing? What inspired her to write a particular work? Where do her ideas originate? and What is the role of the artist in society? Because interviews show where the meanings are by providing the missing links, interviews today have become a valid genre, a literary art form.[1]

In the interviews included in this section of the anthology for literary enrichment and knowledge, author Flora Nwapa recalls her happy childhood in Ugwuta, her ancestral home. She remembers listening to the stories of the local griottes where she learned the community's traditional practices and culture. Nwapa also shares the routines of her studious life at C. M. S. Girls' School in

Elelenwa, near Port Harcourt and the University College, Ibadan, where she mastered the English classics. The "ups" and "downs" of being a woman writer, and publisher of adult and children's books, coming from the horse's mouth, make all three interviews invaluable.

The best of interviews promote a reliable assessment of an author by exploring the recesses of the subconscious. In the beginning of her literary career with the publication of *Efuru*, Flora Nwapa (re)creates in her fiction, a water deity, Ogbuide, as a powerful goddess who gives women wealth but not children. In her last novel, *The Lake Goddess*, written thirty years later, Flora Nwapa characterizes Ogbuide as the "Queen Mother . . . Good Mother,"[2] who gives women power, wealth and children. Sabine Jell-Bahlsen in "An Interview With Flora Nwapa," gets to the bottom of this contradiction: "[My] interview, together with her novels, reveals Nwapa's deep-seated ambivalence, not only toward Western intrusions, but also toward the Lake Goddess, Ogbuide/Uhammiri herself, her values, gifts, ethics, social codes and requirements."[3] Thus Jell-Bahlsen unravels the subconscious layers of Nwapa's thoughts and ideas.

Apart from touching on the subconscious, conversing with the subject also points to her creative muse(s). Theodora Akachi Ezeigbo's interview, "A Chat With Flora Nwapa," provides valuable insight into the different ways Author Nwapa derived her sources of inspiration. Flora Nwapa prides herself on painting authentic heroines of the Nigerian women she listened to and the people she came into contact with during her various roles as wife, mother, teacher, administrator, commissioner, writer and publisher of Nigeria's first feminist press. So naturally she attributes her plots to the vicarious experiences she witnessed as a citizen of Nigeria:

> A story somebody told me had a great impression on me and I started to write *One Is Enough*. In four months, it was done. *Cassava Song* came to me suddenly. A woman, a beggar, came to Oguta and sang in praise of cassava. In my office, I remembered the woman and her song. The urge to write the story of cassava in verse caught my imagination and I finished writing it in two days. *Mammywater* came like that too. Michael Crowther came to see me. I think it was in 1976 in Enugu. He wanted to see Oguta and we went to see the lake. We took a canoe to Urashi

River (which flows into the lake) and there we saw
the gun-boat sunk by the Biafran Navy. The impression I had after that visit to Lake Oguta made me
write *Mammywater.*

The fact that Nwapa made use of such inspirational encounters
added to her success as a writer and her fame as an insightful
author who made history.

But how does a historically significant writer who has made it
see herself? What path(s) does she advise others to take who may
want to follow her footsteps? Would she take that road again? In
her ground-breaking tête à tête with Marie Umeh, "The Poetics of
Economic Independence for Female Empowerment: An Interview
With Flora Nwapa," the distinguished author puts Ugwuta on the
map as one of the few communities in Igboland where the androge-
nous nature of male and female is respected and encouraged:

> . . . In Ugwuta, women have certain rights that
> women elsewhere, in other parts of the country, do
> not have. For instance, in Ugwuta, a woman can
> break the kola nut where men are.[4] If she is old, or
> if she has achieved much or if she has paid the bride
> price for a male relation and that member of the
> family is there, she can break the kola nut. And
> everybody will eat the kola nut. But in certain parts
> of Igboland, a woman is not even *shown* a kola nut,
> not to talk about *breaking* it.

Flora Nwapa attributes her extraordinary life, as Africa's first internationally acclaimed female writer of adult and children's books,
as the publisher of Africa's first womanist press, Tana Press Limited
and Flora Nwapa Books and Company, and as the Mother of
African Women's Writing, to her passion for story telling.[5] As you
read and listen to the following interview- encounters may you be
transformed.

Marie Umeh

Notes

1. See Christopher Silvester's excellent introduction to his book, *The Norton
   Book of Interviews* (New York: Norton, 1996), p. 35.
2. These phrases are taken from Flora Nwapa's last novel, *The Lake Goddess.*
   (Unpublished manuscript) p. 247.
3. Sabine Jell-Bahlsen makes this assertion in her "Interview With Flora
   Nwapa," which appears in this volume.

4. In a greater part of Igboland, women do not break kola nut. Kola nut is a status symbol and male preserve; it symbolizes peace, happiness, trust and friendship. However, according to Flora Nwapa, prominent, influential women in the Ugwuta area of Igboland have this right. Perhaps it is because of the power and influence of the mystical Lake Goddess, Ogbuide. See Marie Umeh's "The Poetics of Economic Independence for Female Empowerment: An Interview With Flora Nwapa," which appears in this volume.
5. Charlotte Chandler points out in her book, *The Ultimate Seduction,* that a vital love for one's work is the main ingredient for success. See her introduction, pp. 20-23.

WORKS CITED

Chandler, Charlotte. *The Ultimate Seduction.* Garden City, New York: Doubleday, 1984.

Nwapa, Flora. *The Lake Goddess.* (Unpublished manuscript)

Silvester, Christopher. *The Norton Book of Interviews.* New York: Norton, 1996.

An Interview with
Flora Nwapa[1]

Sabine Jell-Bahlsen

Flora Nwapa's relationship to her hometown Ugwuta's[2] patron deity, the lake goddess Ogbuide/Uhamiri,[3] evolved over the years. This is evident from Nwapa's novels, as well as from an interview, I conducted with Flora Nwapa, together with Chief Francis Ebiri* of Orsu-Obodo/Ugwuta II, in 1988.

At the time of the interview, I was involved in field research in and around Ugwuta studying the goddess, her worship, and her devotees. In addition, I was preparing the production of a documentary film on Mammy Water.[4] I had been invited by the priestess, Eze Mmiri di Egwu of Orsu-Obodo, during my first visit to Ugwuta in 1978. I was immediately fascinated by this powerful woman and her followers, and embarked on a long journey studying the goddess's worship and rituals. Chief Francis Ebiri has been my research assistant and translator all along, since we first met in Ugwuta in 1978. He is the son of the late Ebiri Obua of Orsu-Obodo, a widely known, highly respected diviner, herbalist, and priest of the lake goddess Uhammiri. Ebiri Obua has healed, initiated, and installed many of the current priests and priestesses of

---

* Chief Francis Ebiri unexpectedly passed away at the age of 50 in his home on September 3, 1996.

Ogbuide/Uhammiri, including the Eze Mmiri. Chief Ebiri Obua has passed on much of his knowledge to his son, Chief Francis Ebiri, who nevertheless is neither a herbalist, nor a diviner, or a priest of the goddess.

At the time of our interview, Flora Nwapa and her husband, Gogo Nwakuche, were living in Enugu. There she had established her publishing company, Tana Press, Ltd. Despite her fame and busy schedule, Nwapa frequently visited her hometown, Ugwuta. The interview was conducted in Nwapa's office at Tana Press in Enugu.

During the 1988 interview, Flora Nwapa's knowledge of the lake goddess, Ogbuide/Uhammiri, and her worship turned out to be somewhat limited. I was surprised to find out that Nwapa's information about the goddess, her worshippers, and her rituals was apparently based primarily on stories and hearsay rather than on first-hand ritual or religious experiences. And Nwapa's first encounter with the Eze Mmiri, a renowned priestess of Uhammiri in Orsu-Obodo, occurred only two years after our interview, when Nwapa was asked by an American professor to interview a real priestess of the lake goddess. In 1990, the Eze Mmiri, a village woman of modest means, told me how she had been honored for the first time by a personal visit from the world-famous writer.

In our interview of 1988, Nwapa revealed her own painful awareness of the extent to which the European impact on her parents' Christian faith and education had divorced herself from Ugwuta's pre-Christian religious beliefs and traditions. As a child, Nwapa was not permitted to attend those activities; her parents and their foreign mentors deemed them "pagan rituals." Ironically, these rites make up the life blood of Ugwuta's culture, touching upon the local people's dearest and most sacred values, as they evoke a host of local gods and goddesses and their rules and rites. Ugwuta society is socially stratified. Ugwuta's ordinary people focus on the goddess of their lake, Ogbuide, Eze Nwanyi, even today. Flora Nwapa belonged to a different class. Her parents had built a plantation and made a fortune during Ugwuta's early heydays, trading in palm oil with John Holt and other European companies. This commercial exchange between European envoys and local enterpreneurs was accompanied by the impositions of European colonial rule and missionary activities. Flora Nwapa was a child of this uneven union. On the one hand, Nwapa benefited directly from her parents' trading relations and the resulting wealth by receiving an excellent education. On the other hand, this gift turned out a mixed blessing endowing her with European values, while also denying her full access to the "university of the village." As a small child, Nwapa spent much time in her "pagan" grandfather's compound. Yet, in 1988, she clearly distinguished herself from the villagers and "their" beliefs. This distinction came with regrets though, as Nwapa pondered about her missed opportunities: she could have established a personal shrine for the goddess Uhammiri/Ogbuide,[5] were it not for her Christian education.

Today, as in Nwapa's youth, the local people of Ugwuta still believe in what Nwapa had heard about in stories but was not permitted to worship herself: Ogbuide, the Lake Goddess. Nwapa recollects a special event when her "pagan" paternal grandfather was performing a sacrifice, most likely to his ancestors. Although the village children would normally partake in all of these local events, this was a special event for Nwapa, because her grandfather did not allow other children to attend. Nwapa herself had sneaked away from her own Christian home to afford participation.

The ritual Nwapa recollects in the inteview took place in her grandfather's compound, that is, in a domestic setting, as is common for sacrifices to the male ancestors. By contrast, rituals for the —female—lake goddess Ogbuide would normally take place either in a priest's, or a priestess's compound, or at special sacred locations outside of the town. It is highly unlikely that Nwapa, a child of a wealthy and distinguished family, would have been able to venture out of town without her parents' knowledge or permission. Indeed, Nwapa did not describe a single ritual event focusing on Uhammiri/Ogbuide.

As a child, Nwapa had picked up a lot of stories about the lake goddess in her grandfather's compound. But as her schooling progressed, so did her rooting in Christianity, while her "pagan" connections dwindled, a process she describes in *Efuru:* "When your parents send you to school, you automatically became a Christian" (103). In a similar vein, Nwapa's last, posthumously published book, *The Lake Goddess,* also vividly recalls how the heroine, Ona, suffers as a student who is drawn to the lake goddess, Ogbuide/Uhammiri, but is trapped at the mercy of a fanatic Christian teacher.[6] In her later years, Nwapa was increasingly aware of the mixed blessing of (her own) Christian education and its propagating of foreign values. She knew that without that education, she might not have become the world-famous author she was. On the other hand, Nwapa also recalls in the interview the pain she felt at the end of the Nigerian civil war, when her townspeople were *ordered in church not to believe,* as they did, that Ogbuide/Uhammiri, the lake goddess had saved them from invading soldiers.

The interview, together with her novels, reveals Nwapa's deep-seated ambivalency, not only towards Christian intrusions, but also against the lake goddess, Ogbuide/Uhammiri herself, her values, gifts, ethics, social codes, and requirements. While embracing the gift of wealth and success, commonly ascribed to Uhammiri, Nwapa questions the idea of the deity's other gift, the gift of children,[7] the blessing cherished most by Ugwuta people (male and female alike).

In *Efuru,* Nwapa raises doubts about the goddess, Uhammiri and her blessings. Some of Nwapa's characters question not only whether Uhammiri has children of her own, but also whether the deity would grant children/fertility, and by extension, whether the gift of children really represents a woman's ultimate desire. This proposition is almost unspeakable to local ears.[8] An ambitious woman endowed with money and higher education, Flora Nwapa strove to achieve more in life than the average Nigerian woman, or a "mother of many." In her understandable and highly contemporary desire for professional success and appreciation beyond the home, Nwapa inadvertently challenges a core value of Ugwuta society, customarily ascribed to the lake goddess.

Troubled by what Nwapa wrote about Mammywater/Uhammiri and children in *Efuru* and in *Mammywater,* I kept repeating this question during the interview. At first Nwapa responded naively, so it seems that "local people did not associate Uhammiri with children." Chief Ebiri, a man born, raised and living locally was surpised, shocked, and gasping in disbelief, when he heard what Nwapa said about "local beliefs." But, when I told Nwapa what *foreigners* had written about Mammy Water—that Mammywater was supposed to be foreign herself—it was Nwapa's turn to gasp in disbelief. At first, she thought that we were referring to a foreign John Holt worker who had "seen Mammywater." But when I went on to explain to her that foreigners had suggested that Mammywater was *herself* alien, Nwapa became very clear and firm. She knew and candidly explained two basic points about Ugwuta's lake goddess, Mammywater: 1) Mammywater's local name is Ogbuide/ Uhammiri. She is a local deity and very authentic. 2) *Water is the life giving thing* (in Africa). As everyone in Ugwuta knows, the water goddess of the lake gives life, not only to plants and animals, but above all to humans in the form of children. The goddess

Uhammiri/Ogbuide is the Goddess of wealth *and* children, originally considered to be one. Nwapa knew this very well, even
though *she chose not to propagate this (cultural) message* in her earlier work.

Ironically, missionaries aiming to destroy local beliefs and the
loyalties of local people in their "pagan" deities would also take
resort to introducing doubts about the merits and powers of these
gods' and goddesses' abilities to give to people the most precious
gift: children. However, Nwapa's deconstruction of the goddess's
gifts of children was personal and genderized in motivation, rather
than consciously serving the needs of a foreign organization, such
as the church. After the Nigerian civil war, the lives of the heroines of Nwapa's novels and of the author herself, temporarily
moved away from Ugwuta and its lake goddess, as *Women are
Different* and *One is Enough* are set in Lagos. Yet, Nwapa's interest
in Ugwuta's Lake Goddess grew, together with her awareness of
the conflicts and crises introduced by foreign beliefs and interventions, even as she had distanced herself from home. By 1988,
at the time of our interview, Nwapa had begun to rediscover local
lore. 1988 may actually have been a turning point in Nwapa's own
life, her interests and inquiries. From the interview, it becomes
evident that Nwapa was reflecting upon her evolving relationship
to the goddess of the lake, Ogbuide/Uhammiri. Shortly afterwards,
Nwapa ventured to interview the Eze Mmiri of Orsu-Obodo, a major
priestess of Uhammiri, in 1990. In 1988, Nwapa was still ambivalent about the goddess and her benefits to women. But she (re)con-
structed the deity's image in her last novel, *The Lake Goddess,*
written after her encounter with the priestess. In this posthumously published work, Nwapa clearly voices her concerns about
the conflicts created between local and foreign values. These conflicts are tearing apart peoples' minds and lives; they also drove
the heroine of *The Lake Goddess,* Ona, the gift of the goddess, to
the brink of madness. Nwapa's growing concern with the conflict
between the double-edged benefits of local custom and its lake goddess to women on the one hand, and the equally double-edged benefit of Christianity and foreign values, on the other, is evident
throughout the 1988 interview.

N: As a child stories fascinated me.
S: Just stories?
N: Yes, just stories. They are mysterious; we listen. I really listen;
I sit down and listen before going to sleep, you know, late at night.

Others would go to sleep but I would listen to the story teller. That's where I learnt about Mammywater.

S: It's from those stories that you learnt about the goddess?

N: I definitely heard about her.

S: You are from Ugwuta?

N: I am from Ugwuta.

S: And of course, in Ugwuta, Mammy Water has a local name, like in so many places.

N: Yes. It's Uhammiri. It's Ogbuide.

I commissioned that painting. (She points at a large painting behind her desk.) I commissioned Uche Okeke to paint it for me.

S: I can see the lady with long hair.

N: That lady is supposed to be Ogbuide.

S: Aha. She has long hair.

N: Yes, it's supposed to be Ogbuide. The lady in the shadow is Ogbuide. But the lady with the necklace and the earrings is supposed to be Efuru.

S: Okay.

N: I commissioned him to paint Efuru in her glory.

S: Is that Efuru from your book, *Efuru?*

N: Efuru, hm. Yes, Efuru in her glory. Hm-mm.

S: When was this painting done?

N: It is nearly ten years old now. Each and every day it makes more meaning to me as the years are passing. I just sit there and then I watch it. Maybe it inspires me.

S: Yes. And I can see her wearing the costume of the Mammy Water worshippers. She is wearing the white dress.

N: Aha. That is what she is wearing.

S: Ogbuide is important for Efuru in your novel.

N: Yes.

S: Can you explain the importance of Ogbuide to Efuru?

N: What is the importance? Efuru from no fault of hers had had two unsuccessful marriages. Her only child dies. Her husband accuses her of adultery. I think that is the one that she could not take. Because in our tradition, the worst thing that can happen to a woman is for her husband to accuse her of adultery. It is a very serious offense. And if she is innocent, she invariably leaves her husband, especially if she is proven innocent. And this is exactly what Efuru did. That is, she did what was expected of her. That is, she went to the shrine and swore that she was not guilty of adultery. This question of adultery came up because of her illness. She became very ill and nobody knew what it was. So people said, "Oh, she is very ill, this is because she did this." But she knew she was innocent. Any man who accuses you of adultery, you would not want to stay with. So when all these things had happened, Ogbuide, the Woman of the Lake, started appearing in her dreams. And she began to think that, oh after all, I am not actually of this world. I think I should do better if I become a worshipper of the Woman of the Lake.

S: So, Ogbuide, or Mammywater, as the English say, could be seen as somebody who will help a woman in her times of need?

N: Aha. Certainly, certainly in times of need.

S: And compensate her for losses?

N: Exactly.

S: And hardship?

N: Exactly. And another factor is that when an Ugwuta woman is prosperous, it is associated with the Woman of the Lake. Wealth is associated with the Woman of the Lake. And she herself thinks that it is not her own doing, that it is the Woman of the Lake who is doing it for her. She begins to worship the Woman of the Lake on her own without anybody telling her to do so.

S: I keep hearing in different places that Mammywater, or Ogbuide is very much associated with wealth....

N: Yes.

S: ... and with children.

N: But then, during the time of Efuru, those I know who worshipped the Woman of the Lake invariably did not have children. They did not have children of their own. I don't know why this was so. Mammywater was not associated with children. In the last pages of my book, *Efuru,* I was saying this because it was something that I could not understand. It is the medical doctor talking to Efuru:

> "Where is Adizua? Any news about Adizua?"
> "I have heard nothing about him. To me he has
> been dead years ago."
> "So he did not return?"
> "No he did not return."
> The doctor shook his head.
> "How is your wife? I have not asked."
> "Oh she is well. I left her in the country of the
> white people."
> "All alone?"
> "She lives with an elderly woman who takes great care
> of her and her two sons."

"That is good. I think I should be going," Efuru says, getting up.

"I think you should consider going back to your husband, Efuru."

"It is not possible."

"Let day break, Efuru."

"Let day break." Efuru slept soundly that night. She dreamt of the Woman of the Lake, her beauty, her long hair and her riches. She had lived for ages at the bottom of the lake and was as old as the lake itself. She was wealthy, she was beautiful. She gave women beauty. But she had no child. She had never experienced the joy of motherhood. Why then did the women worship her?" (281)

That was in my last page, the last sentence of my book.

S: Yes. I remember that.

N: Why did the women worship her, if the women are so preoccupied with bearing children? Our culture demands that a woman must have a child, and if a woman does not have a child, she is not fullfilled. But here is this Woman of the Lake who was never associated with having children deified by the community. Why then do the women worship her when she has no children? Do they worship her because she brings them wealth? Is it just wealth that they want? Or, do they also want children? But then, if they want children, which is so important in their culture, why do they worship her? But at that time, nobody associated the Woman of the Lake with children. No one. No one at the time when *Efuru* was written and when I was growing up.

E: [Gasping in disbelief.]

S: So the historical time of your book setting is ...?

N: It is when I was growing up. That was in the late thirties and early forties.

E: And by the time Efuru became a woman associated with the Woman of the Lake, how old was she? I want to know if perhaps she was above the age of getting pregnant.

N: Yes. I understand your question. The period seems to be 1940-1950.

E: Okay.

N: That is the period I had in mind. I was born in 1931, and at that time, Efuru got married to her husband. Efuru should have been about sixteen. Efuru could not have been younger than fifteen. She was about sixteen.

S: That was the normal time to get married at the time? So she did not go to school; so she could marry early?

E: What was the age when she started having the dispute with her husband?

N: Ah, she must have been a very young person, if she was sixteen when she married [laughs]. The story takes place around 1930-35. Don't you think so?

S: Yes. And she had had a child who died.

N: Yes. So you see that was the period I was writing about. And remember the period was not after the Second World War. The Second World War ended in 1945.

E: Yes.

N: Then there was not much emphasis on wealth and material things as we have it today. There were no cars then, no Mercedes, no radios, no TVs, nothing like that.

S: [Laughs.] Were not children equivalent to wealth in the old days?

N: Yes. You had farms in Ugwuta at the time with wealthy women who traded with UAC, John Holt, etc., in Ugwuta. And that was the time that I was writing about in *Efuru*. You see, *they* never associated the Woman of the Lake with child-bearing.

E: Giving children.

S: At this time wealth is considered money. But then, before that time, wealth was children!

N: Wealth was children, then.

The first important achievement for a woman was to have children. If she did not have children, it was such a handicap. And what she did at that time was to get young people to stay with her. It was slaves before, buying slaves, bringing them in her household and then the slaves would have children. I think this practice was accepted then.

E: Well, she has made the wealth also.

N: Aha, because only a wealthy woman who was barren could buy slaves. That was not possible for the poor. So we are talking of the well-to-do, the rich woman.

S: A change in the notion of wealth has taken place.

N: Yes.

S: In the very old days, somebody would be considered wealthy if she had many children. Then later on, a woman could be wealthy just from money.

N: Yes from money, that is, from material possessions.

S: Today, a person may actually be caught up in a conflict between money and children. Because as you pointed out earlier, somebody may find it hard to do trading and bring up small children.

N: Yes, if you start making money, people would say in Ugwuta even today, "Ah, ah, please, get married first. Have one or two children and then, you can go on to make money. Time passes as they say. Not that the time passes biologically, but they claim that if your mind is all set on wealth, then "children will run away." [She laughs.]

E: I think age *is* a factor in this case.

S: I mean what you are addressing is a very universal problem relating to any nation or country.

N: Exactly.

S: Either making money or getting ahead professionally, and then having children ....

N: Yes.

E: And you see, one other thing is that a woman who is not with her husband, and who is a worshipper of a goddess like Ogbuide might prefer to keep herself reserved and in that case, child-bearing would not come her way.

N: No, [it] would not come her way.

E: I am looking at it from a different angle, a man's point of view. In the case of Efuru....

S: There may be a contradiction between looking for money and success and having children. I mean it is a lot of pressure in a woman's life.

N: Yes.

S: So, is Mammywater the one who sort of compensates or helps a woman to sustain the pressure?

N: It depends on the woman herself. Some say that Mammywater chooses them, like in *Efuru.*

S: Yes.

N: When she goes to bed, she dreams, and all that. She tells her father. Her father says this is the sign. Mammywater has chosen you to be one of her worshippers. So in that case that was what Efuru was doing. Efuru, she was now past child-bearing, remember, and she had already "made it," she was wealthy. All she wanted from Mammywater was that peace of mind. It was just the peace of mind that she wanted. Of course, you have in these modern days the material things. Many women will voluntarily seek the help of Mammywater without feeling that they want to be called. In those days, these people who were called behaved in an extraordinary way.

S: You mean they were deviant from normal women?

N: Yes. When a woman would just suddenly begin to sing Mammywater [songs], something happened and she might begin to dress differently, and make sacrifices to Mammywater. She would worship her in short. Then we know that it is something that is genuine. Today, I don't know how genuine it is for women who say they worship Mammywater.

S: I don't know. But we have been interviewing many different people and heard different stories. Often the person would say that they were very ill, or [to E.] remember the woman at Calabar who was a maid. She would not fullfill her duties. She would just stand and stare. And it was a form of illness that she could not cope with the social demands.

E: Exactly.

S: So she was having dreams and when they finally brought her to a priest, a fully installed person, they would interpret it as being called by Mammywater.

N: Yes.

S: So the belief in the goddess helps them to find their way.

N: Yes, to have peace of mind.

E: Some even consult diviners to find out why they are sick.

N: Yes.

E: If it is proven that they are being called by the mermaid, they will be taken to somebody who can effect a cure, that is a worshipper of the mermaid....

N: Exactly.

E: ... And the priests and priestesses initiate them....

N: Exactly.

E: ... And that is how we found these worshippers whom we interviewed.

N: Like in *Efuru*, after illness, after disapointment, she becomes a worshipper.

E: Yes.

N: After disapointment, illness, and things like that, you want peace of mind. It is not so much wealth, is it? The case is not so.

S: No. It is peace of mind. In the cases we saw, wealth was sort of secondary. Peace of mind is what people were searching for. After peace of mind, wealth joined in and was ascribed to Mammywater from what I have seen.

N: But then you can have cases where you find that somebody comes to you and tells you that if you sacrifice to Mammywater, you are going to be a very wealthy person. And then, if you do it, maybe a door or two will open for you. It can happen.

S: What of yourself now? Mammywater features prominently in so many of your books and stories. We have mentioned *Efuru* and then there is also your children's book, entitled *Mammywater,* and I think she also appears in *Idu.*

N: Oh yes, she appears in *Idu.*

S: Why are you so interested in Mammywater?

N: That is what I was talking about. It is because I have been told these fantastic stories about Mammywater and lots and lots of things about her building under the water. Then there was something about her palm tree that attracted my attention and how she came out and met with human beings. These are the stories that you believed as a child, as you believe in father Christmas. My children do. So that was the sort of thing I believed in. But I would have thought that I would really have been a worshipper....

E: Because you have eventually found peace?

N: [Laughs.] I should have been a worshipper. But maybe it is my Western education that prevented me from being a worshipper. If I did not receive this kind of education....

E: Then you could have been a worshipper, but then....

N: ... But then my Christian background ... because my parents were the children of the first converts. My two grandmothers were Christians. They baptised us almost at the same time. Then they were fanatics about Christianity. We were exposed to so many things. We children were not allowed to do certain things, like talking about Mammywater. My [parents] called her worship "paganism"; they talked of "heathens" .... But then I was lucky because I went to live with my grandmother at a very early age in a huge compound. Not only my mother, but there were so many women who were not Christians. That was a more relaxed atmosphere. There, I was able to listen to all those stories of Mammywater, Okita, and all those things. I could not help hearing all those stories. If I had stayed with my parents, because they were Christians, they would not take interest [condone my participation] in "pagan" rituals. Remember that the "idols" and all those shrines for the local deities were destroyed. It is because of my sensitivity to these stories. Sabine, I cannot explain it, but I am not a worshipper. Yet invariably, Mammywater occurs in my stories. Even in *Never Again.* At the end of the war, let me show you: "Many people were not home yet. There was no home to return to at the end of the war."

S: Aha. You are quoting from *Never Again.*

N: Yes, I am reading from *Never Again*. It deals with the war and Ugwuta people.

S: So again you find people grateful to Uhammiri for....

N: And that is it. I remember that. There was so much about it. There was so much about the deliverance of Uhammiri. Who wants the church? Let me briefly tell you the story. We returned to Ugwuta in 1967 or 1968. By September '68, the federal troops had overrun Owerri and were pushing on to Ugwuta. But then there was an attack from Port Harcourt and one afternoon, we saw ourselves fleeing. We got to Mgbidi, the place was set on fire, then we stayed about 10 kilometers from Mgbidi, and a month, not quite, 3 or 4 days after we fled, we were told that the federal troops had been driven away from Ugwuta. I did not immediately go there, but my husband and my brother went. They reported that everything was gone; that all the people had left the town. It was a battleground and people could not go in. So we had to wait for about 2-3 months, before we went home again. Some waited until the end of the war.

S: People were afraid of the fighting?

N: Yes. We returned to Ugwuta before the war ended. Then, when the war ended, Ugwuta people were full of praise for the Woman of the Lake. They said that the Woman of the Lake was responsible for their coming back to their homes. Because it was only in Ugwuta, that is, within the Biafran territory, that the people were driven from their homes and came back, I think between Ugwuta and one or two other places. The federal troops just landed. Within 24 hours they were wiped out. You see, Ugwuta people attributed this to the powers of Uhammiri.[9] Because to them, it is Uhammiri that protects them. Not only women and children. She protects everybody. There is even a precedent for this in the history of Ugwuta. In the 19th century, some troops came from the Midwest to fight against Ugwuta.

S: From Benin?

N: Yes, from the kingdom of Benin. You know we have a boundary with the Midwest on the other side [of the River Niger]. The story—it could be a legend—has it that the waters of the lake, Uhammiri, parted when the troops were coming. When they got to the middle of the lake, Uhammiri closed up again and all of them drowned. It is just like the parting of the waters of the Red Sea in the bible.

S: That was an earlier invasion from the Benin kingdom?

N: Yes, from the Benin area. Now whether this is true or not, I don't know. But this is a story that has been handed down from generation to generation. And the Ugwuta man believes that no army of occupation coming across the lake ever goes back alive. The Woman of the Lake would destroy that army. Now when the Nigerian federal troops came from Port Harcourt crossing the lake, Ugwuta people said it was the lake that drove the invading army away from Ugwuta. Now in church, the pastor got so angry when he heard this that he did not mince words. He told Ugwuta people, "Either you are a Christian, or not." And I was in church that day.

S: This is appalling. He actually told the people that they could not be Christians if they continued believing in what was dear to them!

N: Exactly.

S: Those fanatics try to alienate people from the very holy grounds—and waters—of their hometown.

N: Exactly.

S: This could create a lot of pressure on a person.

N: Yes exactly. This is the situation. But you find now that as time goes on, Christianity lost out. Christianity is losing its grip on the people.

S: Yes, as people are getting more educated, they will realize the importance of their own culture. And because of education they will no longer fall easy victim to this type of foreign indoctrination.

N: Exactly. So I have not fully made up my mind on my own position, yet.

S: Uhammiri is just there.

N: It's just there. I cannot give an example. I cannot say this is why I am doing it. It is just in me.

S: There are foreigners who have said that Mammywater is a foreign import.

N: That Mammywater is a foreign import?

E: A foreign import.

N: It was during the time of early colonialization. I think [there] is a piece of writing by Chinua Achebe about Mammywater appearing to an UAC official.

S: Oh, you mean appearing to a foreigner?

N: A foreigner.

S: No. That is not what I am referring to. I am referring to some foreign scholars who say that Mammywater *herself* came from abroad. These people claim that the worship of Mammywater itself is a foreign cult similar to Christianity. One of the reasons these authors give is the use of an imported chromolithograph being used in the worship of the goddess. What do you think of that?

N: Oh my God! [Laughs.]

S: How did Mammywater come to Ugwuta?

N: It has been there all the time. It is like any of our gods and goddesses. It is just there. It is connected with water. And *water is the life giving thing.*

S: You are saying that Mammywater is indigenous, that she originated in Ugwuta?

N: I should think so. Very authentic. Very original. Because we have a name for it. It is Ogbuide.

E: And it is not a god of foreigners?

N: And you can see that I don't call her Mammywater in *Efuru.*

S: No [you don't].

N: I either call [her] Uhammiri[10] or Ogbuide.

S: Yes.

N: You see, I used the name *Mammywater,* because I was writing a children's book. And I know that Mammywater is something that will attract children.

S: You will find the same name in Lagos or Ibadan.

N: That's right. Going to different locations Mammywater is a more acceptable name than the highly localized name of Uhamiri, in Nigeria and even in West Africa.

S: Yes, it is used in Ghana and in the Ivory Coast.

N: Yes.

S: The name, Mammywater, is used in many places for the group's localized water spirits, mysterious beings from the water.

N: Yes, water spirits.

E: In our interviews very many people say that Mammywater goes to the market. Is this true, and what does it mean?

N: I have said that the stories I heard as a child depicted Mammywater coming out of the water. Then she goes shopping in the market.

E: In this case she makes herself visible to the people?

N: Yes. It was believed at the time that she makes herself visible. Although nobody has said to me, "Yes, I saw her," or something [like that]. But because of hearsay, this person said, that person said, it is believed that on Nkwo[11] days in Ugwuta, Mammywater will come out, and Mammywater will buy.

S: And of course, Ugwuta's market is right by the water side.[12]

N: Yes.

E: Did you also hear that Mammywater made friends with some

prominent Ugwuta men and women?

N: Yes, yes.

E: And she goes to their houses?

N: Yes, I have heard that also. But then at the time I was writing, at the time I was growing up, Mammywater was not associated with men because she was associated with women. But as time went on, I discovered that it was said that Mammywater is a friend to such and such a person. And that Mammywater used to visit him at night. Do you hear that?

E: Yes, I hear that. Can you also tell us about the relationship between Uhammiri and Urashi?

N: I was trying to do that in one of my stories, a sequence to *Mammywater.* I have not actually succeeded in what I wanted to do. So it has not actually been published yet. We have a legend that Uhammiri or Ogbuide quarreled constantly with Okita. Okita is the water spirit of Urashi. We call it Okita. As a child, I went to my grandfather's, my father's father. My paternal grandfather was not a Christian. He resisted Christianity until the end. My grandmother on the other hand was a fanatic Christian, [so much so] that it broke their marriage.

S: You see what damage it does.

N: Their conflicting religious beliefs broke their marriage. I think he was celebrating [an ancestor]. I remember it very clearly. Either he was feasting his ancestors, or another spirit. And when that happens, a goat is killed and children hang around, and of course, as Christians we were not allowed [to participate]. We were not allowed to go there. But unknown to our parents, I went [to my grandfather's compound]. Our paternal grandfather receives us but not other children. He was so delighted to see us. And there was one man there who was staring and somebody (not my grandfather) said that it was Okita. The man laughed and said it was Okita [Urashi, the river god].

S: Okita was in the crowd?

N: [Laughing] Yes, he was in the crowd. It was a smallish person, you know. He had this white chalk there, and there he was, standing, and this child was telling me, "Look, that's Okita!" [She laughs.] And whether he was the one ... I don't know why he had come to my grandfather who was feeding his ancestors. So there was just constant quarrel between my paternal grandparents; similar to the Woman of the Lake and Okita. I think it has something to do with supremacy. Who should be superior over the other.

S: Are they actually husband and wife?

N: Yes, they are husband and wife.

E: Is Okita not believed to be a human being?

N: I said that Okita is a human being but also a spirit. These beliefs are open to interpretations. I referred to Okita as Urashi in *Mammywater.*

S: Yes, you mentioned it in the book.

N: I mentioned it in the book:

"Great spirit, I had to wait for ten years ... I send people to Urashi to ...."

Listen, I think there is something I mentioned about Urashi and Mammywater. [She turns some pages of her book, *Mammywater.*]

S: I think when they ...

N: Where they meet Okita, the Spirit of the Urashi River, with the eagle feather. This is a fish [pointing to the picture on p. 36]. He uses a fish to fan himself.

S: Aha.

E: Like a messenger?[13]

N: No, it is not a messenger, but [a] priest, a sort of priest. Let me see ...

"Yes, of course, you said we are going to see Urashi," says Deke.

"Yes, to pay our respects. I sometimes go there with distinguished visitors."

"Distinguished visitors! Am I a distinguished visitor?"

"Now, now, you are asking too many questions again ...." (35)

So this is it. You see they quarrel constantly.

S: And you mentioned the dividing line ....

N: Aha, between the lake and the river. You have seen it?

S: Yes, I have seen it.

N: You have seen the muddy and the clear water. These waters are separate, they never mix. And it's been there for ages.

E: And some people say they are husband and wife.

N: Some people say they are husband and wife and they quarrel a lot. They are husband and wife ... of course husband and wife quarrel [laughter]. That is very natural.

S: And one is red and one is white.

N: No, it is not red really.

S: Urashi? Is he not red?

N: No, he is muddy. And another thing is that because the lake's water is clear, deep blue and all that, the lake people look down on the river as small and ugly....

S: What about the animals you mentioned, especially the snake?

N: Yes, you find the snake and the crocodile in the river, but *not* in the lake.

S: Aha?

N: Not in the lake. There are no crocodiles in the lake.

E: We mean to say that there are animals that are a favorite of the lake goddess.

N: There are animals that exist in rivers but not in the lake.

S: So, Uhammiri does not associate with the snake?

N: Yes. But Urashi does. And this is what they believe in.

S: What of the tortoise?

N: Tortoise. I think the tortoise can be found mainly in the river.

S: A more personal question: I think your husband has built a temple at Urashi's sacred grove?

N: Hm.

S: Can you say anything about it?

N: No. I have not seen it. It is believed that his wealth comes from the water deities. [She laughs]

E: Very interesting.

S: From Urashi?

N: No from Uhammiri.

S: From Uhammiri? So why did he build his temple at the site of the old shrine of Urashi on the river, not on the lake, by Uhammiri's shrine? Is it because he is a man?

N: No. He has a hotel at the river. Have you seen it? It has been there for a long time. It is fantastic but not finished. He spent real money on it. It is beautiful, by the Urashi riverside.

S: [To Ebiri] You have mentioned it. It is by the river side.

N: It is supposed to be a tourist attraction.

S: What of yourself now? You are independently wealthy.

N: You cannot say that I am poor.

S & E: No.

N: You are asking whether my wealth is associated with the goddess?

S: Yes, the goddess.

N: But of course you know my father was a wealthy man. Have you seen the plantations we have in Ugwuta?

S: No.

E: I have only taken her to your mother's old building. That is where my junior sister lives.

N: Aha.

S: And we met your brother playing tennis, but that was all.

N: Okay.

S: You were going to tell us something about your writing.

N: When I have a story and the story is clear in my mind, all I know is that I am just writing a story, not wanting to impress anybody, not wanting to put anything down, or write a moral lesson, or whatever. I discover that I write better that way. So that was how I wrote *Mammywater*, spontaneously. I just liked the idea and the stories of Mammywater. I don't know whether or not you know Michael Crowder? Somebody was telling me that he died this year. Michael Crowder was a historian who has been in Nigeria for many years. He worked for the Nigerians and he has written so many books. In 1976-77, Crowder came to Enugu and looked me up and said he wanted to go to Ugwuta. I said Okay. I was not doing anything at the time. I was thinking of setting up a publishing company. So I took him to Ugwuta. He said he wanted to see the lake. So we took a dug-out canoe with an old woman, you know these old women who paddle. And we went there past the lake to Urashi. We went to where the gunboat has sunk. Have you gone there?

S: No, I have not been there.

E: You must see that.

N: So as we were coming back, as I was giving him the landmarks, I said, "As a child we believed that the Woman of the Lake lived there. That is the shrine of the Woman of the Lake. You see the confluence of the river and the lake. You see the confluence of the lake and the muddy water. They never mix." And I just went [to Ugwuta] and came back to [Enugu]. We had lunch and he went back to Ibadan. The day after I returned to Enugu, I took up writing *Mammywater*. And two days after, I just finished writing *Mammywater*. You see, I have never thought about it. After that visit, I began writing and that is how I wrote *Mammywater*. It never occured to me that I was going to write *Mammywater*. So I began to write and I saw what was going on. That is how it was.

S: Just by intuition.

N: Yes.

S: You know my own case is quite the same because I came here as an academic to do research. But Mammywater attracted me.

N: You were drawn to Mammywater.

S: I never dreamt of making a film ...

N: Yes.

S: I am not a filmmaker originally.

N: Aha.

S: I never dreamt about making a film about Mammywater in particular. The idea developed as I learnt more about the Lake Goddess from the people of Ugwuta.

N: Aha.

S: And it grew from there. It is not something that I have consciously planned. Although, of course, by trying to realize it, I had to start planning.

N: Exactly. That is how it is.

S: That is, once you are in it, then... I mean you have to start thinking about it. Once you start writing a book, then you have to start thinking about how to print it.

N: Yes. So that is how it went. That is how my own fascination for the Woman of the Lake grew. It is so alive. And as it is, I should have a little shrine. [She laughs]

S: Or, we could have one together. [Laughter]

E: And I will be the one to initiate you both. [Laughter]

NOTES

1. At Flora Nwapa's office in Enugu on November 4, 1988.

2. Ugwuta is the Igbo name of an Igbo town named Oguta by the British colonial administration. Ugwuta is located in Imo State of southeastern Nigeria. The British name and spelling, Oguta, is still in use by the contemporary Nigerian administration and on road maps. Ugwuta and her sister towns, Orsu-Obodo, Esi-Orsu, Nnebukwu, Nkwesi, Mgbelle, and Izombe form a group of towns known as the Oru. The Oru, Omoko and Onitsha are the riverine Igbo, a division of the Igbo people, an ethic group of approximately 20 million people in Southeastern Nigeria. The Oru speak a dialect of the Igbo language and distinguish themselves through their own cultural peculiarities. On the other hand, the Oru also share certain cultural and linguistic features with the Igbo people as a whole.

3. The lake goddess associated with Ugwuta lake is known by several different names,e.g. Uhammiri and Ogbuide. The goddess's praise name and form of address is Eze Nwanyi, or Eze Mmiri (Water Monarch), a title also given to her priestesses. In a similar vein, the people who worship this goddess use several different names for themselves on the ethnic, community, and individual levels. See also S. Jell-Bahlsen, "Names and Naming: Instances from the Oru Igbo" *Dialectical Anthropology* 13 (1989):199-207.

4. This film was produced in 1989 and published in the USA in 1991 as *Mammy Water: In Search of the Water Spirits in Nigeria.* (Berkely: University of California Extension Center for Media and Independent Learning, 1991). I had lived for a year in Orsu-Obodo/Oguta II, in 1978/9, and have returned there many times since, developing an ongoing relationship with the local community, while pursuing many months of in-depth field research over a prolonged period of time.

5. To this day, it is somewhat puzzling why Nwapa did not establish a shrine for the Lake Goddess, as many Ugwuta people expected her to do,

because they ascribe Nwapa's professional success to *their* Lake Goddess, Ogbuide/Uhammiri. Nwapa also refused the priestess's 1990 request to perform a major sacrifice of a cow to Ogbuide/Uhammiri, because of her Christian background, upbringing, and social standing, as Nwapa herself told me during her visit to New York in 1991. By contrast, and as is common knowledge in Ugwuta, a man from Omoko built a temple for Ogbuide in her sacred grove to publically demonstrate his gratefulness to this deity. In a similar vein, Nwapa's husband has built a temple in appreciation of the goddess's help at the river god Urashi's sacred grove on the river bank.

6. During the colonial days, when Nwapa attended school, the school system of that part of the country was dominated by the missions and various churches. After independence the schools were officially secularized, but many are still heavily influenced and controlled by their original confessional sponsors.

7. See also Jell-Bahlsen (in this anthology) "Flora Nwapa and Uhammiri/Ogbuide, the Lake Goddess: An Evolving Relationship."

8. See also Jell-Bahlsen's "The Concept of Mammy Water in Flora Nwapa's Novels" *Research in African Literatures,* 26, 2 (Summer 1995):30-41 and (in this anthology) "Flora Nwapa and Uhammiri/Ogbuide, Eze Nwanyi, the Lake Goddess: An Evolving Relationship."

9. This belief is still strong in Ugwuta today. My documentary film, *Mammy Water: In Search of the Water Spirits in Nigeria,* presents the on-camera account of an elderly Ugwuta woman, Madame Nwametu, who narrates that Uhammiri "drowned the enemies' gunboat during the [Nigerian civil] war."

10. There are different possible spellings. In *Efuru,* Nwapa uses the spelling Uhamiri. I am following the translations and spellings of Chief Ebiri, who assured me that Uhammiri is locally spelled with "mm" as in the Igbo word, "mmiri" (water).

11. The Igbo week has four market days called Nkwo, Eke, Orie, and Afor in Ugwuta. Markets rotate, as each town holds its market on a different day. Ugwuta's market day is Nkwo.

12. I was here referring to Ugwuta's old market, the Umudei market located directly on the Umudei waterside. This market burnt down some time ago but still exists. A major market in the past, it now appears small when compared to the new market with modern cement stalls built at the opposite end of town, near where the main road enters the town. The new market is in partial session every day, but fully occupied only on Nkwo days.

13. "Messenger" here means "a local deity," as the villagers deferentially refer to their local deities, the Arishi, as "messengers" [of the supreme god, Chi-Ukwu].

A Chat With Flora Nwapa

Theodora Akachi Ezeigbo

It was on a beautiful sun-drenched afternoon in August, 1989 that I interviewed Flora Nwapa in the lovely home of our mutual friend, Nina Mba, a historian and Senior Lecturer at the University of Lagos, who resides in Apapa, Lagos. It was during the period known in Nigeria as the "August break," which marks a lull after the usual downpour of the month of July and before the rains return in September. I had told Nina of my desire to interview Flora and in her characteristic efficient manner, she had gone ahead and arranged the interview for me. I had met Flora at the launching of her book, *One Is Enough* at the Nigerian Institute of International Affairs at Victoria Island, Lagos in 1987. It was there my interest in her works intensified.

On the day of the interview, I arrived and Flora was already there, as beautiful and elegant as ever. Nina left us to get on with the business of the moment. Flora made me feel relaxed and all through the interview, she answered my questions directly, without equivocation. Her tone and comportment were lively and friendly, even when she talked about issues she found distasteful, such as the quality of criticism writers like her were receiving from literary critics. The interview lasted about one hour. Afterwards, Nina rejoined us, bringing some refreshments with her. We talked informally for a while, then I left about two hours later, filled with admiration for Flora whom I had come to know better after the meeting.

E: It is on record that you are a pioneer woman writer in this part of the world. When did you start writing and what was the catalyst that brought about your writing activity?

N: At Archdeacon Crowther Memorial Girls' School, Elelenwa, near Port Harcourt. While at the college, I read a lot and enjoyed stories. I started writing there.

E: What experience inspired your writing of *Efuru*, your first novel?

N: One day, on a drive from Enugu to Onitsha, the idea of *Efuru* came to me along that stretch of road called Agu Awka. I was cruising at a speed of 80 miles per hour. The idea came strongly and began to gestate. I started to write the story down at Onitsha. The urge to write was there and I went on and on. When I eventually finished writing it, I got it typed.

E: You seem to have completed the writing rapidly?

N: You could say that! For me, the idea to begin writing was never consciously planned. It came suddenly. For me, writing was never chosen. Writing chose me. After writing, you don't know if you have written a good story. Sometimes you need courage to send it out; this is especially the case with new writers. After I had written *Efuru*, Chinua Achebe graciously accepted to read it and then sent it to Heinemann. I owe it to him—I mean the publication of the book.

E: That's one of the wonderful things Achebe has done for this country, indeed, for Africa as a whole. He has been instrumental to the discovery of many talents whom he helped to get published. Don't you think the situation would be different if our other successful writers were like him in this regard?

N: Yes, but things are different now, especially with the state of the economy. Publishing books is more difficult today.

E: What about *One Is Enough* and the others you have written in recent times? How did you come about writing them?

N: A story somebody told me had a great impression on me and I started to write *One Is Enough*. In four months, it was done. *Cassava Song* came to me suddenly. A woman, a beggar, came to Oguta and sang in praise of cassava. In my office, I remembered the woman and her song. The urge to write the story of cassava in verse caught my imagination and I finished writing it in two days. *Mammywater* came like that too. Michael Crowther came to see me. I think it was in 1978 in Enugu. He wanted to see Oguta and we went to see the lake. We took a canoe to Urashi River (which flows into the lake) and there we saw the gun-boat sunk by the Biafran Navy. The impression I had after that visit to Lake Oguta made me write *Mammywater*.

E: Many feminist critics at home and abroad have observed the radical trend noticeable in your later novels. What is responsible for this trend?

N: When I start writing. I don't want to prove anything; I don't set out to write a feminist novel. I have an open mind; my aim is to tell a good and an interesting story as honestly as possible. I try to be faithful to my vision. In doing so, a lot happens. Characters are consciously created; midway the characters take over the story. You introduce them and want them to be subservient, but they develop lives of their own and take over the story. Sometimes you may have some people in mind, but they are never the same people. There are changes; hence they are fictional.

E: There is the question of influence which interviewers are always asking writers. Are there particular writers who may have influenced you?

N: Everyone with an inclination or ambition to write must read a great deal. I have read a lot. Chinua Achebe may have influenced me. When I was young, I was intrigued by Ekwensi's *When Love Whispers*. I read many of Hemingway's works and he influenced me positively. For example, his style in *A Farewell to Arms* and *For Whom the Bell Tolls*. Both novels are based on wars and conflict. I was touched by T. S. Eliot's *Murder in the Cathedral*. My uncle introduced me to Bernard Shaw and Oscar Wilde. I have also read fellow women writers like Buchi Emecheta and some African-American writers.

E: Would you say that women writers are neglected?

N: Not really. The problem is that there are not many of us. Having said that, one is often surprised when women writers are asked why they always make women their protagonists. I think this question is discriminatory, for it is never asked of male writers.

E: Have you fared well in the hands of literary critics? Would you say they have been fair to you?

N: I don't bother about them at all. I don't allow them to guide my writing. If a critic or a journalist wishes to interview me about my books, I want to find out if he or she has taken the trouble to read them. If he/she comes equipped with this knowledge, then we can discuss. A point can take us hours to discuss and savor. If critics read and study a writer's books and not base their judgement on hearsay, they would be judicious. There is evidence from what they write that they have not taken the trouble to read the works they write about.

E: I want to draw your attention to your novel, *Idu*. Now, we know that having children and rearing them are very important in our culture, and so it is difficult to understand why Idu would want to die and follow her husband. It's not only that she was pregnant but also had a little son. Wouldn't she want to live to have the baby and care for the children?

N: Human behavior is not always guided by logic. Sometimes things happen in life that seem unnatural. Having lost her husband, certain things that seemed important to Idu before now lost their significance.

E: Don't you see inconsistency in your vision here?

N: No. With a woman like Idu, her reaction was not unusual. She was the type of woman who felt deeply and loved passionately. She and her husband were very close.

E: There is a development in *Efuru* that baffles me too. All those years Efuru was childless, Adizua stayed with her, but after she had a child, he left her. Why was that?

N: There are people who are irresponsible. This is the point in the novel, that Efuru's love was misplaced—a man that was totally unworthy of that profound love was its beneficiary. Moreover, you must realize that sometimes, even in our culture, children cannot keep a man. He could leave the woman even after children are born. Some people are irresponsible, and irresponsibility could worsen at any time in a person's life. Efuru's tragedy is that she gave her love to a worthless man. It is a calamity for such a fate to befall such a splendid woman. Wasn't it better, therefore, that she should in the end decide to worship the goddess of the lake and give the deity her devotion?

E: Your finest and most admirable protagonists in your two novels are virtuous women. Were you reacting against the tendency of our earliest male writers to portray female protagonists who are prostitutes or immoral women like Ekwensi's Jagua Nana, Soyinka's Simi and Aluko's Gloria?

N: They are virtuous but they are not idealized. In the traditional past, there were outstanding women like them. You can't compare them with Amaka in *One Is Enough*.

E: Are your female protagonists meant to be role models either to stress the importance of virtue and industry as seen in *Efuru* and *Idu* or to underscore the need for women to be assertive and self-reliant as illustrated in Amaka?

N: No. I was not out to create role models. These women are delineated just the way they are. A woman can be virtuous and she can be assertive. She can refuse to agree to compromise with a situation or the status quo. It depends on what motivation I have when portraying a character. In *One Is Enough*, I was not out to create a role model in Amaka. I was simply creating a woman who found herself in a corrupt society and struggled to exist there the way she thought best.

E: You have published two volumes of short stories. I have read them and I feel the short story is a genre over which you have good control. But I notice in recent times that you have been writing only novels. Have you thought of writing more short stories?

N: Yes. I love short stories and hope to do more in that area. I like writing them. When I have an idea for a short story, I start and do not stop until I finish—usually it could take two days. If it lasts longer than that, I lose interest.

E: I beg your pardon, but I want to return to the issue of critics. You see, I am a critic myself and I'm worried about the disenchantment many of our writers express concerning critics, especially Nigerian critics. I'd like you to talk freely about the things you think we do wrong.

N: In Nigeria today we don't have good critics. If we do, they will inspire writers to write more. Critics allow sectionalism to mar their work, like writing only about certain people or certain kinds of writing. Good critics could give writers feedback on what people feel about their works. They should not be saying because a certain book is a woman's work, there are certain things she must talk about. Writers thrive on controversy and critics must write about their works to make them create better. However, my view is that a writer should continue to write, develop herself and not bother about critics.

E: How have your works fared in the book market?

N: Quite well. For example, *One Is Enough* sold 25,000 copies in the first year of publication and within the next year it sold 25,000 more.

E: Are you satisfied with your present style of writing, or do you intend to experiment with other modes and techniques?

N: That's the way I write and I prefer it that way. I think of the best way to recreate my vision.

E: Would you, one day, try to recreate experiences in other genres you haven't tried yet—drama for instance?

N: Yes. I'm thinking of writing a play. My son once suggested I do and friends have done so, too. I am thinking seriously of writing a play.

The Poetics of Economic Independence for Female Empowerment: An Interview With Flora Nwapa

Marie Umeh

Flora Nwapa Nwakuche, popularly known as Flora Nwapa, Africa's first internationally acclaimed female novelist and publisher, died of pneumonia on October, 16, 1993 at the age of 62 in Enugu, Nigeria. She was buried at Amede's Court in Ugwuta, her home town. In what was to be my last conversation with Flora Nwapa Nwakuche in December, 1992, in Scarsdale, New York, when she was on tour in the United States, the renowned author spoke not only of the glory she received as the first African woman to be published internationally, but also unashamedly of her position as a writer globally, coming from a formerly colonized state. Very much in tune with her culturally nationalistic view towards her motherland, Nwapa applauded the androgynous nature of Ugwuta society while she decried the "multiple marginality" she experienced with her Euro-Western publisher who regarded her as a "minor writer" and therefore did not bother to print and distribute her books locally and internationally when they were in demand, as they would have if she came from a so-called first-world country. According to Nwapa, Heinemann's placing her in the literary back-waters resulted in the piracy of her books in Africa and the death of her voice globally. As Ama Ata Aidoo once said, when the canon-ical establishment refuses to promote, print, distribute, read and critique your books, they kill you creatively (38). Recognizing her status as "other," Nwapa took it upon herself to publish and dis-tribute her books trans-continentally, and established Tana Press Limited in 1977 for this purpose. It is, therefore, my contention that Nwapa's resistance to the canonical politics of her erasure is behind her distancing herself from the term "feminist" to describe her ideological position in world letters. Certainly Nwapa x-rayed and analyzed her own realities and concluded that sexism is a sec-ondary problem which arises out of race, class and the exploitation of people of color. Hence, she prefers to identify with Alice Walker's term, "womanist," which reflects the African reality of effacement based on ethnic difference, if labelling is at all necessary.

The popular Eurocentric view is that the position of African women is one of subordination to husbands, and the repression of talents outside the domestic realm. However, despite the asymmet-rical nature of some African societies, gifted African women were not deterred from playing significant roles exercised by female lead-ers such as, Moremi, Queen Amina of Zaria, and Olufunmilayo Ransome-Kuti. Similarly, Flora Nwapa contends that if she is con-sidered the doyenne of African female writers, the glory goes to the oral historians and griottes who mesmerized her with stories about the mystical powers of Ogbuide, the mother of the lake, her family members of industrious women and men who served as role mod-els, as well as her penchant toward service and the pursuit of excel-lence. Accordingly, in opposition to the belief that women in the Igbo area of Nigeria do not break kola nut, Nwapa informs us that a woman in Ugwuta society, who has achieved because of her indus-try and persistence, is recognized for her accomplishments and she has the privilege of breaking and sharing kola nut. Of course this *modus operandi* points to the complexity and complementary nature of some Igbo societies in pre-colonial Africa which ensured its kith and kin that a woman who has distinguished herself, would not have her gender mitigated against her. The Ugwuta community is there-fore one of those special communities in Igboland where status and recognition are not biologically based. Ugwuta society, it appears, subscribes to Victor Uchendu's view that "a child who washes his/her hands, eats with elders." Indeed, the unit of analysis is the individual. For her courage in exploiting the complementary sex-role system in Ugwutaland, despite the obstacles pioneers must confront and overcome, Flora Nwapa is certainly a phenomenon.

By breaking the silence of women in Nigerian letters, Flora Nwapa has made a name for herself as a major twentieth-century African woman writer. Since the publication of her first novel, *Efuru* (1966), she had gained an impressive readership in both African and international circles, as well as critical acclaim for her novels: *Idu* (1970), *Never Again* (1975), *One Is Enough* (1981), and *Women Are Different* (1986). Her two collections of short stories are entitled *This Is Lagos & Other Stories* (1971) and *Wives At War & Other Stories* (1975). She has also published a book of poems, *Cassava Song and Rice Song* (1986). Apart from her creative works for an adult readership, Nwapa held the reputation of a fine creator of children's books: *Emeka: Driver's Guard* (1972), *Mammywater* (1979), *My Tana Colouring Book* (1979), *My Animal Number Book (1979)*, *The Miracle Kittens (1980)*, *Journey To Space (1980)*, *and The Adventures of Deke* (1980). Two books of plays, *The First Lady* and *Conversations*, were published in 1993. And her novel, *The Lake Goddess*, will be published posthumously. For her achievements as a writer and publisher of Tana Press Limited and Flora Nwapa Books, she received The Officer of the Order of the Niger (OON) Award in 1983 from the Federal Government of Nigeria and the University of Ife Merit Award for Authorship & Publishing in 1985, to name only two of the distinguished prizes accorded her. Among the many honorary positions she held during her lifetime, she was the President of the Association of Nigerian Authors (1989), a member of P.E.N. International (1991) and the Commonwealth Writers Awards Committee (1992).

With the characterizations of Flora Nwapa's female protagonists in her adult fiction, she complicated female identity as delineated in the literature of Chinua Achebe and his brothers, by critiquing both their gender conventions and power relations between men and women in the homestead. Thus, the female literary tradition she initiated was rooted in resistance, a protest against the one-dimensional images of Nigerian women either as wives, mothers, *femmes fatales* or rebel girls. Although it has not been recognized, Nwapa's work represents a monumental effort to invent an African female personality and attitude and to define an African female subject narrativistically. Indeed, her explorations of the female psyche link her works theoretically and thematically with womanist writers such as Alice Walker, Toni Morrison, Ifeoma Okoye, and Zaynab Alkali, to name only a few, whose aim is not only to present the female point of view but also to subvert patriarchal authority on women in world literary history. Her canonical contribution to Nigerian letters is then a poetics of economic independence and self-reliance for female empowerment. Nwapa sets the record straight by the power of the pen. She actually feminizes Nigerian letters as she realistically fictionalizes the shrewd, ubiquitous market women, energetic female farmers, sagacious wives and mothers, and astute women chiefs and priestesses as an integral part of quotidian existence. It is in her fiction that the enterprising African woman takes a stand and demands her rightful place in the halls of global literary history.

Nwapa would have agreed that African men wield a great deal of power in African society. On the other hand, her honest portrayal of Ugwuta women insists on the complementary nature of the society, beginning with a mixed-gender age grade system, a mystical Lake Goddess, who guarantees women, as well as men, power, prominence and peace. The historian Nina Mba posits: "From the turn of the century until the worldwide depression in 1929, the trade in produce and imported goods with the European firms in Onitsha and riverine areas was dominated by women traders. Omu Okwei, for example, began as a petty agent for the Royal Niger Company in 1904 and ended up as a big transporter, moneylender and property owner" (48). Indeed, woman is something, an achiever, a go-getter in Ugwutaland, not only for her special child-bearing and child-rearing abilities but also for her potential to benefit her community spiritually, educationally, economically and psychologically. As Mgbada, the diviner in Nwapa's last novel, The Lake Goddess, tells Ona's husband: "We believe that the Goddess protects us and inspires us to great heights. We believe that no invader from any part of the world can destroy us. We believe that the deity is a beautiful and ageless woman who is partial to women" (208). Coming from this rich tradition where women paddle canoes up, down, and across Ugwuta Lake transporting passengers and their wares for a nominal fee, where women are leaders in trade and commerce, where a democratic sex-gender system recognizes talent, regardless of one's sex, where confidence and perfection is nurtured in both females and males, is it any wonder that Flora Nwapa was able to touch the hand of Goddess?

UMEH: Congratulations on the publishing and launching of a number of your books with Africa World Press in New Jersey. How does it feel to be so successful. . . . to be Africa's first internationally acclaimed published female novelist in the English language?

NWAPA: Thank you very much, Marie. It feels good. It feels fulfilling.

UMEH: What are the rewards and difficulties of being the first African woman novelist and publisher in Nigeria?

NWAPA: I've had my ups and downs. In 1966, when Heinemann published *Efuru* I did not receive much publicity because Nigeria was in a turmoil. There was a *coup d'etat* in 1966 and the whole system had broken down. And in 1967 I had to go back to Eastern Nigeria where all the Igbos returned from all over Nigeria. The war was fought for thirty months and when we came back we had to start all over. It was at this time when I was a minister in East Central State, which is what Enugu was called in those days, that I continued to write again. But my second novel, *Idu,* was also published by Heinemann in 1970. These are the two books published by Heinemann. After that I thought I should have some African publishing companies distribute my books. That was when Nwamife Publishing Company came out. They were the first to publish *This is Lagos* and *Never Again.* It's been fulfilling. I must say there's been a lot of hard work. There's been a lot of frustration all the same. But the problem in Nigeria is the problem of having a reading public, the people who will appreciate your work. In the 80s and 90s things have been accelerating. But, you discover that not many people can afford to buy books. I was lucky because for the past five years the *West African Examination Council,* called *WAEC,* had *Efuru* on its reading list. It was something that I should have been congratulated for. It should have brought in a lot of money. However, the problem was piracy. Heinemann Publishing Company could not bring out the books on time. Therefore, pirates took over so that writing and publishing didn't make any impact. The school system did not make an impact at all on my earnings.

UMEH: What advice would you give to women who would like to write and publish?

NWAPA: The advice that I would give to them is one, they should read and read and read. You have to be a good reader. And you have to be a good listener to be able to write. I think this is the advice that I would give to them. When you hear people asking writers, "Where do you find the time to do all of this"? I tell them that time has little to do with it. If you have a story to tell, the story is there in you and it will haunt you until you tell it. So, I would advise them to read and listen.

UMEH: What circumstances in your life are responsible for the writing of both your adult novels and your children's books?

NWAPA: I read a lot of books. As a child before I went to high school, I listened to a lot of stories, moonlight stories told by the women in Ugwuta. As a child I would call on anybody who promised to tell me a story. I would sit down and listen. And when I went to high school I had read practically everything that I could find, so that contributed again to my writing. After graduating from Ibadan University in Nigeria and Edinburgh University in London, I came home to Nigeria and I taught for a while. While I was teaching I discovered I had plenty of time on my hands. I didn't know what to do with it. So, I began to write stories about my schooldays. It was in this process that I began the story of *Efuru.* It just happened. I started writing the story of this woman and then I went on and on and I discovered that I had a good story to write and a good story to tell. I continued to write until I finished it. There was nothing in me when I was in school that made me feel I was going to be a writer. It was one of those things that just happened. I didn't have the ambition to say, "Oh, Flora you are going to be a writer so work towards it." No, it didn't happen that way. But having written *Efuru* and having published it, I continued to write. It is difficult to write children's books. But I remember that Christopher Okigbo, in those days when he was working for Cambridge University Press, had asked me to write a children's book. I told him, "OK Chris, I will do something about it." I began to write *Emeka—Driver's Guard,* and I finished it. Chris had died during the war.[1] So, I sent my manuscript to London University Press and they published it in 1971. Now when I had my own publishing company, I decided that I needed good books for my growing children. When I went to the book stores, I didn't see anything that was good for my children. That was when I started writing more children's books. I wrote *Mammywater* in 1979.

UMEH: Flora, what year did you start your publishing company, Tana Press?

NWAPA: Tana Press was opened in 1976 and business started in 1977 and I published *Mammywater* in 1979.

UMEH: Out of all your creative works, which one has given you the most satisfaction? And why?

NWAPA: This is a difficult question to ask, Marie. Similar to Buchi Emecheta, I too feel that books are like your children. It's not easy for one to say that you love one and not the other. They are all good books. I cannot tell you that I like this one or that I like that or, that I prefer this one. It is difficult for me to say.

UMEH: My favorite novel is *One Is Enough*. What was your audience's reaction to *One Is Enough?*

NWAPA: Hmmm. It is hard to say. But let me tell you what happened. *One Is Enough* was published in 1982. And a friend of mine who read the book came to me and told me that *One Is Enough* was a true story of a friend. I did not know her friend. It was three years after this, when I was at a funeral that another friend came along and said, "Look, the lady you wrote about in that story is in this audience." I couldn't believe it. So, the lady came over, shook my hand, and said, "Mrs. Nwakuche, I heard so much about this book. People say it is my story. I haven't read it; so I don't know whether it's my story or not." But she didn't feel bad about it. What I want to say is this, after the war in 1970, things changed a great deal in Nigeria. During the war, that is the Nigerian Civil War, women saw themselves playing roles that they never thought they would play. They saw themselves across the enemy lines, trying to trade, trying to feed their children and caring for their husbands. At the end of the war, you could not restrict them any more. They started enjoying their economic independence. So what they tolerated before the war, they could no longer tolerate. For example, if you discovered that your marriage is not giving you satisfaction or that your in-laws are worrying you because you have not produced a male child or a female child, whatever the case maybe, you can just decide to leave that family and go to the big city. The big city at that time was Lagos, where you were anonymous, where nobody seemed to care what you did for a living. So, I think I wouldn't presume that *One Is Enough* had an impact on Nigerians. I would presume that all *One Is Enough* is about is the story of what is happening in male/female relationships in Nigeria today.

UMEH: You are indeed a prolific writer: eight adult works and seven children's books. Have your books been translated into other languages?

NWAPA: Yes, *Efuru* has been translated into French. Unfortunately up to now, I do not have a copy. *Efuru* has also been translated into the Icelandic language.

UMEH: Which writers do you admire? And outside of your literary foremothers who expressed themselves in the oral tradition, which literary artists have influenced your writings?[2]

NWAPA: I would say that Chinua Achebe influenced me a great

deal. He influenced me in my adult life. But as a young girl in school many writers, such as Ernest Hemingway, and Charles Dickens also influenced me a great deal.[3]

UMEH: In addition to your enthusiasm for story telling, do you have a sense of mission? What is your purpose in writing?

NWAPA: I write because I want to write. I write because I have a story to tell. There is this urge always to write and put things down. I do not presume that I have a mission. If you continue to read my books, maybe you could find the mission. But I continue to write because I feel fulfilled. I feel satisfied in what I'm doing.

UMEH: Are there any autobiographical elements in your creative writing?

NWAPA: None! I am not like Efuru, neither am I like Idu, Neither am I Amaka in any way.

UMEH: What do you think of Leopold Sedar Senghor and Ali Mazrui's statements that "African women have always been liberated"? In other words, is there any truth in the statement?

NWAPA: For me, yes! In Ugwuta women have certain rights that women elsewhere, in other parts of the country, do not have. For instance, in Ugwuta, a woman can break the kola nut where men are.[4] If she is old, or if she has achieved much or if she has paid the bride price for a male relation and that member of the family is there, she can break the kola nut. And everybody would eat the kola nut. But in certain parts of Igboland a woman is not even *shown* a kola nut, not to talk about *breaking it.*

UMEH: Now we know why you're a first—Nigeria's first female novelist, East Central State's first female comissioner, Nigeria's first enterprising female publisher of the nation's first feminist press, etc. It's because Ugwuta traditional society certainly nurtured you into being an independent thinker. It appears that you don't even wince before you perform a task.

NWAPA: Thank you, Marie.

UMEH: The critic, Katherine Frank, in an article entitled, "Women Without Men: The Feminist Novel in Africa" describes you as a radical feminist. What is your opinion of this assessment?

NWAPA: I don't think that I'm a radical feminist. I don't even accept that I'm a feminist. I accept that I'm an ordinary woman who is writing about what she knows.[5] I try to project the image of women positively. I attempt to correct our men folks when they started writing, when they wrote little or less about women, where their female characters are prostitutes and never-do-wells. I started writing to tell them that this is not so. When I do write about women in Nigeria, in Africa, I try to paint a positive picture about women because there are many women who are very, very positive in their thinking, who are very, very independent, and very, very industrious.

UMEH: What do you perceive to be the major ideological difference between male and female writing in Nigeria?

NWAPA: The male writers have disappointed us a great deal by not painting the female character as they should paint them. I have to say that there's been a kind of an ideological change. I think male writers are now presenting women as they are. They are not only mothers; they are not only palm collectors; they are not only traders; but they are also wealthy people. Women can stand on their own. My example is Beatrice in *Anthills of the Savannah* by Chinua Achebe. In that novel, Beatrice stands out. She was the one who really understood what was going on. The men were too ideological. They were not actually down to earth. It was Beatrice who was practical.

UMEH: Certainly, Chinua Achebe's attitude towards women in *Anthills of the Savannah* published in 1988 is a far cry from his portrayal of women in *Things Fall Apart* published in 1959. What do you feel about Elechi Amadi's female character in his latest novel, *Estrangement?*

NWAPA: *Estrangement* is another story which I enjoyed a great deal because Elechi Amadi tried to portray this unfortunate woman in a true picture. He has sympathy for the Nigerian woman. In *Estrangement* you could see that the heroine was treated very, very positively by the author. All her misfortunes were clearly stated. And the way the author portrayed her showed understanding.

UMEH: Africa has produced three Nobel Laureates in literature, Wole Soyinka, Naguib Mahfouz and Nadine Gordimer. What was your immediate reactions to each writer's winning the Nobel Prize for literature?

NWAPA: I was very pleased. I was very excited. I was delighted.

UMEH: Do you recognize Nadine Gordimer as an African writer? Who is an African writer?

NWAPA: Nadine Gordimer is a white South African. She has been writing for a very long time; she has sympathy for the black South African. She is an African writer.

UMEH: More and more people, even those in Muslim countries, are moving away from polygamy and polygyny. But in Nigeria they are still common. What do you think about this, since they are effecting the Nigerian family as a unit?

NWAPA: Well, I think it is the society. It is the age that we are in. I think it's going to pass. It started in the 70s and it is going on and on. There is this stigma on a woman who elects to be single. Mothers bring up their daughters telling them that they have to marry. In my own language we say, "No matter how beautiful one is, if she doesn't get married she's nothing." It's left for us who have received a western education to de-emphasize this tradition. However, you discover that a woman who has gone to college, who is working, who has a profession, who is a lawyer or a doctor, who doesn't have a husband, then she will not mind being a second wife. In fact, polygamy is becoming very fashionable in Nigeria these days among western educated women.

UMEH: Do you think it's right for people to say that every woman should have a husband and a child?

NWAPA: No! However, I'm telling you about the tradition.[6] If you had a child out of wedlock in those days in the community that I grew up in, your child was not legitimate. Nevertheless, things are changing; people are now accepting it. In fact, when I was growing up if a young girl became pregnant we viewed her with horror. The child was not baptized. Now when these things happen the baby is baptized.

UMEH: Nowadays, do you think a woman would elect to have a child, if a husband is not forthcoming, rather than enter into a polygamous marriage?

NWAPA: Many women are doing this, Marie. Many women are saying that they don't want a husband but they want a child.

UMEH: So your female character, Amaka who has twins and refuses to marry the father, in One Is Enough, is prophetic?

NWAPA: I think she is, because she, like many women, has had experiences in her married life, with men generally, which were nothing but war.

UMEH: Do you have a specific message for women? Do you have an ideology or some words of advice?

NWAPA: Yes, I do. I feel that every woman, married or single, must have economic independence. If you look at One Is Enough, I quote a Hausa proverb which says, "a woman who holds her husband as a father dies an orphan."

UMEH: My interpretation of the proverb is that a woman should be economically independent. One should not rely on inheritance or men for survival?

NWAPA: Exactly.

UMEH: Thank you very much.

Acknowledgements

I am most grateful to my friends and mentors who have graciously shared with me Igbo/Ugwuta cultural traditions: Chikwenye Okonjo Ogunyemi, Davidson Umeh, Amede Leslye Obiora and Uzoma Gogo Nwakuche.

Notes

1. The Nigerian Civil War took place from 1967-70 in the Southeastern part of the country.
2. African women literary artists made their mark in oral literature. They are given prominence in the telling of moonlight stories, educating children through the medium of proverbs, riddles, folktales, song, dance, as well as singing praise songs at traditional marriage ceremonies and burial ceremonies, to name only a few of the avenues of artistic expressivity practiced by Ugwuta women.
3. Chinua Achebe, editor of Heinemann's African Writers' Series, read Flora Nwapa's first manuscript *Efuru* and recommended it for publication. Chinua Achebe, author of *Things Fall Apart,* is also considered the Father of the Nigerian novel. So, he has many literary followers. Ernest Hemingway (*The Old Man and The Sea*) and Charles Dickens (*The Adventures of Oliver Twist*) were read by Nigerian school children when Nigeria was under British rule. So, apparently Nwapa studied them.
4. In a greater part of Igboland, women do not break the kola nut. Kola nut is a status symbol and male preserve; it symbolizes peace, good will, trust and friendship. However, according to Flora Nwapa, prominent, influential women in the Ugwuta area of Igboland have this right. Perhaps it is because of the power and influence of the mystical Lake Goddess, Ogbuide.
5. In a conversation with Alison Perry in London, Flora Nwapa describes herself as a womanist, thereby identifying with Alice Walker's term, *womanist.* See Walker's definition of *womanist* in the preface of her book, *In Search of Our Mothers' Gardens.* Also see Chikwenye Ogunyemi's essay entitled, "Womanism: The Dynamics of the Contemporary Black Female Novel in English."
6. Flora Nwapa, in her novels, *One Is Enough* and *Women Are Different,* contends that there are different ways of living one's life fully and fruitfully. . . marriage is not the only way.

Works Cited

Aidoo, Ama Ata. "Unwelcome Pals and Decorated Slaves." *AFA:Journal of Creative Writing.* 1(1982): 34-43.

Awe, Bolanle. *Nigerian Women in Historical Perspectives.* Lagos/Ibadan: Sankore/Bookcraft, 1992.

Davies, Carole Boyce and Anne Adams Graves, eds. *Ngambika: Studies of Women in African Literature.* Trenton, NJ: Africa World Press, 1986.

Davies, Carole Boyce. "Writing Off Marginality, Minoring and Effacement." *Women's Studies International Forum*. 14.4 (1991): 249-263.

Egejuru, Phanuel A. A Book Review of *Power and Powerlessness of Women in West African Orality*. Ed. Raoul Granqvist & Nnadozie Inyama. *Research in African Literatures*. 24.3(Winter, 1993): 148-152.

Inyama, Nnadozie. "The Rebel Girl in West African Literature: Variations on a Folk Theme." *Power and Powerlessness of Women in West African Orality*. Ed. Raoul Granqvist & Nnadozie Inyama. Sweden: Umea University, 1992. 109-121.

Mba, Nina Emma. *Nigerian Women Mobilized: Women's Political Activity in Southern Nigeria, 1900-1965*. Berkeley: Institute of International Studies, 1982.

Mojola, Yemi. "Flora Nwapa." *Perspectives on Nigerian Literature: 1700 to the Present. Volume II*. Ed. Yemi Ogunbiyi. Lagos: Guardian Books, 1988. 122-127.

Nwapa, Flora. *Cassava Song and Rice Song*. Enugu: Tana, 1986.

___ *Efuru*. London: Heinemann, 1966.

___ *Idu*. London: Heinemann, 1970.

___ *The Lake Goddess*. (Forthcoming)

___ *Never Again*. Trenton, NJ: Africa World Press, 1992.

___ *One Is Enough*. Enugu: Tana, 1982.

___ *This Is Lagos and Other Stories*. Trenton, NJ: Africa World Press, 1992.

___ *Wives At War: And Other Stories*. Trenton, NJ: Africa World Press, 1992.

___ *Women Are Different*. Enugu: Tana, 1986.

Ogunyemi, Chikwenye. "Womanism: The Dynamics of the Contemporary Black Female Novel in English." *Revising the Word and the World: Essays in Feminist Literary Criticism*. Ed. Vèvè A. Clark, Ruth-Ellen B. Joeres, and Madelon Sprengnether. Chicago: University of Chicago Press, 1993, 231-248.

Perry, Alison. "Meeting Flora Nwapa." *West Africa*. 3487 (June 18, 1984): 1262.

Sofola, 'Zulu. "Foreword." *Africana Womanism: Reclaiming Ourselves* by Clenora Hudson-Weems. Troy, Mich: Bedford, 1993.

Uchendu, Victor C. *The Igbo of Southeast Nigeria*. Porthsmouth, New Hampshire: Heinemann, 1991.

Walker, Alice. *In Search of Our Mothers' Gardens: Womanist Prose*. San Diego: Harcourt, 1983.

Chronology of Flora Nwapa's Life and Works

1931   • January 13: Born Flora Nwanzuruahu Nkiru Nwapa in Ugwuta, Nigeria to Christopher Ijeoma Nwapa and Martha Onyema Onumonu, both of Ugwuta, Imo State.

1936   • Attends C.M.S. Central School in Ugwuta.

1944   • Attends Archdeacon Crowther Memorial Girls' School, Elelenwa, near Port Harcourt, Nigeria.

1949   • Attends C.M.S. Girls' School, Lagos.

1950   • Receives Cambridge Overseas Senior School Leaving Certificate Grade 2.

1950-51 • Attends Queen's College, Lagos.

1952   • Teaches at Priscilla Memorial Grammar School, Ugwuta. Obtains 'A' Levels in English and History in G.C.E.

1953   • Matriculates at University College, Ibadan (UCI).

1955   • President, Women's Hall Residence, University College, Ibadan, Nigeria.

1957
- Receives B.A. (London) in May, University College Ibadan, Nigeria.
- Attends University of Edinburgh, Scotland.

1958
- Receives Diploma in Education in May, University of Edinburgh, Scotland. Travels to Europe and spends nine weeks in Switzerland.
- Returns to Nigeria and joins Ministry of Education as Woman Education Officer, Inspectorate Division, Calabar.

1959
- Teaches in Queen's School, Enugu.
- A daughter, Ejine, is born June 28th.

1961
- Begins writing the novel, *Efuru*.

1962
- Joins University of Lagos as an Administrative Officer.

1964
- Teaches English and Geography to journalist students at the International Press Institute (IPI) at the University of Lagos.
- Becomes member and Secretary of the Society of Nigerian Authors.
- Promoted to Assistant Registrar in charge of Public Relations at the University of Lagos.

1965
- Receives a grant for nine weeks from the Ford Foundation and State Department U.S.A. to study Public Relations, Fund Raising and Alumni Administration in U.S.A. universities.
- Receives a British Council grant in the United Kingdom for three weeks to study University Administration.

1966
- Publishes *Efuru* by Heinemann Educational Books, earns the distinction of being the first African woman to publish in London and Nigeria's premier woman novelist in the English language.

1967
- Marries Gogo Nwakuche.

1969
- A son, Uzoma, is born April 12th.

1970   • Serves the Executive Council of East Central State, Nigeria as Commissioner of Health and Social Welfare; reunites displaced children from Gabon and Ivory Coast to their families in Nigeria. Closes down numerous orphanages which sprung up during the Nigerian Civil War.
        • Publishes a second novel, *Idu*, by Heinemann Educational Books, London.

1971   • Publishes *This Is Lagos and Other Stories*, Nwamife Publishers Ltd., Enugu, Nigeria. A daughter, Amede, is born October 28th.
        • Accepts political appointment as Commissioner of Lands, Survey and Urban Development, East Central State, Nigeria.

1972   • Publishes Children's book, *Emeka—Driver's Guard*, University of London Press, London.

1973   • Presents paper, "African Women in Literature" on March 9th to the Women's Society at the University of Nigeria, Nsukka (UNN).
        • Attends African Literature Association Conference in Syracuse, New York; presents paper, "African Women in Literature."

1974   • Accepts political appointment as Commissioner of Establishment, East Central State, Nigeria.

1975   • Publishes *Never Again*, a novel, Nwamife Publishers, Ltd., Enugu.
        • Ends political appointment as General Yakubu Gowon's regime ends.

1976   • Visiting Lecturer in Creative Writing at Alvan Ikoku College of Education, Owerri.
        • Registers Tana Press Ltd. and Flora Nwapa Books & Company to publish and print children and adult books.

1977   • Founder and Managing Director, Flora Nwapa Books & Company and Tana Press, Ltd., Enugu, Nigeria.

1979   • Publishes *Mammywater, My Tana Colouring Book*, and *My Animal Colouring Book*, Flora Nwapa Books.

1980   • Publishes *The Miracle Kittens, Journey to Space, The Adventures of Deke*, Flora Nwapa Books.
  • Attends Frankfurt Book Fair and exhibits books published by Flora Nwapa Books and Tana Press Ltd.
  • Receives Certificate of Merit, Nigerian Association of University Women.

1981   • Nigerian Representative at Bologna International Children's Book Fair.
  • Attends African Literature Association Conference in Claremont, California. Vice-President, Association of Nigerian Authors (ANA), Lagos State.
  • Publishes the novel, *One Is Enough*, Tana Press, Ltd. Publishes *My Tana Alphabet Book* and *My Animal Number Book*, Flora Nwapa Books.

1983   • Receives Officer of the Niger (OON) Award from the Nigerian Federal Government.
  • Attends Seminar in Freetown, Sierra Leone; presents paper, "Creative Writing and Publishing for Childrenin Africa."

1984   • Attends First Feminist Book Fair in London. Attends University of Iowa, Iowa City, U.S.A., International Writing Program.
  • Receives Certificate of Honorary Fellow in Writing Award.
  • Lectures at U.S. universities (Ames, Austin, Lincoln, Northwestern) on African Literature.

1985   • Attends International Conference at Michigan State University, East Lansing and presents paper, "The Black Woman Writer and the Diaspora."
  • Attends Kalamazoo College and the University of Michigan at Ann Arbor; presents paper "The Black Woman Writer and the Politics of Publishing."
  • Attends Women's Forum (End of the Decade Conference), Nairobi, Kenya and presents paper, "Women As Guardian of Culture in Africa."

- Attends International Zimbabwe Book Fair, Harare; reads paper, "An African Woman as Publisher." Receives the University of Ife Merit Award on Authorship and Publishing.

1986
- Attends the Second International Feminist Book Fair in Oslo, Norway. Reads paper, "Sisterhood and Survival: The Nigerian Experience."
- Translation of *Efuru* into French.
- Publishes *Women Are Different, Cassava Song & Rice Song and Wives at War & Other Stories*, Tana Press.

1987
- Receives Honarary awards from the Solidra Circle in Lagos and The Octagon Club in Owerri, Nigeria.

1988
- Attends Third International Book Fair in Montreal, Canada and presents paper, "Women's Writing Versus Tradition, Economic, Social and Political Oppression," and participates in a Seminar in Toronto.
- Attends a conference in London organized by *Akina Mama wa Africa* (Solidarity Among African Women.) Presents paper, "Women in Writing and Publishing."

1989
- Participates in Cultural Exchange Program between Nigeria and Switzerland. Translation of short story, "The Chief's Daughter," into Swiss German.
- Visiting Professor in Creative Writing at University of Maiduguri, Nigeria. President, Association of Nigerian Authors (ANA), Bornu State.

1990
- Accepts appointment by President of Nigeria, Ibrahim Babangida, on the Commission on Review of Higher Education in Nigeria.

1991
- Presents Key Note Address "Nwanyibu: Womanbeing & African Literature" at the 17th Annual African Literature Association Conference at Loyola University, New Orleans, Louisiana.
- Attends and presents paper, "Priestesses and Power Among The Riverine Igbo" at the conference, *Queens, Queen Mothers, Priestesses and Power: Case Studies in African Gender*. April 8-11, 1991. The Schomburg

Center for Research in Black Culture, New York:1-9.
- Begins research and writing of Ugwuta priestesses which culminates in the novel, *The Lake Goddess* (unpublished).
- Translation of *Efuru* into Dutch.

1992
- Publishes her creative works, *Never Again, One Is Enough, Women Are Different, This Is Lagos & Other Stories, Wives at War & Other Stories,* in the United States with Africa World Press, Inc., Trenton, New Jersey.
- Receives Medal of Honor from Former Governor Amadi Ikweche of Imo State in Owerri, Nigeria.
- Attends Seminar on Children's Literature, University of Umea, Sweden; presents paper, "Writing and Publishing for Children in Africa."
- Celebrates Black History Month at John Jay College of Criminal Justice, CUNY, and Black Books Plus, New York. Speaks on "African Women in Literature": February 18.
- Attends Fifth International Feminist Book Fair in Amsterdam, Holland; presents paper, "The Image of the New Woman."
- Attends The First International Conference on "Women in Africa and the African Diaspora: Bridges Against Activism and the Academy," at the University of Nigeria, Nsukka (UNN) July 13-18; gives Keynote address at the Plenary Session: "Women and Creative Writing in Africa." A Speaker on panel, Conversations with African Women Writers.
- Invited as one of the judges on the Commonwealth Writer's Prize Award Committee in Nairobi, Kenya.
- Presents paper, "The African Woman Writer" at various universities in the United States. Member of PEN International.
- Gives manuscript *The Lake Goddess*, a novel, to Dr. Chester Mills for editing and computing printing.
- Interviewed by Marie Umeh and leaves manuscript, *The Lake Goddess*, with Marie Umeh and Marvie Brooks for publishing.

1993
- Accepts invitation by Gay Wilentz and appointment to teach Creative Writing at East Carolina University in North Carolina.
- Publication of *The First Lady: A Play*. Enugu: Tana Press.
- Publication of *Conversations [Plays]*. Enugu: Tana Press.
- Dies from pneumonia on October 16th at University of Nigeria Teaching Hospital, Enugu.
- Buried at Amede's Court in Ugwuta, Nigeria on November 20th.

1997
- Translation and publication of *Efuru* in Germany.

Flora Nwapa (1931-1993)
A Bibliography

Brenda F. Berrian

NOVELS

Efuru. London: Heinemann, 1966.

Idu. London: Heinemann, 1970.

Never Again. Enugu: Nwamife, 1975; rpt. Trenton, N.J.: Africa World Press, 1992.

One Is Enough. Enugu: Tana, 1981; rpt. Trenton, N.J.: Africa World Press, 1992.

Women Are Different. Enugu: Tana, 1986; rpt. Trenton, N.J.: Africa World Press, 1992.

The Lake Goddess. (Unpublished manuscript)

The Umbilical Cord. (Unpublished manuscript)

BIOGRAPHY

Golden Wedding Jubilee of Chief and Mrs. C. I. Nwapa, April 20, 1930-April 20, 1980. Enugu: Flora Nwapa Books, 1980.

SHORT STORIES

"My Spoons Are Finished." *Présence Africaine* 63(1967):227-35.

"Idu." *Présence Africaine* 1.4(1968):50-52.

This Is Lagos and Other Stories. Enugu: Nwankwo-Ifejika, 1971. rpt. Trenton,

N.J.: Africa World Press, 1992.

"The Campaigner." *The Insider: Stories of War and Peace from Nigeria.* Ed. Chinua Achebe. Enugu: Nwankwo-Ifejika, 1971. 73-78; rpt. *African Rhythms: Selected Stories and Poems.* Ed. Charlotte Brooks. New York: Washington Square, 1974. 136-55.

"Ada." *Black Orpheus* 3.4(1976):20-30.

"This Is Lagos." *Unwinding Threads: Writing by Women in Africa.* Ed. Charlotte Bruner. London: Heinemann, 1983. 40-48; rpt. *Daughters of Africa: An International Anthology of Words and Writings by Women of African Descent, From Ancient Egyptian to the Present.* Ed. Margaret Busby. New York: Pantheon, 1992. 399-406.

Wives At War and Other Stories. Enugu: Tana, 1980. Rpt. Trenton, N. J.: Africa World Press, 1992.

The New Game and Other Stories. (Unpublished manuscript)

The Silent Passengers and Other Stories. (Unpublished manuscript)

Miri and Other Stories. (Unpublished manuscript)

The Debt and Other Stories. (Unpublished manuscript)

PLAYS

The First Lady: [A Play]. Enugu: Tana, 1993.

"The Sychophants." In *Conversations:* [Plays]. Enugu: Tana, 1993

"Two Women in Conversation." In *Conversations:* [Plays]. Enugu: Tana, 1993.

CHILDREN'S LITERATURE

Emeka—Driver's Guard. London: University of London, 1972.

Mammywater. Enugu: Flora Nwapa Books, 1979.

My Animal Colouring Book. Enugu: Flora Nwapa Books, 1979.

My Tana Colouring Book. Enugu: Flora Nwapa Books, 1979.

The Adventures of Deke. Enugu: Flora Nwapa Books, 1980.

Journey to Space and Other Stories. Enugu: Flora Nwapa Books, 1980.

"The Magic Cocoyam." A short story in *Journey to Space and Other Stories.* Enugu: Flora Nwapa Books, 1980.

The Miracle Kittens. Read It Yourself Series. Enugu: Flora Nwapa Books, 1980.

My Tana Alphabet Book. Enugu: Flora Nwapa Books, 1981.

My Animal Number Book. Enugu: Flora Nwapa Books, 1981.

POETRY

Cassava Song and Rice Song. Enugu: Tana, 1986.

ESSAYS

"Women in Politics." *Présence Africaine* 141.1 (1978):115-21.

"Nigeria—The Woman As A Writer." *Realities* (1985):1, 22-24.

"Sisterhood and Survival: The Nigerian Experience." Paper presented at the Second International Feminist Book Fair, Oslo, Norway, 1986: 1-20.

(Unpublished paper)
"Writers, Printers and Publishers." *Guardian* [Lagos] 17 Aug. 1988:16.
"The Role of Women in Nigeria." 1-17 (Unpublished paper)
"Priestesses and Power Among the Riverine Igbo." Paper presented at the conference, *Queens, Queen Mothers, Priestesses and Power: Case Studies in African Gender*. April 8-11, 1991. The Schomburg Center for Research in Black Culture, New York:1-9.
"Priestesses and Power Among the Riverine Igbo." In *Queens, Queen Mothers, Priestesses, and Power: Case Studies in African Gender*. Ed., Flora Edouwaye S. Kaplan. New York: New York Academy of Sciences, 1997, 415–424.
"Women and Creative Writing in Africa." In *Sisterhood, Feminisms, and Power*. Ed., Obioma Nnaemeka. Trenton, New Jersey: Africa World Press, 1998: 106-118.
"Writing and Publishing for Children in Africa—A Personal Account." In *Preserving the Landscape of the Imagination: Children's Literature in Africa*. Ed., Raoul Granqvist & Jürgen Martini. [*Matatu* 17-18] Amsterdam /Atlanta:Editions Rodopi, 1997:265-275.

CRITICISMS OF FLORA NWAPA'S WORKS

Abanobi, N. N. *Women's Role in the World of Flora Nwapa's Novels*. B.A. thesis, University of Nigeria, Nsukka, 1977-78.
Acholonu, Catherine. *Western and Indigenous Traditions in Modern Igbo Literature*. Düsseldorf: University of Düsseldorf, 1985.
Acholonu, Rose. *Love in Nigerian Fiction*. Ph.D. Dissertation, University of Port Harcourt, 1991.
___ "The Dynamism of Love: Flora Nwapa's *Idu* and *Efuru*." Ed. Rose Acholonu. *Family Love in Nigerian Fiction: Feminist Perspectives*. Owerri, Nigeria: Achisons, 1995:126-148.
___."Flora Nwapa's Heroines: *One Is Enough* and *Women Are Different*." Ed. Rose Acholonu. *Family Love in Nigerian Fiction: Feminist Perspectives*. Owerri, Nigeria: Achisons, 1995:177-92.
___."The Female Predicament in the Nigerian Novel." *Feminism in African Literature: Essays on Criticism*. Ed. Helen Chukwuma. Lagos: New Generation, 1994:38-52.
Achufusi, Ify G. "Feminist Inclinations of Flora Nwapa." *Critical Theory & African Literature Today* 19(1994):101-114.
Adelugba, Segun. "Women in Nigerian Literature." *National Concord* 4 Sept. 1990:5.
Afuba, Ifeanyi. "Positive Feminism Identified in 3 Novels." *The Guardian* [Nigeria] 14 April, 1986:14.
Aidoo, Ama Ata. "THESE DAYS [III]—A Letter to Flora Nwapa." *Research in African Literatures* 26.2(1995):17-21.
Amuta, Chidi Nnanna. *Dissonant Harmony: Art and Social Reality in Literature Based on the Nigerian Civil War*. Ph.D. Dissertation, Obafemi Awolowo University, 1987.
Andrade, Susan Z. "Rewriting History, Motherhood, and Rebellion: Naming an African Women's Literary Tradition." *Research in African Literatures*

21.1(1990):91-110.

Anozie, Sunday O. *Sociologie du Roman Africaine: Réalisme, structure et déter-
mination dans le roman ouest africain.* Paris: Auber Montaigne, 1970. 26,
38, 103-9.

Asanbe, Joseph. *The Place of the Individual in the Novels of Chinua Achebe, T.
M. Aluko, Flora Nwapa and Wole Soyinka.* Dissertation Abstracts
International (DAI) 40(1980):5447A.

___ "The Context of the Nigerian Writer: The Example of Nwapa and
Emecheta." The Writer and the Critic: The 7th Annual Conference Held
at the University of Calabar, Calabar, April 20-23,1983. Ed. K. Echenim.
n.p.: *Modern Languages Association of Nigeria*, n.d. 211-25.

___ "Context of Writer and Audience: Nwapa ad Emecheta." *LARES* 6-7(1984-
85):185-96.

Ashimole, Elizabeth O. "Nigerian Children's Literature and the Challenges of
Social Change." *Children and Literature in Africa.* Ed. Chidi Ikonne, Emelia
Oko and Peter Onwudinjo. Ibadan: Heinemann, 1992:70-81.

Awafinya, Michael. "Flora Nwapa: Nigeria's First Female Novelist." *Sunday
Concord* 11 May 1986:11-12.

Azuonye, Chukwuma. "Folk Stereotypes and the Theme of Marital
Compatibility in the Novels of Flora Nwapa." *NKA* 2(1988): 1-12.

Bala, L. Sasi. "Heroines of Flora Nwapa." *Commonwealth Fiction.* Vol. II. Ed.
R. K. Dhawan. New Delhi: Classical, 1988: 260-72.

Banyiwa-Horne, Naana. "African Womanhood: The Contrasting Perspectives
of Flora Nwapa's *Efuru* and Elechi Amadi's *The Concubine.*" In *Ngambika:
Studies of Women in African Literature.* Ed. Carole Boyce Davies and Anne
Adams Graves. Trenton, NJ: Africa World, 1986:119-30.

Bazin, Nancy Topping. "Weight of Custom, Signs of Change: Feminism in the
Literature of African Women." *World Literature Written in English*
25.2(1985):183-97.

___ "Feminism in the Literature of African Women." *Black Scholar* 20.
3.4(1989):8-17.

Bengu, Sibusiso M. E. "African Cultural Identity and International Relations:
Analysis of Ghanaian and Nigerian Sources 1958-1974." Pref. Roy
Preiswick. Geneva: University of Geneva, 1976; Doctoral Dissertation.
Rpt. *Gods Are Not Our Own.* Pietermarizburg: Shuster and Shuster, 1976.

Berrian, Brenda F. "African Women As Seen in the Works of Flora Nwapa and
Ama Ata Aidoo." *CLA Journal* 25.3(1981):331-39.

___ "The Reinvention of Woman through Conversations and Humor in Flora
Nwapa's *One Is Enough.*" *Research in African Literatures* 26.2(1995):53-67.

___ "In Memoriam: Flora Nwapa (1931-1993)." *Signs: Journal of Women in
Culture and Society* 20.4(1995):996-99.

Boehmer, Elleke. "Stories of Women and Mothers: Gender and Nationalism
in the Early Fiction of Flora Nwapa." *Motherlands: Black Women's Writing
from Africa, the Caribbean and South Asia.* Ed. Susheila Nasta. London:
The Women's Press, 1991. 3-23.

Booth, James. *Writers and Politics in Nigeria.* London: Hodder and Stoughton,
1981:80-81.

Bottcher, Karl-Heinz. *Tradition und Modernitat bei Amos Tutuola und Chinua*

Achebe: Grandzuge der Westafrikanischen Erzahl-literatur Englischer Sprache. Bonn: Verlag Grundmann, 1974.

Brown, Lloyd W. "The African Woman as Writer." *Canadian Journal of African Studies* 9.3(1975):493-501.

___ *Women Writers in Black Africa.* Westport, Conn.: Greenwood, 1981. 122-57.

Bruner, Charlotte. "A Decade for Women Writers." *African Literature Studies: The Present State/L'Etat Present.* Ed. Stephen Arnold. Washington, D.C.: Three Continents, 1985:189-216.

Bryce-Okunlola, Jane. *A Feminist Study of Fiction by Nigerian Women Writers.* Ph.D. Dissertation. Ile-Ife: Obafemi Awolowo University, 1989.

___ "Conflict and Contradiction in Women's Writing on the Nigerian Civil War." *African Languages and Cultures* [London] 4 (1991): 29-42.

___ "Inventing Autobiography: Some Examples From Fiction and Journalism by Nigerian Women Writers." *Seminar on Aspects of Commonwealth Literature: 1989-1990.* London: Institute of Commonwealth Studies, 1990: 1-9.

___ "Motherhood as a Metaphor for Creativity in Three African Women's Novels: Flora Nwapa, Rebeka Njau and Bessie Head." *Motherlands: Women's Writing from Africa, the Caribbean and South Asia.* Ed. Susheila Nasta. London: The Women's Press, 1991:200-18.

Busia, Abena. "Flora Nwapa: A Tribute from the President of the ALA." *ALA Bulletin* 20.1(1994):7.

Cavendish, Luci. "Tract and Fiction." *The Guardian* [London] 6 August 1992:10.

Chikwe, Kema. *Women and New Orientation: A Profile of Igbo Women in History.* Owerri: Prime Time, 1994.

Chukukere, Gloria Chineze. *Gender Voices and Choices: Redefining Women in Contemporary African Fiction.* Enugu: Fourth Dimension, 1995.

___ "Images of Women in Contemporary African Fiction." *Introductory Readings in the Humanities and Social Sciences.* Ed. Gloria Chukukere. Onitsha: University Publishing, 1988:13-29.

___ *The Dilemma of the Woman as Writer and Protagonist in Contemporary African Fiction.* Ph.D. Dissertation, Lagos: University of Lagos, 1984.

Chukwuma, Helen O. Ed. "The Legacy of Flora Nwapa." *Journal of Women's Studies in Africa* 1(1997).

___ *Feminism in African Literature: Essays on Criticism.* Lagos: New Generation, 1994.

___ "Flora Nwapa is Different." *Feminism in African Literature: Essays on Criticism.* Enugu: New Generation, 1994:115-130.

___ "Voices and Choices: The Feminist Dilemma in Four African Novels." *Feminism in African Literature.* Lagos: New Generation, 1994:215-227.

___ "Two Decades of the Short Story in West Africa." *WAACLALS Journal* [Calabar] 1(1989):1-14.

___ "Aspects of Style in Igbo Oral Literature." *Introductory Readings in the Humanities and Social Sciences.* Ed. Gloria Chukukere. Onitsha: University Publishing, 1988:31-52.

___ "Nigerian Female Authors, 1970 to the Present." *Matatu* [Frankfurt]

1.2(1987):23-42.

Condé, Maryse. "Three Female Writers in Modern Africa: Flora Nwapa, Ama Ata Aidoo and Grace Ogot." *Présence Africaine* (Paris) 82(1972):132-43.

Coulon, Virginia. "Women at War: Nigerian Women Writers and the Civil War." *Commonwealth Essays and Studies* 13.1(1990): 1-12.

Coussy, Denise. *Le Roman Nigérian*. Paris: Editions Silex, 1988.

Dathorne, O. R. *African Literature in the Twentieth Century*. Minneapolis: University of Minnesota, 1974. 117-118.

Davies, Carole Boyce. "Motherhood in the Works of Male and Female Writers: Achebe, Emecheta, Nwapa and Nzekwu." *Ngambika: Studies Of Women in African Literature*. Ed. Carole Boyce Davies and Anne Adams Graves. Trenton, NJ: Africa World, 1986:241-56.

___ "Private Selves and Public Spaces: Autobiography and the African Woman Writer" *Neohelicon* 17.2(1992):183-210.

___ "Wrapping One's Self in Mother's Akatado-Cloths: Mother-Daughter Relationships in the Works of African Women Writers." *Sage* 4.2(1987):11-19.

___ "Writing Off Marginality, Minoring, Effacement." *Women's Studies International Forum* 14(1991):249-63.

Duruoha, Iheanyichukwu. "The Igbo Novel and the Literary Communication of Igbo Culture." *African Study Monographs* (Kyoto)12(1991):185-200.

Egbufor, Ada. "Flora Nwapa: Africa's First Lady of Letters."*African Profiles International*. March/April, 1993:34.

Ebeogu, Afam. "Igbo Sense of Tragedy: A Thematic Feature of the Achebe School." *Literary Half-Yearly* 24.1(1983):69-86.

Egejuru, Phanuel A. "Flora, Onyiba Nwanyi." *ALA Bulletin* 20.1 (1994):16-17.

Eko, Ebele, Julius Ogu & Emelia Oko. *Flora Nwapa: Critical Perspectives*. Calabar: University of Calabar Press, 1997.

Emecheta, Buchi. "Nwayi oma, biko nodu nma." *West Africa* 24-30 October 1994: 1831.

Emenyonu, Ernest N. "African Literature Revisited: A Search for Critical Standards." *Revue de Littérature Comparée* 43. 3.4(1974):387-97.

___ "Post-War Writing in Nigeria." *Issue* 3.2(1973):49-51.

___ "Who Does Flora Nwapa Write For?" *African Literature Today* 7(1975):28-33.

___ "The Nigerian Civil War and the Nigerian Novel: The Writer as Historical Witness." *Studies on the Nigerian Novel*. Ibadan: Heinemann, 1991:89-105.

___ "Flora Nwapa: A Pioneer African Female Voice Is Silenced." *ALA Bulletin* 20.1(1994):10-11.

Emenyonu, Patricia T. "The Role of Contemporary Female Nigerian Writers in the Education of Nigerian Youth." *The Literary Criterion* 23.1.2(1988): 216-21. Rpt. *African Literature Comes of Age*. Ed. C. D. Narasimhaiah and Ernest Emenyonu. Mysore: Dhvanyaloka, 1988:216-21.

Ette, Mercy. "Ink Dries in the Pen." *Newswatch*, 8 Nov. 1993:38.

Ezeigbo, Theodora Akachi. *Gender Issues in Nigeria: A Feminine Perspective*. Lagos: Vista Books, 1996.

___ *Fact and Fiction in the Literature of The Nigerian Civil War*. Ojo Town, Nigeria: Unity, 1991.

___ "Traditional Women's Institutions in Igbo Society: Implications for the Igbo Female Writer." *African Languages and Cultures* 3.2(1990):149-65.

___ "Reflecting the Times: Radicalism in Recent Female-Oriented Fiction in Nigeria." *Literature and Black Aesthetics.* Ed. Ernest Emenyonu et. al. Ibadan: Heinemann, 1990. 143-57.

Ezeigbo, Theodora Akachi & Liz Gunner, ed. *The Literatures of War.* [Special Issue.] *African Languages and Cultures* 4.1(1991).

Ezenwa-Ohaeto. "Flora Nwapa: *Cassava Song and Rice Song:* A Book Review." *OKIKE: An African Journal of New Writing* 29(November 1990):106-107.

___ "The Other Voices: The Poetry of Three Nigerian Female Writers." *Canadian Journal of African Studies* 22(1988):662-68.

___ "The Notion of Fulfillment in Flora Nwapa's *Women Are Different."* *Neohelicon* 19.1(1992):323-33.

___ "The Child Figures and Childhood Symbolism in Flora Nwapa's Children's Fiction." *Research in African Literatures* 26.2 (1995):68-79.

___.*Chinua Achebe, The Life: Facing The Frontiers.* Oxford: James Curry, 1997.

Fayemi, Olufunmilayo Oluwafolakemi. *Origins and Evolution of the Short Story Genre in Nigeria: An Analytic Survey.* Ph.D. Dissertation, University of Ibadan, 1990.

Fishburn, Katherine. *Reading Buchi Emecheta: Cross-Cultural Conversations.* Westport, Conn.: Greenwood, 1995.

Frank, Katherine. "Women Without Men: The Feminist Novel in Africa." *African Literature Today* 15(1987):14-34.

___ "Flora Nwapa: Africa's First Woman Publisher." *Africa Now.* 25(May, 1983):61-62.

Galloy, Martine. *La femme africaine dans la littérature nigerienne: Les cas de Flora Nwapa et Cyprian Ekwensi.* Doctorat du 3e cycle, Université de Paris III, 1982.

Gardener, Susan. "The World of Flora Nwapa." *The Women's Review of Books* 11.6(1994):9-10.

Githaiga, A. *Notes on Flora Nwapa's Efuru* . Nairobi: Heinemann, 1978.

Gordimer, Nadine. "Themes and Attitudes in Modern African Writing." *Michigan Quarterly Review* 9(1970):22; rpt. *The Black Interpreters: Notes on African Writing.* Johannesburg: Ravan,1973:20-21.

Holloway, Karla F. C. *Moorings and Metaphors: Figures of Culture and Gender in Black Women's Literature.* New Brunswick, N.J.: Rutgers University, 1992:168-187.

Ibeleme, Emmanuel. "Women Writers Bag Top Fiction Prizes." *New African* 2 (1986):49.

Ikonne, Chidi. "The Folk Roots of Flora Nwapa's Early Novels." *Orature in African Literature Today* 18(1992):96-104.

___ "The Society and Women's Quest for Selfhood in Flora Nwapa's Early Novels." *Kakaki* 3(1983-84):76-91; rpt. *Kunapipi* 6.1(1984):68-78.

Iloegbunam, Chuks. "Flora Nwapa (1931 to 1993)." *West Africa* 15-21 Nov. 1993:2088.

Iweriebor, Ifeyinwa. "Remembering Our Flora: A Beautiful Voice." *African News Weekly* 6.2(January 27, 1995):18 & 27.

James, Adeola. "*Idu*: Flora Nwapa (Review)." *African Literature Today*

5(1971):150-53.

Jarret-Kerr, Martin. "Christian Faith and the African Imagination." *Religion in Life* 41.4(1972):559-60.

Jell-Bahlsen, Sabine. "The Concept of Mammywater in Flora Nwapa's Novels." *Research in African Literatures* 26.2(1995):30-41.

___ *Mammywater: In Search of the Water Spirits in Nigeria.* A Film-Video. Berkeley: University of California Extension Center for Media and Independent Learning, 1991.

___ "The Flora Nwapa I Knew & Loved: A Tribute To Flora Nwapa." *African Profiles International.* February/March, 1994:38.

Jones, Eldred. "Locale and Universe." *The Journal of Commonwealth Literature* 3(1967):127-31.

Kandji, Diouf Fatou. "Des calvaires de la femme africaine dans la création romanesque de Buchi Emecheta, Flora Nwapa, Ngũgĩ Wa Thiong'o et Ahmadou Kourouma." *Bridges: A Senegalese Journal of English Studies* [Dakar] 4(1992):113-33.

Kern, Anita. "Flora Nwapa: *Never Again.*" *World Literature Written in English* 17.1(1978):58-59.

Klima, Vladimir. *Modern Nigerian Novels.* Dissertation Orientales 88. Prague: Oriental Institute of Academic Publishing House of Czezhoslovak Academy of Sciences, 1969.

Larson, Charles. "Whither the African Novel?" *CLA Journal* 13.2 (1969):169.

Laurence, Margaret. *Long Drums and Cannons: Nigerian Dramatists and Novelists 1952-1966.* London: MacMillan, 1968:187-91.

Lazarus, Neil. "Africa." *Longman Anthology of World Literature by Women, 1875-1975.* Ed. Marian Arkin and Barbara Shollar. New York: Longman, 1989:1061-72.

Lee, Sonia. "Le thème de bonheur ches les romancières de l'Afrique occidentale." *Présence Francophone* 29(1986):91-103.

___ "Changes in the Mother Image in West African Fiction." *Neohelicon* 14.2(1987):139-50.

Leopold, Wanda. "Powiesci Ibo [Novels by Ibo Writers]." *Przeglad Socjologiczny* 23(1969):370-87.

Lindfors, Bernth. "Achebe's Followers." *Revue de Littérature Comparée* 48. 3.4(1974):577-78.

___ "Big Shots and Little Shots of the Anglophone African Literature Canon." *Commonwealth Essays and Studies* 14.2(1992):89-97.

___ "The Famous Author's Reputation Test." *Kriteria: A Nigerian Journal of Literary Research* 1.1(1988):25-33.

___ *Long Drums and Canons: Teaching and Researching African Literatures.* Trenton, N. J.: Africa World, 1995.

___ "New Trends in West and East African Fiction." *Review of National Literatures* 11.2(1971):21, 26.

___ *Nigerian Fiction in English: 1952-1967.* Dissertation Abstracts International (DAI) 30(1969):2535A.

___ "Nigerian Novels of 1966." *Africa Today* 14.5(1967):30-31.

Lippert, Anne. *The Changing Role of Women as Viewed in Literature of English and French Speaking Africa.* Ph.D. Dissertation, Indiana University, 1972.

Little, Kenneth. *The Sociology of the Urban Woman's Image in African Literature.* London: MacMillan, 1980:58, 135-38.

Lyonga, Pauline Nalova. *Uhamiri or a Feminist Approach to African Literature: An Analysis of Selected Texts by Women in Oral and Written Literature.* Dissertation Abstracts International (DAI) 46.7(1986):1940A.

McLuskie, Kathleen and Lyn Innes. "Women and African Literature." *Wasafiri* 8(1988):3-7.

Maja-Pearce, Adewale. "Flora Nwapa's *Efuru:* A Study in Mis-placed Hostility." *World Literature Written in English* 25(1985):10-15.

___ *A Mask Dancing: Nigerian Novelists of the Eighties.* London: Hans Zell, 1992:27-29, 153-55, 176, 182.

Mason, Nondita. "Women and Development in Third World Writing." *Populi* (New York) 5.4(1978):45-49.

Mba, Nina E. *Nigerian Women Mobilized: Women's Political Activity in Southern Nigeria, 1900-1965.* Berkeley: University of California, 1982.

___ "Flora Nwapa at 60: Forever First." *Guardian* [Nigeria] 13 Jan. 1991: B6.

Mberekpe, Frank P. *Efuru in Summary and Revision Essay Questions.* Foreword Flora Nwapa. Enugu: Tana, 1987.

Mojola, Yemi. "The Novelist's View of Women in Igbo Traditional Culture: The Example of Flora Nwapa." *Nigeria Magazine* 56. 3.4(1988):25-33.

___ "The Works of Flora Nwapa." *Nigerian Female Writers: A Critical Perspective.* Ed. Henrietta C. Otokunefor and Obiageli Nwodo. Lagos: Malthouse, 1989:19-29.

___ "Flora Nwapa." *Perspectives on Nigerian Literature: 1700 to the Present*, Vol. Two. Ed. Yemi Ogunbiyi. Lagos: Guardian Books, 1988:122-27.

Morrison, Kathleen. *Metaphysical Concepts in West African Prose: Spiritual Significance and Aesthetic Implications.* Ph.D. Dissertation, University of Western Ontario, 1991.

Mugambi, Helen Nabasuta. *The Wounded Psyche and Beyond: Conformity and Marginality in Selected African and Afro-American Novels.* Dissertations Abstracts International (DAI) 50(1989):944A.

___ "Re-creating a Discourse: The Scriptable Novels of Nwapa and Emecheta." *Understanding Women: The Challenge of Cross- Cultural Perspectives.* Ed. Marilyn R. Waldman, Artemis Leontis and Muge Galin. Athens: The Ohio State University, Papers in Comparative Studies 7, 1992:167-79.

Mutiso, Gideon. *Socio-Political Thought in African Literature: Weusi?* New York: Macmillan, 1974:57-58.

—-. "Women in African Literature." *East Africa Journal* 8. 3(1971):7.

Nandakumar, Prema. "An Image of African Womanhood: A Study of Flora Nwapa's *Efuru." Africa Quarterly* 11(1967):136-46; rpt. *Recent Commonwealth Literature.* Vol. 2. Ed. R. K. Dhawan et al. New Delhi: Prestige, 1989:159-72.

Ngcobo, Lauretta. "Black African Women Writers." *Cambridge Journal of Education* 14.3(1984):16-21.

Nichols, Lee. *African Writers at the Microphone.* Washington, D.C.: Three Continents, 1984:98, 105, 149-150, 245.

Njoku, Teresa U. "Womanism in Flora Nwapa's *One Is Enough* and *Women Are Different." Commonwealth Quarterly* 14.39 (1989):1-16.

Nkosi, Lewis. "Women in Literature." *Africa Woman* 6 (1976):36-37.

Nnaemeka, Obioma. "Towards a Feminist Criticism of Nigerian Literature." *Feminist Issues* 9.1(1989):73-87.

___ "Feminism, Rebellious Women and Cultural Boundaries: Rereading Flora Nwapa and Her Compatriots." *Research in African Literatures* 26.2 (1995):80-113.

___ "From Orality to Writing: African Women Writers and the (Re)Inscription of Womanhood." *Research in African Literatures* 25.4(1994):137-157.

___ *Sisterhood, Feminisms, and Power.* Trenton, New Jersey: Africa World Press, 1998.

Nnolim, Charles E. "The Nigerian Tradition in the Novel." *CNIE* 2.2(1983):22-40.

___ "Trends in the Nigerian Novel." *Matatu* 2 (1987): 7-22; rpt. *Literature and National Consciousness.* Ed. Ernest N. Emenyonu et al. Ibadan: Heinemann, 1989:53-65.

___ "Mythology and the Unhappy Woman in Nigerian Fiction." *Critical Theory and African Literature.* Ed. Ernest N. Emenyonu et al. Ibadan: Heinemann Educational Books (Nigeria), 1987: 45-54.

Nutsukpo, Fafa Margaret. *Feminist Trends in the Novels of Flora Nwapa.* Master's Thesis, University of Port Harcourt, 1991.

Nwaegbe, William D. O. *The Problem of Alienation in Modern West African Literature.* Ph.D. Dissertation, University of Nigeria, Nsukka, 1985.

Nwahunanya, Chinyere. *Tragedy in the Anglophone West African Novel.* Ph.D. Dissertation, University of Port Harcourt, 1990.

Nwankwo, Chimalum. "The Igbo Word in Flora Nwapa's Craft." *Research in African Literatures* 26.2(1995):42-52.

Nzewi, Meki. "Ancestral Polyphony." *African Arts* 11.4(1978):74, 92-94.

Odi Assamoi, Georgette. *Le problème de l'éducation dans le roman africain de language anglaise.* Doctorat du 3e cycle, Paul Valéry Université, Montpellier III, 1977.

Ofochebe, O. A. *Destiny and the Individual in Elechi Amadi's The Concubine, Flora Nwapa's Efuru, Onuora Nzekwu's Wand of Noble Wood.* B.A. Thesis, University of Nigeria, Nsukka, 1977-78.

Ogu, J. N. "Creativity and Children's Literature." *Children and Literature in Africa.* Ed. Chidi Ikonne, Emelia Oko and Peter Onwudinjo. Ibadan: Heinemann, 1992:82-87.

Ogundipe-Leslie, Molara. *Re-Creating Ourselves: African Women & Critical Transformations.* Trenton, N. J.: Africa World, 1994.

___ "Dirge to Flora Nwapa." *ALA Bulletin* 20.1(1994):14-15.

___ "The Female Writer and Her Commitment." *Women in African Literature Today* 15(1987):5-13.

Ogunyemi, Chikwenye Okonjo. *Africa WO/MAN Palava: The Nigerian Novel By Women.* Chicago:University of Chicago, 1996:131-182.

___ "Womanism: The Dynamics of the Contemporary Black Female Novel in English." *Signs: Journal of Women in Culture and Society* 11.1(1985):63-80; rpt. *Revising the Word and the World: Essays in Feminist Literary Criticism,* Ed. Vèvè A. Clark, Ruth-Ellen B. Joeres, & Madelon Sprengnether. Chicago: University of Chicago, 1993.

___ "Introduction: The Invalid Dea(r)th, and the Author: The Case of Flora Nwapa, *aka* Professor (Mrs.) Flora Nwanzuruahu Nwakuche." *Research in African Literatures* 26.2(1995):1-16.

Ogunyemi, Chikwenye Okonjo & Marie Umeh, Ed. "Flora Nwapa." Special Issue. *Research in African Literatures* 26.2(1995).

Ojo-Ade, Femi. "Women and the Nigerian Civil War: Buchi Emecheta and Flora Nwapa." *Etudes Germano-Africaines* [Dakar] 6(1988):75-86.

___ "Female Writers, Male Critics." *African Literature Today* 13(1983):158-79.

Okereke, Grace Eche. "Children in the Nigerian Feminist Novel." *Children and Literature in Africa.* Ed. Chidi Ikonne, Emelia Oko and Peter Onwudinjo. Ibadan: Heinemann, 1992:138-149.

___ *The Independent Woman in Selected Novels of Buchi Emecheta, Flora Nwapa, Mariama Bâ and Zaynab Alkali.* Ph.D. Thesis, University of Calabar, 1991.

___ "Language, Gender, Space and Development in the Novels of Selected Igbo Writers." Rockefeller Project, Howard University, Washington, D.C., May, 1995.

Okonkwo, Juliet I. "Adam and Eve: Igbo Marriage in the Nigerian Novel." *Conch* 3(1971):137-51.

___ "The Talented Woman in African Literature." *African Quarterly* 15(1975):35-47.

Ola, Virginia U. "Flora Nwapa and the Art of the Novel." *Medium and Message* [Calabar] 1(1981):91-111.

Oladeji, Niyi. "Women in the Nigerian Novel: Two Novelists' Attitudes to Nigerian Womanhood." *Marang* 4(1983):35-46; rpt. *LARES* 9(1987):116-26.

O'Malley, P. "Recent Nigerian Fiction." *Nigerian Opinion* 3.4 (1967):191.

Oni, Dokun. *Comprehensive Notes Series: Efuru: Summary Explanations Questions & Answers.* JAMB/GCE. Ilesa, Nigeria: Eddok Nigeria Enterprises, 1987.

Onwudiofu, S. N. *Flora Nwapa's Efuru in Notes, Revision Hints, General and Chapter Summaries, Character Sketches, Sample Essay Questions and Answers.* Onitsha: Tabansi, 1988.

Opara, Chioma Amadi. *Towards Utopia: Womanhood in the Fiction of Selected West African Writers.* Ph.D. Dissertation, University of Ibadan, 1987.

___ "Femininity as Cultural Reification: The Female Voice in the Nigerian Novel." *Ogele* [Port Harcourt] 3.1(1988):14-20.

Palmer, Eustace. "Review of *Efuru*." *African Literature Today* 1(1968):56-58.

Pawlikowa-Vilhanowa, Viera. "Women in African Literatures." Ed. Eckhard Breitinger. *Defining New Idioms and Alternative Forms of Expression.* Amsterdam/Atlanta: Rodopi,1996. 167-178.

Peters, Jonathan. "English-Language Fiction from West Africa." *History of Twentieth-Century African Literature.* Ed. Oyekan Owomoyela. Lincoln: University of Nebraska, 1993:9-48.

Peterson, Kirsten Holst. "Unpopular Opinions: Some African Women Writers." *Kunapipi* 7. 2.3(1985):107-20; rpt. *A Double Colonization: Colonial and Post-Colonial Women's Writing.* Ed. Kirsten Holst Peterson and Anna Rutherford. Mundelstrup, Denmark: Dangaroo, 1986:107-22.

___ "Unorthodox Fictions about African Women." The Edward A. Clark Center for Australian Studies, Univ. of Texas, Austin. *International Literature in*

English: Essays on the Major Writers. Ed. Robert L. Ross. New York: Garland, 1991:283-92.

Phillips, Maggie. "Engaging Dreams: Alternative Perspectives on Flora Nwapa, Buchi Emecheta, Ama Ata Aidoo, Bessie Head and Tsitse Dangarembga's Writing." *Research in African Literatures* 25(1994):89-103.

Pifferi, Annisa. *La Donna Nel Romanzo Africano In Lingua Inglese (Letteratura).* Calliano, Italia: Mangrini Editori, 1985.

Ramba, Opa. "African Profiles Honors Nwapa's Memory." *African Profiles International.* February/March, 1994:36-7.

Sample, Maxine J. Cornish. *The Representation of Space in Selected Works by Bessie Head, Buchi Emecheta and Flora Nwapa.* Dissertation Abstracts International (DAI) 51.5(1990):1611A.

___ "In Another Life: The Refugee Phenomenon in Two Novels of the Nigerian Civil War." *Modern Fiction Studies* 37.3(1991):445-54.

Scheub, Harold. "Two African Women." *Revue des Langues Vivantes* 37.5(1971):545-58.

Schipper, Mineke. "Women and Literature in Africa." *Unheard Words.* Ed. Mineke Schipper. London: Allison & Busby, 1984:22-58.

Schmidt, Nancy J. "Children's Books by Well-Known African Authors." *World Literature Written in English* 18.1(1979):114-23.

___ "Nwapa, Flora (Florence Nwanzuruahu Nkiru Nwapa)." *Twentieth-Century Children's Writers.* Ed. Tracy Chevalier. Chicago and London: St. James, 1989:730-31.

Shoga, Yinka. "Women Writers and African Literature." *Afriscope* 3.10(1973):44-45.

Slomski, Genevieve. *Dialogue in the Discourse: A Study of Revolt in Selected Fiction by African Women.* Dissertation Abstracts International (DAI) 47.5(1986):1721A.

Stratton, Florence. *Contemporary African Literature and the Politics of Gender.* New York: Routledge, 1994. 80-130.

___ *Manichean Aesthetics Reconsidered: Contemporary African Literature and the Politics of Gender.* Ph.D. Dissertation, School of Oriental and African Studies, University of London, 1991.

___ "The Shallow Grave: Archtypes of Female Experience in African Fiction." *Research in African Liteatures* 19.2(1988):143-69.

Strong-Leek, Linda. *Excising the Spiritual, Physical, and Psychological Self: An Analysis of Female Circumcision in the Works of Flora Nwapa, Ngũgĩ wa Thiong'o, and Alice Walker.* Dissertation Abstracts International (DAI-A) 55/07 (1995).

Swann, Joseph. "The Heroic, the Ironic, and the Underlying Condition: Interpretations of (African) Life in the Short Stories of Cyprian Ekwensi, Chinua Achebe and Flora Nwapa." *The Story Must Be Told: Short Narrative Prose in the New English Literatures.* Ed. Peter O. Strummer. Würzburg: Konigshausen und Neumann, 1986:47-57.

Taiwo, Oladele. *Culture and the African Novel.* London: Macmillan; New York: St. Martin's, 1976:ix, 5, 156, 183.

___ *Female Novelists of Modern Africa.* London: Macmillan, 1984:47-83.

Ugwu, Anthonia Nnenna. *A Study of Some Aspects of "Nigerian English" in*

Nigerian Prose Fiction. Ph.D. Dissertation, University of Ibadan, 1990.

Uhunmwangho, Amen. "On Feminism: Flora Nwapa and Buchi Emecheta." *Guardian* [Lagos] 30 Mar. 1991: 16; rpt. *Third World First* [Lagos] 2.4(1991):22-24.

Umeh, Marie. "Ivory Befits Her Ankles: In Celebration of Flora Nwanzuruahu Nwapa and Her Legacy." *1994 Annual Selected Papers.* Ed. Kofi Anyidoho, Abena Busia and Anne Adams. *African Literature Association*, 1998:1-20. (Forthcoming)

___ "Finale: Signifyin(g) The Griottes: Flora Nwapa's Legacy of (Re)Vision and Voice." *Research in African Literatures* 26.2 (1995):114-23.

___ "A Tribute to Flora Nwapa." *ALA Bulletin* 20.1(1994):8-9.

___ "The Poetics of Thwarted Sensitivity." *Critical Theory and African Literature.* Ed. Ernest N. Emenyonu. Calabar Studies in African Literature. Ibadan: Heinemann, 1987:194-206.

___ "Towards the Demythification of Female Images in Children's Literature: The Nigerian Challenge." Paper presented at ALA Fifteenth Annual Meeting. L'Université Cheick Anta Diop, Dakar, Senegal, 1989:1-22.

___ "Children's Literature in Nigeria: Revolutionary Omissions." In *Preserving the Landscape of the Imagination: Children's Literature in Africa.* Ed., Raoul Granqvist & Jürgen Martini. [*Matatu* 17-18] Amsteredam/Atlanta: Editions Rodopi, 1997:191-206.

Vivlov, V. N. *Proza Nigerii [Nigerian Prose].* Moscow: Nauka, 1973.

___ "Formirovania nigeriskai literatury (The Formation of Nigerian Literature)." *Vzaimosviazi Afrikanskikh Literatur Cbornik Statel [Interrelationships of African Literature and Literatures of the World].* Ed. Irina Nikiforova. Moscow: Nauka, 1975.

Wastberg, Per. *Afrikas moderna litteratur [Modern African Literature].* Stockholm: Wahlstrom & Widstrand och Nordiska Afrikainstitutet, 1969:98.

Wilentz, Gay. "The Individual Voice in the Communal Chorus: The Possibility of Choice in Flora Nwapa." *ACLALS Bulletin* 7.4 (1986):30-36.

___ *From Africa to America: Cultural Ties That Bind in the Works of Contemporary African and African-American Women Writers.* Dissertation Abstracts International (DAI)47(1987):3421-22A.

___ "Flora Nwapa: *Efuru.*" *Binding Cultures: Black Women Writers in Africa and the Diaspora.* Bloomington: Indiana University, 1992:3-19.

___ "Flora Nwapa." *Twentieth-Century Caribbean and Black African Writers. Dictionary of Literary Biography* 125. Ed. Reinhard Sander and Bernth Lindfors. Detroit: Gale Research, 1993:178-83.

___ "Flora Nwapa [Nwakuche] 1931-1993." *ALA Bulletin* 20.1 (1994):12-13.

___ "Flora Nwapa, 1931-1993." *The Women's Review of Books* 11.6(1994):8.

Zell, Hans. "Publishing in Africa: Writing Wrongs for Children." *West Africa* 7 Mar. 1983:607-10; rpt. in *African Books Publishing Record* 9(1983):13-14.

INTERVIEWS

Anon. *West Africa* 14 July 1972: 891.

Anon. *West Africa* 9 Oct. 1972: 1355.

Adeleke, Segun, Bose Adeogun, and Ben Nwanne. "Interview with Flora Nwapa: Some ANA Members Are Crazy." *Quality Weekly,* 23 Aug.1990:28, 31-35.

Adeniyi, Dapo. "Nigerian Female Writers: Scanty Drops from the Inkpot." *Sunday Times* 22 Sept. 1991:17-18.

___ "I Am Not a Feminist—Flora Nwapa." *Daily Times* (Lagos) 16 Nov. 1991:20.

Agueta, John. "Flora Nwakuche." *Sunday Observer* 18 Aug. 1975:6.

___ "Flora Nwakuche." *Interview with Six Nigerian Writers.* Benin City: The Bendel Newspapers Corporation, 1974:22-7.

___ *Sunday Observer.* 18 Aug. 1975:6.

___ "Nwakuche—Novelist with Concern for Women." *Daily Times* (Lagos) 3 July 1982:7.

Awoyinfa, Michael. "Flora Nwapa: Nigeria's First Female Novelist." *National Concord* 23 Jan. 1989:7.

Chimombo, Steve. "Flora Nwapa Interviewed." *Outlook* 1(1986):1-7.

Cooke, Michael G. "Flora Nwapa: An Interview." *Commonwealth of Letters* 2.1(1990):14-27.

Egbuna, Maureen. "Personality Interviewed." *Cactus* (1980-81):28- 29.

Ezenwa-Ohaeto. "Flora Nwapa." *ALA Bulletin* 19.4(1993):12-18.

Frank, Katherine. "Flora Nwapa, Africa's First Woman Publisher." *Africa Now* May 1983:61-62.

Fraser, Gerald. "A Writer Who Seeks To Reconcile 2 Worlds." *The New York Times* 2 June 1990.

Ibeabuchi, Aloysius. "Nwakuche: Novelist with Concern for Women." *Daily Times* 3 Jul. 1982:7.

James, Adeola. "Flora Nwapa." *In Their Own Voices: African Women Writers Talk.* London: James Currey, 1990:111-17.

Nichols, Lee. "Interview for the VOA with Flora Nwapa, ALA Annual Meeting March 23, 1991." *ALA Bulletin* 17.3(1991):8-9.

Ogan, Amma. "Flora Nwapa: The Stories of Women Come Naturally to Me." *The Guardian* [Lagos] 25 Mar. 1985:4.

Olayebi, Bankole. "We Are Not Feminists." *African Guardian* 18 Sept. 1986:40.

Othman, Abubakar Adamu, Lubasa N'ti Nseendi, Hassan Kurfi, and Adetayo Alabi. "Flora Nwapa: A Swipe at the Ideologies." *The Guardian* [Lagos] 5 Jan. 1991:6.

Perry, Alison. "Meeting Flora Nwapa." *West Africa* No. 3487, 18 June 1984:1262.

Umeh, Marie. "The Poetics of Economic Independence for Female Empowerment: An Interview with Flora Nwapa." *Research in African Literatures* 26.2(1995):22-29.

Uwechue, Austa. "Flora Nwakuche, née Nwapa, A Former Cabinet Minister and One of Africa's Leading Women Writers, Talks to Austa Uwechue." *Africa Woman* 10(1977):8-10.

BIBLIOGRAPHICAL AND BIOGRAPHICAL REFERENCES

Anon. *African Year Book and Who's Who 1977.* London: Africa Journal Ltd., 1976:1272.

Anon. *Women and Women Writers in the Commonwealth.* London : Commonwealth Institute and the National Book League, 1975:34.

Berrian, Brenda F. "Bibliographies of Nine Female African Writers." *Research in African Literatures* 12.2(1981):214-36.

___ *Bibliography of African Women Writers and Journalists.* Washington, D.C.: Three Continents Press, 1985.

___ "An Update: Bibliography of Twelve African Women Writers." *Research in African Literatures* 19.2(1988):229-31.

___ "Bibliography: Flora Nwapa (1931-1993)." *Research in African Literatures* 26.2 (1995):124-29.

Creque-Harris, Leah. "Literature of the Diaspora by Women of Color." *Sage* 3.2(1986):61-64.

Fister, Barbara, ed. *Third World Women's Literatures: A Dictionary and Guide to Materials in English.* Westport CT: Greenwood, 1995:59, 95, 186, 207, 224-25, 249.

Gibbs, James, Comp. *A Handbook for African Writers.* Oxford: Hans Zell, 1986:22-23.

Herdeck, Donald, Comp. *African Authors: A Companion to Black African Writing 1300-1973.* Washington, D. C.: Black Orpheus, 1973:303-04.

Jahn, Janheinz and Claus Dressler, Comps. *Bibliography of Creative African Writing.* Nendeln: Krauss-Thomson, 1971: 165-66.

Jahn, Janheinz, Ulla Schild and Almut Nordmann. *Who's Who in African Literature.* Tübingen: Horst Erdmann Verlag, 1971: 275-76.

Kirkpatrick, D. L., ed. *Twentieth-Century Children's Writers.* New York: St. Martin's, 1983.

Lang, D. M. "African." *The Penguin Companion to Literature.* Harmondsworth: Penguin, 1969.

Lindfors, Bernth, Comp. *Black African Literature in English.* Detroit: Gale Research, 1979:8-10, 355-56.

___ *Black African Literature in English: 1982-1986.* New York: Hans Zell, 1989:32, 305-06.

___ *Black African Literature in English: 1987-1991.* New York: Hans Zell, 1995:439-40.

Mayes, Janis A. Comp. "Studies of African Literatures and Oratures: An Annual Annotated Bibliography, 1987." *Callaloo* 11.4(1988):846-903.

Ramsaran, J. M. *New Approaches to African Literature.* Ibadan: Ibadan University, 1970:35.

Saint-Andre-Utudjian, Eliane. *A Bibliography of West African Life and Literature.* New York: Africana, 1977:44, 81, 83, 93, 107, 118.

Schmidt, Nancy J. Comp. *Supplement to Children's Books on Africa and Their Authors.* New York: Africana, 1979:177.

Sivard, Ruth Leger. *Women: A World Survey.* Washington, D.C. World Priorities, 1985:44.

Skrujat, Ernestyna. *Afryka W Tworczosci Jej Pisarzy [Africa Through the Works of the Authors].* Studium Afrykanistyczne Uniwersytetu Warsawskiego, 1973.

Zell, Hans. *African Books in Print. I.* London: Mansell Information, 1978:325-26.

Zell, Hans and Helen Silver, Comp. *A Reader's Guide to African Literature.* New York: Africana, 1971:38, 163.

Zell, Hans, Carol Bundy and Virginia Coulon, Comps. *A New Reader's Guide to African Literature.* New York: Africana, 1983:158- 59, 439.

Contributors

Julie Agbasiere is a Senior Lecturer in the Faculty of Arts at Nnamdi Azikiwe University in Awka, Nigeria. She is the author of several articles on African Literature in books namely, *Emerging Perspectives on Buchi Emecheta, Children and Literature in Africa, Current Trends in Literature and Language Studies in West Africa, The Humanities and Social Changes, Introductory Readings in the Humanities and Social Sciences, The English Handbook,* and *Dictionnaire des oeuvres littéraires négro-africaines de langue française.* She has also published articles in *Présence Francophone,* the *Journal of Educational Research,* and a book, *Le chemin de l'est* (1996).

Ifi Amadiume is an Associate Professor of Religion and African Studies at Dartmouth College. Her publications include *Male Daughters, Female Husbands: Gender and Sex in An African Society* and *African Matriarchal Foundations: The Igbo Case.* She is also a published and award-winning poet.

Adeline Apena is a Professor of African History and Women's Studies at the Sage Colleges in Troy, New York. She is a Betty Harder McCllelan Fellow in the Humanities and Secretary of New York State African Studies Association. Her publications include

Colonization, Commerce and Entrepreneurship in Nigeria: The Western Delta 1914-1960 (1996) and many published essays in scholarly books and journals.

Susan Arndt is a doctoral candidate at Humboldt University in Berlin, Germany where she also teaches African Literatures. She has presented papers on African oral literatures at various conferences and published essays in books, such as *Emerging Perspectives on Buchi Emecheta*, as well as in scholarly journals globally.

Ada Uzoamaka Azodo teaches French and African Literature in the Department of Foreign Languages and Literatures at Valparaiso University in Indiana. She is the author of *L'imaginaire Dans Les Romans de Camara Laye*. A critical essay, "Work in Gold as Spiritual Journey in Camara Laye's *L'enfant Noir*," appears in *Journal of Religion in Africa* (1994). Her forthcoming work includes a book co-edited with Gay Wilentz, entitled, *Emerging Perspectives on Ama Ata Aidoo*.

Brenda F. Berrian is an Associate Professor and Chair of the Department of Africana Studies at the University of Pittsburgh. She is the editor of *Bibliography of African Women Writers and Journalists* (1985) and *Bibliography of Women Writers from the Caribbean* (1989). Her articles on African-American, African and Caribbean writers appear in scholarly books and journals across the globe.

Marvie Brooks is a Reference Librarian at John Jay College of Criminal Justice, CUNY.

Jane Bryce teaches African Literature in the Department of English at the University of the West Indies in Cave Hill, Barbados. Her published articles appear in *Wasafiri, Motherlands, Unbecoming Daughters of the Empire, Readings in Popular Culture*, and *Writing & Africa*.

Buchi Emecheta is Africa's most prolific female writer today. She has published nineteen books, which have been translated into fourteen languages. Her published essays, along with her *magnum opus, The Joys of Motherhood*, have established her as one of the "big shots" in the Anglophone African writer's canon.

Ernest N. Emenyonu holds a joint appointment in the English Department and the Department of African and Afro-American Studies at Brandeis University in Waltham, Massachusetts. His published books include *Cyprian Ekwensi* (1974), *The Rise of the Igbo Novel* (1978), *African Literature for Schools and Colleges* (1985), *The Essential Ekwensi* (1987), *African Literature Comes of Age*, with C. D. Narasimhaiah (1988), *Studies on the Nigerian Novel* (1991), *Current Trends in Literature and Language Studies in West Africa*, with Charles Nnolim (1994). He has also published three volumes of children's stories: *Bedtime Stories for African Children* (1989), *The Adventures of Ebeleako* (1991) and *Uzo Remembers His Father* (1992). His critical essays have appeared in scholarly journals and books throughout the world. He is the founding editor of the Calabar Studies in African Literature.

Theodora Akachi Ezeigbo is a Senior Lecturer in the Department of English at the University of Lagos, Nigeria. Her publications include *Fact and Fiction in the Literature of the Nigerian Civil War* (1991), *Gender Issues in Nigeria: A Feminine Perspective* (1996), three collections of short stories, *Rhythms of Life, Echoes in the Mind*, and *Rituals and Departures* and two children's books, *The Buried Treasure* and *The Prize*. Her critical essays appear in anthologies and journals globally.

Ezenwa-Ohaeto is a Nigerian poet, short story writer, literary critic and scholar. He is the author of the poetry collections: *Songs of A Traveller, I Wan Be President, Bullets for Buntings* and *Pieces of Madness*; the literary works, *Examination Guide to Alkali's The Stillborn* (1991) and *Chinua Achebe, The Life: Facing the Frontiers* (1997), as well as editing *Making Books Available and Affordable* (1995). He has won several prizes including a short story first prize (1979), a B.B.C. Arts and Africa Award (1981), and the Orphic Lute Poetry Prize (1985). His works have been translated into French, Russian and German. He is currently a Visiting Professor at the University of Bayreuth in Germany.

Naana Banyiwa Horne teaches English, African American Studies, American Ethnic and Minority Literature and Women's Studies at Indiana University, Kokomo.

Chukwuemeka Ike is one of Nigeria's leading novelists who has authored *Toads for Supper* (1965), *The Naked Gods* (1970), *The*

Potter's Wheel (1973), *Sunset at Dawn* (1976), and *The Chicken Chasers* (1980). He has also played many roles in Nigeria as schoolmaster, university administrator (UNN), chief executive of the West African Examinations Council and university professor (Maiduguri). He is currently the President of Nigerian Book Foundation in Awka, Nigeria.

Sabine Jell-Bahlsen is an anthropologist and a filmmaker who teaches at the Rhode Island School of Design in Providence, RI. She has spent much time in Ugwuta nearly every year, since 1978. Her published articles appear in the journals, *Research in African Literatures* and *Dialectical Anthropology* and in several anthologies such as, *Sisterhood, Feminisms, and Power* and *Queens, Queen Mothers, Priestesses and Power*. She has independently produced several ethnographic documentary films, including *Mammy Water: In Search of the Water Spirits in Nigeria and Owu: Chidi Joins the Okoroshi Secret Society*. She is currently working on a book reflecting on her field work in Nigeria.

Mary E. Modupe Kolawole is a Senior Lecturer in Literature and Women's Studies at Obafemi Awolowo University in Ile-Ife, Nigeria. She has published a book, *Womanism and African Consciousness* (1997). Her published articles and book chapters appear in *Nigerian Women in Social Change, Journal of English Studies, Great Themes in African Literature*, and *Wole Soyinka: Collected Essays*.

Nina E. Mba teaches in the Department of History at the University of Lagos. Her first book, *Nigerian Women Mobilized: Women in Southern Nigerian Political History 1900-1965* (1982) is a landmark in the history of Nigerian women. She is the co-author of a forthcoming biography of Funmilayo Ransome-Kuti (University of Illinois, 1996). Among her other publications are essays on Nigerian literature.

Chester St. H. Mills, born in Manchester, Jamaica, is an Associate Professor in the Department of English and Print Journalism at Southern University in New Orleans, Louisiana. He received his B.A. in German, his M.A. in Spanish from the State University of New York at Binghamton and his Ph.D. in Literary Studies from Washington State University.

Fehintola Mosadomi is currently a doctoral student in linguistics at Tulane University. She holds two Masters degrees from the University of Delaware. She has published poetry both in French and English, and articles in Francophone African Studies and linguistics.

Teresa U. Njoku Ph. D., is a professor in the Department of English at Abia State University in Nigeria. She teaches African and English Literatures and has published many articles in Nigerian and international journals. Presently, she is working on a book entitled *The Feminist Temperament in West African Fiction*.

Chimalum Nwankwo teaches African Literature and English at the University of North Carolina in Raleigh. His critical essays appear in books and journals globally. He is also a published and award-winning poet.

Molara Ogundipe-Leslie is a professor of literature and Women's Studies. One of the leading feminist/critical thinkers from Africa, she is a much published poet, literary critic, essayist and author of several books and numerous articles.

Obododimma Oha teaches in the Department of English at the University of Ibadan in Nigeria. He writes on Rhetoric of Conflict, Gender Linguistics and Meaning in Discourse. His essays have appeared in journals and books in Nigeria, Europe, and in the United States. Generally, he doubles as a creative writer and an analyst of discourse.

Emelia C. Oko is a Senior Lecturer in the Department of English at the University of Calabar, Nigeria. Her book, *The West African Novel and Social Evolution*, is published by Nok. Her published articles appear in journals in Nigeria and abroad such as, *The CONCH, Nsukka Studies in African Literatures, and the Encyclopedia of Post-Colonial Literature*. She has edited many issues of *Calabar Studies in African Literature*.

Ifeoma Okoye teaches English at Nnamdi Azikiwe University, in Awka, Nigeria. She is the author of *Behind the Clouds, Men Without Ears* (1984 winner of Association of Nigerian Authors Best Fiction of the Year Prize) and *Chimere*, all published by Longman. She has also written books for children and articles on teaching English.

701

Osonye Tess Onwueme, one of Africa's award winning play-wrights, is currently the first Distinguished Professor of Cultural Diversity and Professor of English at the University of Wisconsin, Eau Claire. Her plays have earned recognition on stage in Africa, the United Kingdom and the United States. In 1985 her play, *The Desert Encroachers*, won the Association of Nigerian Authors Literary Prize in drama; *Tell It to Women* won the 1995 drama prize of the Association of Nigerian Authors (ANA). Her other plays include *A Hen Too Soon, Ban Empty Barn, Mirror for Campus, The Reign of Wazobia, The Artist's Homecoming, Some Day Soon, Legacies, Riot in Heaven* and an anthology of *Three Plays* published by Wayne State University Press, Detroit Michigan.

Shivaji Sengupta is Vice-President of Academic Policy at Boricua College and a Professor of English. He has published many articles on literary criticism and theory in American, British and Indian journals. He has published *A Critical Edition of John Dryden's MacFlecknoe* (1985) and *Absalom and Achitophel* (1991). Professor Sengupta has also published a book of poems, *Jonaki* (1980).

Florence Stratton taught literature at Njala University College in Sierre Leone for nineteen years. Her latest book is *Contemporary African Literature and the Politics of Gender* (1994). She has published numerous articles in scholarly journals such as, *Research in African Literatures.* Currently, she is teaching at the University of Regina in Canada.

Linda Strong-Leek teaches in the Department of English at Florida International University in Miami, Florida.

Marie Umeh teaches Literature of the African World, Western Literature and Writing Composition in the Department of English at John Jay College of Criminal Justice, CUNY. She is the editor of *Emerging Perspectives on Buchi Emecheta.* Her essays have appeared in books and journals in Africa, Europe and the United States.

Gay Wilentz is an Associate Professor of English at East Carolina University. She was a Fulbright Scholar to Nigeria, where she met and worked with Flora Nwapa. Her book, *Binding Cultures: Black Women Writers in Africa and the Diaspora*, examines women's role in the transmission of culture on both sides of the Atlantic. She has published in *College English, African American Review, Research in African Literatures, Twentieth Century Literature* and *MELUS.*

Permissions

Grateful acknowledgements are extended to the following individuals and publishers, who generously allowed us to include their works and photographs in this anthology:

Sabine Jell-Bahlsen, who created the map of Oguta especially for this anthology and who supplied the photograph of Flora Nwapa in her office at Tana Press, Enugu, which appears on the back cover of the book, as well as the photograph of the Priestess of Oguta which appears also within the book.

Theodora Akachi Ezeigbo, who gave us the map of Nigeria created for this anthology.

Nina E. Mba who gave us the photograph of Flora Nwapa for the front cover of the book and others that appear within the book.

Chief Gogo Nwakuche, who gave us permission to print the pictures of his wife, Flora Nwapa Nwakuche, and himself in the two photos that appear in the anthology.

Uzoma Nwakuche, who gave us permission to quote from Flora Nwapa's unpublished manuscript, *The Lake Goddess* (forthcoming).

Ejine Nzeribe, who supplied the photographs of the Nwapa-Nwakuche family which appear throughout the book.

Indiana University Press, Bloomington, which gave us permission to reprint Marie Umeh's conversation with Flora Nwapa, "The Poetics of Economic Independence for Female Empowerment: An Interview with Flora Nwapa," which first appeared in *Research in African Literatures,* 26.2 (Summer, 1995):22-29.

Thunder's Mouth Press, which gave us permission to publish part of Birago Diop's poem, "Spirits," printed in Ellen Conroy's book, *The Negritude Poets.*

The editor has attempted to acknowledge all copyright holders. I shall be very grateful to hear from anyone who has been inadvertently overlooked or incorrectly cited. The necessary changes will be made at the first opportunity.

Index

women, 0-5, 8-10, 12-17, 19-27,
29, 31, 34-36, 41-43, 45-49, 51-
75, 77-78, 82-83, 85-94, 99-100,
104-116, 118, 121-125, 127, 129,
131-141, 143-187, 191-193, 197-
198, 201-203, 205-207, 209-214,
217-219, 221-259, 263-275, 277-
289, 292-293, 295-312, 314-321,
323, 330-333, 341, 343-346, 348-
349, 351, 353, 355-367, 370-371,
373-374, 376-381, 383-384, 387,
390, 392-396, 398-399, 401-402,
404-411, 413-414, 433, 435, 439,
441-444, 446-454, 457-465, 468-
478, 481-488, 491-495, 498, 500-
502, 504-505, 509, 511-521,
523-529, 531-540, 542-545, 547-
553, 555-558, 560-564, 572, 575,
577-579, 583-584, 592, 608, 625-
626, 630-632, 637, 640-645, 648,
651, 657-659, 662-672
women and war, -victimization of
women, 483–492
women and Colonialism and post-
Coloniality, 277–288
Women's Liberation, 365, 373, 383-
384, 563
women's roles, 47, 234, 296, 363,
453, 459
Women's War of 1929, 127
Woolf, Virginia,166, 185, 187, 394,
410
writing, unofficial, 404, 407

yam as king of crops, 502